Degradation of Bioactive Substances:

Physiology and Pathophysiology

Editor

Jens H. Henriksen, M.D., Ph.D.
Chief Physician
Department of Clinical Physiology
Hvidovre Hospital
Hvidovre, Denmark
and
Associate Professor of Clinical Physiology
University of Copenhagen
Copenhagen, Denmark

CRC Press
Boca Raton Ann Arbor Boston London

Library of Congress Cataloging-in-Publication Data

Degradation of bioactive substances : physiology and pathophysiology /
editor, Jens H. Henriksen.
p. cm.
Includes bibliographical references.
Includes index.
ISBN 0-8493-6858-8
1. Metabolism. 2. Bioactive compounds—Metabolism.
3. Biomolecules—Pathophysiology. [1. Metabolic Diseases.]
I. Henriksen, Jens H.
[DNLM: 1. Blood Chemical Analysis. 2. Catecholamines—metabolism.
3. Peptides—metabolism. QY 450 D3221]
QP171.D44 1991
616.07′561—dc20
DNLM/DLC
for Library of Congress 90-2609
CIP

Direct all inquiries to CRC Press, Inc., 2000 Corporate Blvd., N.W., Boca Raton, Florida, 33431.

International Standard Book Number 0-8493-6858-8

Library of Congress Card Number 90-2609
Printed in the United States

This book is dedicated to

my father E. Henriksen,
my wife Dorte,
my daughter Katrine,
my son Ulrik,
and
my friend Karin.

PREFACE

The availability of specific and sensitive assays is a major factor in the sustained explosive expansion of knowledge in biomedical sciences that has been occurring during the past 25 years. A particularly productive field has been the measurement of the level of a number of bioactive substances in plasma. The advances in our understanding of degradation processes and their importance in the dynamics of bioactive material have necessitated this book in order to provide recent information in a comprehensive and accessible way.

The interprelation of data made available by the application of kinetic techniques to physiological and clinical problems may require a high level of mathematical skill even when relatively simple models are used. These problems will be facilitated through the presentation of some fundamental aspects that are reviewed in the first chapters and wherever necessary. The reader of this book does not need any special mathematical or kinetic background. The text is, in fact, intended primarily for those who are working actively in the field of bioactive substances such as neurotransmitters, hormone growth factors, neuro- and biomodulators, signal substances, connective tissue markers, etc.

THE EDITOR

Jens H. Henriksen, M.D., Ph.D., is Chief Physician, Department of Clinical Physiology, Hvidovre Hospital and Associate Professor of Clinical Physiology, University of Copenhagen, Copenhagen, Denmark.

Dr. Henriksen graduated from Christianhavns Gymnasium in 1964, received his B.S. in physiology in 1968, and obtained his M.D. and Ph.D. degrees in 1973 and 1982, respectively, from the University of Copenhagen. He is now 45 years old.

Dr. Henriksen is a member of the Danish Society of Clinical Physiology and Nuclear Medicine, Danish Association for the Study of the Liver, European Association for the Study of the Liver, International Association for the Study of the Liver, and Danish Society for Internal Medicine. He is a fellow of the International College of Angiology and board member of the Society of Theoretical and Applied Therapy. He served as board member of the Society of Clinical Physiology (1978-1982) and of the Danish Association for the Study of the Liver (1983-1988), and was President of the Society of Theoretical and Applied Therapy (1988-1989). He has received three awards: Kommunitetet, Grosserer Vendsyssels Legat, and Gardejer af Stenløse Peder Laurids Pedersens Legat and a large number of research grants from the Danish Medical Research Foundation, the NOVO Foundation, King Christian X Foundation, C. Studsgaards Fund,* Borgens Foundation, The Ebba Celinder Foundation, The Medical Research Foundation, Regions of Copenhagen, the Faroe Islands and Greenland, Lœge Ingrid Munkholm's Foundation, and Max and Anna Friedman's Foundation.

Dr. Henriksen has presented over 40 invited lectures at international meetings. He has published more than 180 research papers and has been the author or co-author of 11 books/book chapters. His current major research interests include hemodynamics and homeostatic mechanisms in chronic liver disease and kinetics of transperitoneal transfer.

* Lœgeforeningens Forskningsfond.

CONTRIBUTORS

Jens Lundbœk Andersen
Department of Medical Physiology C
University of Copenhagen
Copenhagen, Denmark

Lars Juel Andersen
Department of Medical Physiology C
University of Copenhagen
Copenhagen, Denmark

Vicente Arroyo, M.D.
Professor of Medicine
Liver Unit
Hospital Clinic i Provincial
University of Barcelona
Barcelona, Spain

Kirsten D. Bentsen, M.D.
Senior Registrar
Department of Internal Medicine, B
Frederiksberg
Copenhagen, Denmark

Mauro Bernardi, M.D.
Associate Professor
Department of Patologia Medica I
University of Bologna
Bologna, Italy

Peter Bie, M.D., Ph.D.
Associate Professor
Department of Medical Physiology C
University of Copenhagen
Copenhagen, Denmark

Niels Juel Christensen, M.D., Ph.D.
Director
Department of Internal Medicine and Endocrinology
Herlev University Hospital
Herlev, Denmark

Jan Fahrenkrug, M.D., Ph.D.
Department Head
Department of Clinical Chemistry
Bispebjerg Hospital
Copenhagen, Denmark

J. Robert E. Fraser, M.D., F.R.C.P., F.R.A.C.P.
Reader in Medicine
University of Melbourne
Parkville, Victoria, Australia

Steen Gammeltoft, M.D., D.M.Sc.
Associate Professor
Department of Clinical Chemistry
Bispebjerg Hospital
Copenhagen, Denmark

Antonio Gasbarrini, M.D.
Research Fellow
Department of Patologia Medica I
University of Bologna
Bologna, Italy

Giovanni Gasbarrini, M.D.
Professor of Medicine
and
Head
Department of Patologia Medica I
University of Bologna
Bologna, Italy

Alexander L. Gerbes, M.D.
Assistant Professor
Klinikum Grosshadern
Department of Medicine II
University of Munich
Munich, Germany

Jens H. Henriksen, M.D., Ph.D.
Chief Physician
Department of Clinical Physiology
Hvidovre Hospital
Hvidovre, Denmark
and
Associate Professor of Clinical Physiology
University of Copenhagen
Copenhagen, Denmark

Linda Hilsted, M.D.
Senior Registrar
Department of Clinical Chemistry
Bispebjerg Hospital
Copenhagen, Denmark

Jens Juul Holst, M.B., B.S., Ph.D., Sc.D.
Professor
Department of Medical Physiology C
University of Copenhagen
Denmark, Copenhagen

N. M. Hooper, M.D., Ph.D.
Department of Biochemistry
University of Leeds
Leeds, England

A. J. Kenny, M.D., Ph.D.
Department of Biochemistry
and Molecular Biology
University of Leeds
Leeds, England

Torvard C. Laurent, M.D., D.M.Sc.
Professor
Department of Medical
and Physiological Chemistry
University of Uppsala
Uppsala, Sweden

Giorgio La Villa, M.D.
Investigator
Clinica Medica II
University of Florence
Florence, Italy

Fred Nyberg, Ph.D.
Professor
Department of Pharmacology
University of Uppsala
Uppsala, Sweden

Mairead M. T. O'Hare, M.D., Ph.D.
Department of Clinical Chemistry
Rigshospitalet
Copenhagen, Denmark

Jens Møller Rasmussen, M.D.
Registrar
Department of Infectious Diseases
Odense University Hospital
Odense, Denmark

Jens F. Rehfeld, M.D., Ph.D.
Professor
Department of Clinical Chemistry
Rigshopitalet
Copenhagen, Denmark

Peter Schmidt, B.S.
Research Fellow
Department of Medical
Physiology C
University of Copenhagen
Copenhagen, Denmark

Thue W. Schwartz, D.M.Sc.
Professor
Department of Laboratory
Molecular Endocrinology
Rigshospitalet
Copenhagen, Denmark

Sven-Erik Svehag, D.V.M., Ph.D.
Professor
Department of Medical Microbiology
Odense University
Odense, Denmark

B. Teisner, M.D.
Associate Professor
Department of Medical Microbiology
Odense University
Odense, Denmark

Lars Terenius, Ph.D.
Professor
Department of Drug Dependence
Research
Karolinska Institute
Stockholm, Sweden

Franco Trevisani, M.D.
Research Fellow
Department of Patologia Medica I
University of Bologna
Bologna, Italy

Angelika M. Vollmar, Ph.D.
Institute of Pharmacology, Toxicology
and Pharmacy
University of Munich
Munich, Germany

Kjeld Winkler, M.D.
Senior Physician
Department of Clinical Physiology
Hvidovre Hospital
Copenhagen, Denmark

TABLE OF CONTENTS

Chapter 1

DEGRADATION OF BIOACTIVE SUBSTANCES: PHYSIOLOGY AND PATHOPHYSIOLOGY

Jens H. Henriksen

The last 2 decades have seen a substantial increase in the number of identified bioactive substances, and refined analyses now allow reliable determination of the concentration of several of these substances in plasma and tissue. The production, release, regulation, and function of hormones, neurotransmitters/modulators, signal substances, enzymes, plasma factors, tissue compounds, etc. are dealt with extensively in journals and textbooks. In contrast, the degradation of these substances has hitherto attracted much less interest. A number of new investigations on the disposal of circulating bioactive substances has given a more detailed insight into the mechanisms governing the plasma concentration of these materials, which justifies an updated survey.

The kinetics of major groups of bioactive substances are described with respect to physiology, location, and regulation of degradation. Basic aspects of general and specific kinetics of degradation are discussed in Chapters 2 and 3. Owing to their general importance, enzymatic degradation by peptidases and receptor-mediated endocytosis and degradation are described in separate Chapters 4 and 5, respectively. Descriptions of the degradation of specific proteins, peptides, amines, steroids, and polysaccharides are given in individual chapters. Insulin, C-peptide, and proinsulin are not described separately because these peptides are considered in the chapters on whole-body and organ degradation and receptor-mediated kinetics.

Information on the degradation of a number of newly discovered bioactive substances such as endothelin, calcitonin gene-related peptide is either lacking or so scanty that it is premature to discuss their disposal in this book. Nor does it include, for example, prostaglandins, the coagulation system, thyroid and parathyroid hormones, and sex hormones, metabolites in general, or drugs because of the comprehensive body of literature on these topics.

Pathophysiology and its role in diagnostics are considered wherever important, but sometimes only by giving examples of particular diseases.

Chapter 2

KINETICS OF WHOLE-BODY AND ORGAN DEGRADATION

Jens H. Henriksen

TABLE OF CONTENTS

I. INTRODUCTION

Kinetics deals with the relation of mass and concentration with time in the body. Several kinetic methods are available for the determination of interchanges between different parts of the organism and of the final irreversible disposal, either by excretion or by metabolic degradation. Concepts like volume of distribution, plasma clearance, half-life, plasma residence or transit times, rate constants, metabolic clearance rate, and appearance and release rates are used in most kinetic models. The two major types of kinetic designs are (1) *compartmental or multicompartmental models,* and (2) *noncompartmental ("black-box") models.* One or the other is generally applied for kinetic analysis in the multifarious areas of life sciences. Both methodologies may be highly sophisticated, employing advanced mathematical equations and computer programs. However, a number of facets have, on occasion, been misused, misinterpreted, and misunderstood. This chapter will not deal in detail with the many different models, their assumptions, pitfalls, limitations, and advantages, but it will review some basic conceptions with special reference to biological problems and include some examples. Whereas it is not the intention to go deeply into the mathematical aspects, the multiexponential approach to the great majority of compartmental and noncompartmental problems justifies presentation in some detail. In practice, fitting exponentials is not difficult and the availability of standard computer programs has made this approach very popular. As long as this procedure is meant to provide a practical description of curves and not necessarily to describe the biological features closely connected with specific processes, the computer is unlikely to generate too many artifacts.

II. COMPARTMENTAL AND MULTICOMPARTMENTAL MODELS

The concept behind compartmental modeling is to describe a system by a finite number of pools which interact by mass exchange.[1,2] This approach necessarily entails simplification of the processes under investigation. One should, however, choose a model that has a direct analogy to the particular biological system. A limited knowledge of physiology may typically have a significant adverse effect on the specific form and order (i.e., number of pools) of a compartmental model, which will lead to major or minor fallacies.

A *compartment* (or pool) is a delimited amount of material that acts kinetically in a homogeneous manner. In a multicompartmental model, two or more compartments are interconnected so that there is mass exchange among some or all of them (see Figure 1). Exchange of material may occur by indicator or mother substances crossing some physical barrier (e.g., membrane transport or blood transport), or by chemical transformation (e.g., metabolism). *Mamillary models* consist of a central compartment connected with peripheral pools, none of which are connected with each other. In *catenary models,* all the pools are arranged in a chain with each pool connected in a series to its closest neighbors. A compartment is *open* if it leaks to the environment and *closed* if it does not. Compartmental models of biological systems usually have at least one open compartment.

The mathematical aspects of compartmental modeling usually involve writing mass-balance equations for the chosen kinetic structure, followed by quantification of the model from experimental data. Applicability to a physiological system entails a number of assumptions of which two important ones are *linearity* and *stationarity.* Linearity signifies that responses can be superadded; that is, the response to several stimuli is the sum of the separate responses. Stationarity denotes the systemic steady state, i.e. the constancy of all systemic parameters in time, which means that repetition of the same stimulus will produce repetition of the same response. The *rate constant* (k) is the fraction of mass in one compartment transferred into another per unit of time. The *distribution volume* (V_D) is equivalent volume of distribution relative to a reference (plasma) concentration rather than a true volume.

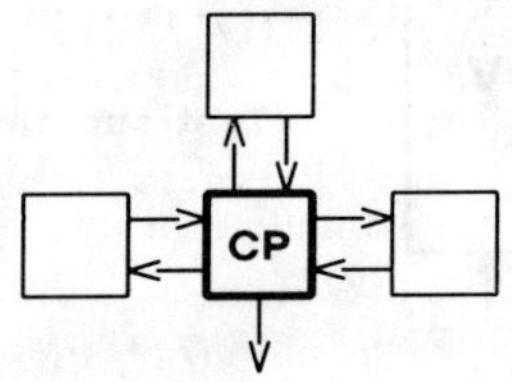

Mammillary system

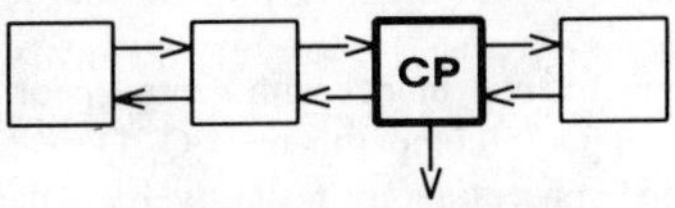

Catenary system

FIGURE 1. Kinetic models. The mamillary system consists of a central pool (CP) connected with peripheral pools, none of which is connected to another. The catenary system consists of pools arranged in series.

Material in a compartment, by definition, must be *homogeneously distributed*; that is, it must have the same concentration throughout the compartment at all times. This means that a multicompartmental model can be a good approximation if the rate of mixing within each compartment is rapid in relation to the rates of transfer among the compartments, but poor if this is not the case. Compartments with concentrations that are always in a fixed proportion to each other may be impossible to distinguish. A common and practical approximation is to "lump" such pools into a single one for analytical purposes, but it is not generally appreciated that lumped pools have useful physiological interpretations. Thus, the compartmental structure may be severely limited by the available data but this does not necessarily restrict the applicability of a compartmental model, provided that it is structured and interpreted appropriately.

Multiexponential solutions for multicompartmental models usually imply the determination of the values of the rate constants and distribution volumes that, in some sense, best fit experimental data; for instance, those that minimize the sum of the squared errors between the data and their corresponding variables in the model. The system is usually 'disturbed' by a known indicator substance and the values of model parameters are derived from the measured output (e.g., an input-output or stimulus-response experiment). Some of the rate constants or pools can sometimes be determined from independent experiments; for example, from the study of subsystems, other test materials, or from published values for analogous systems.

An example of solution of a multicompartmental model follows. A bicompartmental model is illustrated in Figure 2. The rate constants, k_1 and k_2, describe the transfer of the matter under study from compartment 1 to compartment 2 and back, respectively. k_3 describes disposal; that is, all irreversible loss and metabolic degradation of a particular substance. In the present model this takes place exclusively from compartment 2. Assuming linearity and stationarity, differential equations can be based on the principle of conservation of matter:[3]

$$dQ_1/dt = -k_1 \cdot Q_1 + k_2 \cdot Q_2$$

$$dQ_2/dt = k_1 \cdot Q_1 - \left(k_2 + k_3\right) \cdot Q_2$$

$$\left[d\left(Q_1 + Q_2\right)/dt = -k_3 \cdot Q_2\right]$$

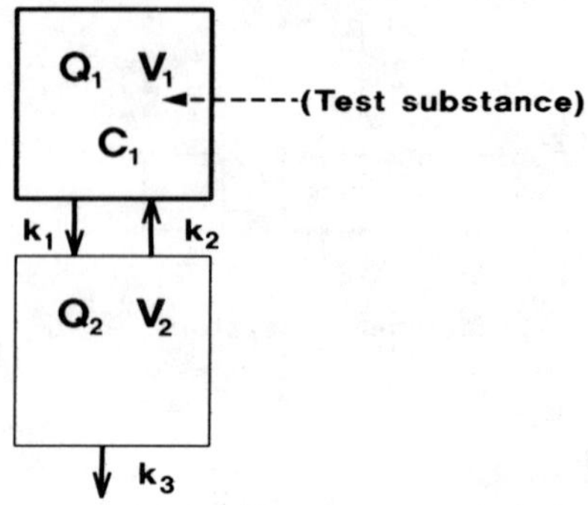

FIGURE 2. Bicompartmental model with central pool (compartment 1) and lumped peripheral pools (compartment 2). Q_1, Q_2, V_1, V_2, C_1, and C_2 are masses, volumes, and concentrations, respectively, in the two pools.

$Q = Q_1 + Q_2$ is the total quantity of test substance. The concentration in compartment 1 is $C_1 = Q_1/V_1$ and the equivalent concentration in compartment 2 is $C_2 = Q_2/V_2$, where V_1 is the volume of compartment 1 (often the plasma space) and V_2 the equivalent volume of compartment 2 (volume of distribution $V_D = V_1 + V_2$). In this case, Q_1, Q_2, k_1, k_2, k_3, V_1, and V_2 are uniquely identifiable from concentration-time measurements in compartment 1. The concentration-time curve is generally described by a sum of exponential functions.

$$Y(t) = A_1 \cdot e^{b_1 \cdot t} + A_2 \cdot e^{b_2 \cdot t} + \ldots .. + A_i \cdot e^{b_1 \cdot t}$$

$$Y(t) = \sum_1^n A_i \cdot e^{b_i \cdot t}$$

In the case of decreasing Y(t), the b_n values are negative and the following equations can be written:

$$Y(0) = \sum_1^n A_i$$

$$\text{Area under } Y(t) = \sum_1^n A_i \big/ b_i = \int_0^\infty Y(t)dt$$

$$\text{Slope: } dY(t)/dt = \sum_1^n A_i \cdot b_i \cdot e^{b_i \cdot t}$$

$$\text{Initial slope} = \left(\sum_1^n A_i \cdot b_i \right) \Bigg/ \sum_1^n A_i$$

$$\text{Mean transit time } \bar{t} = \left(\sum_1^n A_i \big/ b_i \right) \Bigg/ \sum_1^n A_i \quad \text{(residue detection, see Sections A and B.)}$$

$$\text{Mean transit time } \bar{t} = \left(\sum_1^n A_i \big/ b_i^2 \right) \Bigg/ \left(\sum_1^n A_i \big/ b_i \right) \text{(outflow detection, see Sections A and B.}$$

If $A = \sum_1^n A_i$ denotes the fractional concentration of indicator (e.g., μmol/μmol·l $= l^{-1}$) and

$C_1(0) = Q \cdot \sum_1^n A_i$ and $C_1(t) = Q \cdot \sum_1^n A_i \cdot e^{b_i \cdot t}$, then the differential equations can be solved with the following solutions based on *a biexponential* function:

$$V_1 = Q/C_1(0) = 1/(A_1 + A_2)$$

$$V_2 = \frac{V_1^2 \cdot (A_1 + A_2) \cdot (A_1 \cdot b_1 + A_2 \cdot b_2)^2}{A_1 \cdot A_2 \cdot (b_1 - b_2)^2 + (A_1 + A_2)^2 \cdot b_1 \cdot b_2}$$

$$= V_D \cdot k_1/(k_2 + k_3)$$

where

$$k_1 = V_1 \cdot (A_1 \cdot b_1 + A_2 \cdot b_2)$$

$$k_2 = V_1 \cdot A_1 \cdot A_2 \cdot (b_1 - b_2)^2 / k_1 \cdot (A_1 + A_2)$$

$$k_3 = V_1 \cdot (A_1 + A_2) \cdot b_1 \cdot b_2 / k_1$$

In the case of multiexponential functions of a higher order than two, the solutions become much more complex.

If an irreversible removal occurs from compartment 1 (in addition to that of compartment 2), a fourth rate constant (k_4) must be included in the differential equations. Not all the system parameters are then uniquely identifiable from the concentration-time relationship in compartment 1 alone. Other information (overall disappearance from the body, size of distribution volume, concentration-time relationship in compartment 2, etc.) is needed.

The number of exponential terms (n) that can be accurately and meaningfully fitted to a set of data

$$Y(t) = A_1 \cdot e^{b_1 \cdot t} + \ldots\ldots + A_n \cdot e^{b_n \cdot t}$$

is one of the most limiting factors in determining compartmental model order (N, i.e., the number of pools), especially when only one pool is accessible to measurement. The reasons for this are *mathematical* as well as *statistical*. The value of n is rarely greater than three or four for single-input/single-output models, even for large values of N. This means that system compartments that are not directly measured are usually a composite of pools ("lumped" pool) with more or less similar kinetic characteristics (e.g., proportional concentrations) in relation to the measured pool(s).

It is beyond the scope of this survey to describe in detail the many facets of advanced compartmental modeling and computer applications. Readers should consult books and reviews on this specific topic.[1–6]

III. NONCOMPARTMENTAL MODELS

The basic noncompartmental model is illustrated in Figure 3. The most common case is that of a central pool connected with any number of peripheral pools. The *inlet arrow* represents the sum of all *de novo* entries of the substance into the central pool. Likewise, the *outlet arrow* represents the sum of irreversible disposal; that is, metabolism, elimination, degradation, external loss, etc. in the system. Thus, a system with a definite structure is implied in the

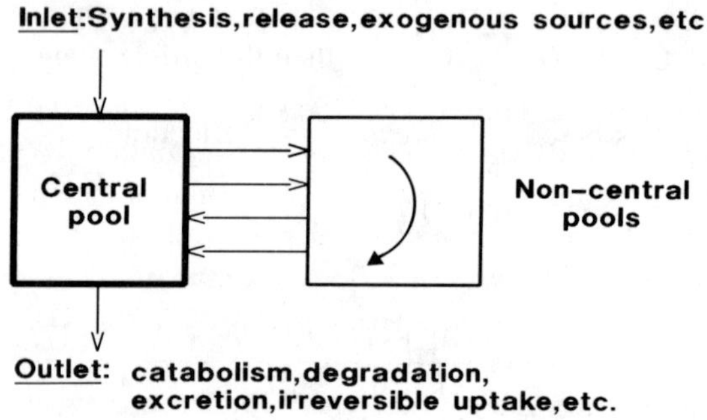

FIGURE 3. Noncompartmental ("black-box") model with central pool, inlet, outlet, exchange to, and recirculation within and from noncentral pools.

noncompartmental model (noncompartmental analysis is not "model independent", as it has sometimes been labeled in the literature). Sources, sinks, recirculation, test sampling, etc. are all explicitly associated with one (central) pool. The sources are equivalent ones that reflect inlets reaching the central pool directly and indirectly; for example, hormones generated from precursors in peripheral tissues (indirect sources). The sinks may also appear elsewhere in the system, for example, hormone disposal in tissues (indirect sinks).

The major conceptual advantage of a noncompartmental model is that any number of recirculations or exchanges can occur with any number of noncentral pools, none of which have to be identified with any physiological structures. The multicompartmental model, on the other hand, requires more specific structures. That mathematical analyses of kinetic data can be accomplished with equations involving integrals rather than differentials may be an extra benefit.

Noncompartmental analysis is generally used to assess steady-state, whole-body variables: plasma clearance rate (Cl), distribution volume (V_D), total body pool size, whole-body mean residence (θ) or transit ($\bar{t}$) times, fractional clearance rate (FCR), and metabolic clearance rate (MCR) (see the following section on whole-body degradation). For determination of the above kinetic variables in noncompartmental analysis, the substance under study must be directly eliminated from the central pool into which the indicator is introduced and measured (e.g., intravenous injection and blood sampling). This means that no or relatively little disposal can occur in or from noncentral (e.g., extravascular) pools. Strictly speaking, this condition can be difficult to meet for a wide variety of physiologically interesting materials that are metabolized intracellularly (e.g., hormones and transmitters). In practice, however, this condition may be met approximately, for instance, in cases where transcapillary exchange rates are rapid compared to overall distribution times and at the same time the noncentral pools are relatively small and minimally "leaky".[7] However, the degree of approximation depends on quantitative as well as structural aspects of the system. A second condition implied by the noncompartmental model is that all sources of the substance under study must be transported to the central pool, that is, the same pool into which the indicator is introduced.

Briefly, the noncompartmental model can provide accurate determination of kinetic variables for systems in which all sources and all sinks are directly associated with, and only with, the central measurement pool. However, the plasma clearance rate can also be accurately assessed when noncentral pool sinks are present if there is only one source leading directly into the central pool. If these conditions are not met, lower bounds or indeterminate underestimates may be obtained for Cl, MCR, V_D, and total body pool size, and FCR may be overestimated by the non-compartmental analysis (Reference 2, and Chapter 18).

Special problems on *flow dependency and limitation, receptor mediated degradation, enzyme limited degradation,* and disposal through *major metabolic pathways* are considered separately in Chapters 3, 4, and 5.

IV. WHOLE-BODY DEGRADATION

In principle, kinetics of whole-body degradation may be assessed by either whole-body detection (residue detection) or by time-varying plasma sampling.[1] Whole-body detection is based on registration of a time-quantity curve for the amount of material retained in the entire organism. This method has a relatively limited application due to practical difficulties but may, in principle, be the most direct way of solving complicated overall kinetic problems. An example of whole-body detection is the assessment of protein metabolism.[8,9] Serum albumin, gamma globulins and other plasma proteins can be labeled with ^{131}I for application in whole-body detection. Whole-body residue of tracer may be determined in one of two ways: (1) direct counting of gamma photons in the entire organism, and (2) subtraction of all external loss (urine, feces) from the amount of tracer initially introduced. It is important to note that for the assessment of radioactivity in the whole-body residue, only the substance under study must contain the label. The plasma sampling method for assessment of whole-body degradation may be a technique which applies (1) as a steady indicator level in plasma, as obtained by continuous infusion, followed by registration of the time course after termination of the infusion, or a (2) type which applies a bolus injection followed by measurement of the plasma disappearance.[10]

A common error in all whole-body kinetics is spending too a short time on plasma sampling or curtailing the registration of other relevant data. Prolonged sampling periods are time-consuming and the concentration may be low and thus difficult to measure. Nevertheless, the area under the plasma curve (concentration-time integral) may depend substantially on the presence or absence of a long "tail" with a relatively small concentration.[1] Similarly, in multiexponential functions the presence of an additional exponential element with a small rate constant may have a heavy impact on the results obtained. Another problem is the representativeness of plasma samples collected from peripheral veins. It is often assumed that a peripheral blood sample is close to the ideal mixture of the blood pool. In the case of rapid elimination with short half-lives and large clearance values, the use of peripheral venous blood samples may lead to substantial overestimation of the true clearance value (even exceeding the size of the cardiac plasma or blood output). The correct value can be obtained by applying arterial (well mixed) samples instead. The fallacy about peripheral blood has also been considered in the chapter on catecholamine disposal. Before discussing the degradation of specific substances, some basic concepts and definitions should be described.

A. KINETIC CONCEPTS AND DEFINITIONS

The *mean transit time* ($\bar{t}$) denotes the average time taken by all molecules of a given material to pass through a specified system.[1] The concept is easiest to understand from a residue detection experiment after a bolus (i.e., single rapid) injection, see Figure 4. If a "finite" number of indicator particles are injected, their passage through the system may be compared to a car race with several individual transit times: t_1, t_2, ... t_i, and $\bar{t}$ is then

$$\bar{t} = \frac{\sum_{1}^{n} t_i}{n}$$

If an "infinite" number of indicator molecules are introduced, t may be expressed as:

$$\bar{t} = \frac{\int_0^\infty m(t)dt}{m_o}$$

where m(t) is the indicator residue mass-time function and m_o the amount of indicator introduced into the system (m_o = m(o)). In other words, $\bar{t}$ is the area under the residue curve per unit mass of indicator introduced.

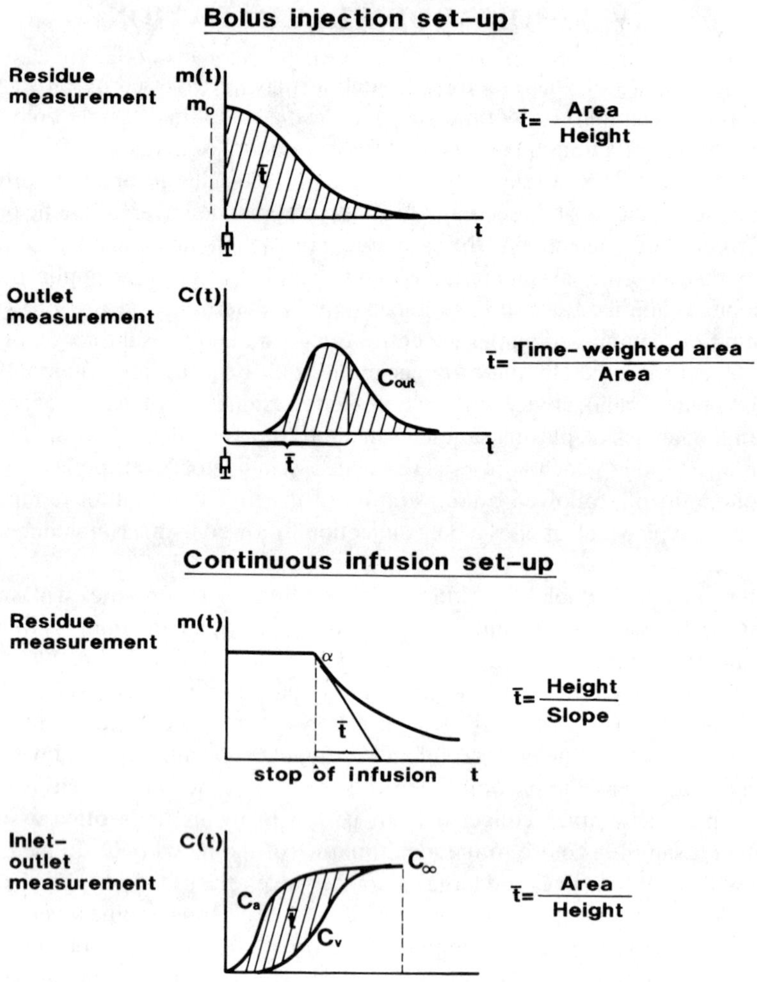

FIGURE 4. Different types of kinetic set-ups. t = mean transit time, m(t) and c(t) = amount and concentration of test substance to time t, respectively (see text).

A fundamental relationship exists between $\bar{t}$, volume (V), and flow (F) through a system:[1,2]

$$V = \bar{t} \cdot F \text{ or } \bar{t} = V/F$$

If the volume is the volume of distribution (V_D), the equation can be written as:

$$V_D = \bar{t} \cdot Cl$$

where Cl is the system clearance. The equation may also occur as:

$$M = \bar{t} \cdot j$$

where M and j are mass (e.g., μmol) and flux, (e.g., $\mu\text{mol} \cdot \text{min}^{-1}$), respectively.

In addition, the mean transit time may be derived from bolus injection and outlet (outflow) detection, or by a continuous influsion experiment either with residue detection or inlet-outlet detection (see Figure 4).

Volume of distribution (V_D) is considered a virtual space (or pool) and is defined by applying the mass balance concept to the amount of indicator (m_o) or endogenous substance (M) relative to its concentration in a reference fluid (plasma) after complete mixing:

$$V_D = m_o / c(\infty) \text{ or } V_D = M/C$$

where c(∞) is the indicator concentration at the steady state and C the concentration of endogenous substance. The *clearance concept* (Cl) also presupposes the existence of a reference fluid. Cl is the volume of this reference fluid (plasma) equivalent to the transfer of the substance across a border (e.g., capillary membrane, cell membrane at metabolic sites, glomerulus, bile canaliculi) per unit time:

$$C1 = j/C$$

where C is the concentration in the reference fluid. The *fractional clearance rate* (FCR) is the clearance rate relative to the distribution volume:

$$FCR = C1/V_D = j/M = \bar{t}^{-1}$$

Catabolism is often designated as the *metabolic clearance rate* (MCR) and the FCR described as the *fractional catabolic rate. Plasma appearance rate* (AR) during the steady state is then given by

$$AR = M \cdot FCR = Cl \cdot C$$

The *extraction fraction* (or *extraction ratio,* E) is, by definition, the fraction of the inflowing substance that does not leave via the outflow (i.e., the fraction that is removed during a single passage of the blood):

$$E = \frac{C_{artery} - C_{vein}}{C_{artery}}$$

The *clearance* is then given by $Cl = \text{plasma flow rate} \times E = j/C_{artery}$ The relationship between Cl, C, blood flow, and metabolic removal is considered in Chapter 3.

Degradation of *albumin* may serve as an example of an endogenous substance with a relatively long body residence time. Albumin kinetics are well described from both noncompartmental and compartmental points of view and they illustrate basic concepts of whole-body turnover.[8-17]

B. WHOLE-BODY KINETICS OF SUBSTANCES WITH A SLOW RATE OF TURNOVER

Albumin molecules are synthesized in liver cells, from which they pass directly into the blood stream.[10,11] Thus, labeled tracer albumin given intravenously occurs in the body in essentially the same route as does endogenous albumin. Whole-body residue can be followed by daily whole-body counting or by subtracting the urinary excretion of the label from the dose injected. This presupposes that the urinary excretion of the label (often ^{131}I or ^{125}I) is much faster than the turnover of albumin. This holds good for iodide if the thyroid uptake is blocked and the urinary excretion of iodide is normal. The resulting residue radioactivity-time curve will be close to that of the residue of the labeled albumin alone.[10] The mean transit time of albumin in the entire organism ($\bar{t}_{WB}$, plasma plus extravascular spaces), which equals total body albumin mass relative to the disposal rate can then be assessed from the area of the whole-body radioactivity time curve divided by the initial height of the curve:

$$\bar{t}_{WB} = \frac{\int_0^\infty m(t)dt}{m_o} = area/height$$

The disposal rate (which, during the steady state, equals the rate of synthesis) can be determined as the ratio between the introduced tracer dose (Q) and the area under the specific activity-time curve [c(t)] in plasma:

$$j = \frac{Q}{\int_0^\infty c(t)dt \Big/ C_p} = C1 \cdot C_p$$

C_p is the plasma concentration of endogenous albumin. Hence, the total amount of albumin in the body (alb_{WB}) is

$$alb_{WB} = j \cdot \bar{t}_{WB}$$

The *intravascular mass of albumin* (alb_{IVM}) is derived from

$$\begin{aligned} alb_{IVM} &= plasma\ volume \cdot C_p \\ &\ or \\ &= Q \cdot C_p / c(o) \end{aligned}$$

where plasma volume [= Q/c(o)] is the initial volume of albumin distribution.

The mass/flux ratio of plasma albumin will express the *mean time of sojourn (or residence)* in the plasma space:

$$\theta_p = \frac{alb_{IVM}}{j} = \frac{\int_0^\infty c(t)dt}{c(o)}$$

θ_p is quite different from $\bar{t}_{WB}$ because the latter refers to the entire passage of labeled albumin through the integral system (whole-body), whereas θ_p describes albumin sojourn in plasma, excluding the duration of its transit (recirculations) in the extravascular spaces, until irreversible loss takes place.

The distribution of albumin between intravascular and extravascular spaces may be expressed as a *distribution ratio* (DR equal to circulating albumin relative to whole-body albumin). DR can be derived from the following equation:

$$j = \frac{alb_{IVM}}{\theta_p} = \frac{alb_{WB}}{\bar{t}_{WB}}$$

and, by rearrangement,

$$DR = \frac{alb_{IVM}}{alb_{WB}} = \frac{\theta_p}{\bar{t}_{WB}} = \frac{\int_0^\infty c(t)dt / c(o)}{\int_0^\infty m(t)dt / m(o)}$$

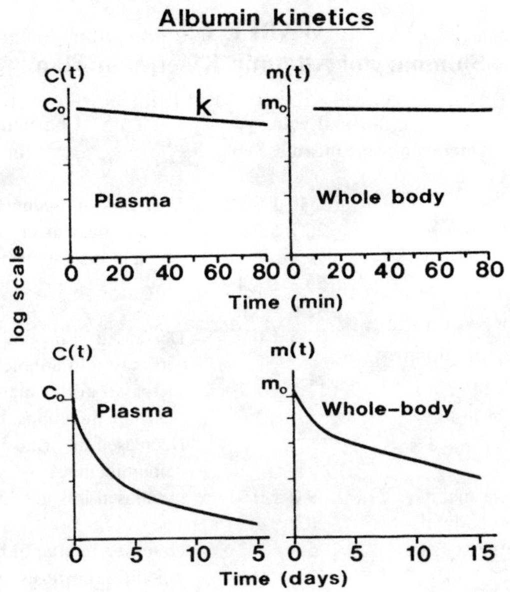

FIGURE 5. Whole-body and plasma curves after i.v. injection of radiolabeled serum albumin. After initial mixing (10 min), the plasma curve reflects transcapillary passage during the next hours, while the whole-body curve shows degradation (top panel). After some days, a steady state between central and noncentral pools has been achieved and the plasma and whole-body curves are parallel in a semilogarithmic plot, the slope reflecting the rate of disposal (metabolic degradation plus external loss, bottom).

From the intravascular space, albumin passes into the extravascular spaces from which it returns into the circulation mainly via the lymph. The rate constant (transcapillary escape rate, k) for this passage can be obtained by the initial slope of the plasma specific activity-time curve after initial mixing in the blood stream, as shown in Figure 5. This value is about 0.05 h^{-1} and may be increased in a number of diseases in which the exchange is accelerated (e.g., arterial hypertension, diabetes, or cirrhosis).[13,15] The amount of albumin (q) that passes through the capillary walls is then:

$$q = alb_{IVM} \cdot k$$

and the mean transit time in plasma (tp) for this passage will be

$$\bar{t}_p = \frac{alb_{IVM}}{q} = l/k$$

The various kinetic constants for albumin are summarized in Table 1. It is clear that most dynamics and disposal rates are slow. Thus, experiments with continuous infusion are impracticable, solely because several weeks may pass before a steady state is reached. With respect to the detection of local degradation of albumin, individual organs cannot be identified by the selective organ venous outflow catetherization technique (the Fick principle, see Section V.A): Supposing the MCR of albumin is in the order of 0.2 ml/min, differences in the arteriovenous concentration of less than, for example, 0.2/600 = 0.03% would be measured, assuming that all the albumin was degraded in, for example, the kidneys, and even smaller arteriovenous differences in the case of widespread degradation. *In vitro* studies of tissue cultures and

TABLE 1
Summary of Albumin Kinetics in Man

Kinetic concept	Normal value (mean ± SD)	Characteristic change in disease
Plasma concentration, μmol/l	540 ± 800[a]	Decreased in malnutrition, cirrhosis, nephrotic syndrome, cancer, post-surgery syndrome, protein loosing gastropathy
Intravascular mass (IVM_{alb}), μmol/kg	26.7 ± 2.0	
Whole body mass, μmol/kg	59.3 ± 4.4	
Extravascular mass, μmol/kg	32.6 ± 3.0	
Distribution ratio	0.45 ± 0.02	Decreased in myxedema, cirrhosis with ascites; increased in uremic patients without ascites
Fractional catabolic rate, % IVM_{alb}/24 h	11.1 ± 1.1	Increased in rheumatoid arthritis, fever, chronic infections, thyroid disorders
Metabolic clearance rate, ml.24 h ·kg	5.1 ± 1.6	Decreased in cirrhosis, protein malnurtrition, albuminemia
Mean sojourn (residence) in plasma, days	9.0 ± 0.9	Decreased in some hypercatabolic states
Mean transit (residence) in the body, days	20.0 ± 1.7	
Transcapillary escape rate, % IVM_{alb}/h	5.9 ± 1.2	Increased in arterial hypertension, diabetes mellitus, cirrhosis, congestive heart failure, uremia, psoriasis, erythrodermia
Transcapillary clearance, ml/24 h · kg	2.7 ± 0.8	
Intravascular mean transit time, h	16.9 ± 3.4	
Extravascular mean transit time, h	20.7 ± 4.3	Increased in myxedema, cirrhosis with ascites
Average number of transcapillar passages of a single plasma albumin molecule before irreversible disposal	12.8 ± 2.8	

Note: Values from author's laboratory. All values are similar to those from other reports (e.g., References 8 to 13, 17, and 85).

[a] Range.

enzyme studies do, in fact, suggest a widespread disposal of albumin.[10,15,16] As seen from the low MCR and transcapillary clearance, there is no blood flow limitation for albumin, not even in organs with very low perfusion (see Chapter 3).

By considering $\bar{t}_p$ in relation to θ_p it can be estimated that an albumin molecule passes the capillary wall on average about $\bar{t}_p/\theta_p = 13$ times before irreversible degradation occurs. An almost constant value of $\bar{t}_p/\theta_p$ in patients with different alterations of k and MCR has led to the hypothesis that the disposal of albumin takes place in relation to its transcapillary passage.[17]

As mentioned above, whole-body residue measurements are difficult to perform. They may, however, be substituted by plasma sampling alone. If one considers the plasma space to be the ultimate pool of albumin before irreversible disposal, the whole-body mean transit time can be expressed as:

$$\bar{t}_{WB} = \frac{\int_0^\infty t \cdot c_p(t)dt}{\int_0^\infty c_p(t)dt}$$

This may also be written as

$$\bar{t}_{WB} = \int_0^\infty t \cdot h(t)\, dt$$

where h(t) is the frequency function of albumin disposal

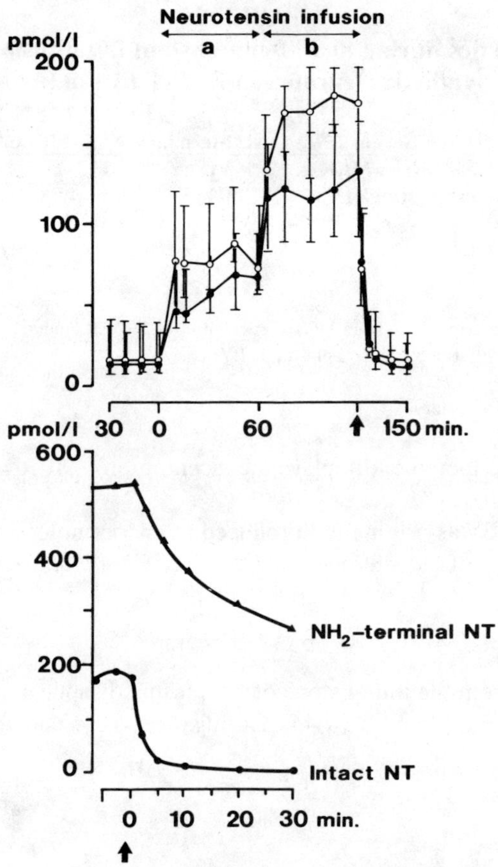

FIGURE 6. Upper panel: Median (± interquartile range) concentrations of intact neurotensin (NT 1-13) in arterial (○-○) and peripheral venous plasma (●-●) during i.v. infusion of 144 (a) and 288 (b) pmol/kg.h in six normal subjects. Lower panel: Median concentrations in arterial plasma of intact neurotensin (NT) (▲-▲) and NH_2-terminal NT (●-●) after stop of continuous i.v. infusion of 288 pmol/kg·h of intact NT 1-13.

$$\bar{t}_{WB} = \int_0^{\infty} t \cdot h(t)\, dt$$

The kinetics of albumin may, as illustrated above, be described by the noncompartmental (black-box) analysis. Compartmental analysis has also been performed by Rieve and co-workers.[9] It is beyond the scope of this survey to give the compartmental description of albumin kinetics but, in both cases, the basic concepts are the same and plasma as well as whole-body curves may be described by the sum of exponential functions.

C. WHOLE-BODY KINETICS OF SUBSTANCES WITH A FAST RATE OF TURNOVER

In principle, substances with a fast removal rate may be studied by the bolus injection technique as well as that of continuous infusion. However, the metabolic clearance rate may be fast enough to cause serious mixing problems in the plasma, and difficulties with accurate timing of the plasma samples may render the bolus injection technique inappropriate. An example is the rapid degradation of catecholamines, which is considered in Chapter 18.

In the *constant infusion* experiment (see Figure 6) the whole-body *metabolic clearance rate* (MCR) can be deduced during the steady state by dividing the infusion rate by the plasma

TABLE 2
Kinetics during and after Constant i.v. Infusion of 288 pmol/kg · h Synthetic Neurotensin [NT (1-13)] in Six Normal Subjects

Sample site	Intact NT (antiserum 759 A4) NH2-terminal $T_{1/2}$ (min)	MCR (ml/kg · min)	VD (ml/kg)	NT (antiserum 72 A9-4) $T_{1/2}$ (min)	MCR (ml/kg · min)	VD (ml/kg)
Artery	1.4	22.5	38.9	10.3	6.4	130.6
Cubital vein	2.3[a]	35.7[a]	118.3[a]	13.5	6.3	140.0

Note: Values given are medians from Holst Pedersen et al.[20] MCR: metabolic clearance rate; V_D: volume of distribution; and $T_{1/2}$: plasma half-life.

[a] $P < 0.01$ compared to artery.

concentration (or by the increment in plasma concentration): MCR = infusion rate/C_p = j/C_p (or = infusion rate/Δc_p).

The concept of MCR was originally introduced in endocrinology by Tait and co-workers in 1962[18] in their studies of aldosterone kinetics. The *volume of distribution* (V_D) is

$$V_D = MCR \cdot \bar{t}_{WB}$$

where $\bar{t}_{WB}$ is obtained from the initial slope of the plasma disappearance curve after termination of the infusion ($\bar{t}_{WB} = t_{1/2}/\ln 2 = t_{1/2} \cdot 1.44$). During the steady state, the appearance rate (AR) can be estimated as

$$AR = MCR \cdot C_p$$

The application of these equations, which are common in a variety of kinetic set-ups, presupposes that the changes in the plasma concentration reflect changes in the entire distribution space. Moreover, the plasma concentration (C_p) must be the concentration of the "metabolic inlet". In the case of a relatively low MCR, the concentration in plasma, as obtained from a peripheral vein, may be close to that of the metabolic site. But if the MCR is high (more than about 0.5 to 1.0 l/min · kg) or the substance under study binds significantly to protein (see below), the application of the plasma concentration from a peripheral vein may cause serious errors. The reason for the high MCR in this case is that samples obtained at that location in addition to whole-body metabolism may also reflect local degradation (see Chapter 18). By using a venous instead of an arterial sample, the MCR will be overrated and may in some cases even erroneously exceed the value of the cardiac plasma output (about 3.5 l/min or 50 ml/min · kg). V_D and $\bar{t}_{WB}$ are also affected (see Table 2). Use of the initial plasma slope after termination of the indicator infusion (residue detection kinetics) should be limited to substances with a rapid transcapillary passage (i.e., whole-body permeability surface area product much greater than MCR); otherwise, $\bar{t}_{WB}$ may be erroneously estimated with a subsequent error in the estimation of V_D. Special problems exist if an endogenous substance is released and degraded locally without the involvement of plasma transport (e.g., neurotransmitters) or if a substantial degradation takes place in the blood itself (red cells or plasma). The kinetic conditions for V_D and MCR are then not fulfilled.

Neurotensin, a 13 amino acid peptide, may serve as an example of the continuous infusion technique for determination of whole-body kinetics. Figure 6 illustrates the concentrations of neurotensin in arterial and cubital venous plasma in a control period and during continuous i.v. infusion of 144 and 288 pmol/kg·h neurotensin 1-13.[20] The figure shows that the plasma levels increase proportionately with the infusion rate, which indicates first-order kinetics.

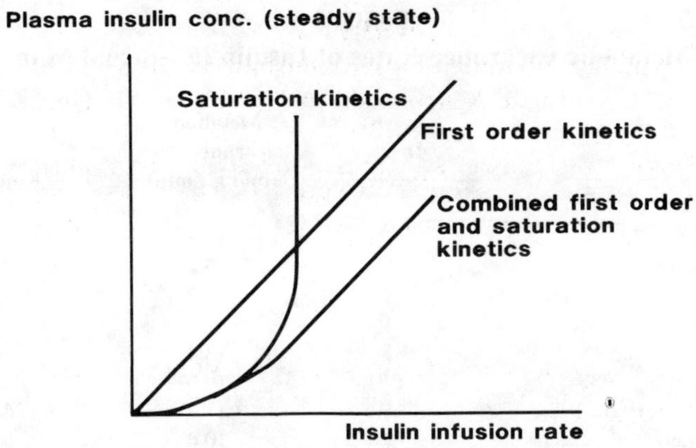

FIGURE 7. Relationship between steady-state plasma concentration of insulin and infusion rate of exogenous insulin in different models of insulin kinetics.

Moreover, there is a substantial difference between the MCR calculated from arterial samples and that from venous samples (see Table 2). The reason is local extraction of neurotensin in the forearm in addition to whole-body disposal of this peptide. The arterio-venous extraction ratio of neurotensin increases with increasing arterial peptide level. This may be due to protein binding of the peptide, allosteric enzyme kinetics, local release, peripheral vasoconstriction, or because neurotensin accelerates its own degradation (see Section V.A). The volume of distribution is assessed as $V_D = MCR \cdot t_{WB}$, the latter factor being obtained from the slope of the initial plasma disappearance after termination of the continuous infusion. This type of kinetics presupposes, as mentioned above, that the disappearance of neurotensin from plasma reflects its disappearance from the entire body (whole-body residue detection). The MCR is so high that limitations in transcapillary passage and plasma flow in tissues with relatively low perfusion may call into question the validity of this assumption.

The immunoreactive site of the molecule is another issue of relevance in the quantitation of peptide degradation. This may be illustrated by the employment of different antisera for the measurement of neurotensin. Antisera 759 A4 and 72 A9-4 react with intact neurotensin.[20] The plasma immunoreactivities during and after constant infusion of biosynthetic neurotensin 1-13 are, however, quite different, (see Figure 6). This is because antiserum 72 A94 reacts not only with intact neurotensin 1-13, but also with NH_2-terminal degradation products (NT 1-12, NT 1-11, NT 1-10, NT 1-8, and NT 1-6). Moreover, its reactivity for, for example, NT 1-8 and NT 1-6 is almost twice that for intact neurotensin.[20] Neurotensin degradation products have longer plasma lives than their mother substance, which is the reason for the slower disappearance of plasma 72 A9-4 immunoreactivity compared with that of 759 A–4 (see also Table 2).

Insulin metabolism may serve as another example of rapid whole-body degradation. Different models have been proposed for the degradation of circulating insulin.[21] According to one model, insulin is removed by a nonsaturable process (first-order kinetics), but to another by a saturable process (zero-order kinetics). A third model describes a combination of the two pathways (see Figure 7). It is now widely accepted that insulin kinetics saturates at supra-physiological concentrations of plasma insulin.[21] A number of variables may affect its whole-body disposal: hepatic extraction, glucose intake, cathecholamines, insulin antibodies, and various diseases. The interaction between insulin molecules and receptors[22] is described in Chapter 5. Insulin does not strictly fulfill formal assumptions for a substance which can be described with a clearance over a wide range of plasma concentrations. Nonetheless, the MCR of insulin has been commonly accepted as a measurement of insulin disposal from plasma; see Table 3.[23-52]

TABLE 3
Metabolic Clearance Rates of Insulin in Normal Man

	Insulin species[a]	Metabolic clearance rate (ml/kg · min)	Kinetics
Genuth, 1972	P	12.3	First order
Sönksen et al., 1973	H	23.4	do
Shervin et al., 1974	P	11.2	do
McGuire et al., 1979	P	12.4	do
Minaker et al., 1982	P	13.4	do
Home et al., 1982	P/H	19.6	do
Ferrannini et al., 1983	P	11.7	do
Del Prato et al., 1983	P	10.6	do
McGulloch et al., 1983	P	17.3	do
Waldhäusl et al., 1983	P/H	9.9	do
Adeniyi-Jones et al., 1983	P/H	32.9	do
Pagano et al., 1983	?	11.0	do
Pagano et al., 1984	?	7.5	do
Gray et al., 1984	B/H	21.1	do
Proietto et al., 1984	P	13.6	do
Vannini et al., 1984	P	19.5	do
Fink et al., 1985	?	18.7	do
Waldhäusl et al., 1985	H	11.1	do
Dimitriadis et al., 1985	P	16.1	do
Nygren et al., 1985	P	16.5	do
Yki-Järvinen et al., 1985	P	11.4	do
DeFeo et al., 1986	P	12.3	do
Cohen et al., 1986	P	24.9	do
Randin et al., 1986	P	12.1	do
Van Haeften et al., 1986	P	17.3	do
Henriksen et al., 1987	EH	15	First order
Frost et al., 1973	H	33.3[b]	Michaelis-Menten
Thorsteinsson et al., 1985	P	27.7[b]	do
Thorsteinsson et al., 1987	P/H	35.2[b]	do
Elleman et al., 1987	P	24.9[b]	do

[a] B: bovine insulin, EH: endogenous human insulin, H: human insulin, and P: porcine insulin.
[b] In studies applying first-order kinetics despite evidence of saturation, the metabolic clearance rate is indicated for the lowest infusion rate. Values in the table are means from References 23 to 52, modified from Reference 21.

V. ORGAN AND TISSUE DEGRADATION

Assessment of the disposal of bioactive substances in single organs (or determination of the fractional share of a specific organ in whole-body disposal) is important in physiology and pathophysiology. Traditionally, the *liver* and *kidneys* have attracted most attention because of their important metabolic and excretory functions. However, *skin, adipose tissue,* and *skeletal muscles* may be equally important owing to their large size. The *lungs* are interesting because they may serve as a filter for the entire cardiac output. In the last decade, a number of enzymes and receptors have been identified in numerous organs and tissues, including the endothelial cells of the *blood vessels.* Methods of assessing disposal in single organs or tissue are (1) elective catheterization of organ veins with sampling of endogenous substances, metabolites, or tracers from the organ inlet and outlet (Fick principle), (2) bolus injection into the arterial inlet followed by elective sampling from the outlet, (3) tissue sampling after indicator

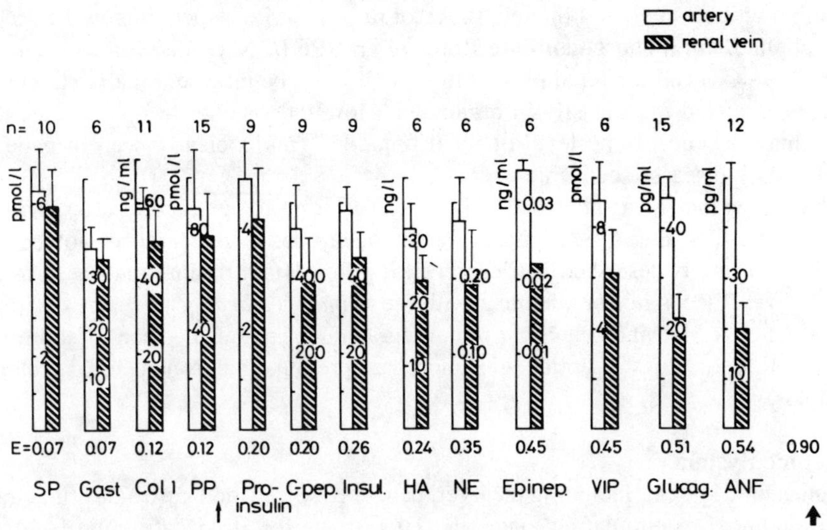

FIGURE 8. Arterio-renal venous extraction of a number of bioactive substances: SP = substance P, gast = gastrine, PP = pancreatic polypeptide, Col1 = fragment of procollagen III propeptide, C-pep = C-peptide, Insul = insulin, HA = hyaluronic acid, Epinep = epinephrine, NE = norepinephrine, VIP = vasoactive intestinal polypeptide, Glucag = pancreatic glucagon, ANF = atrial natriuretic factor. The small arrow denotes the filtration fraction, the large arrow shows the extraction ratio of hippurate. E = extraction ratio.

injection (sink method), (4) measurements before and after exclusion of one or several organs (stopped-flow technique), and (5) excretion measurement (urine and bile).

A. ASSESSMENT OF ORGAN DISPOSAL DURING ELECTIVE CATHERIZATION OF ORGAN VENOUS OUTLET (THE FICK PRINCIPLE)

The arterio-organ venous extraction technique demands selective catheterization of the particular organ vein. A number of organs are accessible to elective venous catheterization: for example, renal veins, hepatic veins, coronary sinus, internal jugular vein, left adrenal vein, and deep muscular veins.[53,54] It may be difficult to obtain blood from the mixed venous outlet of other organs, such as the right adrenal, thyroid, or bone marrow. The arterio-venous *extraction ratio* should always be considered in relation to the plasma flow rate. If the organ venous blood flow is slow, the blood samples must be collected very slowly in order to avoid downstream contamination from larger vessels. Sampling from wedged catheters (e.g., in pulmonary artery or hepatic veins) is difficult. Moreover, arterial and organ venous samples must be taken at exactly the same time because a number of endogenous peptides, hormones, etc. show a pulsatile release with subsequent oscillations in the plasma concentration.[55,56] It is also essential to overcome the catheter deadspace, not only once, but three to five times before collecting the sample.

1. Kidney

The *kidney* is an organ well suited for the extraction method. Both renal veins are easy to catheterize. The left renal vein receives minor inlets from the left adrenal gland and testis/ovary. The blood flow is normally high and well mixed; only in patients with severely reduced unilateral or total kidney function is the perfusion low. Renal blood flow may be assessed by hipurate clearance, ^{133}Xe wash-out, or thermodilution techniques.[57,58] Figure 8 illustrates renal extraction for a number of bioactive substances as determined from the renal vein catetherization method.[48,53,54,59-66]

If the renal elimination of a substance is small, the difference in arterial and renal venous concentration will be small and the relative error may be large, depending on the accuracy of the assay applied. For a constant elimination rate j = RPF ($C_a - C_{rv}$), the arterio-renal venous difference ($C_a - C_{rv}$) and the renal plasma flow (RPF) will be inverse. Small extractions may, therefore, be detected more easily in organs with lower flow rate. In other cases, the renal extraction may depend on the level of renal function. This is clearly seen in patients with unilateral kidney disease (see Table 4).

The renal clearance of a substance can be derived from the extraction ratio and the renal plasma flow rate Cl = E · RPF The general relationship between clearance, flow, extraction, and elimination rate is described in Chapter 3. It is important to note that the *arterio-renal venous clearance* needs not be identical with the *urinary clearance*. Urinary clearance will often be smaller, which indicates that renal degradation has taken place in addition to the urinary excretion. Insulin, C-peptide, glucagon, vasopressin, hyaluronan, and catecholamines are examples.

2. Splanchnic System

The splanchnic system, including the liver, pancreas, spleen, and gastrointestinal tract, may eliminate a number of circulating materials. This is true for atrial natriuretic factor (ANF), hyaluronan, catecholamines, pituitary peptides, and renin, see Figure 9.[20,48,53,54,59,61,66,67] As portal venous blood is difficult to collect, the extraction analysis of the liver itself is complex. Moreover, the dual inlet from the hepatic artery and portal vein adds to the difficulty.

Pancreatic peptides must be given special attention, because they are released directly into the portal blood stream. The hepatic vein catheterization technique can provide substantial information.[48,68–70] The hepatic extraction ratio of insulin may be determined as

$$\left(C_p - C_{hv}\right)_I / C_{p,I}$$

where $C_{hv,I}$ is the plasma concentration of insulin in the hepatic vein and $C_{p,I}$ is the estimated presinusoidal concentration of insulin derived from the equation:

$$C_{p,I} = \left(C_{hv} - C_a\right)_C + C_{a,I}$$

where the suffix C denotes C-peptide and $C_{a,I}$ the arterial plasma concentration of insulin. This equation presupposes (1) an equimolar release of insulin (I) and C-peptide (C) from the pancreatic beta-cells into the portal vein, and (2) no significant degradation of C-peptide through its transheptatic passage. The fraction of endogenously released insulin degraded by the liver (DF) can be determined as

$$DF = 1 - \left(C_{hv} - C_a\right)_I / \left(C_{hv} - C_a\right)_C$$

The derivation of this simple equation may illustrate some general kinetic considerations. The pancreatic release rate of insulin (RR_I) and C-peptide (RR_C) may be expressed as

$$RR_I = PVF \cdot \left(C_{pv} - C_a\right)_I = HPF \cdot \left(C_p - C_a\right)_I$$

and

$$RR_C = PVF \cdot \left(C_{pv} - C_a\right)_C = HPF \cdot \left(C_p - C_a\right)_C$$

where PVF and HPF are portal venous and hepatic plasma flow rates, respectively, and C_{pv} is the peptide concentration in portal venous plasma. It is assumed that portal venous and hepatic arterial plasma are mixed at the sinusoidal inlet, where the concentration is C_p. Equimolar release implies that $RR_I = RR_C$ and thereby, that

TABLE 4
Arterio-Renal Venous Extraction Ratios of Insulin, C-Peptide, Proinsulin, and Glucagon

	Insulin		C-peptide		Proinsulin		Glucagon	
	Controls	Kidney disease[a]	Controls	Kidney disease[a]	Controls	Kidney disease[a]	Controls	Kidney disease[b]
Kidney with higher function	0.26 ± 0.05	0.21 ± 0.07	0.20 ± 0.04	0.20 ± 0.05	0.20 ± 0.06	0.23 ± 0.08	0.55 ± 0.07	0.51 ± 0.08
Kidney with lower function		0.11 ± 0.04[c,d]		0.10 ± 0.06[e,f]		0.01 ± 0.07[d,f]		0.31 ± 0.07[c,d]

Note: Values are given as the mean ± SEM. Controls (n = 9).[48,61]

[a] n = 4.
[b] n = 8.
[c] Significantly different from controls, P < 0.02.
[d] Significantly different from best kidney, P < 0.01.
[e] Significantly different from controls, P < 0.05.

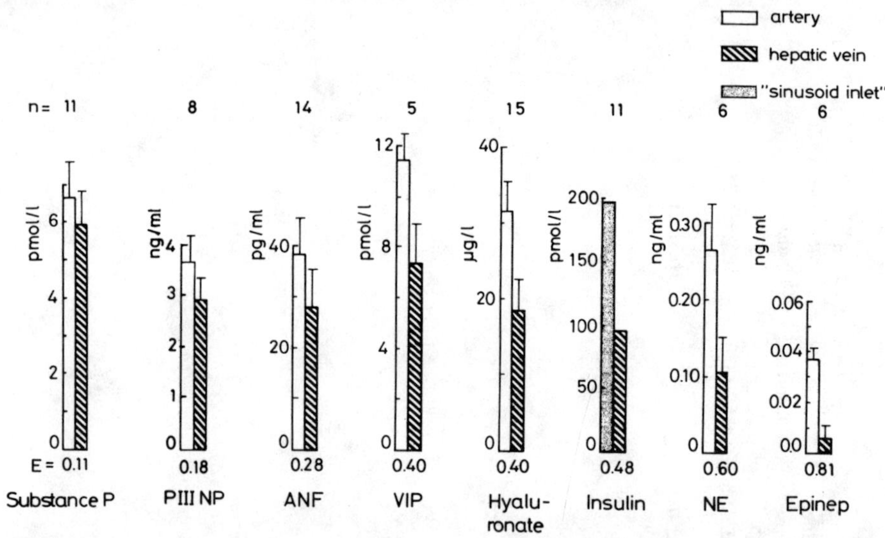

FIGURE 9. Arterior-hepatic venous extraction of a number of bioactive substances. Abbreviations: see legend to Figure 8.

$$HPF \cdot (C_p - C_a)_I = HPF \cdot (C_p - C_a)_C \tag{1}$$

As it is assumed that no significant hepatic degradation of C-peptide takes place, hepatic input equals hepatic output:

$$HPF \cdot C_{p,C} = HPF \cdot C_{hv,C}$$

which implies that $C_{p,C} = C_{hv,C}$. Substituted into Equation 1, the following estimate of $C_{p,I}$ may be derived:

$$C_{p,I} = (C_{hv} - C_a)_C + C_{a,I} \tag{2}$$

The fraction of endogenously released insulin degraded by the liver (DF = hepatic rate of insulin degradation divided by RR_I) can be expressed as

$$DF = \frac{HPF \cdot (C_p - C_{hv})_I}{HPF \cdot (C_p - C_a)_I} = \frac{(C_p - C_{hv})_I}{(C_p - C_a)_I} \tag{3}$$

from Equations 2 and 3, it follows that

$$DF = \frac{(C_{hv} - C_a)_C + C_{a,I} - C_{hv,I}}{(C_{hv} - C_a)_C}$$

$$= 1 - (C_{hv} - C_a)_I / (C_{hv} - C_a)_C$$

During the steady state, the overall release rate will equal the overall removal rate. On this assumption, the extrahepatic metabolic clearance rate of C-peptide may be estimated from the following equation.

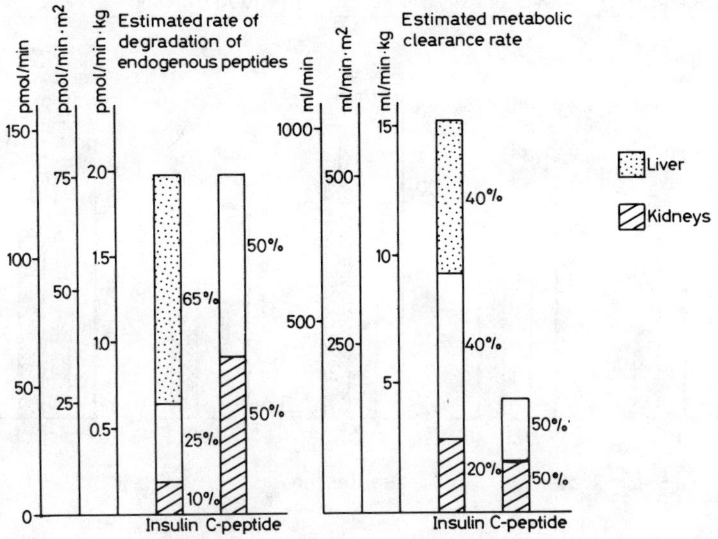

FIGURE 10. Whole-body and organ disposal of endogenous insulin and C-peptide as obtained by the hepatic venous catheterization technique.

$$MCR_C = HPF \cdot (C_{hv} - C_a)_C / C_{a,C}$$

In fasting, supine persons without liver disease, HPF is about 800 ml/min (average value of normal subjects from our laboratory), MCR_C is then on the average 300 ml/min (approximately 4.3 ml/min· kg). This values is close to that obtained by Faber et al.[71] who used a bolus injection, and Polonsky et al.[72] who used constant intravenous infusion of synthetic human C-peptide.

The metabolic clearance rate of insulin (MCR_I) may be assessed as the sum of the hepatic and extrahepatic clearances of that peptide

$$MCR_I = \left[HPF \cdot (C_p - C_{hv})_I / C_{p,I} \right] + \left[HPF \cdot (C_{hv} - C_a)_I / C_{a,I} \right]$$

Estimated in this way MCR_I is about 1000 ml/min (or approximately 15 ml/min·kg or 480 ml/min·m²); see Figure 10 and Table 3.

3. Disposal Affected by Food Ingestion

As illustrated in Figure 11, the splanchnic degradation of atrial natriuretic factor (ANF) may serve as an example of organ removal affected by food intake. The arterio-hepatic venous extraction ratio of ANF decreased somewhat after intake of a test meal.[73] Nevertheless, splanchnic clearance and the disposal rate of ANF increased substantially. This is probably because of the concomitant increase in splanchnic blood flow and oxygen uptake governed by the meal. The implications of alimentary enhancement of splanchnic ANF disposal are unknown.

The hepatic extraction of insulin may depend on the level of glucose output from the liver. Thus, an oral glucose load may increase the plasma insulin response, in part because of a lower hepatic extraction of insulin.[74]

4. Disposal and Organ Function

Degradation of a number of substances depends on the function of the particular organ in question (see Table 4). Compared with healthy organs and organs with a less-affected overall function, extraction may be decreased in pathological conditions with organ insufficiency.

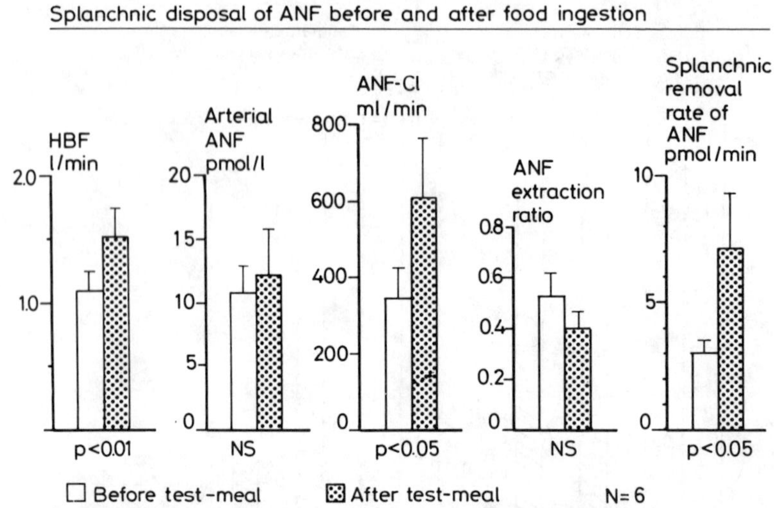

FIGURE 11. Splanchnic disposal of atrial natriuretic factor (ANF) before and after food intake in subjects without liver disease.

This has been considered in the foregoing sections. However, in some cases, disposal seems to be independent of parenchymal function. The explanation may be substantial metabolic activity of endothelial cells or degradation in the plasma itself, which does not depend on the functional level of specific cells in a particular organ. An example is hepatic degradation of atrial natriuretic factor (ANF). In patients with cirrhosis,[60,66] hepatic disposal of ANF is not decreased, although such patients have a reduced hepatocellular function. Examples of other substances degraded in the blood vessels are substance P and hyaluronic acid (HA). Substance P is discussed in Chapter 9, HA is almost exclusively degraded by liver endothelial cells (Reference 63 and Chapter 16). However, in the presence of cirrhosis, the hepatic HA extraction is decreased. This is due to the defenestration of the liver endothelium that takes place in the cirrhotic liver with subsequent restriction in the endothelial uptake of HA molecules. On the other hand, hepatic degradation of renin is normal in patients with cirrhosis. Renin is degraded in liver endothelial cells (see Chapter 6).

The above discussion illustrates that a general prediction as to changed disposal in organs with reduced function is impossible.

5. Disposal Affected by Drugs

A number of drugs can affect degradation and excretion of bioactive substances. It is beyond the scope of this survey to describe drug interaction and pharmacokinetics. The reader is referred to handbooks on this topic.

6. Extraction "Recruitment"

Special problems may arise if one determines a relatively small extraction ratio in the basal condition, and a subsequent increase in the circulating level (e.g., by intravenous infusion) shows that the extraction ratio increases; see Table 5. This is the opposite of the constant extraction ratio of the first-order kinetic experiment and the decreasing extraction ratio under zero-order kinetics. Various explanations for the increasing extraction are

1. A relatively large and/or firm protein binding with only extraction of free substance may be considered a zero offset artifact; that is, addition of a relatively large (protein-bound)

TABLE 5
Extraction Ratio (E) and Plasma Levels of Neurotensin in Brachial Artery and Cubital Vein after Constant Intravenous Infusion of Synthetic Neurotensin (NT 1-13) in Six Normal Subjects

Dose of NT (1-13) infused (pmol/kg·h)	Intact NT (antiserum 759 A4) Artery (pmol/l)	Intact NT (antiserum 759 A4) Vein (pmol/l)	Intact NT (antiserum 759 A4) E (%)	NH2-terminal NT antiserum 72 A9 4) Artery (pmol/l)	NH2-terminal NT antiserum 72 A9 4) Vein (pmol/l)	NH2-terminal NT antiserum 72 A9 4) E (%)
Basal	17	14	10	57	48	10
144	80[a]	62[a]	23[a]	348[a]	292[a]	17
288	183[a]	122[a]	34[a]	557[a]	484[a]	15

Note: Values are medians; E = arterio-cubital venous extraction ratio.[20]

[a] Significantly different from basal values (all $P < 0.05$).

value in the denominator of the extraction equation. By increasing the circulating level, the fraction of protein bound material becomes relatively smaller and, consequently, the calculated extraction ratio increases to a constant value.

2. If the substance under study is released as well as extracted in the same vascular bed, the release rate may be of almost the same size as that of the extraction rate.[75] This gives a small or no significant net extraction in the basic situation (examples are noradrenaline, substance P, and calcitonin gene-related peptide). By increasing the circulating level, the net extraction ratio may increase because of an almost unchanged release rate and a tissue uptake proportional to the plasma concentration.
3. The presence of allosteric metabolic enzyme sites may enhance the catabolic velocity with increasing concentration (analogous to the oxygen uptake curve of reduced hemoglobin).
4. A substance may accelerate its own extraction through receptor mechanisms or metabolic regulation. An example is adrenaline, which is known to enhance its own extraction in peripheral tissues.[76]
5. The blood flow rate may change and thereby change the extraction ratio without a change in the transcapillary flux rate of extracted material. On the other hand, recruitment of capillaries may enhance the "number of metabolic sites" with subsequent increase in extraction.

B. SINGLE INJECTION MULTIINDICATOR TECHNIQUE

The multiindicator technique applies injection of a bolus into the arterial organ inflow, followed by elective sampling from the venous outflow.[77-79] The single injection multiindicator dilution technique involves injection of a reference substance which remains in the bloodstream during a single organ passage. Other test materials are added to the injected volume, depending on the process to be studied: typically, a marker of extracellular volume (sucrose or ^{51}Cr EDTA). The third tracer should be an indicator for the mother substance degraded in the organ. The technique may also provide information on the transcapillary exchange of different material, including that metabolized in the organ. Figure 12 illustrates the single injection multiindicator method for different tracers;[79-82] this technique has been elaborated by Goresky and co-workers.

Indicator dilution curves should be normalized with respect to the amount of tracer bolus injected (see Figure 12). Labeled albumin is suitable for intravascular tracer, as it does not leave the capillaries to any significant degree during a single passage. Its outflow profile is shaped by the distribution of transit times in the large, nonexchanging vessels (arteries and

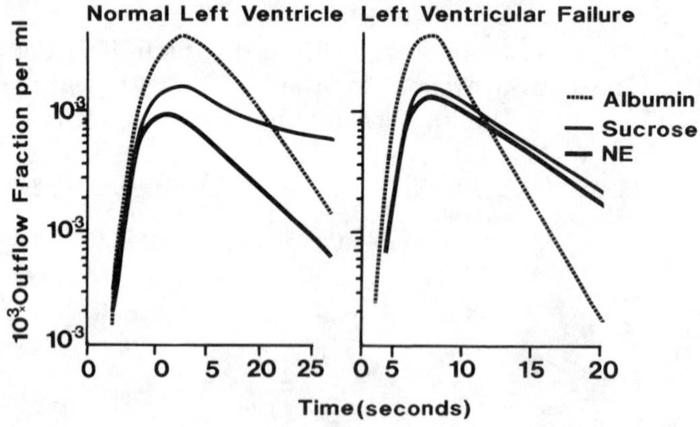

FIGURE 12. Left panel: Normalized coronary sinus outflow curves from a patient with mitral stenosis (normal left ventricular function), and right panel: a patient with arotic stenosis, aortic regurgitation, mitral regurgitation, and left ventricular hypertrophy (left ventricular insufficiency); ^{125}I-albumin, ^{14}C sucrose; ^{3}H norepinephrine (NE). It is seen that a substantial difference is present between the sucrose and NE outflow curves in the patient with normal left ventricle, indicating a major tissue uptake of norepinephrine (redrawn from Reference 82).

veins) and the capillaries in which the exchange of matter takes place. The up-slope and peak of the extracellular tracer curve are smaller than that of the intravascular tracer curve because some of it transverses the capillary barrier and enters the interstitial space. The remaining extravascular tracer transverses the capillaries without leaving. Some part of the indicator that has left the capillaries will return later to these vessels and the blood flow carries it via the veins to the sampling catheter. The extravascular tracer curve then crosses the intravascular tracer curve. If samples were collected for a sufficiently long time, all of the extravascular tracer would be recovered (e.g., the areas under the two curves would be the same). This may be difficult to show in the intact organism owing to recirculation of the tracer. The tracer of the metabolic test substance is handled in a similar fashion at the capillary membrane, but in the interstitial space an additional process occurs; that is, accumulation and degradation by metabolic enzymes and specific receptors (Chapters 4 and 5). The returning component of the metabolic tracer is reduced by these uptake and degradation processes. Whereas the form of the different tracer curves makes it apparent that a significant fraction of the metabolic tracer bolus does not leave the capillary, further analysis is needed in order to obtain a model of capillary transfer and the processes in the surrounding tissue under study. This analysis applies considerations at the single capillary level, introduction of a function describing dispersion of capillaries, and, lastly, modeling to approximate the experimental curves to different model parameters and in some cases to different models. The single capillary model would often be one in which there is bidirectional transfer at the capillary membrane with irreversible removal from the interstitial space by the above-mentioned receptors and metabolic pathways.

Analysis of these events occurring at the capillary level in space and time leads to two-dimensional partial differential equations that must be solved simultaneously. The first equation deals with the conservation of matter and the second with the rate of accumulation of metabolic tracer in the interstitial space. It is beyond the scope of this chapter to go into detail, and the reader is referred elsewhere,[1,77,79,81] but the process involves advanced iterative computer procedures. A number of conditions must be met in order to apply the method of single injection multiindicator outflow detection. Some of these are summarized in Chapter 18. A drawback in the study of human pathophysiology is the fact that catheters must be placed

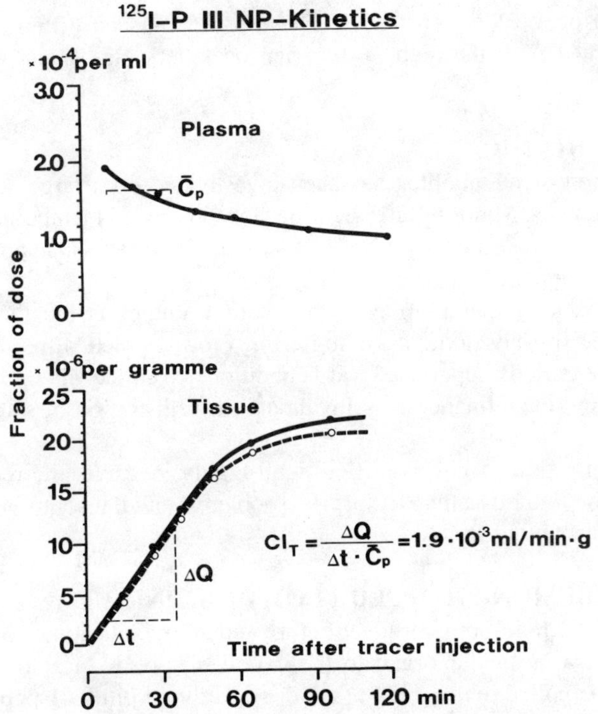

FIGURE 13. Tissue uptake of procollagen III aminoterminal propeptide (^{125}I-P III NP) in skin. The plasma concentration is expressed as the fraction of injected tracer dose per milliliter plasma. Tissue uptake shows fraction of tracer in total tissue (solid line) and fraction of tracer in extravascular tissue (dashed line) as obtained by ^{51}Cr-red cell correction. Note the almost linear initial tissue uptake (from L. T. Jensen et al., unpublished results).

in the arterial inflow and in a mixed equivalent outflow of the organ under study. In principle, the technique is very elegant and has provided important information on degradation and uptake of several test substances in the liver, heart, and brain.

C. TISSUE SAMPLING METHOD

Analysis of tissue specimens obtained after injection of indicator may provide substantial information on preferential uptake and accumulation of the test substance and its metabolites in tissues and organs. Owing to its invasive nature, this method has been applied in man only to persons undergoing subsequent surgical procedures (e.g., cholesterol uptake in the aortic intima,[83,84] uptake in fractured bones, and uptake and degradation in malignant tumors and skin). Rapid uptake of a connective tissue marker is illustrated in Figure 13.

This technique presupposes that the tissue can be considered a "sink" for the test substance; that is, that the presence of accumulated metabolites does not influence the uptake process from blood and interstitial space. Moreover, this method demands the application of intravascular markers (labeled red cells or labeled albumin). Another condition is that the tracer must not leave the tissue before appropriate sampling has taken place. By chemical separation, a number of metabolites may be identified and quantified at a given time after introduction of the test substance. In the case of a rapidly changing plasma concentration, the consequent variation in tissue or organ input may demand some convolution/deconvolution procedures. These are mathematical operations performed in order to reconstruct the tissue curve equivalent to that which would be obtained if the plasma concentration was kept constant by a balanced infusion of the test substance.[1] These procedures are often difficult and should always be performed with the help of experienced kineticists, owing to the risk of major artifacts.

D. EXCRETION STUDIES

Urinary excretion of metabolites has been investigated on a large scale in a number of pharmacokinetic studies. Major progress, for instance in steroid metabolism (Chapter 17), has been achieved by the assessment of metabolites in the urine. This also applies to the study of catecholamine metabolism (Chapter 18).

A major drawback of quantitative urinary output studies is the fact that 24-h urinary sampling is vitiated by substantial sampling error. From a practical point of view, it is very difficult, even under strictly supervised ward conditions, to collect all voided urine. Moreover, urinary retention and incontinence may invalidate the registration of sampling intervals and times.

Quantitative collection of bile is difficult and highly invasive, the reason for which this procedure should be limited in man to specific problems, and the solution of which cannot be obtained by other techniques.

E. ORGAN EXCLUSION (STOPPED-FLOW) TECHNIQUE

Measurements of whole-body turnover before and after exclusion of one organ may show the contribution of a particular organ to the overall disposal. In animal experiments, this technique has been applied in the study of the degradation of pituitary peptides and renin, etc. (see Chapters 6 and 13).

The biological effect of injection of material into the general circulation compared with the effect of equimolar injection into a specific organ artery may illustrate whether or not any disposal takes place into the organ in which the bioactive substance has been introduced (Chapter 9). In this context, it should be borne in mind that distribution into extravascular pools of some of the injected material with a subsequent delayed reentry into the circulation may blunt the biological response, without any presence of irreversible disposal in the organ, which renders the method more or less semiquantitative.

Results from organ exclusion and organ bypassing techniques should always be analyzed very carefully because of the risk of artifacts. Moreover, the surgical trauma and invasive stress may activate a number of neuroendocrine functions with increased release of bioactive substances, thus making the results difficult to interpret.

VI. CONCLUSION

The design of the kinetic approach to degradation should be highly specific in order to solve problems about the disposal of bioactive material. Sometimes, several methods (e.g., compartmental as well as noncompartmental models) provide an equally adequate description of the process under study. In such cases, various approaches may highlight different aspects of the process and thereby yield supplementary information. In other cases, a specific technique should be selected even though other techniques and methods could be applied. A specific technique may be selected based solely on assumptions, system limitations, flow independency, or sampling intervals and sites. The use of exogenous test material in biologically active doses vs. radioactive tracers should also be carefully considered.

No advanced or highly sophisticated kinetic approach can substitute for thorough analysis of the biological processes under study; especially in pathophysiology, organ insufficiency or altered metabolism may have a substantial influence not only on the magnitude of the kinetic variable (metabolic clearance rate, half-life, volume of distribution, and residence time), but also on the kinetic system structure (e.g., total number of compartments, presence of flow limitation, shunts, altered capillary exchange, presence of diffusive barriers, and presence of alternative elimination routes).

ACKNOWLEDGMENTS

This survey has been supported by the NOVO Foundation and Læge Ingrid Munkholms Legat.

REFERENCES

1. **Lassen, N. A. and Perl, W.,** *Tracer Kinetic Methods in Medical Physiology,* Raven Press, New York, 1979.
2. **DiStefano, J. J. and Landaw, E. M.,** Multiexponential, multicompartmental, and noncompartmental modeling. I. Methodological limitations and physiological interpretations, *Am. J. Physiol.,* 246, R651, 1984.
3. **Lightfoot, E. N.,** *Transport Phenomena and Living Systems. Biomedical Aspects of Momentum and Mass Transport,* John Wiley & Sons, New York, 1976.
4. **Bregner, P.-E. E.,** Tracer dynamics and the determination of pool-sizes and turnover factors in metabolic systems, *J. Theor. Biol.,* 6, 137, 1964.
5. **Anderson, D. H.,** *Compartmental Modeling and Tracer Kinetics,* Springer-Verlag, Berlin, 1983.
6. **Carson, E. R., Cobelli, C., and Finkelstein, L.,** *The Mathematical Modeling of Metabolic and Endocrine Systems,* John Wiley & Sons, New York, 1983, 1.
7. **Henriksen, J. H. and Christensen, N. J.,** Plasma norepinephrine in humans: limitations in assessment of whole body norepinephrine kinetics and plasma clearance, *Am. J. Physiol.,* 20, E743, 1989.
8. **McFarlane, A. S.,** Catabolism of plasma proteins, *Lancet,* ii, 131, 1963.
9. **Reeve, E. B. and Bailey, H. R.,** Mathematical models describing the distribution of ^{131}I-albumin in man, *J. Lab. Clin. Med.,* 60, 923, 1962.
10. **Rossing, N.,** *Human Albumin Metabolism Determined with Radioiodinated Albumin,* Munksgaard, Copenhagen, 1971.
11. **McFarlane, A. S.,** Measurement of synthesis rates of liver-produced plasma proteins, *Biochem. J.,* 89, 277, 1963.
12. **Parving, H.-H., Rossing, N., Nielsen, S. L., and Lassen, N. A.,** Increased transcapilary escape rate of albumin, IgG, and IgM after plasma volume expansion, *Am. J. Phsiol.,* 227, 245, 1974.
13. **Hesse, B., Parving, H.-H., Lund-Jacobsen, H., and Noer, I.,** Transcapillary escape rate of albumin and right atrial pressure in chronic congestive heart failure before and after treatment, *Cir. Res.,* 39, 358, 1976.
14. **Henriksen, J. H., Parving, H.-H., Lassen, N. A., and Winkler, K.,** Filtration as the main mechanism of increased protein extravasion in liver cirrhosis, *Scand. J. Clin. Lab. Invest.,* 40, 121, 1980.
15. **Putnam, E.,** *Plasma Proteins,* 2nd ed., Academic Press, New York, 1975.
16. **Honoré, B. and Pedersen, A. O.,** Conformational changes in human serum albumin studied by fluorescence and absorption spectroscopy. Distance measurements as a function of pH and fatty acids, *Biochem. J.,* 258, 199, 1989.
17. **Lassen, N. A., Parving, H.-H., and Rossing, N.,** Filtration as the main mechanism of overall transcapillary protein escape from the plasma, *Microvasc. Res.,* 7, i, 1974.
18. **Tait, J. F., Little, B., Tait, S. A. S., and Flood, C.,** The metabolic clearance rate of aldosterone in pregnant and nonpregnant subjects estimated by both single-injection and constant-infusion methods, *J. Clin. Invest.,* 41, 2093, 1962.
19. **Henriksen, J. H., Schaffalitzky de Muckadell, O. B., and Bülow, J. B.,** Does liver-intestine significantly degrade circulating endogenous substance P in man?, *Scand. J. Gastroenterol.,* 21, 300, 1986.
20. **Pedersen, J. H., Andersen, H. O., Olsen, P. S., and Henriksen, J. H.,** Pharmacokinetics and metabolism of neurotensin in man, *J. Clin. Endocrinol. Metabol.,* 68, 294, 1989.
21. **Thorsteinsson, B.,** Kinetic models for insulin disappearance from plasma in man, *Dan. Med. Bull.,* 37, 143, 1990.
22. **Gammeltoft, S.,** Insulin receptors: binding kinetics and structure-function relationship of insulin, *Physiol. Rev.,* 64, 1321, 1984.
23. **Genuth, S. M.,** Metabolic clearance of insulin in man, *Diabetes,* 21, 1003, 1972.
24. **Sönksen, P. H., Tompkins, C. V., Srivastava, M. C., and Nabarro, J. D. N.,** A comparative study on the metabolism of human insulin and porcine proinsulin in man, *Clin. Sci. Mol. Med.,* 45, 633, 1973.

25. **Shervin, R. S., Kramer, K. J., Tobin, J. D., Insel, P. A., Liljenquist, J. E., Berman, M., and Andres, R.,** A model of kinetics of insulin in man, *J. Clin. Invest.,* 53, 1481, 1974.
26. **McGuire, E. A., Tobin, J. D., Berman, M., and Andres, R.,** Kinetics of native insulin in diabetic, obese and aged men, *Diabetes,* 28, 110, 1979.
27. **Minaker, K. L., Rowe, J. W., Tonino, R., and Pallotta, J. A.,** Influence of age on clearance of insulin in man, *Diabetes,* 31, 851, 1982.
28. **Home, P. D., Massi-Beneditte, M., Shepherd, G. A. A., Hanning, I., Alberti, K. G. M. M., and Owens, D. R.,** A comparison of the activity and disposal of semi-synthetic human insulin and porcine insulin in normal man by the glucose clamp technique, *Diabetologica,* 22, 41, 1982.
29. **Ferrannini, E., Wahren, J., Faber, O. K., Felig, P., Binder, C., and DeFronzo, R. A.,** Splanchnic and renal metabolism of insulin in human subjects: a dose response study, *Am. J. Physiol.,* 244, E517, 1983.
30. **Del Prato, S., Nosadini, R., Tiengo, A., Tessari, P., Avogaro, A., Revisan, R., Valerio, A., Muggeo, M., Cobelli, C., and Toffolog, G.,** Insulin-mediated glucose disposal in type I diabetes: evidence for insulin resistance, *J. Clin. Endocrinol. Metab.,* 57, 904, 1983.
31. **McCulloch, A. J., Home, P. D., Heine, R., Ponchner, M., Hanning, I., Johnston, D. G., Clark, F., and Alberti, K. G. M. M.,** Insulin sensitivity in hyperthyroidism: measurement by the glucose clamp technique, *Clin. Endocrinol.,* 18, 327, 1983.
32. **Waldhäusl, W. K., Bratusch-Marrain, P. R., Vierhapper, H., and Nowotny, P.,** Insulin pharmacokinetics following continuous infusion and bolus injection of regular porcine and human insulin in healthy man, *Metabolism,* 32, 478, 1983.
33. **Adeniyi-Jones, R. O. C., Jones, R. H., Barnes, D. G., Gerlis, L. S., and Sönksen, P. H.,** Porcine and human insulin (NOVO): a comparison of their metabolic and hypoglycaemic activity in normal man, *Diabetes Care,* 6(Suppl. 1), 9, 1983.
34. **Patano, G., Cassader, M., Cavallo-Perin, P., Bruno, A., Ozzello, A., Dall'Omo, A. M., and Imbimbo, B.,** An *in vivo* and *in vitro* study of the mechanisms of prednisone-induced insulin resistance in healthy subjects, *J. Clin. Invest.,* 72, 1814, 1983.
35. **Pagano, G., Cassader, M., Cavallo-Perin, P., Bruno, A., Masciola, P., Ozzello, A., Dall'Omo, A. M., and Foco, A.,** Insulin resistance in the aged: a quantitative evaluation of *in vivo* insulin sensitivity and *in vitro* glucose transport, *Metabolism,* 33, 976, 1984.
36. **Gray, R. S., Cowan, P., Duncan, L. J. P., and Clarke, B. F.,** A comparison of the biological actions and pharmacokinetics of intravenously infused highly purified beef and biosynthetic human insulins in normal man, *Diabet. Metab.,* 10, 188, 1984b.
37. **Proietto, J., Dudley, F. J., Aitken, P., and Alford, F.,** Hyperinsulinaemia and insulin resistance in cirrhosis: the importance of insulin hypersecretion, *Clin. Endocrinol.,* 21, 657, 1984.
38. **Vannini, P., Forlani, G., Marchesini, G., Ciavarella, A., Zoli, M., and Pisi, E.,** The euglycemic clamp technique in patients with liver cirrhosis, *Horm. Metab. Res.,* 20, 1, 1988.
39. **Fink, R. I., Revers, R. R., Kolterman, O. G., and Olefsky, J. M.,** The metabolic clearance of insulin and the feedback inhibition of insulin secretion are altered with aging, *Diabetes,* 34, 275, 1985.
40. **Waldhäusl, W. K., Bratusch-Marrain, P. R., Jensen, P., Kruse, Nowotny, P., and Vierhapper, H.,** Effect of insulin antibodies on insulin pharmacokinetics and glucose utilization in insulin-dependent diabetic patients, *Diabetes,* 34, 166, 1985.
41. **Dimitriadis, G., Baker, B., Marsh, H., Mandarino, L., Rizza, R., Bergman, R., Haymond, M., and Gerich, J.,** Effect of thyroid hormone excess on action, secretion, and metabolism of insulin in humans, *Am. J. Physiol.,* 248, E,593, 1985.
42. **Nygren, A., Adner, N., Sundblad, L., and Wiechel, K.-L.,** Insulin uptake by the human cirrhotic liver, *Metabolism,* 34, 48, 1985.
43. **Yki-Järvinen, H.,** Sex and insulin sensitivity, *Metabolism,* 33, 1011, 1984.
44. **DeFeo, P., Perriello, G., Ventura, M. M., Calcinaro, F., Basta, G., Lolli, C., Gruciani, C., Dell'Olio, A., Santeusanio, F., Brunetti, P., and Bolli, G. B.,** Studies on overnight insulin requirements and metabolic clearance rate of insulin in normal and diabetic man: relevance to the pathogenesis of the dawn phenomenon, *Diabetologia,* 29, 475, 1986.
45. **Cohen, P., Barzilai, N., Barzilai, D., and Karnieli, E.,** Correlation between insulin clearance and insulin responsiveness: studies in normal, obese, hyperthyroid, and Cushing's syndrome patients, *Metabolism,* 35, 744, 1986.
46. **Randin, J.-P., Tappy, L., Scazziga, B., Jequier, E., and Felber, J.-P.,** Insulin sensitivity and exogenous insulin clearance in Graves' disease. Measurement by the glucose clamp technique and continuous indirect calorimetry, *Diabetes,* 35, 178, 1986.
47. **Van Haeften, T. W., Bolli, G. B., Dimitriadis, G. D., Gottesman, I. S., Horwitz, D. L., and Gerich, J. E.,** Effect of insulin antibodies and their kinetic characteristics on plasma free insulin dynamics in patients with diabetes mellitus, *Metabolism,* 35, 649, 1986.

48. **Henriksen, J. H., Tronier, B., and Bülow, J. B.,** Kinetics of circulating endogenous insulin, C-peptide, and proinsulin in fasting nondiabetic man, *Metabolism,* 36, 463, 1987.

49. **Frost, D. P., Srivastava, M. C., Jones, R. H., Nabarro, J. D. N., and Sönksen, P. H.,** The kinetics of insulin metabolism in diabetes mellitus, *Postgrad. Med. J.,* 49, 949, 1973.

50. **Thorsteinsson, B., Fugleberg, S., Feldt-Rasmussen, B., and Binder, C.,** Kinetic models for plasma disappearance of insulin in normal subjects, *Acta Pharmacol. Toxicol.,* 57, 309, 1985.

51. **Thornsteinsson, B., Fugleberg, S., and Binder, C.,** Kinetics of human and porcine insulins in normal and type I diabetic subjects, *Eur. J. Clin. Pharmacol.,* 33, 173, 1987b.

52. **Ellemann, K., Thorsteinsson, B., Fugleberg, S., Feldt-Rasmussen, B., Ortved Andersen, O., Grønbæk, P., and Binder, C.,** Kinetics of insulin disappearance from plasma in cortisone-treated normal subjects, *Clin. Endocrinol.,* 26, 623, 1987.

53. **Henriksen, J. H., Ring-Larsen, H., Kanstrup, I.-L., and Christensen, N. J.,** Splanchnic and renal elimination and release of catecholamines in cirrhosis. Evidence of enhanced sympathetic nervous activity in patients with decompensated cirrhosis, *Gut,* 25, 1034, 1984.

54. **Hollister, A. S., Rodeheffer, R. J., and White, F. J., et al.,** Clearance of atrial natriuretic factor by lung, liver, and kidney in human subjects and the dog, *J. Clin. Invest.,* 83, 623, 1989.

55. **Stagner, J. J., Samols, E., and Weir, G. C.,** Sustained oscillations of insulin, glucagon, and somatostatin from the isolated canine pancreas during exposure to a constant glucose concentration, *J. Clin. Invest.,* 65, 939, 1980.

56. **Paolisso, G., Scheen, A. J., Ouycks, A. S., and Lefebvre, P. J.,** Pulsatile hyperglucagonemia fails to increase hepatic glucose production in normal man, *Am. J. Physiol.,* 252, E1, 1987.

57. **Ring-Larsen, H.,** Renal blood flow in cirrhosis: relation to systemic and portal haemodynamics and liver function, *Scand. J. Clin. Lab. Invest.,* 37, 632, 1977.

58. **Tfelt-Hansen, P., Kanstrup, I.-L., Christensen, N. J., and Winkler, K.,** General and regional haemodynamic effect of intravenous ergotamine in man, *Clin. Sci.,* 65, 599, 1987.

59. **Schütten, H. J., Henriksen, J. H., and Warberg, J.,** Organ extraction of atrial natriuretic peptide (ANP) in man. Significance of sampling site, *Clin. Phys.,* 7, 125, 1987.

60. **Henriksen, J. H., Schütten, H. J., Bendtsen, F., and Warberg, J.,** Circulating atrial natriuretic peptide (ANF) and central blood volume (CBV) in cirrhosis, *Liver,* 6, 363, 1986.

61. **Petersen, K. F. and Henriksen, J. H.,** Endogenous glucagon in different vascular beds: the role of the kidney in degradation of circulating glucagon in man, *Diabetes Nutrition Metab.,* 2, 119, 1989.

62. **Bentsen, K. D., Henriksen, J. H., and Laurent, T. C.,** Circulating hyaluronate: concentration in different vascular beds in man, *Clin. Sci.,* 71, 161, 1986.

62a. **Bentsen, K. D., Henriksen, J. H., Bendtsen, F., Hørslev-Petersen, K., and Lorenzen, I.,** Splanchnic and renal extraction of circulating type III procollagen aminoterminal propeptide in patients with normal liver function and in patients with alcoholic cirrhosis, *Hepatology,* 11, 957, 1990.

63. **Henriksen, J. H., Bentsen, K. D., and Laurent, T. C.,** Splanchnic and renal extraction of circulating hyaluronan in patients with alcoholic liver disease, *J. Hepatol.,* 6, 158, 1988.

64. **Henriksen, J. H., Schwartz, T. W., and Bülow, J. B.,** Endogenous pancreatic polypeptide in different vascular beds: relationship to release and degradation in subjects with normal and decreased kidney function, *Metabolism,* 35, 542, 1986.

65. **Crozier, I. G., Nicholls, M. G., Ikram, H., Espiner, E. A., Yandle, T. G., and Jans, S.,** Atrial natriuretic peptide in humans. Production and clearance by various tissues, *Hypertension,* 8, 11, 1986.

66. **Ginés, P., Jimenez, W., Arroyo, V., Navasa, M., Lopez, C., Tito, L., Serra, A., Bosch, J., Sanz, G., Rivera, F., and Rodés, J.,** Atrial natriuretic factor (ANF) in cirrhosis, plasma levels, cardiac release and splanchnic extraction, *Hepatology,* 6, 636, 1988.

67. **Vierhapper, H., Gasic, S., Nowotny, P., and Waldhaüsel, W.,** Splanchnic disposal of human natriuretic peptide in humans, *Metabolism,* 37, 973, 1988.

68. **Waldhäusl, W., Bratusch-Marrain, P., Gasic, S., Korn, A., and Nowotny, P.,** Insulin production rate following glucose ingestion estimated by splanchnic C-peptide output in normal man, *Diabetologia,* 17, 221, 1979.

69. **Bratusch-Marrain, P. R., Waldhäusl, W. K., Gasic, S., and Hofer, A.,** Hepatic disposal of biosynthetic human insulin and procine C-peptide in humans, *Metabolism,* 33, 151, 1984.

70. **Tranberg, K.-G.,** Hepatic uptake of insulin in man, *Am. J. Physiol.,* 237, E-509, 1979.

71. **Faber, O. K., Hagen, C., Binder, C., Markussen, J., Naithani, V. K., Blix, P. M., Kuzuya, H., Horwitz, D. L., and Rubinstein, A. H.,** Kinetics of human connecting peptide in normal and diabetic subjects, *J. Clin. Invest.,* 62, 197, 1978.

72. **Polonsky, K. S., Licionio-Paixao, J., Given, B. D., Pugh, W., Rue, P., Galloway, J., Karrizon, T., and Frank, B.,** Use of biosynthetic human C-peptide in the measurement of insulin secretion rates in normal volunteers and type I diabetic patients, *J. Clin. Invest.,* 77, 98, 1986.

73. **Henriksen, J. H., Bendtsen, F., and Gerbes, A. L.,** Splanchnic removal of atrial natriuretic factor (ANF) in man: enhancement after food intake, *Metabolism,* 39, 553, 1990.
74. **Madsbad, S., Kehlet, H., Hilsted, J., and Tronier B.,** Discrepancy between plasma C-peptide and insulin response to oral and intravenous glucose, *Diabetes,* 32, 436, 1983.
75. **Brown, M. J., Jenner, D. A., Allison, D. J., and Dollery, C. T.,** Variations in individual organ release of noradrenaline measured by an improved radioenzymatic technique; limitations of peripheral venous measurements in the assessment of sympathetic nervous activity, *Clin. Sci.,* 61, 585, 1981.
76. **Astrup, A., Bülow, J., Christensen, N. J., Madsen, J., and Quaade, F.,** Facultative thermogenesis induced by carbohydrate: a skeletal muscle component mediated by epinephrine, *Am. J. Physiol.,* 13, E226, 1986.
77. **Goresky, C. A.,** A linear method for determining liver sinusoidal and extravascular volumes, *Am. J. Physiol.,* 204, 626, 1963.
78. **Goresky, C. A., Cronin, R. F. P., and Wangel, B. E.,** Indicator dilution measurements of extravascular water in the lungs, *J. Clin. Invest.,* 48, 487, 1969.
79. **Goresky, C. A., Bach, G. G., Cousineau, D., Schwab, A. J., Rose, C., Lee, S., and Goresky, S.,** Handling of tracer norepinepherine by the dog liver, *Am. J. Physiol.,* 256, G-107, 1989.
80. **Cousineau, D., Goresky, C. A., and Rose, C. P.,** Decreased basal cardiac interstitial norepinephrine release after neuronal uptake inhibition in dogs, *Circ. Res.,* 58, 859, 1986.
81. **Cousineau, D., Rose, C. P., and Goresky, C. A.,** Labeled catecholamine uptake in the dog heart: interactions between capillary wall and sympathetic nerve uptake, *Circ. Res.,* 47, 329, 1980.
82. **Rose, C. P., Burgess, J. H., and Cousineau, D.,** Tracer norepinephrine kinetics in coronary circulation of patients with heart failure secondary to chronic pressure and volume overload, *J. Clin. Invest.,* 76, 1740, 1985.
83. **Stender, S. and Hjelms, E.,** *In vivo* influx of free and esterified plasma cholesterol into human aortic tissue without atherosclerotic lesions, *J. Clin. Invest.,* 74, 1871, 1984.
84. **Stender, S. and Zilversmit, D. B.,** *In vivo* influx, tissue esterification and hydrolysis of free and estified plasma cholesterol in the cholesterol-fed rabbit, *Biochim. Biophys. Acta,* 663, 674, 1981.
85. **Parving, H.-H.,** Microvascular permeability to plasma proteins in hypertension and diabetes mellitus in man — on the pathogenesis of hypertensive and diabetic microangiopathy, *Dan. Med. Bull.,* 22, 217, 1975.

Chapter 3

KINETICS OF ELIMINATION FROM THE BLOOD: PRINCIPLES AND PROCEDURES

Kjeld Winkler

TABLE OF CONTENTS

I. INTRODUCTION

Studies of the elimination of substances present in the blood are essential for the understanding of the biological behavior of bioactive materials. These may be regulatory compounds, metabolites, drugs, and toxins, and be derived from endogenous or exogenous sources. Such studies are also valuable in quantitating the effects of disease on the function of the organs responsible for their removal or synthesis. In itself, measurement of the blood concentration level or of changes therein is of rather limited interest for our biological understanding and, therefore, more sophisticated methods have to be used. One of the widely employed concepts in this topic is 'kinetics' (from Greek: movement). In the medical sciences, this usually means a description of the (directional) movements of substances between various compartments of the body. It encompasses the processes by which these movements are caused, and usually describes not only directions, but also the rates of change. This is achieved by a model; that is, a picture in space and time by which the changes occur. This quantitative (mathematical) description is based on structural, physiological and biochemical knowledge of the ways and means by which organ elimination or synthesis is performed. The kinetic description gives information about the time spent by a substance in a given compartment (the 'residence' or 'sojourn' time) and, further, about the parameters of the models usually stated in meaningful physiological terms (clearance, volumes of compartments, and rate of synthesis) to make them generally applicable and understandable.

The complete insight requires quantification of all the processes which determine the concentration in the circulation which usually is the only component accessible for study. In the following, some general principles governing the type of reasoning will be described.

Elimination depends on many cooperating factors, usually making the total description very complex:

1. The properties of the eliminating organ(s)
2. Their site in the circulation — series (lung) or parallel (liver)
3. Total and regional blood flow rates
4. The microcirculatory arrangement, including its capillary permeability
5. Distribution in extravascular compartments
6. Resynthesis or recirculation of the substance already eliminated

Whether elimination occurs in a given organ or at various sites depends on the chemical structure, degree of protein binding, and the properties of the eliminatory organs.

In the following discussion, the role of the liver will be used as a paradigm because of its dominant role in removal processes and because its kinetics are relatively well studied.

II. KINETIC MODELS IN GENERAL

The purpose of kinetics is to describe — and thereby to predict — how changes in blood concentrations by physiologic, biochemical, or pathological processes are brought about.

To measure an elimination curve from the blood ('the central delivery system') after an exogenous load, and to fit it to some heuristic mathematical function (exponential, parabolic, hyperbolic, or polynomial are frequently used) is not difficult, as many functions adequately describe sections of the curves and their parameters, but they do not explicitly tell where and how elimination is brought about. Thus, from the shape of a blood elimination curve, one can only make limited conclusions of the kinetics and mistakes can be found, especially when extrapolation procedures (e.g., to obtain areas) are applied.

It is more profitable, but also more demanding, to apply the inverse procedure: to determine the shape of the blood removal curve (e.g., the relations of the removal rate to blood concentration) by using the available physiological-biochemical knowledge to construct a model as a picture of the mechanisms of removal, and to describe how the parameters are determined by macro- and microanatomic and physiologic circulations, spaces, and removal apparatus.

In modeling of liver kinetics, two different perspectives in the description of the processes have dominated: the *compartmental* and the *single capillary*. In the *compartmental model,* the removal from the blood volume into other compartments is described by the rate constants ('k') between the compartments (in series, parallel, or combinations), the size of the compartments, and organ removal. It is generally assumed that complete instantaneous mixing is at hand. The differential equations are usually more simple than those of the other group of models and standard computer solutions from actual data are often available. The most frequently used model in pharmacokinetics may serve as an example. Another is that of Richards[1] which describes the removal of organic anionic dyes from blood via liver cells into bile. By extending the models and trying to make them more physiologically realistic (e.g., by using Michaelis-Menten type of kinetics to describe the removal instead of first-order removal) their solutions become much more difficult and numerical analyses used as exact solutions may become impracticable. More complex models of this type have been proposed by Winkler[2] and Forker.[3]

In general, the compartmental models are more easy to understand and handle, but frequently they do not offer a complete description of the system under study. In this connection, it should be stressed that the compartments often do not correspond to physiologically measurable spaces of the body, as, for example, when the substance is protein-bound extravascularly (drugs) or in the liver (dyes excreted in bile). They become operational, and can be described as "apparent"; that is, the abstract larger volume in which the concentration is similar to that of the blood. The other concept is the *single capillary model* (in hepatology: 'sinusoidal'). The primary element considered is a tiny, transverse section of the capillary where cellular exchange occurs, integrated throughout its length and thereby to the entire organ. Contrary to the simple compartmental model, it includes the effects on elimination of the concentration gradient found in the microcirculatory arrangement and, thereby, also the influence of the organ perfusion which, in greater or lesser ways, determines the rate of cellular removal of substances. Thus, the assumption of complete mixing in each compartment is replaced by a more realistic concept of the capillary-cellular functional unit, giving a more complex result compared with the test tube condition ("well-mixed jar"). The mathematical and computational solutions may be very heavy; surveys may be found on general capillary physiology,[4] on tracer kinetics,[5] and regarding the liver.[6] The above definitions of classes of models may not completely cover those used by other authors.[6] Black-box analysis, using only organ input and output measurements, is also termed "noncompartmental".[5] The *compartmental models* in the present context assume some further properties of the box; for example, that mixing is complete and the rates of transfer between various compartments are characteristic for that process.

The models thus describe single organ functions and the basic types assume identity of each capillary unit. They can be extended — again for the purpose of realism — to include various types of distribution functions to those parameters of the model where heterogeneity of capillary anatomy and perfusion, and heterogeneity of cellular function (e.g., along the liver sinusoids) is expected.[7,8] This of course adds to complexity but covers experimental facts more satisfactorily. By and large, these distributed models describe the removal efficiency (removal rate at a given blood concentration) of the inherent structure of the organs in the body. A general aspect of these more elaborate models is that, by identical blood concentrations below saturation, the removal rate is smaller than that occurring during *in vitro* conditions. This difference may depict the biological efficiency of the organs *in situ,* but quantitative means to express this interesting parameter are not available at present. It should be added that, to describe removal from the entire body, the two types of model have to be joined. Solutions of various types of models can be found in References 3, 6, and 9 through 11, the last study describing the kinetics of urea synthesis by the liver.

III. REGIMES OF ELIMINATION

In this section, a more detailed discussion on the dependence of the rate of elimination on the blood concentration follows in order to amplify the preceding section on modeling, and serves as a prelude to proper experimental planning. The reader should note that this discussion is based on a discrete model and that the description may be different using other model concepts.

A model describing the hepatic elimination kinetics of galactose may serve as an example. It supposes that elimination by the liver is solely dependent on a system (the rate-limiting step in the removal) which follows simple Michaelis-Menten kinetics. The relations *in vivo* are, however, different from the completely mixed *in vitro* conditions. In an organ with a given main direction of the blood flow in the exchange vessels (sinusoids), a concentration gradient along the direction of flow exists, created by the eliminatory cells. To be able to determine the average or integral concentration which determines the actual rate of elimination, it must be calculated by a model, as actual measurement is usually impossible.

The general relation between the removal rate v, (e.g., mol/min), input blood concentration c_i, liver blood flow F, and the enzyme kinetic constants V_{max} (maximal removal rate), and Km (Michaelis constant) according to this model[12] is

$$v = F \cdot c_i \left[1 - e^{-\left(V_{max} - v\right)/\left(F \cdot K_m\right)} \right] \tag{1}$$

From experimentally determined connected values between c_i, v, and F, V_{max} and K_m can be determined from this "general rate expression" by nonlinear regression analysis,[9] and the system can be described, if the various assumptions of the model are valid. This formulation has been mathematically generalized to contain other types of elimination.[13] The expression has some intrinsic interest as it describes the relation of biochemistry (enzyme kinetics) to physiology (perfusion in an organ).

During certain restrictive conditions which can be mathematically and experimentally defined, various *regimes of elimination* can be discerned. These are dependent on the blood concentration (and the other constants in Equation 1) which theoretically may be varied from infinity to zero. The successive restrictions of these conditions should be carefully distinguished.

A. HIGH BLOOD CONCENTRATIONS

At very high blood concentrations, the maximal rate of elimination is found (*the saturated regime*), as all eliminatory sites are saturated.* The influence of blood flow, concentration, and K_m on elimination rate, as found in the general rate expression, disappears. For many substances, the high concentrations necessary for saturation cannot be achieved, due to unwanted hemodynamic or toxic effects. By again decreasing blood concentration, the *general regime* is reached. This regime is characterized by a complex dependence of the rate of elimination on all variables of Equation 1, included in the blood concentration. This regime is not easily suited for simple kinetic studies.

* When C_i increases, the general expression of Equation 1 approximates the well-known Michaelis-Menten expression: $v = V_{max} \cdot c_i/(c_i + K_m)$ and when $C_i >> K_m$, the saturated regime: $v = V_{max}$ is reached.

B. LOW BLOOD CONCENTRATIONS (CLEARANCE)

During this circumstance, more convenient kinetic expressions appear.[14] Low concentration of the *general clearance regime* is defined by the magnitude of the parameters of the system; that is, if the actual rate of elimination (v) is small in relation to the maximal (V_{max}), which appears from Equation 1, then

$$v/c_1 = F \cdot \left[1 - e^{-V\max/(F \cdot K_m)}\right] \tag{2}$$

The left side of this equation represents a clearance (elimination rate relative to concentration; unit: volume/time), and its size depends, in a rather simple way, on blood flow and the kinetic constants (i.e., the property of the removal system). The magnitude of the exponential constant of Equation 2 depends on the amount of removal apparatus (V_{max}), and of the flow, as well as on the system's K_m. In this context, K_m has a wider (operative) interpretation than in biochemistry, due to the more complex *in vivo* situation ('apparent', 'as seen from the blood'). For this reason, the K_m value of the intact organ may be different from that of the enzymes directly responsible for removal and can be perceived as a symbol of an eliminatory intensity of the organ.

At a fixed rate of flow, it appears that clearance is independent of the blood concentration and, thus, is a useful expression of the process one wants to describe (cf. the clearance notions in renal physiology).

Two clearance concepts can now be introduced. The clearance of Equation 2 can be regarded as a *systemic clearance* expressing the actual rate of removal from the body with the mixed pool concentration c_i as reference. The *intrinsic clearance* is considered below. If, from the present model, the hepatic outflow concentration (c_o) is introduced the general clearance expression (Equation 2) can be transformed into Equation 3 (for derivation, see Reference 12).

$$v/\hat{c} = V_{max}/K_m \tag{3}$$

where $\hat{c}$ is the logarithmic average: $(c_i - c_o)/(\ln c_i - \ln c_o)$. The logarithmic average concentration is a concept which is not always easy to grasp. The notion can be extended to cover all regimes[7,12] and the general regime of Equation 1 becomes $v = V_{max} \cdot \hat{c}/(\hat{c} + K_m)$.

This is the expression for the undistributed model described above, and its similarity to the Michaelis-Menten kinetics of the *in vitro* system should be noted. $\hat{c}$ can be considered as the average sinusoidal concentration. As mentioned, its determination requires measurement of the outflow concentration. One property of this expression is that blood flow does not enter the kinetic description (the connection between removal rate and $\hat{c}$, in general, is flow independent). Another characteristic of $\hat{c}$ is that it is the only concentration which, in the intact organ, makes a Lineweaver-Burk plot a straight line in this model.

It is seen from Equation 3 that the clearance so obtained (*the intrinsic clearance* which requires determination of c_o by hepatic venous catheteriztion) is solely dependent on V_{max}/K_m (i.e., a measure of the liver eliminatory function *per se*).

The systemic clearance is of special importance in pharmacokinetics, and its dependence of liver blood flow accounts for the difference in peripheral blood concentrations obtained after equal doses of drugs given perorally or intravenously ('first pass effect'). The first pass effect is explained by the liver being intercalated between the site of absorption and the general volume of distribution. If a substance is extracted completely by the liver (e.g., certain bile acids or NH_4^+), all of the substance absorbed or secreted to the portal vein tributaries is removed and nothing reaches the remaining part of the body. This presumes that no porto-systemic collaterals are present. The intrinsic clearance is useful for measuring the liver function and the logarithmic average concentration has been particularly important in capillary tube modeling defining the 'best' average capillary concentration to describe elimination.[7] It should, however, be stressed again that controversy exists regarding various models and their interpretation.[10,15-20]

IV. SPECIAL CLEARANCE REGIMES

The general clearance regime can be further developed into two special cases, their classification depending on special numerical combinations of the parameters of Equation 1, (i.e., of the properties of the system).

If the capacity of the removing system (V_{max}) is small in relation to the flow and K_m (if, in Equation 2, $V_{max}/F \cdot K_m << 1$), the clearance becomes

$$v/c_i = V_{max}/K_m \quad (4)$$

The clearance ('enzyme limited regime') is now independent of the perfusion rate. Thus, clearance directly describes the cellular elimination and, if it is supposed that K_m is a fixed value, the clearance becomes proportional to V_{max} or to the amount of enzyme present in the liver.

If, on the other hand, V_{max} is relatively large (if $V_{max}/F \cdot K_m >> 1$), the clearance becomes

$$v/c_i = F \quad (5)$$

The removal, described by this clearance ('the flow-limited regime') is solely determined by and equal to the rate of blood flow to the liver, and not upon the removing mechanisms. Substances with these properties thus can be used for measurements of blood flow without specific hepatic venous sampling (like paraaminohippuric acid in the kidney). Clearance doubles when flow is increased to its double value.

A distinction between *flow-limited* and *flow-dependent* clearance should be made. The removal rate of substances with properties of the general clearance regime is influenced by changes in blood flow, but in a nonproportional way (see Equation 2). The dye Indocyanine Green, used for determinations of the hepatic blood flow, belongs to this category, and if its clearance is used for blood flow determinations or if changes in clearance is used to indicate identical changes in blood flow, systematic underestimation of flow results.[21] The flow-limited category, like ethanol in small amounts, may be used for flow measurements in the normal state. But the caveat mentioned above should always be considered in other conditions, like organ cellular damage in liver disease, where V_{max} declines and the flow-limited condition therefore no longer holds true.

Thus, each substance behaving according to this model has its own clearance regime, but it should be observed that its classification depends on the actual combination of the three constants which may be altered independently physiologically (e.g., a decrease in hepatic blood flow by exercise or a vasodilatation during ethanol administration caused by the effect on the conversion product acetate liberating adenosine[22]), pharmacologically (competition), or by disease. If the actual blood concentration is increased above the limit which defines the clearance regime, the calculated clearance declines alone for this reason, and the clearance concept becomes invalid.

In the following section, examples of the use of regimes other than that of clearance will be shown. The general regime (with blood concentrations above those defining the clearance conditions but below those saturating the removal system) is not well suited for practical measurements, but see Reference 9. The saturated regime may be well suited, as exemplified by the classical Widmark method, for measuring the maximal elimination rate of ethanol.

V. EXPERIMENTAL PROCEDURES

By experiments in the present connection is understood determination of model parameters in animals, man and more specifically in patients. The design and performance of such studies, whether aimed at individuals (e.g., for prediction of correct dosage of drugs) or at studies to acquire new knowledge is strongly dependent on the choice of the type of models discussed in the previous section. In practice, compartmental or capillary models and included combinations thereof are useful, and studies using mixed blood concentration curves or organ extraction methods can be selected. The methods may be applied to steady-state conditions (thereby minimizing mixing problems) or to concentration-time dependent procedures.

A. ORGAN EXTRACTION

Many pitfalls in the use of organ extractions are encountered when used for kinetic information. The extraction* is dependent on the input blood concentration and, when the input load becomes high, the extraction may become immeasurable for analytical reasons. Furthermore, the extraction varies with the blood flow, which — in case of the liver — has a large, individual range. Therefore, flow determinations should be included so that the actual elimination rate is determined. This point of view is especially valid when relative contributions of various organs to each other or to total body removal is determined by different methods. It should be remembered that the catheterization procedures used to obtain mixed outflow blood may give substantial errors due to lack of representativity.

Two special problems must be mentioned here. The first is that it must be taken into consideration — also in modeling — if a substance inside the red blood cells is not in fast equilibrium with the plasma phase during the sinusoidal transit time.[23] A special pathophysiologic problem is the existence of intraorgan vascular shunts which bypass functioning tissue, and are much discussed in cirrhosis of the liver (intrahepatic shunts), but are probably also important in other organ disease. It has some similarity to the problem of recruitment or decruitment of normal capillaries. The model discussed here does not take these shunts into account (but see Reference 24). Their effect is that the measurable venous outflow concentration is different from that of the real capillary outflow (c_o in the formulae). From this difference the expression 'effective hepatic blood flow' has been coined; that is, the part of total organ flow which participates in the exchange of substances for elimination by perfusing active hepatocytes. Many studies misuse substances which normally have a very high extraction for measuring these shunts in pathologic cases, by assuming that a decreased extraction is solely due to admixture of shunted blood (at inflow concentration) to the capillary outflow, thereby ignoring the fact that decreased extraction as well can be caused by abnormal cellular function. Correctly, the shunts have to be specifically determined, as has been done in cirrhotic animals and man.[25,26] Such measurements are difficult in man, but highly wanted due to the pathogenetically important toxic substances of intestinal origin bypassing normal hepatic removal, and their effects on drug removal in liver diseases.

The following conclusions can safely be made from a measurement of the extraction (example, see Reference 27).

* 'Extraction': organ inflow *minus* outflow concentration. This is sometimes confounded with 'extraction fraction', 'extraction ratio', or 'extraction percentage', which is the extraction divided by the inflow concentration, expressed as a fraction or multiplied by 100. 'Blood flow' or 'perfusion' means blood flow rate (e.g., in milliliters per minute). 'Perfusion coefficient' is blood flow per unit of tissue weight, usually per 100 g, or better per gram. Within the clearance regime, the extraction fraction is constant, independent of blood concentration, and in the saturated regime the extraction is constant, assuming that blood flow is constant. In the case of production by an organ, like renin from the kidney, the inverse concept, ('the production fraction', $(c_o - c_i)/c_i$) is of more ambiguous meaning, as production may not — contrary to elimination — be directly dependent on the inflow concentration. It may be used for calculation of a clearance ('synthetance') if production is proportional to inflow concentration, as is the case with urea formation from amino acids in man.[11]

1. Extraction zero (insignificantly different from zero): no organ extraction, unless extraction is counterbalanced by synthesis, (e.g., norepinephrine in kidney)
2. Extraction equals inflow concentration (outflow concentration zero): the elimination is flow limited (i.e., all substance supplied to the organ is removed)
3. The extraction is substantial; without further knowledge of the eliminatory regime during which it is determined, the magnitude of the pure figure is the only information acquired; the extent of information is widened if the data are secured within the clearance regime, as has been discussed earlier

The extraction method can be extended by using stepwise, increasing steady-state concentrations to obtain sets of connected values of elimination rate and appropriate concentrations (c_i, c_o), and to apply nonlinear regression analysis or some type of linear transformation to obtain the parameters of the model, as, for example, is preferred for hepatic galactose elimination[9] and hepatic urea synthesis.[11] These methods are time consuming and, for this reason alone, less applicable to man. The multiple indicator dilution technique has been used for elimination kinetics in the series of eminent studies by Goresky and collaborators.[10,28] The same types of kinetic information as described here can be obtained, but its use in man is difficult as it requires catheterization of the portal vein or the proper hepatic artery. It is especially difficult in patients with chronic liver disease, as the general sinusoidal model has to be extended with an extra capillary membrane ('alcoholic capillarization' of hepatic sinusoids)[26] and with separate determination of the amount of intrahepatic shunting. The technique can be applied to many organs (brain, lung, and heart) but cannot study the integral elimination of the body.

B. THE USE OF PERIPHERAL BLOOD CONCENTRATION IN KINETICS

These procedures use various compartmental models and are an alternative to the organ extraction methods, being more simple to perform and interpret and able to give kinetic information adequate for many purposes. However, they are not always as profound as that of organ extraction studies in that they do not give any prediction of flow effects. The methods require an increase in the blood concentration (a tolerance test) or the use of compartmental tracer methods[29] (glucose or bile acid turnover). The concentration/time course is used for the kinetic study employing a model, as described previously. Several methodologic problems in these studies must be discussed.

1. The Concept of Strategic Regions

This concept deals with the choice of a suitable blood concentration for the purpose of getting the maximal kinetic information with a minimal bias from other processes like those of distribution discussed below.

Certain parts of the entire blood concentration span can be delimited where one of the processes of elimination becomes preponderant or, for practical purposes, supreme. An example is the choice of the 3 to 5-h section of the ^{51}Cr-EDTA elimination curve used for determination of glomerular filtration rate.[30] Unfortunately, many substances have eliminitory patterns which do not allow such regions to be discerned ('soft models'); but, in some of these cases, the kinetic constants can be derived by advanced statistical analysis (the soft models, however, often showing up with huge standard errors of the estimates or lack of convergence of the iterations).

An example of this concept is shown in Figure 1. In the case of ethanol, the strategic regions can be selected on the model basis as discussed above, and the dose given adjusted to the purpose. The entire elimination curve of ^{131}I-human serum albumin can be similarly divided into several sections for quite different physiologic purposes: transit time in circulation, transcapillary escape rate, or albumin degradation rate. The correct model of substances

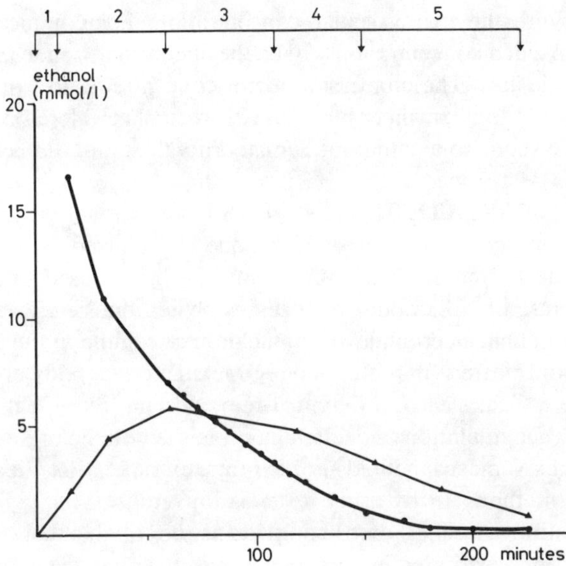

FIGURE 1. The concept of strategic regions. Full circles (●) denote arterial blood ethanol concentration and full triangles (▲) denote ascitic fluid ethanol concentration. The method involves single injection of ethanol in a subclinical dose to a patient with cirrhosis and ascites (2.6 l). Kinetic parameters and model are similar to that described in text. Strategic regions include: 1. Injection. 2. Phase dominated by extravascular distribution. 3. Linear section for measurement of maximal hepatic elimination rate of ethanol by compartmental model. 4. General regime determined by all parameters. K_m can be determined here by the sinusoidal model. The seemingly narrow range of concentration for ethanol in this regime can easily be handled experimentally by the method of stepwise steady-state infusions. 5. Flow-limited regime. Hepatic blood flow equals ethanol blood clearance. Note the slow distribution in the ascitic fluid corresponding to a time delay of about 40 min. The first part of the ascitic curve is the strategic region for measurement of the transperitoneal transfer ('PS'-product = 34 ml/min).

eliminated via liver into bile requires more complex analysis as, for example, shown in Reference 3, because selection of discrete strategic regions is difficult.

VI. METHODOLOGY

A. UNITS

Great care should be taken that correct units of kinetic expressions are used in calculations and are clearly defined in publications. They should be unambiguous and physiologically relevant. This means they should be expressed in the system (medium) in which they react, and not necessarily where they are actually measured. As an example, hepatic and renal flow rates are usually measured in and, therefore, expressed as plasma values. They should be recalculated by means of the hematocrit value to whole blood, the medium where changes in organ flow rates (e.g., by vasoactive hormones) occur.

Similarly, the volume of distribution of substances should be the solvent into which their chemical activity is expressed. If glucose kinetics is studied in the pig, the blood concentration is usually measured. As pig erythrocytes are impermeable to glucose, blood concentrations should be recalculated to plasma water concentration. The last mentioned example is in dynamic equilibrium with the extravascular concentration. For complete plasma protein-bound substances, the plasma concentration will be the choice. The same applies to body fluid concentrations like K_m values. The conversion factor from plasma to extravascular water is approximately 1.05 to 1.1, and for whole blood to extravascular water about 1.15, depending mainly on the plasma protein concentration, hematocrit value, and red cell water content.

B. STEADY/UNSTEADY-STATE TECHNIQUES

For some purposes, the constant infusion technique is preferred, as for determination of glomerular filtration rate by inulin and, if sufficient time has elapsed,* the measurement of elimination rate is unbiased by distribution processes. When the steady state is reached, the concentration is maintained at a certain level (in the general regime or the clearance regimes) or is linearly rising or decreasing (the saturated regime). The total body elimination then can be calculated either as a clearance or a maximal removal rate, the last mentioned by using suitable corrections for accumulation or decumulation in the volume of distribution. This can be directly determined by standard methods as zero-time extrapolation, area-dose calculation, isotope tracer-dilution methods, or by using a standard average value.

The blood concentration to be measured should ideally be equal to that of the entire volume of distribution (apart from the eliminating organs); that is, the mixed venous concentration or, for practical purposes, the arterial concentration. In organ systems like the splanchnic, the hepatic input concentration may however differ from that value if, for example, intestinal absorption or recirculation (bile acids after digestion), metabolism or synthesis in prehepatic splanchnic organs occur, as is especially likely with many hormones.[31] During single injection techniques (seconds or a few minutes), the problem of extravascular diffusion becomes important, as the initial part of the elimination curve is not solely caused by the elimination proper (see Figure 1) and as concentration differences occur in various parts of the volume of distribution. Even if during a certain time, depending on the properties of the substance, the capillary permeability, and the size of the compartment(s), the initial distribution has occurred, the extravascular concentration remains higher than that of the mixed venous blood, as the substance reenters the circulation during the declining blood concentration caused by the ultimate elimination. This means that blood sampling from a peripheral vein frequently is not optimal; this is especially so when elimination from blood is fast in relation to transcapillary exchange rate. The use of 'arterialized' venous blood cannot always be recommended.[32] In quasi-steady-state studies, the average body concentration should be used, but it can only be calculated by more difficult methods (see e.g., Reference 33). The veno-arterial difference may be larger in disease with fluid retention, especially if compartments like ascites with a slow rate of exchange are present (Figure 1). It should also be remembered that veno-arterial differences are not identical in various sections of the circulation, being locally dependent on its special capillary permeability and the regional blood flow, and also if removal or synthesis occur. For substances with a clearance of the magnitude of that of the cardiac output, the mixing time in the blood cannot be ignored, and the arterial sampling site becomes the solely usable one. When this is ignored, large differences in clearances calculated from venous and arterial blood may be found, as, for example, more than 50% with neurotensin.[34] A special problem in modeling and experiment occurs when the substance is removed in the blood itself, particularly when this removal is fast. This happens with adenosine[35] where the half-time in blood is less than 1 s, corresponding to a turnover clearance of greater than 20 times the cardiac output. This is prohibitive for the types of measurements discussed here.

* The delimitation of the concentration-time curve to exclude the distribution phase is routinely done graphically from the general knowledge of the time of distribution; for more accurate studies, special methods exist (e.g., by including the extravasation process in a compartmental model, as is usually done in pharmacokinetics).

C. VERIFICATION OF MODELS

There exists a gap between the feasibility of constructing kinetic models and their verification, and this gap should be closed by some experimental support before being used for actual purposes. This is necessitated because more than one model may depict the same biological process and because of the assumptions (e.g., approximations) implied in model solutions.

A simple way to check a model for calculation of the hepatic maximal removal rate is to ensure within the strategic region chosen, that the arterio-hepatovenous difference is constant after a bolus injection. If the problem concerns a clearance, the logarithm of the difference should be linearly decreasing (or the extraction fraction be constant).

A more detailed analysis may be preferable if more complicated models are used. Experiments can be performed in which the given load or blood flow[18] is changed or another type of study made; and by the changed conditions, the parameters of the model should then be unaffected. By this effort, many insufficient models will be disclosed and may have to be extended.

Studies may further describe the kinetics in a more comprehensive way by using some suitable linear transformation, for example, the Lineweaver-Burk plot (linear relation between inverse concentration and inverse rate of elimination) in a wide range of concentrations, or use the more direct methods of nonlinear regression. The statistical errors of the parameter solution should be acceptably low in relation to the known analytical error. Statistical analysis of the residual variation may help to detect systematic deviations from the model concept. Another way is to identify the similarity to experiments by computer simulations on resulting concentration time curves using reasonable estimates of the parameters. A good test of the stability of the model and its computational solution is to add error terms (see Reference 3).

VII. CONCLUDING REMARKS

Studying the ways by which the organism or single organs handle various substances has, in the last three decades, added much to the understanding of these systems. Mathematical, statistical, and computational methods as well as experimental procedures have proliferated and are becoming rather complex, and perhaps a bit difficult to grasp for those doing physiologic and pathophysiologic studies. It seems for the moment that the advanced model makers may be ahead of their experimental support and that the experimental part should receive more attention and research in the future. Important improvements will be required by including, for example, the unsolved problem of describing the role of protein binding in determining elimination.[36,37] Furthermore, receptor kinetics, which is within the reach of measurement (radiolabeled antibodies), could also be added. The physiologic realism should be maintained because an excess of adjustable parameters make the model solutions ambiguous. Kinetic studies are, if correctly performed, powerful tools in our understanding of homeostasis in health and disease.

ACKNOWLEDGMENTS

The basic ideas expressed in this survey have emerged from fruitful discussions and cooperation with L. Bass (Department of Mathematics, University of Queensland, Brisbane, Australia), N. Tygstrup, and S. Keiding (Medical Department A, Rigshospitalet, University of Copenhagen, Denmark). The study of Figure 1 is an unpublished observation by J. H. Henriksen, Department of Clinical Physiology, Hvidovre Hospital.

REFERENCES

1. **Richards, T. G.,** The plasma concentration of bromosulphalein (B. S. P.) after single intravenous injections in normal and abnormal human subjects, in *The Biliary System,* Taylor, W., Ed., Blackwell, Oxford, 1965, 567.
2. **Winkler, K.,** The kinetics of elimination of bromsulphalein in man after single injection and continuous infusions, in *The Biliary System,* Taylor, W., Ed., Blackwell, Oxford, 1965, 551.
3. **Forker, E. L. and Luxon, B.,** Hepatic transport kinetics and plasma disappearance curves: distributed modeling versus conventional approach, *Am. J. Physiol.,* 235, E648, 1978.
4. **Crone, C. and Levitt, D. G.,** Capillary permeability to small solutes, in *Handbook of Physiology,* Section 2, Vol. IV, Renkin, E. M. and Michel, C. C., Eds., Williams & Wilkins, Baltimore, 1984, chap. 10.
5. **Lassen, N. A. and Pearl, W.,** *Tracer Kinetic Methods in Medical Physiology,* Raven Press, New York, 1979.
6. **Goresky, C. A. and Groom, A. C.,** Microcirculatory events in the liver and the spleen, in *Handbook of Physiology,* Section 2, Vol. IV, Renkin, E. M. and Michel, C. C., Eds., Williams & Wilkins, Baltimore, 1984, chap. 15.
7. **Bass, L.,** Heterogeneity within observed regions: physiologic basis and effects on estimation of rates of biodynamic processes, *Circulation,* 72, 47, 1985.
8. **Groothuis, G.,** *Acinar heterogeneity of hepatocytes in transport functions,* Thesis, Rijksuniversiteit te Groningen, Groningen, 1982.
9. **Keiding, S., Johansen, S., Winkler, K., Tønnesen, K., and Tygstrup, N.,** Michaelis-Menten kinetics of galactose elimination by the isolated perfused pig liver, *Am. J. Physiol.,* 230, 1302, 1976.
10. **Goresky, C. A., Bach, G. G., and Rose, C. P.,** Effects of saturating metabolic uptake on space profiles and tracer kinetics, *Am. J. Physiol.,* 244, G215, 1983.
11. **Vilstrup, H. and Skovgaard, L. T.,** Kinetics of hepatic alanine uptake and urea synthesis in pigs, *Am. J. Physiol.,* 254, G602, 1988.
12. **Bass, L., Keiding, S., Winkler, K., and Tygstrup, N.,** Enzymatic elimination of substrates flowing through the intact liver, *J. Theor. Biol.,* 61, 393, 1976.
13. **Johansen, S. and Keiding, S.,** A family of models for the elimination of substrate in the liver, *J. Theor. Biol.,* 89, 549, 1981.
14. **Winkler, K., Bass, L., Keiding, S., and Tygstrup, N.,** The physiologic basis for clearance measurements in hepatology, *Scand. J. Gastroenterol.,* 14, 439, 1979.
15. **Bass, L.,** Physiologically based models and strategic experiments in hepatic pharmacology, *Biochem. Pharmacol.,* 37, 1425, 1988.
16. **Bass, L.,** Convection-dispersion modeling of hepatic elimination, *J. Pharm. Sci.,* 75, 321, 1986.
17. **Roberts, M. S. and Rowland, M.,** Hepatic elimination-dispersion model, *J. Pharm. Sci.,* 74, 585, 1985.
18. **Keiding, S. and Steiness, E.,** Flow dependence of propranolol elimination in perfused rat liver, *J. Pharmacol. Exp. Ther.,* 230, 474, 1984.
19. **Pang, K. S. and Rowland, M.,** Hepatic clearance of drugs. III. Additional experimental evidence supporting the 'well-stirred' model, using metabolite (MEGX) generated from lidocaine under varying hepatic blood flow rates and linear conditions in the perfused rat liver *in situ* preparation, *J. Pharmacokin. Biopharm.,* 5, 681, 1977.
20. **Wagner, J. G.,** Commentary: relationship among the venous equilibration ('well-stirred') model, the sinusoidal perfusion ('parallel-tube') model, and a specific two-compartment open model, *Drug Metab. Disposition,* 13, 119, 1985.
21. **Kanstrup, I.-L. and Winkler, K.,** Indocyanine green plasma clearance as a measure of changes in hepatic blood flow, *Clin. Physiol.,* 7, 51, 1987.
22. **Carmichael, F. J., Saldivia, V., Varghese, G. A., Israel, Y., and Orrego, H.,** Ethanol-induced increase in portal blood flow: role of acetate and A_1- and A_2-adenosine receptors, *Am. J. Physiol.,* 255, G417, 1988.
23. **Goresky, C. A., Bach, G. G., and Nadeau, B. E.,** Red cell carriage of label, *Circ. Res.,* 36, 328, 1975.
24. **Bass, L., Winkler, K., and Tygstrup, N.,** Arterial-venous concentration differences and the quantification of intrahepatic shunts, in *Microcirculation of the Alimentary Tract,* Koo, A., Lam, S. K., and Smaje, L. H., Eds., World Scientific, Singapore, 1983, 157.
25. **Huet, P.-M., Belgiorno, J., Brault, A., Giroux, L., and Rocheleau, B.,** Assessment of intrahepatic shunts in cirrhotic animals, in *Cirrhosis of the Liver: Methods and Fields of Research,* Tygstrup, N. and Orlandi, F., Eds., Elsevier, Amsterdam, 1987, 91.
26. **Huet, P.-M., Goresky, C. A., Villeneuve, J.-P., Marleau, D., and Lough, J. O.,** Assessment of liver microcirculation in human cirrhosis, *J. Clin. Invest.,* 70, 1234, 1982.
27. **Henriksen, J. H., Schwartz, T. W., and Bülow, J. B.,** Endogenous pancreatic polypeptide in different vascular beds: relationship to release and degradation in subjects with normal and decreased kidney function, *Metabolism,* 35, 542, 1986.

28. **Goresky, C. A., Bach, G. G., Cousineau, D., Schwab, A. J., Rose, C., Lee, S., and Goresky, S.,** Handling of tracer norepinephrine by the dog liver, *Am. J. Physiol.,* 256, G107, 1989.
29. **Hetenyi, G., Perez, G., and Vranic, M.,** Turnover and precursor-product relationships of nonlipid metabolites, *Physiol. Rev.,* 63, 606, 1983.
30. **Brøchner-Mortensen, J.,** *Routine Methods and their Reliability for Assessment of Glomerular Filtration Rate in Adults,* Lægeforeningens Forlag, Copenhagen, 1978, 21.
31. **Henriksen, J. H., Ring-Larsen, H., and Christensen, N. J.,** Hepatic intestinal uptake and release of catecholamines in alcoholic cirrhosis. Evidence of enhanced hepatic intestinal sympathetic nervous activity, *Gut,* 28, 1637, 1987.
32. **Astrup, A., Simonsen, L., Bülow, J., and Christensen, N. J.,** Measurement of forearm oxygen consumption: role of heating the contralateral hand, *Am. J. Physiol.,* 255, E572, 1988.
33. **Bass, L., Bracken, A. J., and Burden, C. J.,** New approaches to uptake by heterogeneous perfused organs: from linear to saturation kinetics, in *Lecture Notes in Biomathematics,* Lambrecht, R. M. and Rescigno, A., Eds., Springer-Verlag, Berlin, 1983, 120.
34. **Petersen, J. H., Andersen H. Ø., Olsen, P. S., and Henriksen, J. H.,** Pharmacokinetics and metabolism of neurotensin in man, *J. Clin. Endocrinol. Metab.,* 68, 294, 1989.
35. **Möser, G. H., Schrader, J., and Deussen, A.,** Turnover of adenosine in plasma of human and dog blood, *Am. J. Physiol.,* 256, C799, 1989.
36. **Bass, L. and Pond, S. M.,** The puzzle of rates of cellular uptake of protein-bound ligands, in *Pharmacokinetics: Mathematical and Statistical Approaches to Metabolism and Distribution of Chemicals and Drugs,* Pecile, A. and Rescigno, A., Eds., Plenum Press, London, 1988, 245.
37. **Nunes, R., Kiang, C.-L., Sorrentino, D., and Berk, P. D.,** Albumin-receptor uptake kinetics do not require an intact lobular architecture and are not specific for albumin, *J. Hepatol.,* 7, 293, 1988.

Chapter 4

PEPTIDASES INVOLVED IN THE METABOLISM OF BIOACTIVE PEPTIDES

A. John Kenny and Nigel M. Hooper

TABLE OF CONTENTS

I. INTRODUCTION

Biologically active peptides are synthesized within membrane-bound compartments as larger precursors which are usually fully processed before their release by exocytosis. This chapter is not concerned with the peptidases involved with intracellular processing of precursor forms, but only with the metabolism of the peptides after release, a process in which cell-surface peptidases play a major role, though other mechanisms should not be excluded. In the case of atrial natriuretic peptide, it has been postulated that a class of 'silent' receptors (the C receptor) may be physiologically important in the clearance of this peptide. Thus far, this is the only peptide for which such a mechanism has been described and, generally, there is no evidence that neuropeptides, once released, can be taken up and metabolized by intracellular mechanisms, as would be the case for some non-peptide transmitters, such as γ-amino butyric acid. Peptide inactivation, leading to termination of the signal, is analogous to that for acetylcholine in which a cell-surface hydrolase plays the key role. Acetylcholinesterase is specific in this process, but the cell-surface hydrolases concerned with biologically active peptides do not usually exhibit a high degree of specificity for a particular peptide. Instead, it seems that a limited number of membrane peptidases, each exhibiting the topology of an ectoenzyme (i.e., asymmetrically anchored in the membrane, with the active site exposed at the extracytoplasmic surface) suffices to inactivate a large number of peptides. The mechanism of hydrolysis of biologically active peptides will depend heavily on the precise location of the membrane peptidase and the peptides that are released or circulate to that site.

In this chapter, we shall first describe the properties of the cell-surface peptidases that have been isolated and studied in some detail. We will then examine how these peptidases may act in concert when the membrane is presented with a susceptible peptide substrate and examine their tissue distribution and mode of anchorage and, finally, we shall consider how they may function physiologically in a selected number of different systems.

II. PROPERTIES OF MEMBRANE PEPTIDASES

The topological criteria for a peptidase acting at the cell surface have been mentioned above. Here we shall summarize the structural and catalytic properties of eighteen peptidases (Table 1), the majority of which can be firmly designated as ectoenzymes. There are, however, a few other peptidases with less convincing credentials as regards their membrane location, but for which claims have been made in regard to a role in neuropeptide metabolism. These are also discussed.

TABLE 1
Properties of the Peptidases Described in This Chapter

Enzyme	Topology[a]	Specificity	Active site	Specific inhibitors	cDNA sequence available	Ref.
Endopeptidase-24.11 EC 3.4.24.11	E	↓ –○–○–●–○–○– (hydrophobic)	Zn^{2+}	Phosphoramidon, thiorphan	Yes	1
Endopeptidase-24.15 EC 3.4.24.15	?	↓ –○–○–●–○–○– (hydrophobic or or basic)	Zn^{2+}?	N-[1(R,S)-carboxy-2-phenylethyl]-Ala-Ala-Phe-pAB	—	2
Endopeptidase-2 or meprin	E	↓/↓ –○–○–●–○–○– (aromatic)	Zn^{2+}	—	—	3
Angiotensin converting enzyme EC 3.4.15.1	E	↓ –○–○–●–● (nonspecific)	Zn^{2+}	Captopril, enalaprilat, lisinopril	Yes	4
Aminopeptidase N EC 3.4.11.2	E	↓ ●–○–○–○– (nonspecific)	Zn^{2+}	Actinonin, amastatin, bestatin	Yes	5
Aminopeptidase A EC 3.4.11.7	E	↓ ●–○–○–○– (Glu/Asp)	Ca^{2+}	Amastatin	—	6
Aminopeptidase P EC 3.4.11.9	E	↓ ○–●–○–○– (Pro)	Zn^{2+}	(Enalaprilat), (L155,212)	—	7
Aminpeptidase W EC 3.4.11.-16	E	↓ ○–●–○–○– (Trp)	Zn^{2+}	Amastatin, bestatin	—	8
Carboxypeptidase M	E	↓ –○–○–○–● (Arg/Lys)	Zn^{2+}?	MGTA GEMSA	Yes	9
Carboxypeptidase N EC 3.4.17.3	P	↓ –○–○–○–● (Lys/Arg)	Zn^{2+}	MGTA GEMSA	Yes	9
Carboxypeptidase P EC 3.4.17.-	E	↓ –○–○–●–○ (Pro,Ala,Gly)	Zn^{2+}	—	—	10
Dipeptidyl peptidase IV EC 3.4.14.5	E	↓ ○–●–○–○– (Pro,Ala)	Ser	Diisopropyl-fluorophosphate	Yes	11
Membrane dipeptidase EC 3.4.13.11	E	↓ ●–● (nonspecific)	Zn^{2+}	Cilastatin	Yes	12
Pyroglutamyl peptidase II (thyroliberinase) EC 3.4.19.-	E?	↓ Glp-His-ProNH_2	?	—	—	13
γ-Glutamyltransferase EC 2.3.2.2	E	↓ ●–○–○–○– (γGlu)	—	AT-125	Yes	14
N-acetylated-α-linked acidic dipeptidase	E?	↓ N-acetyl-Asp-Glu	?	Quisqualate	—	15
Neurotensin degrading enzyme	?	↓ –○–○–●–○–○– (Pro)	?	—	—	16
Substance P degrading enzyme	C	↓/↓ –○–○–●–○–○– (Phe)	?	—	—	3

[a] E = eckoenzyme; P = plasma; C = cytoplasmic; ? = uncertain.

A. ENDOPEPTIDASE-24.11

Endopeptidase-24.11 (enkephalinase; neutral metallo-endopeptidase) is a glycoprotein of subunit M_r 85,000 to 95,000. The slight variation in size of endopeptidase-24.11 between tissues is almost certainly due to differences in the extent of glycosylation.[17] This has been confirmed following the isolation and sequencing of the cDNAs for the rat brain and kidney enzymes which are identical except for two nucleotides — differences which are probably cloning artifacts.[18] The human and rabbit cDNA sequences have also been elucidated[19,20] and show a high degree of similarity with each other and with the sequence for the rat enzyme. The human enzyme is a single polypeptide chain of 742 amino acids. There is a short N-terminal cytoplasmic domain (20 residues) and a single membrane-spanning hydrophobic domain (23 residues) which represents the uncleaved signal sequence (see Section VI.A). Endopeptidase-24.11 usually exists in the membrane as a noncovalently associated dimer with one atom of Zn^{2+} bound per subunit which is essential for catalytic activity.[21,22] Thus, the enzyme is sensitive to chelating agents such as EDTA and 1,10-phenanthroline.

The specificity of endopeptidase-24.11 is directed towards peptide bonds involving the amino groups of hydrophobic residues (–X–Y–, where Y is Phe, Leu, Val, Tyr, Ile, Ala, or Trp), provided this residue is not C-terminal or penultimate at the N-terminus. This rule is applicable to the great majority of peptides studied, but there is one well-substantiated irregularity, in the case of atrial natriuretic peptide, where an Arg–Ser bond was hydrolyzed.[23] The optimum substrate sequence for endopeptidase-24.11 has been elucidated[24] as

P_3	P_2	P_1		P'_1	P'_2	P'_3
Phe	Phe	Ala	↑	Phe	Leu	Ala
Ala		Gly		Leu		

(bond cleaved)

Because of the extended active site, amino acid residues distant from the scissile bond can have a marked influence on the rates of hydrolysis of substrates by the enzyme. Thus, endopeptidase-24.11 has the specificity to hydrolyze a wide range of natural peptide substrates (Figure 1 and Table 2). Only a few of these peptides (atrial natriuretic factor, bradykinin, chemotactic peptide, enkephalins, gastrin, neurotensin, and substance P) have been shown to be cleaved *in vivo* or *in situ* by endopeptidase-24.11.[1]

The substrate specificity of endopeptidase-24.11 is similar to that of the bacterial enzyme thermolysin. However, it should be noted that thermolysin prefers a bulky hydrophobic residue in the P_1 position and a small hydrophobic residue in the P_2 position.[34] The active site of endopeptidase-24.11 has been probed with chemical reagents,[35–37] as well as more recently by site-directed mutagenesis[38] and hydrophobic cluster analysis of the primary sequence.[39] These studies have revealed that the Zn^{2+}-binding and catalytic residues present in thermolysin are conserved in endopeptidase-24.11.

The similarities in the substrate specificities of endopeptidase-24.11 and thermolysin led to the observation that the natural product inhibitor of the bacterial enzyme, phosphoramidon,[40] is a potent inhibitor of the mammalian enzyme (Table 3). The earliest synthetic inhibitor developed was thiorphan[43] which has a mercapto function to coordinate to the active-site Zn^{2+} (Table 3). The selectivity of thiorphan for endopeptidase-24.11 was increased by retro-inversion of the peptide bond to produce retrothiorphan.[44] Another synthetic inhibitor, kelatorphan,[46] is equipotent with thiorphan for endopeptidase-24.11, but also inhibits aminopeptidase N (Table 3) and a dipeptidyl aminopeptidase[65] and, thus, completely protects enkephalins from degradation in brain tissue. Kelatorphan also completely prevents the metabolism of neurokinins A and B by pig striatal synaptic membranes.[66] Although a number

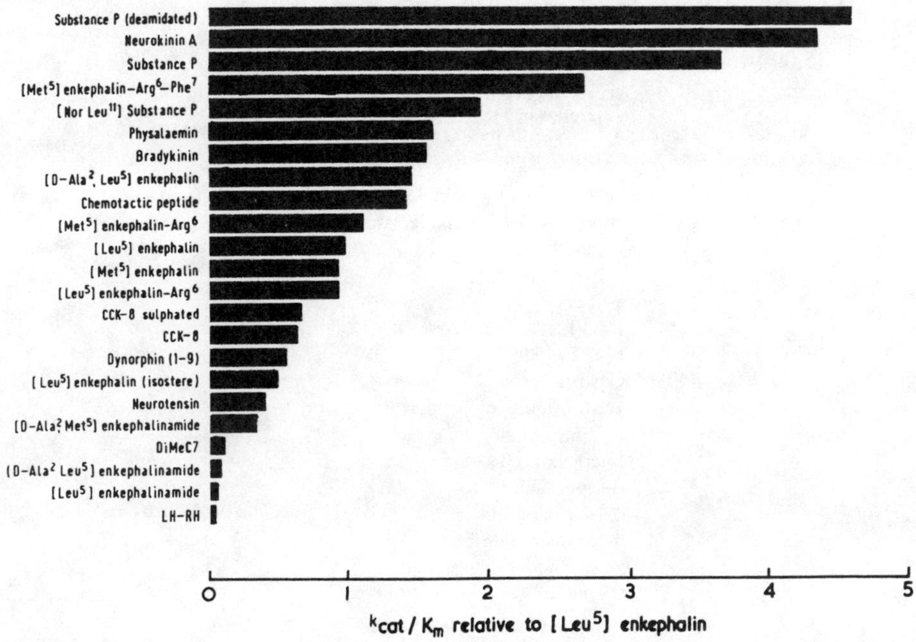

FIGURE 1. Kinetics ofd hydrolysis of bioactive peptides and their analogues by endopeptidase-24.11. The data are expressed as k_{cat}/K_M to [Leu^5] enkephalin and are taken from References 25, 27, 28, and 36. (Modified from Reference 29). Other peptides hydrolyzed by endopeptidase-24.11 for which kinetic data are not available are listed in Table 2.

of other endopeptidase-24.11 inhibitors have been designed,[67,68] phosphoramidon is still one of the most potent and selective inhibitors available.

B. ENDOPEPTIDASE-24.15 FAMILY

Endopeptidase-24.15 is predominantly a cytosolic protein, although there appears to be a membrane-bound component. In the brain, about 20% of the total activity is membrane bound, possibly associated with synaptosomal membranes,[69,70] whereas, in the testes, only about 8% of the total activity is membrane bound.[2] Whether this membrane-bound component is associated with the plasma membrane and has the topology of an ectoenzyme, remains to be determined. The enzyme was first extracted and partially purified from rat brain[71] and has recently been purified from rat testes.[2] Endopeptidase-24.15 is a monomer of M_r 70,000 and is sensitive to inhibition by chelating agents; a Zn^{2+} atom possibly being present at the active site. The enzyme hydrolyzes peptide bonds involving the carboxyl function of hydrophobic or basic residues and the active site has a minimum requirement for five amino acid residues. The specificity is markedly affected by groups in both the P_1 and P'_3 positions, bulky hydrophobic groups being preferred in both cases.[2]

Endopeptidase-24.15 has been shown to hydrolyze a number of natural peptides including dynorphin A 1-8, β-neoendorphin, bradykinin, neurotensin, and luliberin. In the hypothalamus and pituitary, the degradation of luliberin is primarily initiated by endopeptidase-24.15.[72] A selective inhibitor of endopeptidase-24.15 has been synthesized, *N*-[1-(*RS*)-carboxy-2-phenylethyl]Ala–Ala–Phe–pAB[48] (see Table 3).

Recently, it has been shown that the activity attributed to endopeptidase-24.15 is identical with that of the mammalian enzyme Pz-peptidase[73,74] which has been reported to be involved in connective-tissue matrix turnover. The identity of endopeptidase-24.15 with the cysteine peptidase endo-oligopeptidase A[75,76] is unclear, some differences in properties might be

TABLE 2
Other Peptides Hydrolyzed by Endopeptidase-24.11

Angiotensin I
Angiotensin II
Angiotensin III
Atrial natriuretic peptide
Brain natriuretic peptide
Dynorphin 1-13
β-Endorphin
γ-Endorphin
[Met^5]-enkephalin-Arg^6-Gly^7-Leu^8
FMRFamide
Gastrin
Gastrin-releasing peptide
Insulin B chain
Interleukin 1β
β-Lipotropin 61-69
α-Neoendorphin
β-Neoendorphin
Neurokinin B
Oxytocin
Vasoactive intestinal polypeptide

Note: Data taken from References 1 and 30 to 33.

attributable to species variation, although the recent observation that the activity of endo-oligopeptidase A and Pz-peptidase are attributable to the same enzyme[77] would imply that endo-oligopeptidase A and endopeptidase-24.15 are the same protein.

C. ENDOPEPTIDASE-2 FAMILY

The kidneys of rats and mice contain an endopeptidase activity towards the insulin B chain which is not inhibited by phosphoramidon. This additional endopeptidase in rodents has been termed endopeptidase-2 in rats[78] and meprin in mice.[79,80] Endopeptidase-2 and meprin are catalytically and immunologically very similar, if not identical.[81] Endopeptidase-2 in rat kidney is a microvillar membrane-bound enzyme with a complex tetrameric structure.[82] When released by papain, apparently dissimilar subunits (M_r 74,000 and M_r 80,000, respectively) were observed. However, this appears to be the result of papain action since the detergent-solubilized form has a single subunit (M_r 80,000). These subunits are disulphide linked, forming a tetramer, a feature which is uncommon among the microvillar ectoenzymes. Only membrane dipeptidase is similarly disulphide linked.[83] Endopeptidase-2 is a Zn^{2+} metalloprotein, strongly inhibited by chelating agents and phosphate. No specific inhibitor has thus far been identified.

A polyclonal antibody raised to the rat enzyme precipitates the 'meprin' activity from mouse kidney extracts and has been successfully used to purify a sample of meprin by immunoaffinity chromatography. Although both the rat and mouse preparations hydrolyzed ^{125}I-insulin B chain and other assay substrates, there were interesting quantitative differences: the mouse enzyme hydrolyzed azocasein ten times faster and Bz–Tyr–pAB ten times more slowly than rat endopeptidase-2. Both had identical tissue distributions in an immunohistochemical study: in kidney, the justamedullary proximal tubules and, in the gastrointestinal tract, the entire lining was stained, from the duodenum to the rectum.[81]

Endopeptidase-2 has a chymotrypsin-like specificity in that it hydrolyzes Bz–Tyr–pAB but has, like endopeptidase-24.11, an extended activity site.[82] The enzyme is capable of

TABLE 3
Peptidase Inhibitors

Peptidase	Inhibitor	Ki (nM)	I_{50} (μM)	Recommended conc (μM)	Ref.
Endopeptidase-24.11	Phosphoramidon	2	0.013	1	41, 42
	Thiorphan	4.7	0.013	1	42, 43
	Retrothiorphan	6.0	—	1	44
	CPAB	70	—	100	45
	Kelatorphan	1.4	0.002	1	46, 47
Endopeptidase-24.15	N-[1-(R,S)-carboxy-2-] phenylethyl]-Ala-Ala Phe-pAB	800	—	400	48
Angiotensin-converting enzyme	Captopril	1.7	0.021	1	49, 50
	Enalaprilat	0.2	0.003	1	50, 51
	Lisinopril	0.1	0.011	1	50, 51
	L155,212	—	0.003	1	50
Aminopeptidase N	Actinonin	—	1	100	52
	Amastatin	19	0.05	1[a]	33, 53
	Bestatin	4100	3.03	100	53
	Kelatorphan	7000	0.71	100	46, 47
Aminopeptidase A	Amastatin	250	—	10	54
Aminopeptidase P	Enalprilat	—	2.17	100	55
	L155,212	—	0.33	10	55
Aminopeptidase W	Amastatin	—	2	100[a]	8
	Bestatin	—	6	100[a]	8
Carboxypeptidase M	MGTA	—	0.3	10	56, 57
Carboxypeptidase N	GEMSA	1500	—	100	58
	MGTA	5	—	10	59
Dipeptidyl peptidase IV	Diisopropylfluoro-phosphate	—	10	100[a]	60
	Diprotin A	—	3.2	100	61
Membrane dipeptidase	Cilastatin	700	0.11	10	62, 63
γ-Glutamyltrans-peptidase	AT-125	0.6 m*M*	—	—	64
N-acetylated-α-linked acidic dipeptidase	Quisqualate	—	0.48	100	15

[a] Preincubation (30 to 60 min) required for maximal inhibition.

hydrolyzing several natural peptides including luliberin, substance P, bradykinin, neurotensin, [Met^5]enkephalin–Arg^6–Phe^7, and angiotensin II.[82,84,85]

Endopeptidase-2 is also closely related, if not identical, to another metalloenzyme, PABA peptide hydrolase.[3,85] PABA peptide hydrolase is a glycosylated, integral membrane protein present in human small intestine.[86–89] It is a disulphide-linked dimer or tetramer with a similar substrate specificity as endopeptidase-2. PABA peptide hydrolase catalyzed the hydrolysis of angiotensins I and II, bradykinin, insulin B chain, oxytocin, and substance P.[88] The slight structural differences, (e.g., in size and extent of glycosylation) between endopeptidase-2 and PABA peptide hydrolase most probably reflect species differences.

D. ANGIOTENSIN CONVERTING ENZYME

Angiotensin converting enzyme (peptidyl dipeptidase A) is an extensively glycosylated membrane protein of subunit M_r 170,000 to 190,000. As with endopeptidase-24.11, the variations in size between tissues appears to be due to differences in the extent of glycosylation.[50] The cDNAs for human and mouse angiotensin converting enzymes have recently been sequenced.[90,91] The sequence for the human endothelial enzyme codes for a

TABLE 4
Substrate Specificity of Angiotensin-Converting Enzyme

Peptide structure	Cleavage product	Examples	Ref.
Free C-terminus	C-terminal dipeptide ↓ -V-W-X-Y unless X = Pro or Y = Asp, Glu	Angiotensin I, bradykinin, etc.	100
	C-terminal tripeptide ↓ -V-W-X-Y when X = Pro or Trp	Angiotensin II, des-Arg[9]-bradykinin, cholecystokinin-8, [Leu[15]]-gastrin-11-17	101, 102
Amidated C-terminus	C-terminal dipeptide amide	Substance P, [Leu[5]]-enkephalinamide, [Met[5]]-enkephalinamide	50, 103
	C-terminal tripeptide-amide	Substance P, luliberin	103, 104
Blocked N-terminus	N-terminal tripeptide	Luliberin	104

Note: Modified from Erdos, E. G. and Skidgel, R. A., *Lab. Invest.*, 56, 345, 1987.

single polypeptide chain of 1306 residues, beginning with a 29-amino acid signal peptide which is cleaved in the mature enzyme. There is a hydrophobic sequence of amino acids near the C-terminus which probably traverses the lipid bilayer[90,92,93] (see Section VI.A). The mouse enzyme is 83% homologous to human ACE at both the nucleic acid and amino acid levels. A soluble form of angiotensin converting enzyme is present in plasma, cerebrospinal fluid, and other body fluids (see Section VI.C).

The amino acid sequence of angiotensin converting enzyme reveals a high degree of homology between two large domains, each of which bears a putative active site.[90] However, it has been claimed that only one atom of Zn^{2+} is present per molecule of enzyme,[94] suggesting that only one of the two domains is catalytically active, probably the C-terminal domain, by comparison with the testicular enzyme.[95,96] This Zn^{2+} is essential for activity and, thus, chelating agents inhibit the enzyme. Chloride ions (0.1 to 0.3 *M*) are usually required for full activity, although the hydrolysis of bradykinin proceeds in the absence of chloride.[97] The testicular enzyme is a distinct form which has an identical catalytic specificity to the pulmonary enzyme.[98] The cDNA for the testicular form of angiotensin converting enzyme has recently been sequenced[95] and reveals that this form of the enzyme is translated from a shorter mRNA corresponding to the 3´ half of the endothelial transcript.

Angiotensin converting enzyme was originally classified as a peptidyl dipeptidase, removing C-terminal dipeptides from angiotensin I and bradykinin. However, more recently, endopeptidase and peptidyl tripeptidase activities have been attributed to the enzyme (Table 4). Several natural peptides are known to be hydrolyzed by angiotensin converting enzyme *in vitro* (Table 5).

The involvement of angiotensin converting enzyme in the renin-angiotensin system (Section V.B) has led to the enzyme becoming an important target for the treatment of hypertension (see Chapter 6). The development of a number of potent and selective inhibitors of angiotensin converting enzyme has been based on the similarity of its active site with that of two related zinc metalloenzymes, carboxypeptidase A and thermolysin.[68,106] The first clinically successful inhibitor of angiotensin converting enzyme was the thiol compound, captopril, in which the SH group coordinated the Zn^{2+}.[49] Since then, numerous other inhibitors have been developed, the most commonly used being the carboxyl-based inhibitors enalaprilat and lisinopril, where the carboxylate group coordinates the Zn^{2+}.[51,107] The latter inhibitor has proved valuable as a ligand for affinity purification of angiotensin converting enzyme.[50,51]

TABLE 5
Peptides Hydrolyzed by Angiotensin-Converting Enzyme

Peptide	k_{cat}/K_m ($mM^{-1}\cdot min^{-1}$)	Ref. (for k_{cat}/K_m)
Angiotensin I	125	105
Angiotensin II	—	—
Bradykinin	1250	105
des-Arg^9-bradykinin	—	—
BzGly-His-Leu	3.6	104
Cholecystokinin-8	0.29	102
Dynorphin 1-6	—	—
Dynorphin 1-8	—	—
[Leu^5]-enkephalin	0.7	105
[Met^5]-enkephalin	3.5	105
[Leu^5]-enkephalinamide	—	—
[Met^5]-enkephalinamide	—	—
[Met^5]-enkephalin-Arg-Phe	—	—
[Met^5]-enkephalin-Arg-Gly-Leu	—	—
[Leu^{15}]-gastrin (11-17)	0.27	102
Boc-[Leu^{15}]-gastrin (14-17)	0.27	102
Luliberin	1.3	104
β-Neoendorphin	—	—
Neurotensin	2.4	105
Substance P	9.0	105
Substance P (deaminated)	637	105

E. AMINOPEPTIDASES

1. Aminopeptidase N

Aminopeptidase N is an integral membrane glycoprotein of subunit M_r 160,000. The cDNAs coding for rat and human aminopeptidase N have recently been sequenced.[5,108,109] Like endopeptidase-24.11, aminopeptidase N is anchored in the membrane by an uncleaved hydrophobic signal sequence at the N-terminus (Section VI.A). The bulk of the protein (967 amino acids in total) is located at the cell surface. There is a catalytic domain typical of other Zn^{2+} metallopeptidases which shows some homology with aminopeptidase N from *E. coli.*

Aminopeptidase N has a broad specificity, releasing N-terminal amino acids from unblocked di-, tri-, and oligopeptides, but leucinamide is not readily hydrolyzed. Hydrolysis is most rapid when the N-terminal amino acid is Ala, although Phe, Leu, Tyr, and Arg are hydrolyzed at intermediate rates. Glu, Asp, and Pro are hydrolyzed very slowly.[110] Increasing chain length greatly enhances substrate hydrolysis. Aminopeptidase N is the major activity releasing the Tyr^1 from the enkephalins.[111,112] Aminopeptidase N probably also plays a role in the metabolism of cholecystokinin-8[113] and neurokinin A.[114,115]

A number of general aminopeptidase inhibitors are effective on aminopeptidase N (Table 3). Actinonin[52] has the greatest selectivity for aminopeptidase N, although it also partially inhibits endopeptidase-24.11 and angiotensin converting enzyme.[116] Amastatin and bestatin[53] are also valuable inhibitors of aminopeptidase N.

2. Aminopeptidase A

Aminopeptidase A is a dimeric glycoprotein in pig kidney of two subunits of M_r 170,000 each.[6] Compared with the other metallopeptidases, it is unusual in that it contains one atom of Ca^{2+} at its active site. Its best substrates are those with the acidic residues Asp or Glu at the N-terminus, the hydrolysis being activated by Ca^{2+} in the assay. Basic residues are hydrolyzed somewhat more slowly.[6] As well as being sensitive to chelating agents, aminopeptidase A is inhibited by amastatin.[54] It is thought to be involved in the removal of the N-terminal Asp from angiotensin II to form angiotensin III.

3. Aminopeptidase P

In comparison with others in the group, aminopeptidase P is a poorly characterized enzyme, due mainly to difficulties encountered in solubilizing the enzyme from other membrane components.[117] This problem has recently been overcome with the observation that aminopeptidase P is a glycosylphosphatidylinositol-anchored ectoenzyme (see Section VI.A) and can be readily released from the membrane by bacterial phosphatidylinositol-specific phospholipase C.[118] The enzyme is localized in kidney microvilli, intestinal brush borders, and lung tissue.[119–121] Aminopeptidase P has now been solubilized from pig kidney membranes with phospholipase C and purified to virtual homogeneity.[7] The peptidase is a glycoprotein of subunit M_r 91,000. It is partially inhibited by chelating agents such as EDTA and 1,10-phenanthroline, and its activity is stimulated by Mn^{2+} (optimal at 4 m*M)* and Co^{2+}. The enzyme contains one atom of Zn^{2+} per subunit.[55]

Aminopeptidase P appears to be highly specific for X-Pro-Y- peptides, including Gly–Pro–Hyp, although the dipeptide Gly-Pro is also hydrolyzed.[117] Thus, aminopeptidase P may play a role in metabolizing peptides with an N-terminal X–Pro– sequence such as bradykinin and substance P.[121,122] Unlike the other aminopeptidases, aminopeptidase P is not inhibited by actinonin, amastatin, or bestatin. However, and quite surprisingly, the angiotensin converting enzyme inhibitors enalaprilat and its bicyclic lactam, thiaproline derivative (L155, 212), are strong inhibitors of aminopeptidase P[55] (Table 3). Other angiotensin converting enzyme inhibitors such as captopril and lisinopril were without effect.

4. Aminopeptidase W

Aminopeptidase W is unique among the cell-surface peptidases in being discovered by affinity chromatography using a monoclonal antibody generated to pig kidney microvillar membranes.[123] The isolated antigen is an integral membrane glycoprotein of M_r 130,000, existing as a dimer of two identical subunits. A long search for hydrolyzable substrates was rewarded when it was found to split Leu–Trp. Although the enzyme contains one Zn^{2+} per subunit, it is relatively resistant to chelating agents.[8]

Aminopeptidase W hydrolyzes short peptides and exhibits maximal rates towards dipeptides in which the P_1' residue is aromatic (Trp, Phe, or Tyr). Glu–Trp is the best substrate thus far identified.[8] The enzyme is inhibited by amastatin and bestatin (Table 3), but not by actinonin.[8] A fluorometric assay has now been described.[124]

F. CARBOXYPEPTIDASES

1. Carboxypeptidase M

Carboxypeptidase M activity has been detected in membranes from placenta, kidney, and lung[56,59] and was originally believed to be a membrane-bound form of carboxypeptidase N. However, the enzyme has been purified and shown to be structurally, catalytically, and immunologically distinct from carboxypeptidase N.[57] Human carboxypeptidase M is an integral membrane glycoprotein of M_r 62,000. Recently, the cDNA for this enzyme has been sequenced.[125] The 439-amino acid protein contains a cleaved N-terminal signal sequence and a C-terminal hydrophobic region which may act as the transmembrane anchor (see Section VI.A). By comparison with carboxypeptidases A and B, it is apparent that all the active-site residues including the Zn^{2+} ligands are conserved in carboxypeptidase M. The peptidase is inhibited by chelating agents and markedly activated by Co^{2+} ions.[57]

Carboxypeptidase M has a very similar substrate specificity to carboxypeptidase N, in that it cleaves C-terminal basic residues (Arg or Lys) from dipeptides and oligopeptides. However, carboxypeptidase M cleaves Arg faster than Lys compared with carboxypeptidase N, and has a much higher esterase activity relative to its peptidase activity.[9] The enzyme hydrolyzes the C-terminal amino acids from the biological substrates dynorphin A(1-13), [Met^5]enkephalin-Lys^6, [Met^5]enkephalin-Arg^6, bradykinin, and [Leu^5]enkephalin-Arg^6.[57]

No specific inhibitor of carboxypeptidase M is available, although the more general carboxypeptidase inhibitors MGTA (2-mercaptomethyl-3-guanidino-ethylthiopropanoic acid)[56] and GEMSA (guanidino-ethylmercaptosuccinic acid) are relatively potent (Table 3).

2. Carboxypeptidase N

Carboxypeptidase N (kininase I) is a large soluble plasma glycoprotein of M_r 280,000 in humans. It is tetrameric in structure with two, high molecular weight glycosylated subunits (M_r 83,000) which have no activity, and two lower molecular weight unglycosylated subunits (M_r 48,000) which contain the active site.[126] These four subunits are held together by noncovalent interactions. Recently, the cDNA for the small active subunit of human carboxypeptidase N has been cloned and sequenced.[127] There is a cleaved N-terminal signal sequence of 20 residues, followed by the 438 amino acids of the mature protein. As with carboxypeptidase M, the amino acid residues involved in binding the Zn^{2+} and participating in the active site are strictly conserved in carboxypeptidase N. This carboxypeptidase is also inhibited by chelating agents and activated by Co^{2+}.[9]

Carboxypeptidase N has a preference for C-terminal Lys over Arg. Several endogenous peptide and protein substrates are cleaved by carboxypeptidase N including kinins, anaphylatoxins, fibrinopeptides, [Leu^5]enkephalin-Arg^6, [Met^5]enkephalin-Arg^6, [Met^5]enkephalin-Lys^6, and bradykinin.[9,59,128] The general carboxypeptidase inhibitors MGTA and GEMSA also inhibit carboxypeptidase N (Table 3).

3. Carboxypeptidase P

Carboxypeptidase P is an integral membrane glycoprotein of subunit M_r 135,000.[10] The enzyme is present in both renal and intestinal brush borders.[119,129] Carboxypeptidase P exists as a dimer with one Zn^{2+} per subunit; it is inhibited by chelating agents and requires Mn^{2+} for maximum activity.[130]

Carboxypeptidase P cleaves C-terminal amino acids from tri- and oligopeptides when the penultimate residue is Pro, although Ala and Gly are also tolerated. Peptides such as angiotensins II and III are potential substrates for carboxypeptidase P.[10] No specific inhibitor of the enzyme has yet been reported.

G. DIPEPTIDYL PEPTIDASE IV

Dipeptidyl peptidase IV is an integral membrane glycoprotein of subunit M_r 130,000 and exists in the membrane as a noncovalently linked dimer, each subunit containing a single active-site serine residue.[60] Recently, the cDNA sequence for rat liver dipeptidyl peptidase IV has been elucidated.[11] The cDNA codes for a 767-residue polypeptide. As with endopeptidase-24.11 and aminopeptidase N, the uncleaved N-terminal signal sequence constitutes the membrane anchor in the mature protein (Section VI.A). Dipeptidyl peptidase IV is unique among the cell-surface peptidases in being a serine peptidase.

Dipeptidyl peptidase IV releases dipeptides from the N-terminus of susceptible peptides when the penultimate residue is Pro (or Hyp) or Ala.[131] The only natural peptide which has been shown to be hydrolyzed by dipeptidyl peptidase IV is substance P.[132] Bradykinin, by virtue of a second Pro residue at position 3, is resistant to hydrolysis.

Dipeptidyl peptidase IV is inhibited by the serine-protease inhibitor diiso propyl-fluorophosphate.[60] The bacterial product, diprotin A, is also a potent inhibitor of the enzyme[61] (Table 3). Also, some 'suicide' inhibitors have been synthesized[133] and used in functional studies.[134]

H. MEMBRANE DIPEPTIDASE

Membrane dipeptidase (renal or microsomal dipeptidase; dehydropeptidase I) is a glycoprotein of M_r 47,000 in the pig. It is anchored in the membrane by a covalently attached glycosyl-phosphatidylinositol moiety[135,136] (Section VI.A). The human enzyme is considerably larger in size (M_r 59,000) but the difference in M_r between these two species is due almost entirely to more extensive glycosylation of the human enzyme.[137] Membrane dipeptidase exists as a disulphide-linked dimer,[83] each subunit containing one Zn^{2+} at the active site.[138] Recently, the cDNA sequence for pig kidney membrane dipeptidase has been isolated and sequenced.[235] This predicts a cleaved N-terminal signal sequence and a C-terminal hydrophobic region typical of glycosyl-phosphatidylinositol anchored proteins. The amino acid sequence displays no similarity with any known mammalian Zn^{2+}-peptidase or bacterial β-lactamase sequence.

As its name implies, membrane dipeptidase hydrolyzes only dipeptides, but it has a very broad specificity including substrates in which the C-terminal residue has the D-configuration.[139,140] Membrane dipeptidase may play a role in the metabolism of such naturally occurring substances as the glutathione derivative cystinyl-*bis*-glycine, leukotriene D_4,[140,141] and β-lactam antibiotics such as imipenem.[142] Membrane dipeptidase may also play a secondary role in the metabolism of dipeptide products resulting from the primary cleavage of peptides by other peptidases.

Membrane dipeptidase is inhibited by chelating agents and by dithiothreitol. The specific inhibitor, cilastatin (Table 3), was originally developed to protect the β-lactam antibiotic, imipenem, from hydrolysis in the kidney (Section VII.C).[63] This inhibitor can also be used as a ligand for affinity purification of membrane dipeptidase.[62]

I. PYROGLUTAMYL PEPTIDASE II

Pyroglutamyl peptidase II (thyroliberinase) is an ectoenzyme of M_r 230,000. It is a metalloenzyme and is thus distinct from the cysteine enzyme pyroglutamyl peptidase I (EC 3.4.19.3) present in the cytosol.[143] Pyroglutamyl peptidase II activity is highest in the brain, where it is enriched in synaptosomal membranes, with lower levels in lung, liver, and serum.[13,143–145]

Pyroglutamyl peptidase II has a restricted specificity in hydrolyzing only the Glp–His bond of thyroliberin and related synthetic peptides. The same bond in luliberin is not cleaved.[13] It is probably premature, however, to describe this enzyme as a thyroliberin-specific peptidase or thyroliberinase.

J. γ-GLUTAMYL TRANSPEPTIDASE

γ-Glutamyl transpeptidase exists in the membrane as a glycosylated heterodimer, of subunit M_r 30,000 and 50,000. The cDNA sequence for γ-glutamyl transpeptidase has been isolated.[146,147] This predicts a single chain of 568 amino acids which is subsequently processed to the two subunits. The uncleaved signal sequence on the heavy subunit doubles as the membrane anchor (Section VI.A).

γ-Glutamyl transpeptidase cleaves γ-Glu from the N-terminus of oligopeptides, its principal substrate being glutathione. The enzyme will also transfer the γ-Glu group to a variety of amino acids and dipeptides.[148] A nonreversible inhibitor, AT-125 (Table 3), of the enzyme has been described.[64]

K. OTHER PEPTIDASES

1. N-Acetylated-α-Linked Acidic Dipeptidase

The endogenous acidic dipeptide, N-acetyl-Asp-Glu, is a putative neurotransmitter candidate which is metabolized subsequent to its release from nerve endings.[149] The activity responsible for this metabolism has been termed N-acetylated-α-linked acidic dipeptidase (NAALA dipeptidase). This activity is present in synaptosomal preparations from brain and appears to be membrane bound; however, it has yet to be purified. Its distribution is restricted to nervous tissue and kidney.[15,150] NAALA dipeptidase is inhibited by chelating agents, suggesting that it is a metalloenzyme. The aminopeptidase inhibitors bestatin and puromycin substantially inhibited NAALA dipeptidase, although the most potent inhibitor was the excitatory amino acid agonist, quisqualic acid[15] (Table 3). NAALA dipeptidase appears to be selective for N-acetylated α-linked acidic dipeptides as inferred from inhibitor studies, and has a very low K_m value (540 n*M*) for N-acetyl-Asp–Glu.[15] This activity is distinct from membrane dipeptidase (Section II.H) since purified kidney membrane dipeptidase failed to hydrolyze N-acetyl-Asp–Glu and was not inhibited by quisqualic acid.[136]

2. Neurotensin-Degrading Endopeptidase

Neurotensin-degrading endopeptidase was first identified in, and subsequently purified from, rat brain synaptic membranes.[151,152] This peptidase is a monomer of M_r 70,000, and is sensitive to chelating agents indicative of a metalloenzyme.[152] 'Neurotensin-degrading' enzyme has a high affinity for neurotensin (K_m = 2.6 μ*M*), but also cleaves the related peptides xenopsin, neuromedin N and [Lys^8,As n^9]neurotensin-(8-13).[152] Several other natural peptides were not hydrolyzed. The primary specificity of this peptidase appears to be directed towards the sequence –Pro–X– (where X is an aromatic amino acid), although more extensive studies with synthetic peptides are required. An immunochemical study has shown that the peptidase also has a wide distribution in peripheral tissues.[16]

3. Substance P-Degrading Enzyme

Substance P-degrading enzyme has a predominantly cytoplasmic location, although about 35% of the total human brain homogenate activity is associated with the membrane fraction.[153] Similar, if not identical, activities have been identified in the nervous tissue of monkey and rat and rat anterior pituitary.[154–156] In spite of the name, it is now stated to have a broader specificity.[157] The intracellular location is far from clear, but it appears to be either mitochondrial or lysosomal and, thus, not an ectoenzyme. Substance P-degrading enzyme is a monomer of M_r 36,000 to 70,000 and does not contain carbohydrate.[155] The partially purified activity was sensitive to chelating agents and hydrolyzed substance P at the Gln^6–Phe^7, Phe^7–Phe^8, and Phe^8–Gly^9 bonds.[153] Although this enzyme has been suggested to be identical with endopeptidase-2 on the basis of a common pattern of hydrolysis of substance P,[157] the two enzymes are strikingly different in topology, membrane location, size, oligomeric structure, carbohydrate content, and specificity on other substrates.[3]

III. HOW DO THESE PEPTIDASES ACT IN CONCERT?

The information obtained from detailed studies on purified enzymes is important in delineating the bond specificity of the peptidase and generating the quantitative data necessary to assess their kinetics and the effect of inhibitors. However, physiological conditions are far removed from those usually selected for such studies. Two major differences exist: (1) the surface membrane will contain multiple peptidase activities, and (2) the concentration of the peptide substrates will be orders of magnitude lower than those necessary for *in vitro* determinations. The effect of multiple activities is easily investigated using a membrane preparation rich in peptidases, such as renal microvillar membranes, in which many of the peptidases are major proteins in the membrane. In the presence of such a battery of peptidases, most peptides will undergo virtually complete hydrolysis, by which stage the contributions of individual enzymes will have been obscured. What is more relevant is the ordered action of the battery of peptidases, particularly the identity of the enzyme which initiates the attack. Under appropriate conditions, it is possible to stop the hydrolysis at an early stage so that the initial peptide fragments can be identified. Specific inhibitors may then be used to confirm the identity of the peptidase initiating the attack. This approach is illustrated in Figure 2 where a number of peptides have been incubated with pig kidney microvilli.[33,158] The principal enzyme

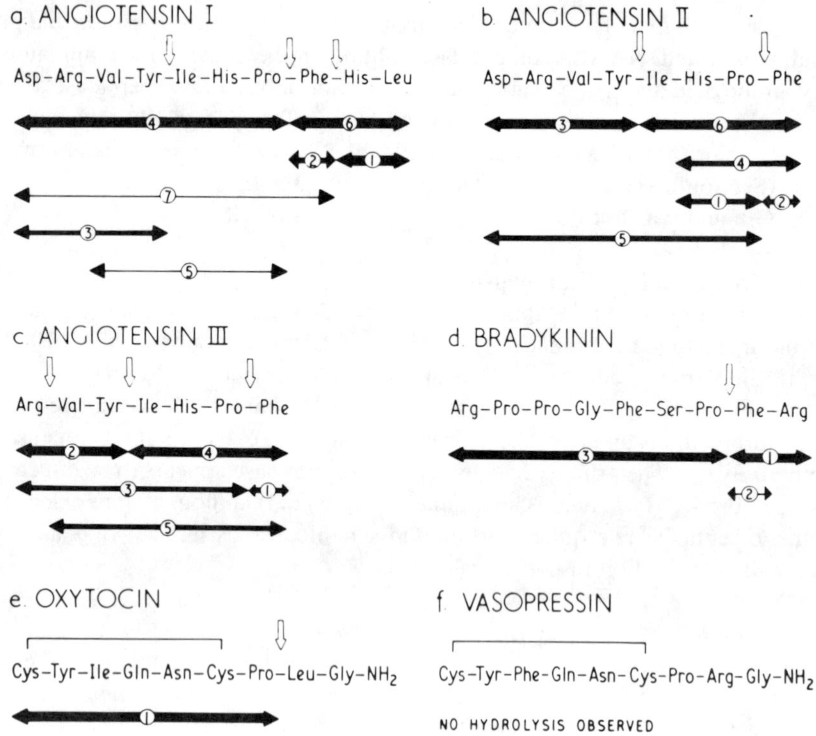

FIGURE 2. Metabolism of peptides by pig kidney microvillar membranes. The figure shows the initial points of attack when the peptides were incubated with pig kidney microvilli. The horizontal arrows define the peptide fragments characterized by amino acid analyses, after separation by HPLC and the thickness of the arrow relates to the yield of that peptide. The open arrows indicate the bonds hydrolyzed. By including specific peptidase inhibitors in parallel incubations, it was possible to assess the relative contributions of the principal peptidases. (See text for details.) (From Stephenson, S. L. and Kenny, A. J., *Biochem. J*, 241, 237, 1987. With permission.)

initiating the hydrolysis of the angiotensins, bradykinin, oxytocin, atrial, and brain natriuretic peptides is endopeptidase-24.11. Angiotensin converting enzyme has a minor role in hydrolyzing the Pro–Phe bond of bradykinin and the formation of His–Leu from angiotensin I is partly the result of this enzyme hydrolyzing the Phe–His bond, but most of this dipeptide arises from the combined attack of endopeptidase–24.11 (generating Phe–His–Leu) and aminopeptidase N (removing Phe). A carboxypeptidase attack, most probably carboxypeptidase P, also plays a role in hydrolyzing the C-terminal Pro–Phe bond in angiotensins II and III. The initial point of attack on atrial natriuretic peptide is the Cys^7–Phe^8 bond. It is noteworthy that this cleavage, in common with the initial attack on many other peptides, is an inactivating step. A similar experimental approach has revealed that the initial attack on atrial natriuretic peptide is essentially dependent on endopeptidase-24.11.[158]

Rat microvilli contain an additional enzyme, endopeptidase-2.[78,81,82] Its repertoire of susceptible peptides overlaps with that of endopeptidase-24.11 but, with the exception of luliberin (which is an extremely poor substrate for endopeptidase-24.11)[25] most peptides are very much more efficiently degraded by endopeptidase-24.11 (Table 6[85]). Phosphoramidon fully inhibits the hydrolysis of luliberin, bradykinin, and substance P by pig microvilli, but is only partially effective for rat microvilli and the residual activity (Figure 3) is a measure of the contribution of endopeptidase-2 in these conditions. However, endopeptidase-2 has a much more restricted tissue distribution than endopeptidase-24.11 and is notably undetectable

TABLE 6
Comparison of Endopeptidase-2 and Endopeptidase-24.11 in Hydrolyzing Three Neuropeptides

Peptide	Endopeptidase-2 K_m (μ*M*)	Endopeptidase-2 k_{cat} (min^{-1})	Endopeptidase-2 k_{cat}/K_m (min^{-1}·μ*M*$^{-1}$)	Endopeptidase 24.11 K_m (μ*M*)	Endopeptidase 24.11 K_{cat} (min^{-1})	Endopeptidase 24.11 k_{cat}/K_m (min^{-1}·μ*M*$^{-1}$)
Luliberin	270	669	2.48	755	840	1.1
Bradykinin	325	87.7	0.27	92	6360	69
Substance P	86	180	2.09	32	5060	159

From Stephenson, S. L. and Kenny, A. J., *Biochem. J.*, 255, 45, 1988.

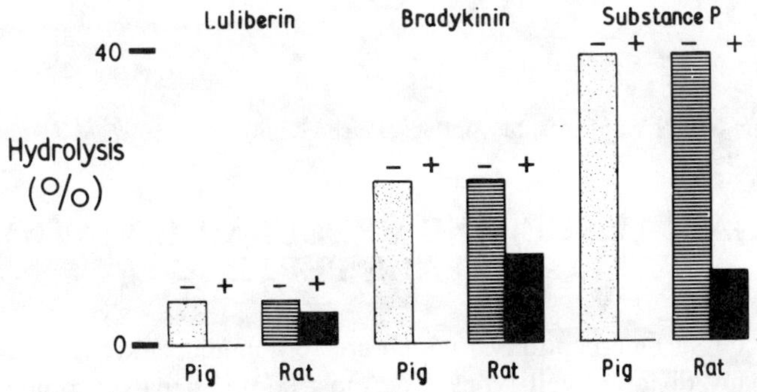

FIGURE 3. Hydrolysis of peptides by pig and rat kidney microvilli. Since rat renal microvilli contain an additional enzyme (endopeptidase-2) which is lacking in pig, rabbit, and human kidneys, this experiment was designed to assess the contribution of this peptidase to the hydrolysis of three peptides. The addition of a cocktail of inhibitors (to block endopeptidase-24.11, aminopeptidases N, A, W, and dipeptidyl peptidase IV) abolished hydrolysis in each case by pig microvilli. The residual activity in the rat membrane is assumed to be the contribution of endopeptidase-2, that is, substantial for luliberin, equal for bradykinin, but minor for substance P. +, –: with and without the inhibitors. (From Stephenson, S. L. and Kenny, A. J., *Biochem. J.*, 255, 45, 1988. With permission.)

in the CNS[81] so that, on many cell surfaces, endopeptidase-24.11 may be the only endopeptidase capable of initiating an attack. Moreover, many peptides appear to be much better substrates for endopeptidase-24.11 than they are for endopeptidase-2. One such peptide is atrial natriuretic peptide[158] and, although endopeptidase-2 can hydrolyze it slowly, it appears to make no contribution to its hydrolysis by rat renal microvilli.[85]

Some degree of caution is necessary in extrapolating the results of these experiments to *in vivo* conditions, since the high concentration of substrate (50 to 250 μ*M) in vitro* might well have revealed only peptidases exhibiting high K_m values. In the case of bradykinin, it was possible to examine this point by repeating the incubation with 20-n*M*[^{3}H]bradykinin. The main labeled product had the same retention time as bradykinin 1-7 and the formation of this peptide was inhibited by 1 μ*M* phosphoramidon.[33] This argues against the existence of a second endopeptidase, with much lower K_m, which might be dominant in physiological conditions. It should also be noted that the crucial role of endopeptidase-24.11, rather than angiotensin converting enzyme, in the inactivation of bradykinin by renal microvilli does not imply that the latter enzyme lacks this role *in vivo*. The crucial consideration is that on the surfaces of the endothelium of the pulmonary vasculature angiotensin converting enzyme is the only enzyme present capable of removing the C-terminal dipeptide (see Section VII.B).

TABLE 7
Distribution of Endopeptidase-24.11 in the Pig

Tissue (homogenates)	Endopeptidase-24.11 (ng/mg of protein)
Kidney (cortex)	5000
Lymph node (piglets/adults)	750/1400
Chondrocytes (isolated from articular cartilage)	650
Jejunum	400
Ileum	130
Duodenum	65
Salivary glands	55
Pancreas	25
Adenohypophysis	25
Adrenals	20
Lung	5
Skin	2.5

From Gee, N. S., Bowes, M. A., Buck, P., and Kenny, A. J., *Biochem. J.*, 228, 119, 1985.

IV. ORGAN, TISSUE, AND CELLULAR LOCATIONS OF CELL-SURFACE PEPTIDASES

The group of cell-surface peptidases which are so abundant in renal microvilli is widely represented on many different cell types. The most comprehensive immunological survey, using an immunoradiometric assay (i.r.m.a.)[159] supported by immunohistochemistry[116] has been made for endopeptidase-24.11 in the pig; and since this enzyme seems to be important in initiating the degradation of a number of peptides, its multiple locations will be described in more detail. Table 7 gives the results of the i.r.m.a. of pig tissues in which the unexpected finding was the abundance of endopeptidase-24.11 in lymphoid tissue, where it appears to be expressed by reticular cells.[160,161] However, subsequent work[162] has not shown similarly high concentrations in lymph nodes from rat, rabbit, or human. The circulation of lymphocytes and certain anatomical features of lymph nodes are unusual in the pig[163] and the high endopeptidase-24.11 content in this species may possibly relate to these differences.

The immunohistochemical survey (Table 8) was more extensive in tissue coverage than that of the less-sensitive i.r.m.a. and offers some clues regarding the likely function of endopeptidase-24.11 at certain locations in terminating peptide signals. What is striking is the very distinctive distribution of the enzyme, being very widespread, but certainly not ubiquitous. For example, only the gonadotrophs among the five cell types in the adenohypophysis were stained;[164] only a minority of neurons in certain discrete regions of the brain express the enzyme and only one zone, that synthesizing aldosterone, of the adrenal cortex was stained. The expression of endopeptidase-24.11 in some cell lineages varies with the stage of differentiation and in malignant transformation. A striking example, which is discussed below, is revealed by work on CALLA (the common acute lymphocytic leukemia antigen).

V. MEMBRANE PEPTIDASES AS CLUSTER DIFFERENTIATION (CD) ANTIGENS

A. CALLA (CD-10) AND ENDOPEPTIDASE-24.11

CALLA, the common acute lymphocytic leukemia antigen, first described in 1975[165] and recently reviewed,[166] is expressed on lymphoblasts from most patients with the common type

TABLE 8
Cells Expressing Endopeptidase-24.11

Epithelia (brush borders)
Renal proximal tubule
Enterocyte
Syncytial trophoblast
Choroid plexus
Epithelia (other types)
Cells lining reproductive tract (male and female)
Acinal cells (prostate, mucous glands, lung)
Duct (pancreas)
Steroid endocrine cells
Leydig cells (testis)
Interstitial cells (ovary)
Zona glomerulosa (adrenal)
Other endocrine cells
Gonadotrophs (adenohypophysis)
Connective tissue
Reticular cells (lymph nodes)
Fibroblasts (human skin in culture)
Chondrocytes (articular cartilage)
Central nervous system
Some neurons in corpus striatum
Peripheral nervous system
Some Schwann cells
Blood
Pre B cells (bone marrow)
$CALLA^+$ cells (common acute lymphocytic leukemia)
Polymorphonuclear leukocytes

of ALL (acute lymphocytic leukemia). CALLA was characterized as a 100-kDa glycosylated polypeptide. Antibodies against CALLA have been used extensively in the diagnosis and therapy of lymphoid malignancies. CALLA had also been reported on some other cells and was shown to be abundant in kidney and intestine. Until very recently, no activity had been ascribed to CALLA, its identification being solely by immunological criteria. Two groups[167,168] have now cloned and sequenced the cDNA encoding CALLA and shown it to be an integral transmembrane protein of 750 amino acids. Letarte et al.[168] observed that the predicted amino acid sequence was identical to that of human endopeptidase-24.11. It has been confirmed that purified CALLA has endopeptidase-24.11 activity.[169,170] The presence of high circulating levels of endopeptidase-24.11 may have a dramatic effect on the clearance of peptides from the blood. The expression of endopeptidase-24.11 in progenitor B cells[171] and transformed cells (but its absence from most circulating B lymphocytes) as well as in some gliomas and melanomas[172,173] (but absent from the normal cell types) awaits an explanation. However, its association with populations of rapidly dividing cells would be consistent with a role in terminating peptide or growth factor signals that may affect cell proliferation in either a stimulatory or inhibitory way.[174]

B. CD13 AND AMINOPEPTIDASE N

CD13 had been defined by monoclonal antibodies as a 150-kDa glycoprotein present on cells of granulocytic and monocytic lineages and is expressed by leukemic blasts in a high proportion of cases of acute myeloid leukemia and, in a few cases, of acute lymphoid leukemia (in which it has an unfavorable prognostic significance). The full cDNA derived sequence for CD13 has now been reported by Look et al.[175] who also noted its identity with that of aminopeptidase N which was published a few months earlier.[108]

C. CD26 AND DIPEPTIDYL PEPTIDASE IV

Dipeptidyl peptidase IV, the only serine peptidase in the group and known to be a marker for $CD4^+$ T cells[134,176] was recently cloned. Its sequence shows homology with gp110, present in the bile canaliculus membrane, and has been reported to be identical to CD26.[11,117]

VI. MEMBRANE ANCHORAGE, RELEASE, AND PLASMA ACTIVITY OF PEPTIDASES

A. MEMBRANE ANCHORAGE

Structural characterization has revealed that cell-surface peptidases can be anchored to the plasma membrane by at least three distinct structures: (1) an N-terminal transmembrane polypeptide, (2) a C-terminal transmembrane polypeptide, or (3) a C-terminal glycosyl-phosphatidylinositol (G-PI) moiety (Figure 4). Historically, a comparison of the molecular properties of the proteinase-solubilized form with the detergent-solubilized form of certain peptidases (aminopeptidases N and A, dipeptidyl dipeptidase IV) revealed the position of the membrane anchor within the protein.[178,179] The proteinase-form, as well as having a smaller M_r by between 3000 to 5000 than the detergent-form, also possessed a different N-terminal amino acid sequence, thus implying that the hydrophobic membrane anchor was located at the N-terminus. When the detergent-solubilized form of endopeptidase-24.11 was purified and partially sequenced, a hydrophobic stretch of amino acids near the N-terminus was revealed.[180] This sequence, which displayed considerable homology to the N-terminal cytoplasmic tail sequence in sucrase-isomaltase, was suggested to represent an uncleaved signal peptide.

This observation was subsequently confirmed when the complete amino acid sequences for these peptidases was deduced from cDNA cloning. For endopeptidase-24.11,[18,19] aminopeptidase N[108] and dipeptidyl peptidase IV,[11] as well as γ-glutamyltransferase,[146,147] it is now apparent that the membrane anchor is the uncleaved N-terminal hydrophobic signal sequence.

The mode of membrane anchorage of angiotensin converting enzyme is clearly different from the above peptidases. Partial amino acid sequences revealed that both the detergent- and proteinase-solubilized form of this peptidase have an identical N-terminal sequence,[92] suggesting that the site of membrane attachment is at the C-terminus. With the elucidation of the cDNA sequence for angiotensin converting enzyme,[90] which predicts a cleaved N-terminal signal sequence and a putative hydrophobic transmembrane spanning region near the C-terminus, this suggestion appears to be confirmed. The recently obtained cDNA sequence for carboxypeptidase M[125] indicates that this peptidase may be similarly anchored by a hydrophobic polypeptide near the C-terminus. However, in neither of these peptidases is there conclusive evidence that the hydrophobic polypeptide predicted by the cDNA sequence constitutes the membrane anchor in the mature protein.

Two other cell-surface peptidases are anchored through their C-terminal residue by a covalently attached G-PI moiety.[181–184] This novel type of membrane anchorage has been identified in a functionally and evolutionarily diverse group of cell-surface proteins, including proteins involved in hydrolytic activity, cell adhesion, transmembrane signaling, complement regulation, and protective surface coat. The C-terminal amino acid of the protein is linked to a glycan chain of three mannose residues via an ethanolamine phosphate. The reducing end of the glycan chain is coupled to a nonacetylated glucosamine residue which, in turn, is linked to the inositol ring of a phosphatidylinositol moiety. Thus, the only part of the structure which interacts with the membrane lipids are the fatty acids in the phosphatidylinositol moiety and no part of the structure interacts with the cytosol (Figure 4). The presence of a G-PI anchor can be identified by a number of criteria other than direct chemical identification of the anchor components. Both membrane dipeptidase and aminopeptidase P are selectively released from kidney microvillar membranes by bacterial phosphatidylinositol-specific phospholipase

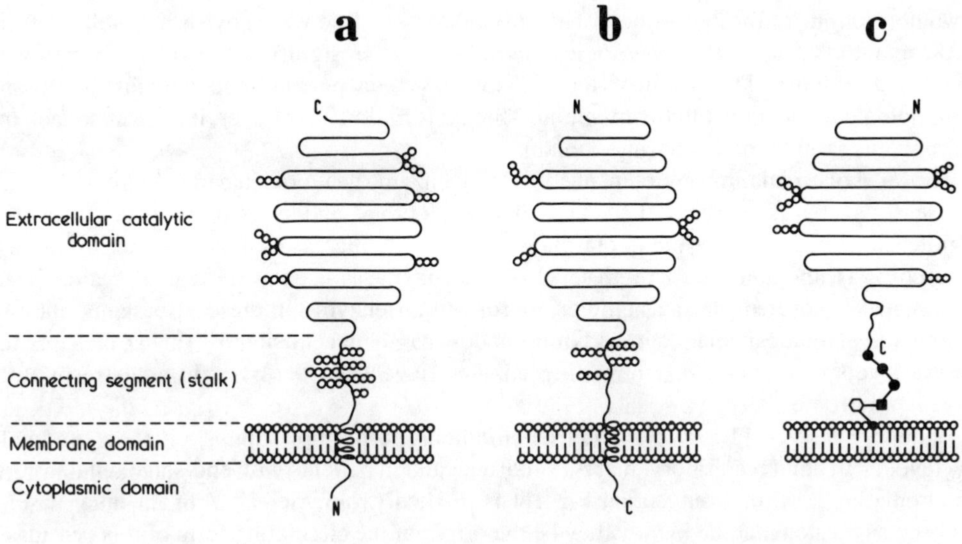

FIGURE 4. Membrane anchorage of cell-surface peptidases. (a) N-terminal transmembrane polypeptide anchor (uncleaved signal peptide) (e.g., endopeptidase-24.11, aminopeptidase N, aminopeptidase A, dipeptidyl peptidase IV, and γ-glutamyltransferase; (b) C-terminal transmembrane polypeptide anchor (e.g., angiotensin-converting enzyme, and carboxypeptidase M); and (c) C-terminal G-PI moiety (e.g., membrane dipeptidase and aminopeptidase P).

C[118,135] and by a G-PI-specific phospholipase D in plasma,[83,185] and both peptidases possess the cross-reacting determinant,[7,136] all characteristic features of G-PI-anchored proteins.

B. RELEASE

As indicated in Table 1, most of the peptidases described in this chapter exist predominantly in a membrane-bound form. However, for some of these enzymes, a soluble form is present in plasma and other body fluids (see Section VI.C). Angiotensin converting enzyme exists in a soluble form in plasma and this form of the enzyme is thought to originate from the vascular endothelium.[186] The plasma enzyme may result from the transcription of an mRNA with a different 3′ end to that for the membrane-bound enzyme; however, a distinct mRNA species was not detected in human vascular endothelial cells.[90] Alternatively, a post-translational modification may account for the soluble form of the enzyme in plasma and other body fluids. In support of this, an EDTA-sensitive hydrolase has been identified in several tissues, including lung, which may be involved in releasing membrane-bound angiotensin converting enzyme in a soluble form.[55,92] No other specific release mechanisms have yet been identified for other transmembrane polypeptide anchored peptidases.

One of the functions of a G-PI membrane anchor may be to allow selective release of the protein in a soluble form.[183] To this end, activities have been identified which can cleave G-PI anchors. These include a G-PI-specific phospholipase C present in rat liver membranes[187] and a G-PI-specific phospholipase D located predominantly in plasma.[188,189] The latter activity has been shown to release membrane-bound membrane dipeptidase and aminopeptidase P in a soluble, active form.[83,185] However, under normal physiological conditions, the precise roles of these G-PI cleaving activities are unclear.

C. PLASMA ACTIVITY

Of the peptidases described in this chapter, carboxypeptidase N is the only one whose primary location is in plasma (Table 1). This peptidase is synthesized in the liver and released into the circulation where its role most probably is to act on potent peptides which may have been released (or escaped) from surrounding tissues and, thus, protect the organism from the unwanted actions of these peptides.[9] The only other peptidase which exists as a soluble form in plasma under normal physiological conditions at a significant level is angiotensin converting enzyme. This form of the enzyme has been purified and is catalytically and immunologically identical to the membrane-bound form.[190,191] However, it appears to lack the hydrophobic membrane anchoring domain.[55]

The level of circulating angiotensin converting enzyme is altered in certain clinical states.[99] In diseases such as sarcoidosis, Gaucher's disease, leprosy, and hyperthyroidism, the level of angiotensin converting enzyme in plasma is significantly increased above normal. In the case of sarcoidosis, the abnormally high level of the enzyme is used as a diagnostic aid.[192] The source of the elevated plasma angiotensin converting enzyme in these diseases is unclear, although in sarcoidosis the afflicted lymph nodes may be responsible.[193] Lower than normal plasma levels of angiotensin converting enzyme have been observed in adult respiratory distress syndrome.[194]

Endopeptidase-24.11 is also found as a circulating form in plasma, albeit at an extremely low level.[195] In adult respiratory distress syndrome and in patients with end-stage renal failure, the circulating level of endopeptidase-24.11 is markedly increased.[194,196] In the latter case, it has been suggested that the kidney may be the origin of the circulating form of this peptidase. The role, if any, of these circulating forms of angiotensin converting enzyme and endopeptidase-24.11 in these pathophysiological situations is unclear. Further studies may well reveal that other, predominantly membrane-bound, peptidases are also increased in these or other disease situations.

VII. PHYSIOLOGICAL ROLES OF MEMBRANE PEPTIDASES

Although experiments with purified enzymes and membrane fractions are an essential prerequisite to understanding the physiological role of an enzyme, the steps from simple *in vitro* preparations to more complex systems involving intact cells, tissues, and whole animals pose many problems. The availability of a few specific, but not toxic, inhibitors has offered one useful approach. The synthesis of peptide analogues which might resist hydrolysis of the putative peptidase offers another way forward. Most of the studies have focused on angiotensin converting enzyme in relation to the renin angiotensin system and on endopeptidase-24.11 in relation to the enkephalins, substance P, and atrial natriuretic peptide, and this section will concentrate on these two peptidases. Mention will also be made of membrane dipeptidase and its role in metabolizing administered β-lactam antibiotics.

A. ENDOPEPTIDASE-24.11

Within the pharmaceutical industry, there have been two waves of intense interest in this enzyme. The first may be designated the 'enkephalinase' period when the search was for inhibitors which might provide a new class of analgesics. The second, focusing on atrial natriuretic peptide, giving hope that an inhibitor might be of value in the treatment of hypertension, has not yet run its course. Both approaches suffer from a flawed hypothesis: that the enzyme is, for all practical purposes, peptide-specific. The very wide range of cell-types expressing this ectoenzyme and the large repertoire of susceptible peptides must raise serious doubts about the basis of this approach. Moreover, the possibility that the C-receptor may have a role in the clearance of atrial natriuretic peptide further complicates the story.

1. 'Enkephalinase'

It is an historical accident that the enkephalins were the first to evoke interest in the metabolic fate of neuropeptides. The first report in 1978 that the initial attack on [Met^5]enkephalin by brain preparations was at the Gly^3–Phe^4 bond was by Craves et al.[197] However, the identity of the enzyme responsible for this hydrolysis was a matter of considerable confusion in the next 4 or 5 years, in part because the removal of the C-terminal dipeptide (Phe–Met) was assumed to be due to the action of a peptidyl dipeptidase resembling, but not identical with, angiotensin converting enzyme.[198] The possibility that it was an endopeptidase attack was first suggested by Blumberg et al.[199] and this concept gained support as a result of work from various laboratories showing that phosphoramidon and certain peptide hydroxamates inhibited striatal enkephalinase[200] and by the finding that a partially purified pituitary membrane endopeptidase hydrolyzing Bz–Gly–Arg–Arg–Leu–2NA, which had certain properties in common with the renal enzyme, also hydrolyzed the Gly–Phe bond of enkephalins.[201] These authors noted that a similar activity was detectable in brain regions reported to be rich in enkephalinase. However, the unambiguous demonstration of the identity of the renal and brain enzymes required immunological and other experiments in which the purified renal enzyme could be compared with the activity in striatal membranes,[42,202] culminating in the purification to homogeneity (24,000-fold) of the brain enzyme by immunoaffinity chromatography.[17] Shortly thereafter, similar immunological evidence was reported from another laboratory.[45]

Some light has been thrown on the metabolism of endogenous peptides in the central nervous system (CNS) by the use of tissue slices which release some of the peptide in response to a depolarizing signal, such as K^+. The peptide released may be quantified by radioimmunoassay and, if the recovery of the peptide is increased by the presence of, say, an endopeptidase-24.11 inhibitor in the perfusing medium, it may be argued that this enzyme has an *in vivo* role in the metabolism of the peptide. Endopeptidase-24.11 inhibitors increased the recovery of [Met^5]enkephalin twofold, but the addition of bestatin augmented the effect, pointing to a contribution from aminopeptidases.[203] The recovery of substance P from slices of substantia nigra was increased five-fold by phosphoramidon and was unaffected by enalaprilat; kelatorphan gave a ten-fold increase[204] — a response greater than that previously reported.[205]

Within the CNS, detailed mapping of endopeptidase-24.11 also provides clues as to its function. It is concentrated within the corpus striatum and striatonigral pathway,[206,209] regions rich in peptides, and at light microscopic level endopeptidase-24.11-rich patches (striosomes and some striatal efferents) exhibited co-localization with substance P and, in some instances, with [Leu^5]enkephalin[206] and it is known from primary cultures that the enzyme is a marker for some striatal neurons.[210] Electron microscopy using an immunoperoxidase technique has demonstrated the enzyme on axonal membranes and boutons synapsing with dendrites.[211] All these observations are consistent with a role in terminating some neuropeptide signals, but co-localization of a defined peptide in synaptic vesicles with the membrane endopeptidase has yet to be demonstrated.

2. Atrial Natriuretic Peptide (ANP) Metabolism

There is good evidence that ANP is rapidly hydrolyzed by endopeptidase-24.11 and that the initial attack is at the Cys^7–Phe^8 bond — a step which opens the ring and is known to inactivate the peptide (see Reference 212 and Chapter 7.) It is also clear that among the battery of peptidases in renal microvillar membranes, it is the endopeptidase which initiates the attack.[158,213] The presence of endopeptidase–24.11 in the zona glomerulosa cells of the adrenal cortex and epithelial cells of the choroid plexus,[116,214] sites at which ANP is known to exert effects, is consistent with a role for the enzyme in terminating the signal. In line with this hypothesis is the observation that more potent long-acting analogues of ANP, (e.g., one lacking the first six residues[215]) would be resistant to hydrolysis at the Cys–Phe bond (the initial site of attack), yielding ANP′. This inactive form of ANP has been detected in human coronary sinus[216] and the metabolic products sampled a few seconds following the injection of ^{125}I–ANP into the renal artery included α–ANP′, both observations consistent with an *in vivo* role for endopeptidase-24.11. However, unambiguous evidence that administration of endopeptidase-24.11 inhibitors has diuretic and natriuretic effects by potentiating endogenous ANP is lacking (or at least difficult to show consistently.[217]) It has been proposed that another mechanism involving receptor-mediated metabolism competes with the endopeptidase for the clearance of ANP from plasma.[218] The effects of ANP on cells are mediated by cGMP and the receptor involved is now known from cDNA cloning to be one type of guanylate cyclase.[219,220] It is a transmembrane protein, the cytoplasmic aspect of which resembles the soluble guanylate cyclase and the cell-surface domain is the ANP receptor, and shows strong homology to another ANP receptor, the C receptor, cloned earlier.[221] Since no second messenger has been identified when ANP binds to this C receptor, it has been argued that it is a 'silent', clearance receptor leading to internalization of the peptide and subsequent degradation within lysozomes. However, there seem to be wide variations in the relative abundance of the two receptors; for example, in aortic smooth muscle, almost all the receptors are of the C-type, while in the testicular Leydig cells, all the receptors are of the B type, i.e., the guanylate cyclase linked receptor. One ANP analogue C-ANF4-22, containing only 15 amino acid residues, being truncated at N- and C-termini and lacking 5 residues within the ring, has no ANP actions but binds to the C-receptor with high affinity.[218,222] *In vivo,* this analogue had little effect on the plasma clearance of ^{125}I-ANP, but did appear to reduce the volume of distribution and hence reduced the metabolic clearance rate, suggesting that the C-receptor plays an important role. However, the possible role of C-ANF$_{4-12}$ as a substrate of endopeptidase-24.11 and, hence, as a competitor with ANP for hydrolysis, has yet to be determined. In other sites such as Leydig cells and the CNS where C-receptors are lacking, ANP metabolism may be dependent on the endopeptidase.

B. ANGIOTENSIN CONVERTING ENZYME

Angiotensin converting enzyme is probably best known for its physiological role in the renin-angiotensin system (RAS), see Chapter 6. It was, however, several years after its discovery in horse plasma[223] before it became evident that angiotensin converting enzyme has a critical role in the RAS. Although the enzyme had been shown to cleave angiotensin I and thus generate angiotensin II (hence its name),[224] it was only when the lung was discovered to be the most important site for the generation of circulating angiotensin II[225,226] and that angiotensin converting enzyme is abundant in this tissue[227] that the connection was made. Because of its ectoenzyme topology and endothelial location,[228] membrane-bound angiotensin converting enzyme can act on circulating angiotensin I. Independently and at the same time, as well as generating a potent vasopressor, angiotensin converting enzyme was shown to inactivate the vasodilator, bradykinin.[229] Thus, the same enzyme has a dual role in blood pressure homeostasis. It is because of this role in blood pressure homeostasis that inhibitors of angiotensin converting enzyme have been developed and successfully used to treat hypertension and congestive heart failure.[49,107,230]

More recently, angiotensin converting enzyme has been implicated in other physiological and pathophysiological processes. Angiotensin converting enzyme has been localized to both the male and female reproductive systems. In males, there is a distinct form of the enzyme in testis (see Section II.D) which is under hormonal control.[231,232] The role of angiotensin converting enzyme in this tissue is unclear. It is probably separate from the RAS system, although other components of this system have recently been localized to the testis, and suggestions put forward include a role in male fertility, the modulation of steroidogenesis, and/or regulation of cell growth and differentiation.[4] Angiotensin converting enzyme and all the other components of the RAS have been detected in ovaries and it has been suggested that they may have paracrine functions in this organ.[4]

It has also been postulated that angiotensin converting enzyme has a role in immunity, as the enzyme is thought to mediate or modulate inflammation. As described earlier (Section II.D), angiotensin converting enzyme can hydrolyze a variety of biologically important peptides including substance P, the enkephalins, and luliberin. The co-localization of angiotensin converting enzyme and substance P to a striato-nigral pathway[233,234] led to the suggestion that the enzyme is involved in the metabolism of this neuropeptide. However, the presence of endopeptidase-24.11 in this same pathway[207] and the ability of inhibitors of endopeptidase-24.11, but not inhibitors of angiotensin converting enzyme to protect substance P released from brain slices from hydrolysis[204] appear to rule out a significant role of angiotensin converting enzyme in substance P metabolism in the substantia nigra. The endogenous neuropeptide substrate(s) for angiotensin converting enzyme in the striato-nigral pathway has yet to be identified.

C. MEMBRANE DIPEPTIDASE

During the search for a β-lactam antibiotic with activity against a broad spectrum of bacteria, a natural product of Streptomyces, thienamycin, was discovered. This compound was unfortunately rather unstable in concentrated solution or as a solid; so an amidine derivative, imipenem, was developed which has substantially improved antibacterial properties. However, when imipenem was administered to humans and other animals, although the serum half-life was very high, urinary recoveries were extremely low. Further investigation revealed that imipenem was rapidly metabolized in the kidneys and the enzyme responsible was found to be the brush-border membrane dipeptidase.[142] This observation led to the development of cilastatin to inhibit membrane dipeptidase, and when co-administered with imipenem, high urinary recovery of the antibiotic was observed.[63] Thus, membrane dipeptidase appears to play a physiological role in metabolizing β-lactam compounds[62] and is the first example of a mammalian β-lactamase.

VIII. SUMMARY AND CONCLUSIONS

In this chapter, we have attempted to describe those cell-surface peptidases that may have a role in metabolizing bioactive peptides in mammalian systems. It is clear that certain of these ectoenzymes, endopeptidase-24.11 and angiotensin converting enzyme, are extremely well characterized in their structural and catalytic properties, and they have been identified as playing central roles in some defined physiological processes. However, for the majority of the peptidases described in this chapter, precise physiological functions have yet to be ascribed, mainly because we are still ignorant of their natural substrates. In general, there is good evidence that in a number of tissues many, but not necessarily all, bioactive peptides will be rapidly and completely hydrolyzed by this battery of cell-surface peptidases acting in concert.

The recent observations that endopeptidase-24.11, aminopeptidase N, and dipeptidyl peptidase IV exist as cluster differentiation antigens, open up new areas for research and pose further questions concerning their (patho)physiological role(s) in growth and differentiation of normal and transformed cells. Other questions are raised by the discovery that membrane dipeptidase and aminopeptidase P possess glycosyl-phosphatidylinositol membrane anchors. This complex anchor structure has been proposed to have functions in addition to retaining a protein at the cell surface. These include the selective release of the protein from the membrane by the action of phospholipases, the property of a higher degree of lateral mobility in the membrane, the provision of an address marker for the intracellular sorting of the protein and the possibility for it to function in transmembrane signaling. Whether any of these functions of the glycolipid anchor relates to the physiological roles of these two peptidases remains to be seen.

Over the coming years, the use of specific and selective inhibitors and antibodies and the availability of cDNA sequence data will undoubtedly further our understanding of the role of peptidases in both physiological and pathophysiological processes.

ABBREVIATIONS

ANF	Atrial natriuretic factor
ANP	Atrial natriuretic peptide
Bz	Benzoyl
CALLA	Common acute lymphocytic leukemia antigen
CD	Cluster differentiation
CNS	Central nervous system
EDTA	Ethylenediamine tetraacetic acid
GEMSA	Guanidino-ethylmercapto-Succinic acid
Glp	Pyroglutamic acid
G-PI	Glycosyl-phosphatidylinositol
HPLC	High-performance liquid chromatography
Hyp	Hydroxyproline
i.r.m.a.	Immunoradiometric assay
MGTA	2-Mercaptomethyl-3-guanidinoethylthiopropanoic acid
2NA	2-Naphthylamide
NAALA	N-acetylated-α-linked acidic dipeptidase
pAB	4-Aminobenzoate
PABA	4-Aminobenzoic acid
RAS	Renin angiotensin system

REFERENCES

1. **Erdos, E. G. and Skidgel, R. A.,** Neutral endopeptidase-24.11 (enkephalinase) and related regulators of peptide hormones, *FASEB J.,* 3, 145, 1989.
2. **Orlowski, M., Reznik, S., Ayala, J., and Pierotti, A. R.,** Endopeptidase-24.15 from rat testes. Isolation of the enzyme and its specificity toward synthetic and natural peptides, including enkephalin-containing peptides, *Biochem. J.,* 261, 951, 1989.
3. **Stephenson, S. L. and Kenny, A. J.,** Sorry, not one of the family, *Biochem. J.,* 259, 622, 1989.
4. **Ehlers, M. R. W. and Riordan, J. F.,** Angiotensin-converting enzyme: new concepts concerning its biological role, *Biochemistry,* 28, 5311, 1989.
5. **Malfroy, B., Kado-Fong, H., Gros, C., Giros, B., Schwartz, J.-C., and Hellmiss, R.,** Molecular cloning and amino acid sequence of rat kidney aminopeptidase M: a member of a super family of zinc-metallohydrolases, *Biochem. Biophys. Res. Commun.,* 161, 236, 1989.
6. **Danielson, E. M., Norén, O., Sjostrom, H., Ingram, J., and Kenny, A. J.,** Proteins of the kidney microvillar membrane. Aspartate aminopeptidase: purification by immunoadsorbent chromatography and properties of the detergent- and proteinase-solubilized forms, *Biochem. J.,* 189, 591, 1980.
7. **Hooper, N. M., Hryszko, J., and Turner, A. J.,** Purification and characterization of pig kidney aminopeptidase P. A glycosyl-phosphatidylinositol anchored ectoenzyme, *Biochem. J.,* 267, 509, 1990.
8. **Gee, N. S. and Kenny, A. J.,** Proteins of the kidney microvillar membrane. Enzymic and molecular properties of aminopeptidase W, *Biochem. J.,* 246, 97, 1987.
9. **Skidgel, R. A.,** Basic carboxypeptidases: regulators of peptide hormone activity, *Trends Pharmacol. Sci.,* 9, 299, 1988.
10. **Hedeager-Sorensen, S. and Kenny, A. J.,** Proteins of the kidney microvillar membrane. Purification and properties of carboxypeptidase P from pig kidneys, *Biochem. J.,* 229, 251, 1985.
11. **Ogata, S., Misumi, Y., and Ikehara, Y.,** Primary structure of rat liver dipeptidyl peptidase IV deduced from its cDNA and identification of the NH_2-terminal signal sequence as the membrane-anchoring domain, *J. Biol. Chem.,* 264, 3596, 1989.
12. **Turner, A. J. and Hooper, N. M.,** Phosphatidylinositol-glycan-tailed membrane proteins: the biochemistry of glycolipid anchors, *Biochem. Soc. Trans.,* 17, 660, 1989.
13. **Wilk, S.,** Neuropeptide-specific peptidases: does brain contain a specific TRH-degrading enzyme?, *Life Sci.,* 39, 1487, 1986.
14. **Semenza, G.,** Anchoring and biosynthesis of stalked brush border membrane proteins: glycosidases and peptidases of enterocytes and renal tubuli, *Annu. Rev. Cell Biol.,* 2, 255, 1986.

15. **Robinson, M. B., Blakely, R. D., Couts, R., and Coyle, J. T.,** Hydrolysis of the brain dipeptide N-acetyl-L-aspartyl-L-glutamate. Identification and characterization of a novel N-acetylated α-linked acidic dipeptidase activity from rat brain, *J. Biol. Chem.,* 262, 14498, 1987.
16. **Checler, F., Barelli, H., and Vincent, J.-P.,** Tissue distribution of a novel neurotensin-degrading metallopeptidase. An immunological approach using monospecific polyclonal antibodies, *Biochem. J.,* 257, 549, 1989.
17. **Relton, J. M., Gee, N. S., Matsas, R., Turner, A. J., and Kenny, A. J.,** Purification of endopeptidase-24.11 ('enkephalinase') from pig brain by immunoadsorbent chromatography, *Biochem. J.,* 215, 519, 1983.
18. **Malfroy, B., Schofield, P. R., Kuang, W.-J., Seeburg, P. H., Mason, A. J., and Henzel, W. J.,** Molecular cloning and amino acid sequence of rat enkephalinase, *Biochem. Biophys. Res. Commun.,* 144, 59, 1987.
19. **Devault, A., Lazure, C., Nault, C., Moual, H. L., Seidah, N. G., Chrétien, M., Kahn, P., Powell, J., Mallet, J., Beaumont, A., Roques, B. P., Crine, P., and Boileau, G.,** Amino acid sequence of rabbit kidney neurtral endopeptidase-24.11 (enkephalinase), *EMBO J.,* 6, 1317, 1987.
20. **Malfroy, B., Kuang, W.-J., Seeburg, P. H., Mason, A. J., and Schofield, P. R.,** Molecular cloning and amino acid sequence of human enkephalinase (neutral endopeptidase), *FEBS Lett.,* 229, 206, 1988.
21. **Kerr, M. A. and Kenny, A. J.,** The purification and specificity of a neutral endopeptidase from rabbit kidney brush border, *Biochem. J.,* 137, 477, 1974.
22. **Fulcher, I. S. and Kenny, A. J.,** Proteins of the kidney microvillar membrane. The amphipathic forms of endopeptidase purified from pig kidneys, *Biochem. J.,* 211, 743, 1983.
23. **Vanneste, Y., Michel, A., Dimaline, R., Najdovski, T., and Deschodt-Lanckman, M.,** Hydrolysis of α-human atrial natriuretic peptide *in vitro* by human kidney membranes and purified endopeptidase-24.11. Evidence for a novel cleavage site, *Biochem. J.,* 254, 531, 1988.
24. **Hersh, L. B. and Morihara, K.,** Comparison of the subsite specificity of the mammalian neutral endopeptidase-24.11 (enkephalinase) to the bacterial neutral endopeptidase thermolysin, *J. Biol. Chem.,* 261, 6433, 1986.
25. **Matsas, R., Kenny, A. J., and Turner, A. J.,** The metabolism of neuropeptides. The hydrolysis of peptides, including enkephalins, tachykinins and their analogues, by endopeptidase-24.11, *Biochem. J.,* 223, 433, 1984.
26. **Turner, A. J., Matsas, R., and Kenny, A. J.,** Are there neuropeptide-specific peptidases?, *Biochem. Pharmacol.,* 34, 1347, 1985.
27. **Hooper, N. M., Kenny, A. J., and Turner, A. J.,** The metabolism of neuropeptides. Neurokinin A (substance K) is a substrate for endopeptidase-24.11 but not for peptidyl dipeptidase A (angiotensin converting enzyme), *Biochem. J.,* 231, 357, 1985.
28. **Connelly, J. C., Skidgel, R. A., Schulz, W. W., Johnson, A. R., and Erdos, E. G.,** Neutral endopeptidase-24.11 in human neutrophils: cleavage of chemotactic peptide, *Proc. Natl. Acad. Sci. U.S.A.,* 82, 8737, 1985.
29. **Turner, A. J., Matsas, R., and Kenny, A. J.,** Endopeptidase-24.11 and neuropeptide metabolism, *Biochem. Soc. Trans.,* 13, 39, 1985.
30. **Goetzl, E. J., Sreedharan, S. P., Turck, C. W., Bridenbaugh, R., and Malfroy, B.,** Preferential cleavage of amino- and carboxy-terminal oligopeptides from vasoactive intestinal polypeptide by human recombinant enkephalinase, *Biochem. Biophys. Res. Commun.,* 158, 850, 1989.
31. **Turner, A. J., Hooper, N. M., and Kenny, A. J.,** Metabolism of neuropeptides, in *Mammalian Ectoenzymes,* Kenny, A. J. and Turner, A. J., Eds., Elsevier, Amsterdam, 1987, chap. 8.
32. **Vogt-Schnaden, M., Gagelmann, M., Hock, D., Herbst, F., and Forssmann, W. G.,** Degradation of porcine brain natriuretic peptide (pBNP-26) by endopeptidase-24.11 from kidney cortical membranes, *Biochem. Biophys. Res. Commun.,* 161, 1177, 1989.
33. **Stephenson, S. L. and Kenny, A. J.,** Metabolism of neuropeptides. Hydrolysis of the angiotensins, bradykinin, substance P and oxytocin by pig kidney microvillar membranes, *Biochem. J.,* 241, 237, 1987.
34. **Pozgay, M., Michaud, C., Liebman, M., and Orlowski, M.,** Substrate and inhibitor studies of thermolysin-like neutral metalloendopeptidase from kidney membrane fractions. Comparison with bacterial thermolysin, *Biochemistry,* 25, 1292, 1986.
35. **Beaumont, A. and Roques, B. P.,** Presence of a histidine at the active site of the neutral endopeptidase-24.11, *Biochem. Biophys. Res. Commun.,* 139, 733, 1986.
36. **Jackson, D. G. and Hersh, L. B.,** Reaction of neutral endopeptidase-24.11 (enkephalinase) with arginine reagents, *J. Biol. Chem.,* 261, 8649, 1986.
37. **Bateman, R. C. and Hersh, L. B.,** Evidence for an essential histidine in neutral endopeptidase-24.11, *Biochemistry,* 26, 4237, 1987.
38. **Devault, A., Sales, V., Nault, C., Beaumont, A., Roques, B., Crine, P., and Boileau, G.,** Exploration of the catalytic site of endopeptidase-24.11 by site-directed mutagenesis. Histidine residues 583 and 587 are essential for catalysis, *FEBS Lett.,* 231, 54, 1988.
39. **Benchetrit, T., Bissry, V., Mornon, J. P., Devault, A., Crine, P., and Roques, B. P.,** Primary structure homologies between two zinc metallopeptidases, the neutral endopeptidase-24.11 (enkephalinase) and thermolysin, through clustering analysis, *Biochemistry,* 27, 592, 1988.

40. **Suda, H., Aoyagi, T., Takeuchi, T., and Umezawa, H.,** A thermolysin inhibitor produced by actinomycetes: phosphoramidon, *J. Antibiot.*, 26, 621, 1973.
41. **Kenny, A. J.,** Proteinases associated with cell membranes, in *Proteinases in Mammalian Cells and Tissues,* Barrett, A. J., Ed., Elsevier/North-Holland Biomedical Press, Amsterdam, 1977, 393.
42. **Fulcher, I. S., Matsas, R., Turner, A. J., and Kenny, A. J.,** Kidney neutral endopeptidase and the hydrolysis of enkephalin by synaptic membranes show similar sensitivity to inhibitors, *Biochem. J.*, 203, 519, 1982.
43. **Roques, B. P., Fournié-Zaluski, M. C., Soroca, E., Lecomte, J. M., Malfroy, B., Llorens, C., and Schwartz, J.-C.,** The enkephalinase inhibitor thiorphan shows antinociceptive activity in mice, *Nature (London),* 288, 286, 1980.
44. **Roques, B. P., Lucas-Soroca, E., Chaillet, P., Costentin, J., and Fournié-Zaluski, M. C.,** Complete differentiation between enkephalinase and angiotensin-converting enzyme inhibition by retrothiorphan, *Proc. Natl. Acad. Sci. U.S.A.*, 80, 3178, 1983.
45. **Almenoff, J. and Orlowski, M.,** Biochemical and immunological properties of a membrane-bound metalloendopeptidase: comparison with thermolysin-like kidney neutral metalloendopeptidase, *J. Neurochem.*, 42, 151, 1984.
46. **Fournié-Zaluksi, M.-C., Chaillet, P., Bouboutou, R., Coulaud, A., Cherot, P., Waksman, G., Costentin, J., and Roques, B. P.,** Analgesic effects of kelatorphan, a new highly potent inhibitor of multiple enkephalin degrading enzymes, *Eur. J. Pharmacol.*, 102, 525, 1984.
47. **Fournié-Zaluski, M.-C., Coulaud, A., Bouboutou, R., Chaillet, P., Devin, J., Waksman, G., Costentin, J., and Roques, B. P.,** New bidentates as full inhibitors of enkephalin-degrading enzymes: synthesis and analgesic properties, *J. Med. Chem.*, 28, 1158, 1985.
48. **Chu, T. G. and Orlowski, M.,** Active site directed *N*-carboxymethyl peptide inhibitors of a soluble metalloendopeptidase from rat brain, *Biochemistry,* 23, 3598, 1984.
49. **Cushman, D. W., Cheung, H. S., Sabo, E. F., and Ondetti, M. A.,** Design of potent competitive inhibitors of angiotensin converting enzyme. Carboxyalkanoyl and mercaptoalkanoyl amino acids, *Biochemistry,* 16, 4384, 1977.
50. **Hooper, N. M. and Turner, A. J.,** Isolation of two differentially glycosylated forms of peptidyl dipeptidase A (angiotensin converting enzyme) from pig brain: a re-evaluation of their role in neuropeptide metabolism, *Biochem. J.*, 241, 625, 1987.
51. **Bull, H. G., Thornberry, N. A., and Cordes, E. G.,** Purification of angiotensin-converting enzyme from rabbit lung and human plasma by affinity chromatography, *J. Biol. Chem.*, 260, 2963, 1985.
52. **Takeuchi, T.,** Production of actinonin, an inhibitor of aminopeptidase M_1, by actinomycetes, *J. Antibiot.*, 38, 1629, 1985.
53. **Rich, D. H., Moon, B. J., and Harbeson, S.,** Inhibition of aminopeptidases by amastatin and bestatin derivatives. Effect of inhibitor structure on slow-binding processes, *J. Med. Chem.*, 27, 417, 1984.
54. **Aoyagi, T., Tobe, H., Kojima, F., Hamada, M., Takeuchi, T., and Umezawa, H.,** Amastatin, an inhibitor of aminopeptidase A, produced by actinomycetes, *J. Antibiot.*, 31, 636, 1978.
55. **Hooper, N. M. and Turner, A. J.,** unpublished data, 1989.
56. **Palmieri, F. E., Bausback, H. H., Churchill, L., and Ward, P. E.,** Kinin and enkephalin conversion by an endothelial, plasma membrane carboxypeptidase, *Biochem. Pharmacol.*, 35, 2749, 1986.
57. **Skidgel, R. A., Davis, R. M., and Tan, F.,** Human carboxypeptidase M., Purification and characterization of a membrane-bound carboxypeptidase that cleaves peptide hormones, *J. Biol. Chem.*, 264, 2236, 1989.
58. **Fricker, L. D.,** Neuropeptide biosynthesis: focus on the carboxypeptidase processing enzyme, *Trends Neurosci.*, 8, 210, 1985.
59. **Skidgel, R. A., Johnson, A. R., and Erdos, E. G.,** Hydrolysis of opioid hexapeptides by carboxypeptidase N. Presence of carboxypeptidase in cell membrane, *Biochem. Pharmacol.*, 33, 2371, 1984.
60. **Kenny, A. J., Booth, A. G., George, S. C., Ingram, J., Kershaw, D., Wood, E. J., and Young, A. T.,** Dipeptidyl peptidase IV, a kidney brush border serine peptidase, *Biochem. J.*, 157, 169, 1976.
61. **Umezawa, H., Aoyagi, T., Ogawa, K., Naganawa, H., Hamada, M., and Takeuchi, T.,** Diprotins A and B, inhibitors of dipeptidyl aminopeptidase IV, produced by bacteria, *J. Antibiot.*, 37, 422, 1984.
62. **Campbell, B. J., Forrester, L. J., Zahler, W. L., and Burks, M.,** β-lactamase activity of purified and partially characterized human renal dipeptidase, *J. Biol. Chem.*, 259, 13486, 1984.
63. **Kahan, F. M., Kropp, H., Sundelof, J. G., and Birnbaum, J.,** Thienamycin: development of imipenem-cilastatin, *J. Antimicrob. Chemother.*, 12, (Suppl. D), 1, 1983.
64. **Gardel, S. J. and Tate, S. S.,** Affinity labeling of γ-glutamyl transpeptidase by glutamine antagonists, *FEBS Lett.*, 122, 171, 1980.
65. **Waksman, G., Bouboutou, R., Devin, J., Bourgoin, S., Cesselin, F., Hamon, M., Fournié-Zaluski, M. C., and Roques, B. P.,** *In vitro* and *in vivo* effects of kelatorphan on enkephalin metabolism in rodent brain, *Eur. J. Pharmacol.*, 117, 233, 1985.
66. **Turner, A. J. and Hooper, N. M.,** Metabolism of tachykinins by cell-surface peptidases, in *Substance P and Neurokinins,* Henry, J. L., Courture, R., Cuello, A. C., Pelletier, G., Quirion, R., and Regoli, D., Eds., Springer-Verlag, New York, 1987, 18.

67. **Schwartz, J.-C., Costententin, J., and Lecomte, J.-M.,** Pharmacology of enkephalinase inhibitors, *Trends Pharmacol. Sci.*, 6, 472, 1985.
68. **Thorsett, E. D. and Wyvratt, M. J.,** Inhibition of zinc peptidases that hydrolyse neuropeptides, in *Neuropeptides and Their Peptidases,* Turner, A. J., Ed., Ellis Horwood Series in Biomedicine, Chichester, England, 1987, chap. 12.
69. **Acker, G. R., Molineaux, C., Clejan, L. A., and Orlowski, M.,** Endopeptidase-24.15 in synaptosomal membranes generates enkephalins from endorphins, dynorphin (1-8) and Met-enkephalin-Arg-Gly-Leu, *Fed. Proc.*, 45, 1719, 1986.
70. **Acker, G. R., Molineaux, C., and Orlowski, M.,** Synaptosomal membrane-bound form of endopeptidase-24.15 generates Leu-enkephalin from Dynorphin 1-8, α and β-neoendorphin and Met-enkephalin from Met-enkephalin-Arg^6-Gly^7-Leu^8, *J. Neurochem.*, 48, 284, 1987.
71. **Orlowski, M., Michaud, C., and Chu, T. G.,** A soluble metalloendopeptidase from rat brain. Purification of the enzyme and determination of specificity with synthetic and natural peptides, *Eur. J. Biochem.*, 135, 81, 1983.
72. **Molineaux, C. J., Lasdun, A., Michaud, C., and Orlowski, M.,** Endopeptidase-24.15 is the primary enzyme that degrades luteinizing hormone releasing hormone both *in vitro* and *in vivo*, *J. Neurochem.*, 51, 624, 1988.
73. **Barrett, A. J. and Tisljar, U.,** The activities of 'Pz-peptidase' and 'endopeptidase-24.15' are due to a single enzyme, *Biochem. J.*, 261, 1047, 1989.
74. **Tisljar, U. and Barrett, A. J.,** Purification and characterization of Pz-peptidase from rabbit muscle, *Arch. Biochem. Biophys.*, 274, 138, 1989.
75. **Camargo, A. C. M., Oliveira, E. B., Toffoletto, O., Metters, K. M., and Rossier, J.,** Brain endo-oligopeptidase A, a putative enkephalin converting enzyme, *J. Neurochem.*, 48, 1258, 1987.
76. **Cicilini, M. A., Ribeiro, M. J. F., De Oliveira, E. B., Mortara, R. A., and Camargo, A. C. M.,** Endooligopeptidase A activity in rabbit heart: generation of enkephalin from enkephalin containing peptides, *Peptides,* 8, 945, 1988.
77. **Tisljar, U., Camargo, A. C. M., Costa, C. A., and Barrett, A. J.,** Activity of Pz-peptidase and endo-oligopeptidase are due to the same enzyme, *Biochem. Biophys. Res. Commun.*, 162, 1460, 1989.
78. **Kenny, A. J., Fulcher, I. S., Ridgwell, K., and Ingram, J.,** Microvillar membrane neutral endopeptidases, *Acta Biol. Med. Germ.*, 40, 1465, 1981.
79. **Beynon, R. J., Shannon, J. D., and Bond, J. S.,** Purification and characterization of metalloendopeptidase from mouse kidney, *Biochem. J.*, 199, 591, 1981.
80. **Bond, J. S., Shannon, J. D., and Beynon, R. J.,** Certain mouse strains are deficient in a kidney brush border metalloendopeptidase activity, *Biochem. J.*, 209, 251, 1983.
81. **Barnes, K., Ingram, J., and Kenny, A. J.,** Proteins of the kidney microvillar membrane. Structure and immunochemical properties of rat endopeptidase-2 and its immunohistochemical localization in tissues of rat and mouse, *Biochem. J.*, 264, 335, 1989.
82. **Kenny, A. J. and Ingram, J.,** Proteins of the kidney microvillar membrane. Purification and properties of the phosphoramidon-insensitive endopeptidase ('endopeptidase-2') from rat kidney, *Biochem. J.*, 245, 515, 1987.
83. **Hooper, N. M and Turner, A. J.,** Ectoenzymes of the kidney microvillar membrane. Isolation and characterization of the amphipathic form of renal dipeptidase and hydrolysis of its glycosyl-phosphatidylinositol anchor by an activity in plasma, *Biochem. J.*, 261, 811, 1989.
84. **Bond, J. S., Butler, P. E., and Beynon, R. J.,** Metalloendopeptidases of the mouse kidney brush border: meprin and endopeptidase-24.11, *Biochem. Biomed. Acta,* 45, 1515, 1987.
85. **Stephenson, S. L and Kenny, A. J.,** The metabolism of neuropeptides. Hydrolysis of peptides by the phosphoramidon-insensitive rat kidney enzyme, 'endopeptidase-2', and by rat microvillar membranes, *Biochem. J.*, 255, 45, 1988.
86. **Sterchi, E. E., Green, J. R., and Lentze, M. J.,** Non-pancreatic hydrolysis of *N*-benzoyl-L-tyrosyl-*p*-aminobenzoic acid (PABA-peptide) in the human small intestine, *Clin. Sci.*, 62, 557, 1982.
87. **Sterchi, E. E., Green, J. R., and Lentze, M. J.,** Non-pancreatic hydrolysis of *N*-benzoyl-L-tyrosyl-*p*-aminobenzoic acid (PABA-peptide) in the rat small intestine, *J. Pediat. Gastroenterol. Nutr.*, 2, 539, 1983.
88. **Sterchi, E. E., Naim, H. Y., Lentze, M. J., Hauri, H.-P., and Fransen, J. A. M.,** *N*-benzoyl-L-tyrosyl-*p*-aminobenzoic acid hydrolase: a metalloendopeptidase of the human intestinal microvillus membrane which degrades biologically active peptides, *Arch. Biochem. Biophys.*, 265, 105, 1988.
89. **Sterchi, E. E., Naim, H. Y., and Lentze, M. J.,** Biosynthesis of *N*-benzoyl-L-tyrosyl-*p*-aminobenzoic acid hydrolase: disulphide-linked dimers are formed at the site of synthesis in the rough endoplasmic reticulum, *Arch. Biochem. Biophys.*, 265, 119, 1988.
90. **Soubrier, F., Alhenc-Gelas, F., Hubert, C., Allegrini, J., John, M., Tregear, G., and Corvol, P.,** Two putative active centres in human angiotensin I-converting enzyme revealed by molecular cloning, *Proc. Natl. Acad. Sci. U.S.A.*, 85, 9386, 1988.
91. **Bernstein, K. E., Martin, B. M., Edwards, A. S., and Bernstein, E. A.,** Mouse angiotensin-converting enzyme is a protein composed of two homologous domains, *J. Biol. Chem.*, 264, 11945, 1989.

92. **Hooper, N. M., Keen, J., Pappin, D. J. C., and Turner, A. J.,** Pig kidney angiotensin converting enzyme. Purification and characterization of amphipathic and hydrophilic forms of the enzyme establishes C-terminal anchorage to the plasma membrane, *Biochem. J.*, 247, 85, 1987.
93. **Hooper, N. M. and Turner, A. J.,** Characterization and isolation of the membrane anchor of angiotensin converting enzyme, *Biochem. Soc. Trans.*, 17, 660, 1989.
94. **Das, M. and Soffer, R. L.,** Pulmonary angiotensin-converting enzyme. Structure and catalytic properties, *J. Biol. Chem.*, 250, 6762, 1975.
95. **Lattion, A.-L., Soubrier, F., Allegrini, J., Hubert, C., Corvol, P., and Alhenc-Gelas, F.,** The testicular transcript of the angiotensin I-converting enzyme encodes for the ancestral, non-duplicated form of the enzyme, *FEBS Lett.*, 252, 99, 1989.
96. **Cumin, F., Vellaud, V., Corvol, P., and Alhenc-Gelas, F.,** Evidence for a single active site in the human angiotensin I-converting enzyme from inhibitor binding studies with [^{3}H]RU44403: role of chloride, *Biochem. Biophys. Res. Commun.*, 163, 718, 1989.
97. **Dorer, F. E., Kahn, J. R., Lentz, K. E., Levine, M., and Skeggs, L. T.,** Hydrolysis of bradykinin by angiotensin-converting enzyme, *Circ. Res.*, 34, 824, 1974.
98. **El-Dorry, H. A., Bull, H. G., Iwata, K., Thornberry, N. A., Cordes, E. H., and Soffer, R. L.,** Molecular and catalytic properties of rabbit testicular dipeptidyl carboxypeptidase, *J. Biol. Chem.*, 257, 14128, 1982.
99. **Erdos, E. G. and Skidgel, R. A.,** The angiotensin I-converting enzyme, *Lab. Invest.*, 56, 345, 1987.
100. **Soffer, R. L.,** Angiotensin-converting enzyme and the regulation of vasoactive peptides, *Ann. Rev. Biochem.*, 45, 73, 1976.
101. **Sheikh, I. A. and Kaplan, A. P.,** Studies of the digestion of bradykinin, Lys-bradykinin, and des-Arg^9-bradykinin by angiotensin converting enzyme, *Biochem. Pharmacol*, 35, 1951, 1986.
102. **Dubreuil, P., Fulcrand, P., Rodriguez, M., Fulcrand, H., Laur, J., and Martinez, J.,** Novel activity of angiotensin converting enzyme: hydrolysis of cholecystokinin and gastrin analogues with release of the amidated COOH-terminal dipeptide, *Biochem. J.*, 262, 125, 1989.
103. **Yokosawa, H., Endo, S., Ogura, Y., and Ishii, S.-I.,** A new feature of angiotensin-converting enzyme in the brain: hydrolysis of Substance P, *Biochem. Biophys. Res. Commun.*, 116, 735, 1983.
104. **Skidgel, R. A. and Erdos, E. G.,** Novel activity of human angiotensin I converting enzyme: release of the NH_2- and COOH-terminal tripeptides from the luteinizing hormone-releasing hormone, *Proc. Natl. Acad. Sci., U.S.A.*, 82, 1025, 1985.
105. **Erdos, E. G. and Skidgel, R. A.,** Structure and functions of human angiotensin I converting enzyme (kininase II), *Biochem. Soc. Trans.*, 13, 42, 1985.
106. **Patchett, A. A. and Cordes, E. H.,** The design and properties of N-carboxyalkyldipeptide inhibitors of angiotensin converting enzyme, *Adv. Enzymol.*, 57, 1, 1985.
107. **Patchett, A. A., Harris, E., Tristram, E. W., Wyvratt, M. J., Wu, M. T., Taub, D., Peterson, E. R., Ikeler, T. J., ten Broeke, J., Payne, L. A., Ondeyka, D. L., Thorsett, E. D., Greenlee, W. J., Lohr, N. S., Hoffsommer, R. D., Joshua, H., Ruyle, W. V., Rothrock, J. W., Aster, S. D., Maycock, A. L., Robinson, F. M., Hirschmann, R., Sweet, C. S., Ulm, E. H., Gross, D. M., Vassil, T. C., and Stone, C. A.,** A new class of angiotensin-converting enzyme inhibitors, *Nature (London)*, 288, 280, 1980.
108. **Olsen, J., Cowell, G. M., Konigshofer, E., Danielsen, M., Moller, J., Laustsen, L., Hansen, O. C., Welinder, K. G., Engberg, J., Hunziker, W., Spiess, M., Sjostrom, H., and Norén, O.,** Complete amino acid sequence of human intestinal aminopeptidase N as deduced from cloned cDNA, *FEBS Lett.*, 238, 307, 1988.
109. **Watt, V. M. and Yip, C. C.,** Amino acid sequence deduced from a rat kidney cDNA suggests it encodes the Zn-peptidase aminopeptidase N, *J. Biol. Chem.*, 264, 5480, 1989.
110. **McDonald, J. K. and Barrett, A. J.,** *Mammalian Proteases: A Glossary and Bibliography*, Vol. 2, Academic Press, London, 1986.
111. **Matsas, R., Stephenson, S. L., Hryszko, J., Kenny, A. J., and Turner, A. J.,** The metabolism of neuropeptides: phase separation of synaptic membrane preparations with Triton X-114 reveals the presence of aminopeptidase N, *Biochem. J.*, 231, 445, 1985.
112. **Giros, B., Gros, C., Solhonne, B., and Schwartz, J. C.,** Characterization of aminopeptidases responsible for inactivating endogenous [Met^5]enkephalin in brain slices using peptidase inhibitors and antiaminopeptidase M antibodies, *Mol. Pharmacol.*, 29, 281, 1986.
113. **Matsas, R., Turner, A. J., and Kenny, A. J.,** Endopeptidase-24.11 and aminopeptidase activity in brain synaptic membranes are jointly responsible for the hydrolysis of cholecystokinin octapeptide (CCK-8), *FEBS Lett.*, 175, 124, 1984.
114. **Nau, R., Schafer, G., Deacon, C. F., Cole, T., Agoston, D. V., and Conlon, J. M.,** Proteolytic inactivation of Substance P and neurokinin A in the longitudinal muscle layer of guinea pig small intestine, *J. Neurochem.*, 47, 856, 1986.
115. **Hooper, N. M.,** Metabolism of Neuropeptides by Cell-Surface Peptidases, Ph.D. thesis, University of Leeds, 1987.

116. **Kenny, A. J., Stephenson, S. L., and Turner, A. J.,** Cell surface peptidases, in *Mammalian Ectoenzymes,* Kenny, A. J. and Turner, A. J., Eds., Elsevier, Amsterdam, 1987, 169.
117. **Dehm, P. and Nordwig, A.,** The cleavage of prolyl peptides by kidney peptidases. Partial purification of a "X-prolyl-aminopeptidase" from swine kidney microsomes, *Eur. J. Biochem.,* 17, 364, 1970.
118. **Hooper, N. M. and Turner, A. J.,** Ectoenzymes of the kidney microvillar membrane. Aminopeptidase P is anchored by a glycosylphosphatidylinositol moiety, *FEBS Lett.,* 229, 340, 1988.
119. **Kenny, A. J., Booth, A. G., and Macnair, R. D. C.,** Peptidases of the kidney microvillus membrane, *Acta Biol. Med. Ger.,* 36, 1575, 1977.
120. **Lasch, R., Koelsch, R., Laahoff, A. M., and Hartrodt, B.,** Is the proline-specific aminopeptidase P of the intestinal brush border an integral membrane enzyme?, *Biomed. Biochem. Acta,* 45, 833, 1986.
121. **Orawski, A. T., Susz, J. P., and Simmons, W. H.,** Aminopeptidase P from bovine lung: solubilization, properties, and potential role in bradykinin degradation, *Mol. Cell. Biochem.,* 75, 123, 1987.
122. **Mentlein, R.,** Proline residues in the maturation and degradation of peptide hormones and neuropeptides, *FEBS Lett.,* 234, 251, 1988.
123. **Gee, N. S. and Kenny, A. J.,** Proteins of the kidney microvillar membrane. The 130 kDa protein in pig kidney, recognized by monoclonal antibody GK5Cl, is an ectoenzyme with aminopeptidase activity, *Biochem. J.,* 230, 753, 1985.
124. **Jackson, M. C., Choudry, Y., Bourne, A., Woodley, J. G., and Kenny, A. J.,** A fluorimetric assay for aminopeptidase W, *Biochem. J.,* 253, 299, 1988.
125. **Tan, F., Chan, S.-J., Steiner, D. F., Schilling, J. W., and Skidgel, R. A.,** Molecular cloning and sequencing of the cDNA for human membrane-bound carboxypeptidase M. Comparison with carboxypeptidases A, B, H and N, *J. Biol. Chem.,* 264, 13165, 1989.
126. **Levin, Y., Skidgel, R. A., and Erdos, E. G.,** Isolation and characterization of human plasma carboxypeptidase N (kininase I), *Proc. Natl. Acad. Sci. U.S.A.,* 79, 4618, 1982.
127. **Gebhard, W., Schube, M., and Eulitz, M.,** cDNA cloning and complete primary structure of the small, active subunit of human carboxypeptidase N (kininase I), *Eur. J. Biochem.,* 178, 603, 1989.
128. **Sheikh, I. A. and Kaplan, A. P.,** Studies of the digestion of bradykinin, lysyl-bradykinin, and kinin-degradation products by carboxypeptidases A, B and N, *Biochem. Pharmacol.,* 35, 1957, 1986.
129. **Auricchio, S., Greco, L., de Vizia, B., and Buoncore, V.,** Peptidylaminopeptidase and carboxypeptidase activities of the brush border of rabbit small intestine, *Gastroenterology,* 75, 1073, 1978.
130. **Dehm, P. and Nordwig, A.,** The cleavage of prolyl peptides by kidney peptidases. Isolation of a microsomal carboxypeptidase from swine kidney, *Eur. J. Biochem.,* 17, 372, 1970.
131. **Bella, A. M., Erickson, R. H., and Kim, Y. S.,** Rat intestinal brush border membrane dipeptidyl aminopeptidase IV: kinetic properties and substrate specificities of the purified enzyme, *Arch. Biochem. Biophys.,* 218, 156, 1982.
132. **Heymann, E. and Mentlein, R.,** Liver dipeptidyl peptidase IV hydrolyses substance P, *FEBS Lett.,* 91, 360, 1978.
133. **Fischer, G., Demuth, H. U., and Barth, A. N.,** *O*-diacylhydroxylamines as enzyme activated inhibitors for serine proteinases, *Pharmazie,* 38, 249, 1983.
134. **Schon, E., Demuth, H.-U., Eichmann, E., Horst, H.-J., Korner, I.-J., Kopp, J., Mattern, T., Neubert, K., Noll, F., Ulmer, A. J., Barth, A., and Ansorge, S.,** Dipeptidyl peptidase IV in human T lymphocytes. Impaired induction of interleukin 2 and gamma interferon due to specific inhibition of dipeptidyl peptidase IV, *Scand. J. Immunol.,* 29, 127, 1989.
135. **Hooper, N. M., Low, M. G., and Turner, A. J.,** Renal dipeptidase is one of the membrane proteins released by phosphatidylinositol-specific phospholipase C, *Biochem. J.,* 244, 465, 1987.
136. **Littlewood, G. M., Hooper, N. M., and Turner, A. J.,** Ectoenzymes of the kidney microvillar membrane. Affinity purification, characterization and localization of the phospholipase C-solubilized form of renal dipeptidase, *Biochem. J.,* 257, 361, 1989.
137. **Hooper, N. M., Keen, J. N., and Turner, A. J.,** Characterization of the glycosyl-phosphatidylinositol anchored human renal dipeptidase reveals that it is more extensively glycosylated than the pig enzyme, *Biochem. J.,* 265, 429, 1990.
138. **Armstrong, D. J., Mukhopadhyay, S. K., and Campbell, B. J.,** Physicochemical characterization of renal dipeptidase, *Biochemistry,* 13, 1745, 1974.
139. **Campbell, B. J.,** Renal dipeptidase, *Methods Enzymol.,* 19, 722, 1970.
140. **Kozak, E. M. and Tate, S. S.,** Glutathione-degrading enzymes of microvillus membrane, *J. Biol. Chem.,* 257, 6322, 1982.
141. **McIntyre, T. and Curthoys, N. P.,** Renal catabolism of glutathione. Characterization of a particulate rat renal dipeptidase that catalyzes the hydrolysis of cisteinylglycine, *J. Biol. Chem.,* 257, 11915, 1982.
142. **Kropp, H., Sundelof, J. S., Hajdu, R., and Kahan, F. M.,** Metabolism of thienamycin and related carbapenem antibiotics by the renal dipeptidase, dehydropeptidase-I, *Antimicrob. Agents Chemother.,* 22, 62, 1982.

143. **O'Connor, B. and O'Cuinn, G.,** Localization of a narrow specificity thyroliberin hydrolyzing pyroglutamate aminopeptidase in synaptosomal membranes of guinea-pig brain, *Eur. J. Biochem.*, 144, 271, 1984.
144. **O'Connor, B. and O'Cuinn, G.,** Purification of and kinetic studies on a narrow specificity synaptosomal membrane pyroglutamate aminopeptidase from guinea-pig brain, *Eur. J. Biochem.*, 150, 47, 1985.
145. **Scharfmann, R., Morgat, J.-L., and Aratan-Spire, S.,** Presence of a particulate thyrotropin-releasing hormone-degrading pyroglutamate aminopeptidase activity in rat liver, *Neuroendocrinology*, 49, 442, 1989.
146. **Coloma, J. and Pitot, H. C.,** Characterization and sequence of a cDNA clone of γ-glutamyltranspeptidase, *Nucleic Acid Res.*, 14, 1393, 1986.
147. **Laperche, Y., Bulle, F., Aissani, T., Chobert, N.-N., Aggerbeck, M., Hanoune, J., and Guellaen, G.,** Molecular cloning and nucleotide sequence of rat kidney γ-glutamyl transpeptidase cDNA, *Proc. Natl. Acad. Sci. U.S.A.*, 83, 937, 1986.
148. **Tate, S. S. and Meister, A.,** Interaction of γ-glutamyl transpeptidase with amino acids, dipeptides and derivatives and analogs of glutathione, *J. Biol. Chem.*, 249, 7593, 1974.
149. **Blakely, R. D., Ory-Lavollée, L., Thompson, R. C., and Coyle, J. T.,** Synaptosomal transport of radiolabel from *N*-acetyl-aspartyl-[^{3}H]-glutamate suggests a mechanism of inactivation of an excitatory neuropeptide, *J. Neurochem.*, 47, 1013, 1986.
150. **Blakely, R. D., Robinson, M. B., Thompson, R. C., and Coyle, J. T.,** Hydrolysis of the brain dipeptide *N*-acetyl-L-aspartyl-L-glutamate: subcellular and regional distribution, ontogeny, and the effect of lesions on *N*-acetylated-α-linked acidic dipeptidase activity, *J. Neurochem.*, 50, 1200, 1988.
151. **Checler, F., Emson, P. C., Vincent, J.-P., and Kitabgi, P.,** Inactivation of neurotensin by rat brain synaptic membranes. Cleavage at the Pro^{10}–Tyr^{11} bond by endopeptidase-24.11 (enkephalinase) and a peptidase different from proline-endopeptidase, *J. Neurochem.*, 43, 1295, 1984.
152. **Checler, F., Vincent, J.-P., and Kitabgi, P.,** Purification and characterization of a novel neurotensin-degrading peptidase from rat brain synaptic membranes, *J. Biol. Chem.*, 261, 11274, 1986.
153. **Lee, C.-M., Sandberg, B. E. B., Hanley, M. R., and Iversen, L. L.,** Purification and characterization of a membrane-bound Substance P-degrading enzyme from human brain, *Eur. J. Biochem.*, 114, 315, 1981.
154. **Nyberg, F., Le Greves, P., Sundquist, C., and Terenius, L.,** Characterization of Substance P (1-7) and (1-8) generating enzyme in human cerebrospinal fluid, *Biochem. Biophys. Res. Commun.*, 125, 244, 1984.
155. **Probert, L. and Hanley, M. R.,** The immunocytochemical localization of 'Substance-P-degrading enzyme' within the rat spinal cord, *Neurosci. Lett.*, 78, 132, 1987.
156. **Horsthemke, B., Leblanc, P., Kordon, C., Wattiaux-de Coninck, S., Wattiaux, R., and Bauer, K.,** Subcellular distribution of particle-bound neutral peptidases capable of hydrolyzing gonadoliberin, thyroliberin, enkephalin and substance P, *Eur. J. Biochem.*, 139, 315, 1984.
157. **Probert, L. and Hanley, M. R.,** Still degrading after all these years, *Biochem. J.*, 259, 621, 1989.
158. **Stephenson, S. L. and Kenny, A. J.,** The hydrolysis of α-human atrial natriuretic peptide by pig kidney microvillar membranes is initiated by endopeptidase-24.11, *Biochem. J.*, 243, 183, 1987.
159. **Gee, N. S., Bowes, M. A., Buck, P., and Kenny, A. J.,** An immunoradiometric assay for endopeptidase-24.11 shows it to be a widely distributed enzyme in pig tissues, *Biochem. J.*, 228, 119, 1985.
160. **Bowes, M. A. and Kenny, A. J.,** Endopeptidase-24.11 in pig lymph nodes: purification and immunocytochemical localization in reticular cells, *Biochem. J.*, 236, 801, 1986.
161. **Bowes, M. A. and Kenny, A. J.,** An immunohistochemical study of endopeptidase-24.11 and aminopeptidase N in lymphoid tissues, *Immunology*, 60, 247, 1987.
162. **Lycett, A. and Kenny, A. J.,** unpublished data, 1989.
163. **Binns, R. M.,** Organization of the lymphoreticular system and lymphocyte markers in the pig, *Vet. Immunol. Immunopathol.*, 3, 95, 1982.
164. **Barnes, K. and Kenny, A. J.,** Endopeptidase-24.11 in the adenohypophysis of the pig is localized in the gonadotrophic cells, *Peptides*, 9, 55, 1988.
165. **Greaves, M. F., Brown, G., Rapson, N. T., and Lister, T. A.,** Antisera to acute lymphoblastic leukemia cells, *Clin. Immunol. Immunopathol.*, 4, 67, 1975.
166. **Tucker, W. L. and McCormack, R. T.,** The common acute lymphoblastic leukemia antigen (CD10) — emancipation from a functional enigma, *Blood*, 73, 625, 1989.
167. **Shipp, M. A., Richardson, N. E., Sayre, P. H., Brown, N. R., Masteller, E. L., Clayton, L. K., Ritz, J., and Reinherz, E. L.,** Molecular cloning of the common acute lymphoblastic leukemia antigen (CALLA) identifies a type II integral membrane protein, *Proc. Natl. Acad. Sci. U.S.A.*, 85, 4819, 1988.
168. **Letarte, M., Vera, S., Tran, R., Addis, J. B. L., Onizuka, R. J., Quackenbush, E. J., Jongeneel, C. V., and McInnes, R. R.,** Common acute lymphocytic leukemia antigen is identical to neutral endopeptidase, *J. Exper. Med.*, 168, 1244, 1988.
169. **Shipp, M. A., Vijayaraghavan, J., Schmidt, E. V., Masteller, E. L., d'Adamio, L., Hersh, L. B., and Reinherz, E. L.,** Common acute lymphoblastic leukemia antigen (CALLA) is active neutral endopeptidase 24.11 ("enkephalinase"): direct evidence by cDNA transfection analysis, *Proc. Natl. Acad. Sci. U.S.A.*, 86, 297, 1989.

170. **Jongeneel, V., Quackenbush, E. J., Ronco, P., Verroust, P., Carrel, S., and Letarte, M.,** Common acute lymphoblastic leukemia antigen expressed on leukemia and melanoma cell lines has neutral endopeptidase activity, *J. Clin. Invest.*, 83, 713, 1989.
171. **Delia, D., Cattoretti, G., Bonati, A., Villa, S., de Braud, F., and Buscaglia, M.,** Detection of the common acute lymphoblastic leukemia antigen (CALLA) on B cells from human fetal tissues. A multiple phenotypic characterization, *Clin. Exp. Immunol.*, 59, 305, 1985.
172. **Carrel, S., de Tribolet, N., and Gross, N.,** Expression of HLA-DR and common acute lymphoblastic leukemia antigens on glioma cells, *Eur. J. Immunol.*, 12, 354, 1982.
173. **Carrel, S., Schmidt-Kessen, A., Mach, J.-P., Heumann, D., and Girardet, C.,** Expression of common acute lymphoblastic leukemia antigen (CALLA) on human malignant melanoma cell lines, *J. Immunol.*, 130, 2456, 1983.
174. **Kenny, A. J., O'Hare, M. J., and Gusterson, B. A.,** Cell-surface peptidases as modulators of growth and differentiation, *Lancet*, 2, 785, 1989.
175. **Look, A. T., Ashmun, R. A., Shapiro, L. H., and Peiper, S. C.,** Human myeloid plasma membrane glycoprotein CD13 (gp150) is identical to aminopeptidase N, *J. Clin. Invest.*, 83, 1299, 1989.
176. **Mentlein, R., Heymann, E., Scholz, W., Feller, A. C., and Flad, H. D.,** Dipeptidyl peptidase IV as a new surface marker for a subpopulation of human T-lymphocytes, *Cell. Immunol.*, 89, 11, 1984.
177. **Ulmer, A. M., Mattern, T., Feller, A. C., Heymann, E., and Flad, H. D.,** TII 19-4-7 and 4EL1C7 but not B1.19.2 (all clustered in CDW26) bind to dipeptidyl peptidase IV (DPP IV), 7th Int. Cong. Immunol., Abstracts, Gustav Fischer Verlag, Stuttgart, 1989, p. 151.
178. **Kenny, A. J. and Booth, A. G.,** Microvilli: their ultrastructure, enzymology and molecular organization, *Essays Biochem.*, 14, 1, 1978.
179. **Kenny, A. J. and Maroux, S.,** Topology of microvillar membrane hydrolases of kidney and intestine, *Physiol. Rev.*, 62, 91, 1982.
180. **Fulcher, I. S., Pappin, D. J. C., and Kenny, A. J.,** The N-terminal amino acid sequence of pig kidney endopeptidase-24.11 shows homology with pro-sucrase-isomaltase, *Biochem. J.*, 240, 305, 1986.
181. **Ferguson, M. A. J. and Williams, A. F.,** Cell surface anchoring of proteins via glycosyl-phosphatidylinositol structures, *Annu. Rev. Biochem.*, 57, 285, 1988.
182. **Lolw, M. G.,** Biochemistry of the glycosyl-phosphatidylinositol membrane protein anchors, *Biochem. J.*, 244, 1, 1987.
183. **Low, M. G.,** Glycosyl-phosphatidylinositol: a versatile anchor for cell surface proteins, *FASEB J.*, 3, 1600, 1989.
184. **Turner, A. J., Ed.,** *Molecular and Cell Biology of Membrane Proteins: Glycolipid Anchors of Cell-Surface Proteins,* Ellis Horwood, Chichester, England, 1990.
185. **Hooper, N. M. and Turner, A. J.,** Hydrolysis of the glycosyl-phosphatidylinositol anchors of renal microvillar peptidases by plasma phospholipase D, *Biochem. Soc. Trans.*, 17, 885, 1989.
186. **Stewart, T. A., Weare, J. A., and Erdos, E. G.,** Purification and characterization of human converting enzyme (kininase II), *Peptides*, 2, 145, 1981.
187. **Fox, J. A., Soliz, N. M., and Saltiel, A. R.,** Purification of a phosphatidylinositol-glycan specific phospholipase C from liver plasma membranes: a possible target of insulin action, *Proc. Natl. Acad. Sci. U.S.A.*, 84, 2663, 1987.
188. **Davitz, M. A., Hereld, D., Shak, S., Krakow, J., Englund, P. T., and Nussenzweig, V.,** A glycan-phosphatidylinositol-specific phospholipase D in human serum, *Science*, 238, 81, 1987.
189. **Davitz, M. A., Hom, J., and Schenkman, S.,** Purification of a glycosyl-phosphatidylinositol-specific phospholipase D from human plasma, *J. Biol. Chem.*, 264, 13760, 1989.
190. **Das, M., Hartley, J. L., and Soffer, R.,** Serum angiotensin-converting enzyme. Isolation and relationship to the pulmonary enzyme, *J. Biol. Chem.*, 252, 1316, 1977.
191. **Lanzillo, J. J. and Fanburg, B. L.,** Angiotensin I converting enzyme from human plasma, *Biochemistry*, 16, 5491, 1977.
192. **Lieberman, J.,** Elevation of serum angiotensin converting enzyme (ACE) level in sarcoidosis, *Am. J. Med.*, 59, 365, 1975.
193. **Silverstein, E., Friedland, J., Lyons, H. A., and Gourin, A.,** Elevation of angiotensin-converting enzyme in granulomatous lymph nodes and serum in sarcoidosis: clinical and possible pathogenic significance, *Ann. N. Y. Acad. Sci.*, 278, 498, 1976.
194. **Johnson, A. R., Coalson, J. J., Ashton, J., Larumbide, M., and Erdos, E. G.,,** Neutral endopeptidase in serum samples from patients with adult respiratory distress syndrome. Comparison with angiotensin-converting enzyme, *Am. Rev. Respir. Dis.*, 132, 1262, 1985.
195. **Almenoff, J., Teirstein, A. S., Thornton, J. C., and Orlowski, M.,** Identification of a thermolysin-like metallopeptidase in serum: activity in normal subjects and in patients with sarcoidosis, *J. Lab. Clin. Med.*, 103, 420, 1984.

196. **Deschodt-Lanckman, M., Michaux, F., De Prez, E., Abramowicz, D., Vanherweghem, J.-L., and Goldman, M.,** Increased serum levels of endopeptidase 24.11 ('enkephalinase') in patients with end-stage renal failure, *Life Sci.*, 45, 133, 1989.
197. **Craves, F. B., Law, P. Y., Hunt, C. A., and Loh, H. H.,** The metabolic disposition of radiolabeled enkephalins *in vitro* and *in situ*, *J. Pharmacol. Exp. Ther.*, 206, 492, 1978.
198. **Swerts, J. P., Perrisot, R., Malfroy, B., and Schwartz, J.-C.,** Is 'enkephalinase' identical with angiotensin converting enzyme?, *Eur. J. Pharmacol.*, 53, 209, 1979.
199. **Blumberg, S., Vogel, Z., and Altstein, M.,** Inhibition of enkephalin-degrading enzymes from rat brain and of thermolysin by amino acid hydroxamates, *Life Sci.*, 28, 301, 1981.
200. **Hudgin, R. L., Charlson, S. E., Zimmerman, M., Mumford, R., and Wood, P. L.,** Enkephalinase: selective peptide inhbitors, *Life Sci.*, 29, 2593, 1981.
201. **Almenoff, J., Wilk, S., and Orlowski, M.,** Membrane bound pituitary metalloendopeptidase: apparent identity to enkephalinase, *Biochem. Biophys. Res. Commun.*, 102, 206, 1981.
202. **Matsas, R., Fulcher, I. S., Kenny, A. J., and Turner, A. J.,** Substance P and [Leu]enkephalin are hydrolyzed by an enzyme in pig caudate synaptic membranes that is identical with the endopeptidase of kidney microvilli, *Proc. Natl. Acad. Sci. U.S.A.*, 80, 3111, 1983.
203. **Chaillet, P., Marcus-Collado, H., Costentin, J., Yi, C. C., de la Baume, S., and Schwartz, J. C.,** Inhibition of enkephalin metabolism by, and antinociceptive activity of, bestatin an aminopeptidase inhibitor, *Eur. J. Pharmacol.*, 86, 29, 1983.
204. **Littlewood, G. M., Iversen, L. L., and Turner, A. J.,** Neuropeptides and their peptidases: functional considerations, *Neurochem. Int.*, 12, 383, 1988.
205. **Mauborgne, A., Bourgoin, S., Bendiel, J. J., Hirsch, M., Berthier, J. L., Haman, M., and Cesselin, F.,** Enkephalinase is involved in the degradation of endogenous substance P released from slices of rat substantia nigra, *J. Pharm. Exp. Therm.*, 243, 674, 1987.
206. **Matsas, R., Kenny, A. J., and Turner, A. J.,** An immunohistochemical study of endopeptidase-24.11 ("enkephalinase") in the pig nervous system, *Neuroscience*, 18, 991, 1986.
207. **Barnes, K., Matsas, R., Hooper, N. M., Turner, A. J., and Kenny, A. J.,** Endopeptidase-24.11 is striosomally ordered in pig brain and, in contrast to aminopeptidase N and peptidyl dipeptidase A ('angiotensin converting enzyme'), is a marker for a set of striatal efferent fibres, *Neuroscience*, 27, 799, 1988.
208. **Pollard, H., de la Baume, S., Bouthenet, M.-L., Schwartz, J. C., Ronco, P., and Verroust, P.,** Characterization of two probes for the localization of enkephalinase in rat brain: [^{3}H]thiorphan and a ^{125}I-labeled monoclonal antibody, *Eur. J. Pharmacol.*, 133, 155, 1987.
209. **Waksman, G., Hamel, E., Fournié-Zaluski, M. C., and Roques, B. P.,** Autoradiographic comparison of the distribution of the neutral endopeptidase 'enkephalinase' of μ and δ opioid receptors in rat brain, *Proc. Natl. Acad. Sci. U.S.A.*, 83, 1523, 1986.
210. **Matsas, R. and Kenny, A. J.,** Imunocytochemical localization of endopeptidase-24.11 in cultured neurons from pig striatum, *Neuroscience*, 31, 237, 1989.
211. **Barnes, K., Turner, A. J., and Kenny, A. J.,** Electronmicroscopic immunocytochemistry of pig brain shows that endopeptidase-24.11 is localized in neuronal membranes, *Neurosci. Lett.*, 94, 64, 1988.
212. **Bergey, J. L., Kotler, D., Delaney, N. G., and Cushman, D. W.,** Factors affecting the vasorelaxant activity of atrial natriuretic peptides, *Fed. Proc. Fed. Am. Soc. Exp. Biol.*, 46, 1296, 1987.
213. **Kenny, A. J. and Stephenson, S. L.,** Role of endopeptidase-24.11 in the inactivation of atrial natriuretic peptide, *FEBS Lett.*, 232, 1, 1988.
214. **Bourne, A., Barnes, K., Taylor, B. A., Turner, A. J., and Kenny, A. J.,** Membrane peptidases in the porcine choroid plexus and other cell membranes in contact with the cerebrospinal fluid, *Biochem. J.*, 259, 69, 1989.
215. **Schiller, P. W., Maziak, L. A., Nguyen, T. M. D., Godin, J., Garcia, R., De Lean, A., and Cantin, M.,** Superactive analogs of the atrial natriuretic peptide, *Biochem. Biophys. Res. Commun.*, 143, 499, 1987.
216. **Yandle, T., Crozier, I., Nichols, G., Espiner, E., Carne, A., and Brennan, S.,** Amino acid sequence of atrial natriuretic peptides in human coronary sinus plasma, *Biochem. Biophys. Res. Commun.*, 146, 832, 1987.
217. **Stephenson, S. L. and Kenny, A. J.,** unpublished data, 1989.
218. **Maack, T., Suzuki, M., Almeida, F. A., Nussenzveig, D., Scarborough, R. M., McEnroe, G. A., and Lewicki, J. A.,** Physiological role of silent receptors of atrial natriuretic factor, *Science*, 238, 675, 1987.
219. **Chinkers, M., Garbers, D. L., Chang, M.-S., Lowe, D. G., Chin, H., Goeddel, D. V., and Schulz, S.,** A membrane form of guanylate cyclase is an atrial natriuretic peptide receptor, *Nature (London)*, 338, 78, 1989.
220. **Schulz, S., Chinkers, M., and Garbers, D. L.,** The guanylate cyclase/receptor family of proteins, *FASEB J.*, 3, 2026, 1989.
221. **Fuller, F., Porter, J. G., Arfsten, A. E., Miller, J., Schilling, J. W., Scarborough, R. M., Lewicki, A., and Schenk, D. B.,** Atrial natriuretic peptide clearance receptor. Complete sequence and functional expression of cDNA clones, *J. Biol. Chem.*, 263, 9395, 1988.
222. **Scarborough, R. M., McEnroe, G. A., Arfsten, A., Kang, L.-L., Schwartz, K., and Lewicki, J. A.,** *D-*

amino acid-substituted atrial natriuretic peptide analogs reveal novel receptor recognition requirements, *J. Biol. Chem.,* 263, 16818, 1988.

223. **Skeggs, L. T., Marsh, W. H., Kahn, J. R., and Shumway, N. P.,** The existence of two forms of hypertension, *J. Exp. Med.,* 99, 275, 1954.
224. **Skeggs, L. T., Kahn, J. R., and Shumway, N. P.,** The preparation and function of the hypertensin-converting enzyme, *J. Exp. Med.,* 103, 295, 1956.
225. **Ng, K. K. F. and Vane, J. R.,** Conversion of angiotensin I to angiotensin II, *Nature (London),* 216, 762, 1967.
226. **Ng, K. K. F. and Vane, J. R.,** Fate of angiotensin I in the circulation, *Nature (London),* 218, 144, 1968.
227. **Cushman, D. W. and Cheung, H. S.,** Concentrations of angiotensin-converting enzyme in the tissues of the rat, *Biochim. Biophys. Acta,* 250, 261, 1971.
228. **Caldwell, P. R. B., Seegar, B. C., Hsu, K. C., Das, M., and Soffer, R. L.,** Angiotensin-converting enzyme: vascular endothelial localization, *Science,* 191, 1050, 1976.
229. **Erdos, E. G. and Yang, H. Y. T.,** An enzyme in microsomal fraction of kidney that inactivates bradykinin, *Life Sci.,* 6, 569, 1967.
230. **Johnston, C. I.,** Angiotensin converting enzyme inhibitors — the balance sheet, *Med. J. Aus.,* 148, 488, 1988.
231. **Strittmatter, S. M., Thièle, E. A., De Souza, E. G., and Snyder, S. H.,** Angiotensin-converting enzyme in the testis and epididymis: differential development and pituitary regulation of isozymes, *Endocrinology,* 117, 1374, 1985.
232. **Velletri, P. A.,** Testicular angiotensin I-converting enzyme, *Life Sci.,* 36, 1597, 1985.
233. **Strittmatter, S. M., Lo, M. M. S., Javitch, J. A., and Snyder, S. H.,** Autoradiographic localization of angiotensin-converting enzyme in rat brain with [^{3}H]captopril. Localization to a striatonigral pathway, *Proc. Natl. Acad. Sci. U.S.A.,* 81, 1599, 1984.
234. **Strittmatter, S. M. and Snyder, S. H.,** Angiotensin converting enzyme immunohistochemistry in rat brain and pituitary gland: correlation of isozyme type with cellular localization, *Neuroscience,* 21, 407, 1987.
235. **Rachel, E., Hooper, N. M., James, P., Semenza, G., Turner, A. J., and Mantei, N.,** cDNA cloning and expression in *Xenopus laevis* oocytes of pig renal dipeptidase, a glycosyl-phosphatidylinositol-anchored ectoenzyme, *Biochem. J.,* 271, 755, 1990.

Chapter 5

RECEPTOR-MEDIATED ENDOCYTOSIS AND DEGRADATION OF POLYPEPTIDE HORMONES, GROWTH FACTORS, AND NEUROPEPTIDES

Steen Gammeltoft

TABLE OF CONTENTS

I. OVERVIEW

Receptor-mediated endocytosis and intracellular degradation of polypeptide hormones, growth factors, and neuropeptides represents the major pathway in the breakdown of these bioactive substances *in vivo*. Three steps are involved: (1) capture by specific receptors on the cell surface, (2) translocation across the plasma membrane, and (3) intracellular proteolysis. Receptors which mediate endocytosis of polypeptide hormones, growth factors, and neuropeptides are also involved in signal transduction whereby the ligand-receptor interaction is translated into activation of intracellular pathways leading to a specific cellular response. Thus, intracellular signaling and endocytosis of receptors are integrated events, although their molecular mechanisms are different. Three general mechanisms of cellular signal transduction of membrane receptors are relatively well understood: (1) coupling of receptors to effectors by GTP-binding proteins (G-proteins), (2) coupling through activation of a tyrosine-specific protein kinase that is intrinsic to the receptor molecule, and (3) ligand-gated ion channels. All three pathways involve a series of intracellular events including activation of enzymes by protein phosphorylations or dephosphorylations.

One mechanism of endocytosis of molecules at the plasma membrane has been recognized: assembly of receptors into coated pits of the membrane followed by budding and formation of an endocytotic coated vesicle. The coat proteins consist of clathrin and two adaptor proteins which recognize the cytoplasmic portion of the membrane receptor. Hormone and growth factor receptors are diffusely distributed on the cell surface and are not trapped in coated pits unless they are occupied with ligand. Other membrane receptors that are primarily involved in endocytosis of their ligands, like low-density lipoprotein (LDL), transferrin, and asialoglycoproteins and mannose-6-phosphate (Man-6-P)-containing proteins, move spontaneously to coated pits and enter cells continuously even in the absence of ligand.

The systems of receptor-mediated endocytosis vary with respect to the fate of ligand and receptor which may follow one of four routes. The classic pathway is degradation of ligand and recycling of receptor. Peptide hormones as well as LDL and asialoglycoproteins are dissociated from the receptor within the endocytotic and prelysosomal vesicles after acidification to pH about 4.5. The ligand is sequentially degraded by specific proteases and less-specific aminopeptidases or endopeptidases, whereas the receptor is recycled back to the plasma membrane. A second pathway has been described for growth factor receptors like the epidermal growth factor (EGF) receptor which are transported to the lysosomes where both receptor and ligand are degraded. Two pathways of endocytosis do not lead to proteolysis of receptor or ligand. The complex of transferrin and its receptor is not dissociated within the endosome, where iron is stripped from transferrin and the apo-transferrin/receptor complex is recycled to the cell surface. Finally, receptor-mediated endocytosis is involved in transport of intact polymeric immunoglobulin A (IgA) and immunoglobulin M (IgM) across epithelial surfaces such as liver cells and mammary cells. Although endocytosis of a variety of extracellular proteins like hormones, growth factors, neuropeptides, and plasma proteins show similarities in the molecular mechanism, significant variations in the intracellular routes and fates exist.

This review is limited to the molecular and cellular mechanism of receptor-mediated endocytosis and degradation of polypeptide hormones, growth factors, and neuropeptides. The concept of receptor-mediated endocytosis was first formulated in 1974 by Goldstein and Brown[1,2] to explain the sequential surface binding, internalization, and intracellular degradation of LDL. It has since been extended to other substances including regulatory polypeptides.[3–5]

II. RECEPTORS FOR POLYPEPTIDE HORMONES, GROWTH FACTORS, AND NEUROPEPTIDES

A. RECEPTOR STRUCTURE

Receptors for polypeptide hormones, growth factors, and neuropeptides are integral

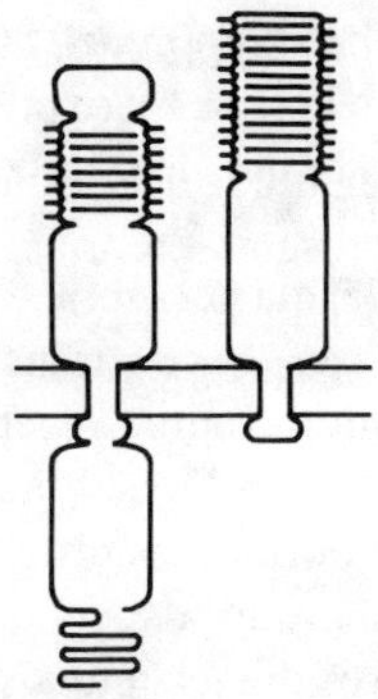

FIGURE 1. Plasma membrane receptor prototypes. The epidermal growth factor receptor (EGFR) represents a prototype of a signaling receptor with an extracellular ligand-binding domain, a disulfide cross-linked domain (hatched), a transmembrane domain, and a large cytoplasmic domain involved in signal transduction. The low density lipoprotein receptor (LDLR) represents a prototype of a transport receptor with an extracellular ligand-binding domain, a disulphide-linked domain (hatched), a transmembrane domain, and a small cytoplasmic domain involved in intracellular targeting. (Redrawn after Reference 16).

membrane proteins and the prototype is composed of extracellular, transmembrane, and cytoplasmic domains which are involved in different functions: ligand binding, receptor aggregation, intracellular signaling, and internalization. Figure 1 shows a schematic drawing of the EGF receptor and for comparison a model of the LDL receptor. The extracellular domains of these receptors are similar, with a cysteine-rich disulfide-linked region, whereas the cytoplasmic domains are very different. The EGF receptor has a large intracellular portion with protein tyrosine kinase activity involved in intracellular signaling,[6,7] whereas the intracellular portion of the LDL receptor is small, reflecting that it has no signal function but mediates endocytosis via interaction with clathrin.[2]

Receptor structures have been determined from the full-length nucleotide sequences of mRNA molecules encoding membrane receptors for hormones, growth factors, and neurotransmitters. The first membrane receptors to be sequenced were the nicotinic acetylcholine receptor[8] and the EGF receptor.[9] These receptor proteins were purified and the aminoterminal amino acid sequence determined. An oligonucleotide probe was synthesized for screening of a cDNA library derived from a tissue or cell expressing mRNA for the receptor. Positive clones were selected by hybridization under stringent conditions and the cDNA insert of the phage was sequenced and the amino acid sequence was deduced. After this pioneering work, three other techniques for cloning and sequencing receptors have been applied. First, homology screening, where receptors with similar sequences are detected by screening the cDNA library with a known receptor cDNA probe under low stringency conditions. One example is the human EGF receptor 2 (HER-2) or erbB-2 protein which was cloned and sequenced by screening with a cDNA probe for the viral oncogene v-erbB.[10] The sequence was similar to that of the rat cellular oncogene c-neu.[11,12] The ligand of this receptor has not yet been identified and it has, accordingly, been designated an orphan receptor. Second, the polymerase chain reaction (PCR) technique has been applied in homology screening of receptor cDNAs. Based on a presumed sequence similarity in a receptor, synthetic oligonucleotide primers have been used for amplification of a particular segment of the uncloned specific cDNA. Hereby, a specific probe was generated for screening of cDNA libraries under stringent conditions. An example of application of this technique is the cloning of the thyroid-stimulating hormone (TSH) receptor.[13] Finally, expression cloning of receptors by transfection of mRNA in a cell and screening for ligand-induced effect: this approach has been used for cloning of neurotransmitter receptors like the substance K receptor by expression of mRNA in a frog oocyte and screening of tachykinin-induced membrane depolarization followed by amplification and sequencing of cDNA.[14]

Based on the primary structure, plasma membrane receptors can be classified into five groups: receptor protein tyrosine kinases, seven transmembrane segment receptors, single transmembrane segment receptors, multisubunit receptors, and ligand-gated ion channels (Table 1). For completeness, the intracellular DNA binding receptors have also been included in the table. Receptors for polypeptide hormones, growth factors, and neuropeptides differ in their structure and are represented in all groups except the ion channel receptors. This suggests that the structure-function relationship of the receptors shows significant variation concerning the ligand binding and signal transduction.

B. SIGNAL TRANSDUCTION

Three major pathways in membrane receptor signal transduction have been recognized: (1) activation of protein tyrosine kinases intrinsic to receptors, (2) coupling of receptors to effector molecules via GTP-binding proteins (G-proteins), and (3) ligand-gated ion channels.[53–56] The different receptor families can be classified into these three groups, but some receptors, like the protein tyrosine kinase family, are also linked to phospholipase C via G-proteins, suggesting that they can activate several signal transduction pathways.[57] Receptor protein tyrosine kinases are activated by polypeptide growth factors like EGF, insulin-like growth factors (IGF), fibroblast growth factor (FGF), platelet-derived growth factor (PDGF), and macrophage-colony-stimulating factor (M-CSF), as well as the hormone insulin.[56] Several studies have indicated that protein tyrosine kinase activation is involved in signal transduction. These include site-directed mutagenesis of receptors in which the kinase has been inactivated by mutation of the ATP-binding site. Cellular transfection and expression of mutant receptors showed that growth factors and insulin failed to induce cellular growth or metabolic responses, although the ligand binding and receptor concentration were normal.[53,56,57] Secondly, cellular injection of specific antibodies which inhibit receptor kinase activity interferes with insulin action.[58,59] Finally, conditions with insulin resistance have been associated with decreased tyrosine kinase activity of the insulin receptor.[58–60]

Various cellular substrates of the receptor protein tyrosine kinases have been demonstrated, including membrane and cytoplasmic proteins with molecular weights ranging from 15,000 to 185,000. The identity and function of the proteins which are phosphorylated on tyrosine in response to growth factors and insulin has not been revealed, but it is possible that they are protein kinases or phosphatases which phosphorylate or dephosphorylate proteins on serine and threonine residues. At present, it seems most likely that receptor protein tyrosine kinases initiate cellular phosphorylation cascades which lead to the ultimate responses.[56–60]

The seven transmembrane segment receptors are coupled to G-proteins which activate or inhibit adenylate cyclase, phosphodiesterase, and phospholipase C. As a result of the enzyme activation or inhibition, the cellular concentrations of second messengers like cyclic AMP, inositol triphosphate, and diacylglycerol are altered. Cyclic AMP activates the cyclic AMP-dependent protein kinase which phosphorylates other protein kinases and phosphatases on serine and threonine which, in turn, regulates enzymes involved in cellular metabolism. Phospholipase C catalyzes the formation of inositol triphosphate and diacylglycerol by cleavage of phosphoinositols. Inositol triphosphate stimulates release of calcium from cellular organelles into the cytoplasm where it is involved in many cellular processes like secretion, contraction, and metabolism. Diacylglycerol activates protein kinase C which regulates several cellular molecules by phosphorylation on serine and threonine.[54,61]

The ligand-gated ion channel family includes neurotransmitter receptors where binding of the ligand results in opening of the ion channel and cellular uptake of ions like Na^+, or Cl^-

TABLE 1
Structural Classification of Receptors

Protein Tyrosine Kinase Family

One segment
- Epidermal growth factor (9)
- Insulin (15, 16)
- Insulin-like growth factor I (17)
- Fibroblast growth factor (18)

Two segment
- Platelet-derived growth factor: A and B (19, 20)
- Macrophage-colony-stimulating factor (colony-stimulating factor-1) (21)

Seven Transmembrane Segment Family

Rhodopsin (22)
Epinephrine: α-1, α-2, β-1, and β-2 adrenergic (23)
Acetylcholine: M1, M2, M3, and M4 muscarinic (24)
Serotonin: 1A, 1B, 1C, 2 (25)
Dopamine: D2 (26)
Angiotensin (27)
Substance K (14)
Substance P (28)
Neuromedin (29)
Luteinizing hormone (30)
Choriongonadotropin (30)
Thyroid-stimulating hormone (13)

Guanylate Cyclase Receptor Family

Atrial natriuretic peptide (31)

Single Transmembrane Segment Family

Nerve growth factor (32)
Low-density lipoprotein (33)
Transferrin (34)
Asialoglycoproteins (35)
Polymeric immunoglobulin A/immunoglobulin M (36)
Mannose-6-phosphate/insulin-like growth factor II (37—39)
Tumor necrosis factor (40)

Multisubunit Family

Erythropoietin (41)
Interleukin-2 (42)
Interleukin-4 (43)
Interleukin-6 (44)
Interleukin-7 (45)
Granulocyte-macrophage-colony-stimulating factor (46)
Growth hormone (47)
Prolactin (48)

Ligand-Gated Ion Channel Family

Acetylcholine: nicotinic (8)
γ-aminobutyric acid (49)
Glycine (50)
Glutamate: kainate (51)

TABLE 1 (continued)
Structural Classification of Receptors

DNA Binding Family (52)

Estrogen
Glucocorticoid
Progesterone
Aldosterone
Vitamin D
Retinoic acid: α, β, and γ
Thyroid hormone: α and β

Note: Receptors are designated by their ligand. Receptor subtypes which are different gene products are indicated with letters and numbers. Representative references are enclosed in parentheses.

leading to changes in membrane potential and cellular effects.[55] Other receptor families, like the single transmembrane segment family and the multisubunit family, have yet unsolved signal transduction mechanisms. It is possible that receptors like the nerve growth factor receptor, as well as the interleukin receptors, are noncovalently associated with protein kinases and that protein phosphorylations on either serine or tyrosine are involved in their signal transduction.[62,63]

III. RECEPTOR-MEDIATED ENDOCYTOSIS

Receptors for hormones, growth factors, neuropeptides, and plasma proteins are internalized from the cell surface to intracellular organelles. The receptor-mediated endocytosis serves at least three different functions: (1) cellular degradation of the regulatory polypeptide (hormone, growth factor, or neuropeptide), whereby the signal mediated by the receptor is terminated and the substance cleared from the extracellular space; (2) cellular desensitization to regulatory substances by reduction of the receptor concentration on the cell surface; and (3) cellular uptake of nutrients (e.g., cholesterol, Fe^{2+}) or plasma proteins (e.g., asialoglycoproteins or phosphomannosylated proteins) which are transported to specific intracellular organelles for degradation or utilization. The cellular phenomenon of receptor-mediated endocytosis was discovered about 15 years ago, but the molecular mechanisms have only been elucidated recently. In these studies, protein engineering techniques with site-directed mutagenesis of receptors and functional studies of mutant receptors transfected into cell lines have been applied. In addition, naturally occurring mutations in patients with defects in receptor internalization have cast light on the molecular events in receptor-mediated endocytosis. The mechanisms will be discussed in more detail below.

Receptor-mediated endocytosis of extracellular ligands should be distinguished from other processes of internalization of extracellular material like phagocytosis and fluid-phase pinocytosis. Although the three processes are similar in that extracellular particles, substances, or fluid are internalized within an invagination of the plasma membrane, the molecular mechanisms involved, the intracellular sorting, and the final destination in cellular organelles are different. Phagocytosis describes the internalization of particulate matter visible by light microscopy (i.e., larger than 0.1 to 0.2 μm in diameter like bacteria, virus, or immune complexes). Phagocytosis involves attachment of the particle to the cell surface and internalization. Both steps require a specific molecular interaction between the particle and the phagocytic cell and is followed by delivery of the material to lysosomes for degradation.[64]

Fluid-phase pinocytosis is used to describe the process wherein smaller substrates (including insoluble particles or solutes) enter the cell in the fluid content of the endocytic vesicle.

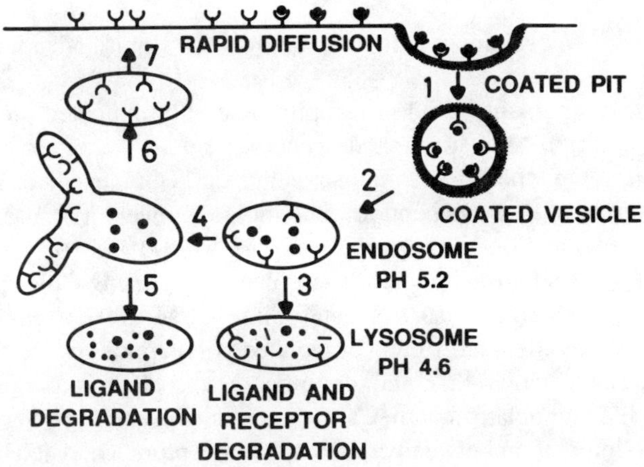

FIGURE. 2. Model of the cellular mechanism of receptor-mediated endocytosis. 1. Receptors bind their ligand and aggregate in coated pits, which bud off to a coated vesicle. 2. Coated vesicles lose their clathrin coat and form acidified endosomes where ligands are dissociated from receptor. 3. Endosomes fuse with lysosomes where ligand and receptor are degraded. 4. Endosomes form multivesicular bodies. 5. Prelysosomal vesicles are formed and the ligand is degraded. 6. Vesicles with unoccupied receptors are formed. 7. Recycling of receptors in the plasma membrane. (Redrawn after Reference 7).

The molecular mechanisms involved are not understood, but it seems most likely that fluid-phase endocytosis is a constitutive process, the function of which is not completely clear. The internalized fluid and its content may be transcytosed to the opposite surface of a polarized cell or be delivered to the lysosome for degradation. Alternatively, the fluid-phase pinocytosis is primarily involved in turnover of membrane lipid and/or protein.[64]

A. CELLULAR MECHANISM OF RECEPTOR-MEDIATED ENDOCYTOSIS

The cellular mechanism of receptor-mediated endocytosis of extracellular ligands involves binding to the receptor on the cell surface, lateral diffusion of receptor-ligand complexes in the membrane, and aggregation in clathrin-coated pits (Figure 2). The receptor and ligand are internalized in a clathrin-coated vesicle which is transformed into an uncoated vesicle termed endosome or receptosome. These vesicles have been shown to have a pH of 5.0 to 6.5 and to be capable of acidification due to the activity of a vacuolar H^+-ATPase which exchanges cations in the endosomal lumen with H^+. The pH can be lowered to 4.5 to 5.0 in the endosomes which fuse with the lysosomes. These endosomes have been termed prelysosomes. In the acid environment, the peptide ligand dissociates from the receptor and is subsequently degraded by specific proteases and unspecific proteases. In some cases, receptors also undergo degradation in the lysosome, while other receptors return to the plasma membrane by recycling in endosomal vesicles.[2–5,64–68]

1. Receptor Aggregation in Clathrin-Coated Pits

Endocytosis by coated pits has been known for more than 20 years, and in 1975, Pearse discovered that the coat is composed of clathrin.[69] Coated pits occupy approximately 2% of the cell surface area, but this figure varies between different cell types from 3.8% in human fibroblasts to 0.4% in adipocytes.[70,71] Receptors for plasma proteins like transferrin and LDL are prelocated in coated pits in the absence of ligand,[2,65] and it has been reported that 60 to 95% of LDL receptors are present in coated pits of human fibroblasts.[70] In contrast, receptors for polypeptide hormones and growth factors are located outside coated pits and their aggregation is induced by the ligand.[4,67,71] It should be added that receptor-mediated endocytosis may also occur via a clathrin-independent pathway, although its significance remains to be established.[65]

The molecular interactions involved in receptor assembly in coated pits have been elucidated recently. In addition to clathrin and receptor proteins, another component termed "adaptor" is present.[73] Two types of adaptors have thus far been identified. The HA-II adaptor consists of a heterodimer of two polypeptides with molecular weight of 100,000, an α-adaptin and a β-adaptin, and two additional polypeptides of M_r 50,000 and 16,000, respectively, and is largely restricted to plasma membrane-coated pits. In the *trans*-Golgi network, a related HA-I adaptor is present which consists of a heterodimer of M_r 100,000 polypeptides, but a γ-adaptin and a β^1-adaptin, with two different associated polypeptides of M_r 47,000 and 19,000, respectively. Studies have shown that adaptors interact with clathrin and form an inner layer of the coat between the outer clathrin lattice and the vesicle bilayer.[68] Fewer studies have been directed towards the interaction between adaptors and receptors. One study indicated that the mannose-6-phosphate (Man-6-P) receptor in solution interacted with adaptors and could be assembled into coats in a stoichiometric fashion.[74]

For several receptors, their cytoplasmic portions contain information which determines their transport in the cell and, in particular, whether they aggregate in coated pits.[73] This applies to the LDL receptor, the transferrin receptor, the poly IgA/IgM receptor and the EGF receptor. For each of these, it has been shown that specific domains and amino acid residues are necessary for concentrating the receptor in coated pits and, therefore, its subsequent endocytosis. Furthermore, covalent modifications like phosphorylation on serine or tyrosine residues may be involved in the regulation of internalization. One example is the EGF receptor where protein tyrosine kinase activity and autophosphorylation is involved in endocytosis and cellular trafficking of the receptor.[75] A mutant kinase-inactive EGF receptor is internalized, but not down-regulated by degradation in the lysosome. Instead, the receptor is recycled to the membrane, indicating that the cytoplasmic portion of the EGF receptor contains signals for intracellular transport. Others have identified a sequence in the C-terminal end of the EGF receptor required for EGF-dependent internalization, down-regulation, and degradation of receptors.[76] At present, the exact nature of these signals remains to be elucidated.

2. Receptor Recycling

Internalized receptors usually recycle from the endosome compartment to the plasma membrane. This is characteristic for polypeptide hormones, LDL, transferrin, Man-6-P-proteins, asialoglycoprotein, and some growth factor receptors. Exceptions are EGF, PDGF, NGF, and interleukin-2 (IL-2) receptors which are degraded in the lysosome. The recycling kinetics show significant differences among various receptors. The intracellular portion of the recycling circuit takes 15 min or less for receptors like the LDL receptor, transferrin receptor, and Man-6-P receptor.[66] For polypeptide hormone receptors like the insulin receptor, the recycling circuit may take about 1 to 5 h.[77,78] The plasma membrane phase can be as short as a few minutes, depending on the location of the receptor on the cell surface in or outside clathrin-coated pits and whether the ligand is present. In many cases, the receptor has been found to recycle at a basal rate even in the absence of added ligand. The basal rates of recycling depend on whether the receptors are integrated into coated pits. LDL-receptors in fibroblasts are, for example, localized preferentially in coated pits, even in the absence of ligand, and recycle rapidly.[2,66,79] Insulin receptors move into the coated microdomains of the cell surface only when associated with ligand and their basal recycling is low.[4,5] The basal recycling may also reflect the continuous nonspecific fluid-phase pinocytosis of extracellular fluid and solutes whereby 50 to 200% of the cell-surface area ~ plasma membrane is internalized per hour.[64,66]

3. Intracellular Routes

It appears that all endocytic receptors enter cells in the same coated pits and are delivered

TABLE 2
Endocytosis of Receptors: Four Intracellular Pathways

Receptor Recycling and Ligand Degradation
Insulin
Glucagon
Somatostatin
Follicle-stimulating hormone
Luteinizing hormone
Chorionic gonadotropin
Insulin-like growth factor I
Insulin-like growth factor II
Transforming growth factor β
Erythropoietin
Vasoactive intestinal polypeptide
Low-density lipoprotein
Asialoglycoproteins
α_2-Macroglobulin
Receptor Degradation and Ligand Degradation
Epidermal growth factor
Platelet-derived growth factor
Nerve growth factor
Interleukin-2
Growth hormone
Receptor Recycling and Ligand Transport
Transferrin (to plasma membrane)
Mannose-6-phosphate-containing proteins (to Golgi)
Receptor and Ligand Transcytosis
Polymeric immunoglobulin A and immunoglobulin M

to the same acidified endosomes.[2,66] Thereafter, the pathways diverge. The receptor-ligand complex may follow one of four routes, which are described below and summarized in Table 2. The sorting of receptors is already initiated at the cell surface since different receptors are internalized independently of each other.[80]

Receptor recycling and ligand degradation is the classic pathway described for LDL, asialoglycoproteins, α_2-macroglobulin, as well as for insulin and pituitary hormones. Within the endosome, the ligands dissociate from their receptors, apparently as a result of the drop in pH to about 4.5.[66] The ligand is carried further to lysosomes where it is degraded. The receptor leaves the endosome via incorporation into the membrane of a vesicle that buds from the endosome surface. These recycling vesicles fuse with the plasma membrane where receptors can bind the extracellular ligand. This route allows reuse of receptors and one LDL receptor can make up to 150 trips through the endosome without losing its function, and mediate the uptake of hundreds of ligands during its usual lifespan of 10 to 30 h.[2]

Receptor degradation and ligand degradation has been described in greatest detail for EGF. After the EGF/receptor complex reaches the endosome, both components are degraded, probably as a result of subsequent co-transport to the lysosome.[3] Although EGF is dissociated from its receptor at the acidic pH in the endosome, the receptor does not return to the cell surface, but rather, it is carried further into lysosomes and degraded. The PDGF, NGF, and IL-2 receptors follow the same route, whereas other growth factor receptors like IGF-I and

TGF β-receptors recycle. In certain cells, however, one population (approximately 20%) of EGF receptors escape degradation and recycle.[3]

Recycling of receptor and transport of ligand has been described for transferrin receptors. In the endosome, the transferrin/receptor complex is not dissociated, but at pH 5 iron is stripped from transferrin. The apo-transferrin/receptor complex returns to the cell surface. The recycling transferrin receptor leaves the endosome by a network of membrane tubules and vesicles that leads it back to the membrane. On the cell surface at neutral pH, apo-transferrin dissociates from the receptor which is now free to bind another molecule of iron-containing transferrin.[81–83]

The Man-6-P receptor is involved in transport of newly synthesized lysosomal hydrolases from the Golgi to the lysosome. A small proportion (approximately 10 to 20%) of Man-6-P receptors are present on the cell surface and mediates endocytosis of extracellular lysosomal enzymes. The Man-6-P-containing ligand is dissociated from the receptor in the endosome at acid pH and transported to the Golgi, whereas the receptor is recycled back to the cell surface.[84,85] Like the LDL receptor, the transferrin and Man-6-P receptors repeat their cycles more than 100 times every 15 to 20 min.[80,85]

In polarized cells, transport of both receptor and ligand across the cell from one surface to another has been described for the receptor that carries the polymeric immunoglobulins IgA and IgM. In a wide variety of epithelial cells, the receptor is expressed on the basolateral surface. Here, ligand can bind and the receptor (with or without ligand) is endocytosed. The receptor is then transcytosed in vesicles to the apical surface. The extracellular, ligand-binding domain of the receptors is cleaved off and released into the apical medium, together with the ligand.[36]

The delineation of four routes for disposal of receptors and ligands implies that cells have multiple mechanisms for sorting receptors after they enter the cell. These mechanisms must be regulated so as to allow different cells to process the same receptor by different routes or a single cell to process the same receptor by different routes at different times. The cloning and sequencing of cDNA of a large number of receptors have enabled various laboratories to directly analyze the structural domains on the receptor involved in intracellular sorting and targeting after endocytosis. A common mechanism seems to be covalent modifications of the receptor by phosphorylation on serine and tyrosine residues.

IV. ENDOCYTOSIS OF POLYPEPTIDE HORMONES, GROWTH FACTORS, NEUROPEPTIDES, AND PLASMA PROTEINS

Receptor-mediated endocytosis has been described for a number of polypeptide hormones, growth factors, and neuropeptides. Endocytosis of regulatory peptides and receptors may serve two functions: (1) degradation of ligand and arrest of ligand-induced signal, and (2) degradation of receptor and cellular desensitization to the ligand. The receptor-ligand interaction is characterized by high affinity and specificity, and the receptor-mediated endocytosis of extracellular substances is efficient and selective and this pathway is the major factor in determining the biological half-life of polypeptide hormones, growth factors, and neuropeptides. The receptor signal is induced at the cell surface and the internalized receptor-ligand complex is still active in signaling, but the dissociation and degradation of the ligand in the endosomal vesicle will deactivate the receptor and arrest the signal. Reduction of the cell-surface receptor concentration following internalization will desensitize the cell to the active substance. For recycling receptors like insulin receptors, the desensitization will be transient, whereas degradation of receptors like the EGF receptor leads to a long-term reduction in cellular responsiveness.

Although the different groups of receptors (receptor tyrosine kinases, G-protein associated receptors, single transmembrane segment receptors, and multisubunit receptors) show overall

TABLE 3
Endocytosis of Receptors: Ligands

Hormones
Insulin
Glucagon
Somatostatin
Growth hormone
Chorigonadotropin
Follicle-stimulating hormone
Luteinizing hormone
Growth Factors
Epidermal growth factor
Platelet-derived growth factor
Insulin-like growth factor I
Insulin-like growth factor II
Nerve growth factor
Erythropoietin
Interleukin-2
Transforming growth factor β
Neuropeptides
Vasoactive intestinal polypeptide
Plasma Proteins
Low-density lipoprotein
Transferrin
Asialoglycoproteins
Mannose-6-phosphate-containing proteins

similarities in the phenomenon of endocytosis, the molecular mechanisms involved and the fate of internalized receptors are different. Thus, autophosphorylation of receptor tyrosine kinases is involved in internalization and determines intracellular targeting, whereas serine phosphorylation of the cytoplasmic portion of the single transmembrane receptors leads to internalization. In the following, the receptor-mediated endocytosis of the polypeptide hormones, growth factors, and neuropeptides, as well as plasma proteins listed in Table 3, will be reviewed.

A. HORMONES

1. Insulin

The insulin receptor is a disulfide-linked heterotetramer consisting of two extracellular α-subunits which bind insulin and two β-subunits with a transmembrane domain and cytoplasmic protein kinase activity.[15,16] The binding kinetics and specificity of the insulin receptor, as well as the kinase activity, have been characterized in great detail.[58–60,86] Receptor-bound insulin is rapidly internalized after its initial localization on the cell surface. The endocytosis of insulin has been demonstrated by morphological and biochemical techniques in different cell types *in vitro* and *in vivo*. The morphological studies, the subcellular localization of ^{125}I-labeled insulin or insulin-ferritin conjugates was analyzed by electron microscopy after injection of insulin *in vivo* or incubation of cells *in vitro*. In these studies, the label was initially associated with the plasma membrane of hepatocytes or adipocytes, but after a few minutes, increasing amounts of grains were localized intracellularly over endosomes, Golgi-like or lysosome-like structures. After 2 to 5 min, there was a fivefold increase and by 30 to 60 min there was a tenfold increase in the ratio of grains associated with intracellular vesicles.[87–89]

In biochemical studies, the distribution of cell-associated ^{125}I-labeled insulin was analyzed by subcellular fractionation of the tissue or cells. After *in vivo* injection of a pulse of ^{125}I-labeled insulin, analysis of the liver-associated ^{125}I-activity showed that the labeling of the plasma membrane fraction was highest 90 s after injection. From 4 to 8 min, the highest amounts of label were contained in high-density and low-density microsomal fractions. Attempts to identify the nature of the fractions by use of biochemical markers, including specific enzyme activities or internalized substances, showed that they did not belong to the Golgi network or lysosomes.[90,91] Similar observations were obtained by incubation of isolated rat adipocytes with ^{125}I-insulin.[92] Thus, it has become widely accepted that after the initial localization to the plasma membrane, the insulin molecules are translocated to the endosomal intracellular compartment. It is controversial whether insulin or its metabolites are transferred to lysosomes to any quantitative degree.[5]

Internalization of surface-bound ^{125}I-insulin was also demonstrated by the increase in the amount of cell-associated radioactivity resistant to acid washing at pH 3 or treatment with trypsin or pronase. After 2 min of incubation of hepatocytes or adipocytes with ^{125}I-insulin, 75 to 90% of the radioactivity was present on the cell surface, but this amount decreased to 20 to 50% after more than 30 min.[93-95]

The fate of internalized insulin has been studied in suspensions of isolated hepatocytes or adipocytes as well as in perfused intact livers by extraction or release of cell-associated radioactivity and analysis by gel filtration. Exposure of cells to ^{125}I-labeled insulin was followed by a lag period of 5 to 10 min before degradation products of small molecular weight and soluble in trichloroacetic acid appeared in the incubation or perfusion buffer. Cell extracts contained degradation products as well as intact insulin. The appearance of degradation products was completely inhibited by addition of an excess of cold insulin as well as antibody to the insulin receptor, indicating that receptor-mediated endocytosis of insulin is involved in the degradation process.[93-97]

The molecular mechanism of insulin degradation is only partly resolved. Studies with inhibitors of intracellular vesicular transport and acidification like chloroquine and monensin, protease inhibitors like sulfhydryl agents and divalent cation chelators, and oxidative energy inhibitors like 2,4-dinitrophenol and cyanide suggest that energy-dependent internalization and acidification of endosomal vesicles driven by a proton pump are crucial, and that thiol protease and metalloprotease activities are involved in the degradation of insulin.[98]

An insulin-specific protease (EC 3.4.22.11) has been purified from skeletal muscle and liver[99] and, recently, the amino acid sequence was determined from the cDNA structure by cloning of the protease from erythrocytes.[100] The number of amino acids is 954 and the molecular weight is about 110,000. The sequence includes a signal peptide showing that it is a secretory or membrane-bound protein. The enzyme degrades insulin with K_m of about 100 nmol/1 and initial cleavage sites between Leu A13–Tyr A14 and Tyr B16–Leu B17 have been reported.[101] It seems likely that less-specific exopeptidases and endopeptidases are active in the cellular proteolysis of insulin following the initial cleavage by the insulin protease.[101] The subcellular localization of the insulin protease has been determined in fractionation studies and was found both in the plasma membrane and the cytosol fraction.[99] Its role in cellular insulin degradation has been further substantiated in studies where pinocytosis of a specific antibody to the insulin protease inhibited insulin degradation.[102]

The majority of endocytosed insulin is degraded but, in addition, a significant portion (approximately 25%) is released as intact peptide.[103,104] The latter pathway has been referred to as transcytosis in endothelial cells[105] and as retroendocytosis in peripheral target cells.[103] In endothelial cells, this pathway may be involved in the transport of insulin from the capillary to the interstitial space, whereas the role of the retroendocytotic pathway is not understood, but it may be important in preserving the biologic signal for insulin action.

Under physiological conditions, the insulin receptor number on the cell surface remains constant, indicating that the receptor is rapidly recycled to the plasma membrane after internalization.[103,104] Brief exposure of cells to insulin induced endocytosis of receptors with $t_{1/2}$ less than 10 min, whereafter the receptors reappeared on the cell surface with $t_{1/2}$ less than 20 min.[78] Long-term incubation of cells with insulin resulted in a concentration-dependent down-regulation of receptors with a lag period of 4 to 6 h, whereafter the receptor number declined with $t_{1/2}$ equal 2 to 3 h. The molecular mechanism of insulin-induced down-regulation is not understood, but increased receptor degradation in the lysosomes may be involved.[106]

Recent studies with transfection of insulin receptor cDNA and expression of receptors in fibroblasts indicate that endocytosis requires a receptor with an active tyrosine kinase. Kinase-defective receptor mutants do not undergo endocytosis.[107] Insulin receptors with active kinase, but with mutations of the regulatory autophosphorylation sites Tyr 1146, 1150, and 1151, show decreased internalization[108] whereas truncation of the C-terminal end does not alter internalization.[109] This indicates that the kinase is implicated in endocytosis of the insulin receptor, either by inducing a conformational change or by covalent modifications of tyrosine residues. It should be added that the insulin receptor kinase remains insulin-sensitive and activated following endocytosis, indicating that internalized receptors may still be involved in intracellular signaling.[110,111]

In conclusion, a large amount of experimental evidence can be formulated in the following sequence of events. Binding of insulin to its receptor on the cell surface is followed by rapid endocytosis of the insulin-receptor complex in endosomal vesicles. Acidification of the vesicle leads to dissociation of insulin from the receptor and its subsequent degradation by the insulin protease and unspecific peptidases. The unoccupied receptor is recycled to the plasma membrane. A minor part of endocytosed insulin is released as intact peptide, a pathway which has been termed retroendocytosis or transcytosis.

2. Glucagon

Glucagon binds to high-affinity receptors which are coupled to G-proteins and adenylate cyclase in target cells.[112] Following exposure to cells, the hormone is degraded via a receptor-mediated mechanism involving endocytosis of glucagon/receptor complex and lysosomal proteolysis of glucagon. Subcellular fractionation and autoradiographic electron microscopical studies have shown that ^{125}I-glucagon-derived radioactivity is localized in lysosomes and Golgi elements of hepatocytes.[113,114] The receptor is not down-regulated by glucagon, indicating that it is recycled back to the plasma membrane. The cell is only temporarily desensitized to glucagon. Thus, the function of receptor-mediated endocytosis of glucagon is to inactivate the hormone, which is supported by the finding that the biologically important NH_2-terminal region of the hormone is modified by limited proteolytic cleavage during cellular metabolism of receptor-bound glucagon.[115]

3. Somatostatin

Somatostatin inhibits secretion by a large variety of pituitary, pancreatic, and gastrointestinal target cells.[116] One of the best documented functions is the control of growth hormone secretion by the pituitary gland. In a model system of a clonal strain of rat pituitary tumor cells, GH_4C_1 cells, somatostatin binds to high-affinity membrane receptors and inhibits growth hormone secretion.[117] Receptor-bound [^{125}I-Tyr-1] somatostatin on GH_4C_1 cells remained acid sensitive for up to 2 h at 37°C, indicating that the peptide stays at the cell surface.[118] In contrast, receptor-bound EGF became rapidly insensitive to acid treatment, indicating that EGF is internalized in the pituitary cell line like other cell types. Other hormones like thyrotropin-releasing hormone, bombesin, and insulin were also internalized by GH_4C_1 cells. Interestingly, about 50% of receptor-bound [^{125}I-Tyr-1] somatostatin was degraded to ^{125}I-tyrosine after 2 h at 37°C compared with greater than 90% of ^{125}I-EGF. Lysosomal inhibitors like chloroquine, ammonium chloride, and leupeptin had no effect on the amounts of receptor-bound and degraded somatostatin. Based on these data, it may be suggested that, in contrast to other peptide hormones and growth factors, receptor-bound somatostatin does not undergo endocytosis but is degraded at the cell surface by membrane proteases.[118] It remains possible, however, that the difference in cellular processing of somatostatin can be explained by experimental factors like the pituitary tumor cell line, the Tyr-1-somatostatin analogue, or the experimental design. Further studies are required to elucidate these questions.

4. Growth Hormone

Growth hormone binds to specific receptors in various target cells including hepatocytes, adipocytes, pancreatic β-cells, and mononuclear cells.[119–121] Cloning of the cDNA encoding the rabbit and human growth hormone receptor have shown that it is a single-chain, 620-amino-acid protein with a large cytoplasmic region, which has no protein tyrosine kinase domain.[47] Current data on the intracellular processing of growth hormone in hepatocytes and adipocytes suggest the following course of events: growth hormone is rapidly bound to receptors on the plasma membrane (within 1 min) and is rapidly internalized. At steady state (60 min), approximately half of the cell-associated growth hormone is intracellular. A sorting event appears to direct internalized growth hormone into one of two pathways, degradative or nondegradative, for processing. About 75% of growth hormone is targeted for degradation, while the remaining 25% traverses the faster nondegradative route and is released intact. The degradation of internalized growth hormone takes place after fusion of endosomes with the lysosome where hormone and receptor are degraded. This fraction of receptors does not recycle since growth hormone induces a decline in binding capacity in the presence of cycloheximide. The degradative processing of growth hormone (binding, internalization, degradation, and release) requires a time span of 1 h at 37°C. The processing of growth hormone via the nondegradative route is less well defined. The release of intact hormone is independent of temperature, energy, and pH, but is modulated by the extracellular concentration of growth hormone. This suggests that growth hormone receptors can cycle in endosomes defective in the acidification mechanism, or that endosomes may fuse with the plasma membrane before acidification occurs. Accordingly, it is likely that the ligand-receptor complex avoids dissociation during the traverse of this pathway before the hormone is released on the cell surface.[120,122]

5. Follicle-Stimulating Hormone, Luteinizing Hormone, and Chorionic Gonadotropin

The pituitary and placental gonadotropins [follicle-stimulating hormone (FSH), luteinizing hormone (LH), and chorionic gonadotropin (CG)] bind to membrane receptors in their target tissues. Isolation of the cDNA for the rat luteal LH-CG receptor revealed a protein with two domains: an extracellular domain of 341 residues and a 333-residue domain containing seven transmembrane segments. This membrane-spanning region displays sequence similarity with all the numbers of the G protein-coupled receptor family.[30] Receptor-mediated endocytosis of FSH, LH, and CG has been studied in their target cells: granulosa cells and luteal cells from the ovarian follicle and Sertoli cells from the testes.[123-129] The binding of FSH to receptors on the surface of rat granulosa cells was followed by a slow internalization which was maximal after 6 to 8 h. Degradation products occurred in the medium after a lag period of 6 h and increased gradually up to 24 h.[123] Inhibitors of endolysosomal acidification and proteases had no effect on the rate of FSH internalization, whereas the degradation was significantly reduced. Furthermore, no down-regulation of FSH receptors occurred during 24 h of incubation. These data suggest that FSH is degraded slowly in an endosomal or lysosomal compartment, whereas the receptor is recycled to the cell surface. The retarded rate of ligand degradation was also demonstrated in studies involving the processing of human LH and CG in rat and ovine luteal cells.[124-128] In contrast, ovine LH was internalized much faster in ovine luteal cells[126,127] and the degradation of FSH by Sertoli cells was rapid and proceeded within 40 min of incubation.[129] A number of experimental reasons can account for the differences in cellular processing of FSH and LH by target cells, including hormone species differences, cell preparations, and experimental design, but it is also possible that the remarkably slow degradation kinetics are intrinsic to these particular hormone receptors and target cells.

B. POLYPEPTIDE GROWTH FACTORS

Polypeptide growth factors are a group of regulatory peptides which have a role in the regulation of cell growth and differentiation in multicellular organisms. The family of growth factors has increased rapidly during the last decade and more than 50 different polypeptides have been characterized completely or partially by their sequence, structure, biosynthesis, receptors, and actions. The amino acid sequences show great variations and the molecular weights vary from about 5000 to 30,000. Based on their physiological actions, the growth factors can be divided into four groups according to their cellular target. Hemopoietic growth factors include the interleukins 1, 2, 4, 5, and 6 with primary activities on lymphoid cells, the granulocyte-macrophage colony-stimulating factors (G-CSF, M-CSF, GM-CSF, and interleukin-3) active on neutrophils, macrophages, eosinophils, megakaryocytes and mast cells, and erythropoietin active in the production of erythroid cells. Epithelial growth factors include EGF, TGFα, IGF-I, and IGF-II which stimulate growth of epithelial cells in various organs like the skin, intestine, and exocrine glands. Mesenchymal growth factors include PDGF, FGF, IGF-I, IGF-II, and TGFβ which act as mitogens and differentiation factors on cells in connective tissue and parenchymal organs. Neurotrophic factors are NGF, FGF, and IGF-II, which are active during development and regeneration of the nervous system. Polypeptide growth factors are synthesized by many different cells which essentially all have specific high-affinity receptors for these peptides. They induce many regulatory actions in a wide variety of cells, and polypeptide growth factors are thus fundamental regulatory molecules acting by both autocrine and paracrine mechanisms. Their cellular mechanism of action including receptor binding, signal transduction, and receptor-mediated endocytosis and degradation has been studied extensively.

1. Epidermal Growth Factor

EGF is a single-chain polypeptide of 53 amino acid residues with 3 disulfide bonds required for its biological activity.[3] The EGF receptor is a single-chain polypeptide of 1186 amino acid residues, with 1 membrane-spanning region and a cytoplasmic tyrosine kinase domain.[9] EGF stimulates cell division and metabolism of a variety of epithelial cells.[3] Receptor-mediated endocytosis and degradation of regulatory polypeptides was first described for EGF in 1976 by Carpenter and Cohen.[130] They postulated that subsequent to the initial binding of EGF to specific plasma membrane receptors, the EGF-receptor complex is internalized and the growth factor ultimately degraded in lysosomes. These conclusions were based on the following observations obtained by incubation of fibroblasts with ^{125}I-EGF at 37°C: (1) cell-bound ^{125}I-EGF is rapidly degraded to [^{125}I]iodo-tyrosine, (2) bound EGF becomes inaccessible to trypsin and antibodies to EGF, (3) exposure of fibroblasts to EGF results in apparent loss of plasma membrane receptors for EGF, and (4) the phenomena were temperature- and energy-dependent and the degradation of EGF was inhibited by lysosomal inhibitors.[3] These biochemical observations were confirmed in studies where internalization was directly visualized by use of a highly fluorescent rhodamine derivative of EGF.[131,132]

Internalization and degradation of EGF was confirmed in studies of other cell types, and it became clear that the receptor-mediated endocytosis of EGF is different from that of polypeptide hormones like insulin. Following its ligand-induced internalization, the EGF receptor is degraded in a lysosomal vesicle leading to depletion of receptors from the cell surface; whereas the insulin receptor is recycled and its receptor concentration on the cell surface kept constant.[7,104] This difference may reflect the functional difference between the two peptides. EGF initiates a number of effects which lead to cell division, a long-term response during which the cell remains desensitized to EGF. In contrast, insulin stimulates short-term metabolic responses during which the cell remains sensitive to insulin. It should be added, however, that a target cell may be desensitized to insulin by receptor down-regulation following chronic exposure to insulin for greater than 24 h.[106]

The molecular mechanism of receptor-mediated endocytosis of EGF has been studied in great detail. Biochemical analysis of the EGF receptor has established that the receptor tyrosine kinase is a multidomain allosteric enzyme which is involved in multiple functions.[7,133] The availability of cloned cDNA sequences encoding the EGF receptor has enabled more refined analysis of receptor functions by mutations of different receptor domains and construction of chimeric receptors.[7,53,133] These studies have demonstrated that a functional tyrosine kinase and phosphorylated receptor are essential for normal cellular routing and degradation of the EGF receptor in lysosomes. In contrast, endocytosis followed by ligand degradation and recycling of the EFG receptor to the cell surface appears to be independent of the tyrosine kinase activity. Thus, kinase-negative mutants of the EGF receptor are not down-regulated and degraded in response to EGF, but are instead recycled to the cell surface for reutilization. The kinase-defective mutant behaves like the unoccupied wild-type EGF receptors which are internalized and recycled continuously.[75] Recent electron microscopic studies have confirmed that wild-type and kinase-defective EGF receptors are sorted in multivesicular endosomes.[134] It seems likely that the kinase activity serves the active role of specifically targeting the EGF receptor for degradation, probably by phosphorylation of the receptor itself as well as specific, yet unidentified, substrates involved in receptor trafficking. Studies of EGF receptor mutants with deletions of the cytoplasmic domain have suggested that specific sequences are also involved in receptor trafficking to lysosomes.[76] In addition, the EGF receptor kinase is active following its rapid translocation from the plasma membrane to intracellular vesicles, indicating that the endocytosed receptor may be involved in signaling.[135]

In conclusion, EGF receptors are internalized and recycled continuously under normal conditions in the absence of the ligand, whereas EGF-activated receptors are trafficked to lysosomes for degradation of both ligand and receptor resulting in down-regulation of cell-surface receptors. A functional receptor tyrosine kinase is also involved in the targeting of the receptor to lysosomes and serves a double function: signal transduction in cell proliferation and cellular desensitization to EGF.

2. Platelet-Derived Growth Factor

PDGF is a potent mitogen for a variety of mesenchymal cells like fibroblasts and smooth muscle cells. PDGF is a disulfide-bonded heterodimer of two similar, but distinct polypeptide chains: A and B chains with a molecular weight of 28,000 to 31,000.[136] PDGF homodimers have also been described and the three isomeric forms (PDGF-AB, PDGF-AA, and PDGF-BB) show differences in receptor binding affinity and biological actions.[137] The actions of PDGF are induced by interaction with two types of receptors, the B-type and the A-type, which recognize the three PDGF isomers with a characteristic specificity: the B-type PDGF receptor binds only PDGF-BB and PDGF-AB, whereas the A-type PDGF receptor binds all three isomers.[137] Both forms of PDGF receptor show intrinsic tyrosine kinase activity. The A- and B-types of PDGF receptors are different gene products which show a high degree of sequence homology.[19,20]

Receptor-mediated endocytosis of PDGF has been described for the B-type receptor and follows the same pathway as described for EGF. After ligand binding, the PDGF/receptor complex is clustered in coated pits at the cell surface and subsequently internalized.[138,139] The ligand is degraded in the lysosome. The PDGF receptor is down-regulated, and recycling does not seem to occur since reappearance of receptors on the cell surface requires intact protein synthesis.[140] In contrast to the EGF receptor, internalization and degradation of PDGF receptors may not involve PDGF-stimulated kinase activity because various kinase-deficient mutants were down-regulated to the same extent as the wild-type PDGF receptor.[57] In conclusion, endocytosed PDGF receptor is transported to the lysosome where ligand and receptor are degraded, leading to arrest of the mitogenic signal and desensitization to PDGF.

3. Insulin-Like Growth Factors

Insulin-like growth factors, IGF-I and IGF-II, are single-chain polypeptides of molecular weight 7649 and 7471, respectively, which are similar to insulin.[86,141] IGF-I and IGF-II interact with two types of surface receptors which are present on all cell types in vertebrates. The IGF-I receptor is a tetrameric protein similar to the insulin receptor which is composed of two α-subunits and two β-subunits. The β-subunit has intrinsic tyrosine kinase activity and its amino acid sequence is similar to other receptor and oncogene tyrosine kinases and is 80% homologous with the insulin receptor.[17] In many cell types, IGF-I and IGF-II induce their cellular effects by interaction with the IGF-I receptor and activation of the tyrosine kinase.[141–143] IGF-I and IGF-II stimulate cell division and metabolism during fetal development, adolescence, and adulthood.[144] Internalization of the IGF-I receptor leads to degradation of the ligand, whereas the receptor is recycled back to the cell surface.[105,145-147] In neuronal precursor cells, the internalization occurs with a $t_{1/2}$ of about 20 min and at steady state about 60% of cell-associated IGF-I is intracellular (Figure 3). In a pulse-chase experiment IGF-I is released from the cells as degradation products whereas cell-associated IGF-I is mainly intact (Figure 4).

Structural and biochemical evidence has shown that the IGF-II receptor is identical to the mannose-6-phosphate (Man-6-P) receptor.[37–39,85,148] The Man-6-P receptor has been implicated in the transport of lysosomal hydrolases carrying Man-6-P recognition marker from the Golgi apparatus to the lysosome.[84] About 10 to 20% of the Man-6-P receptors are present on the cell surface and mediate endocytosis of phosphomannosylated proteins and IGF-II.[85] It is not clear whether the Man-6-P/IGF-II receptor is also involved in intracellular signaling, but recent studies indicate that both IGF-II and Man-6-P may induce cellular responses in various cell types including neuronal cells.[148,149] As shown in Figure 3, binding of IGF-II to the Man-6-P/IGF-II receptor is followed by a rapid endocytosis ($t_{1/2}$ = 5 to 10 min) and, after 1 h, about 80% of cell-associated IGF-II was intracellular.[145] IGF-II is degraded and released as amino acids, whereas cell-associated IGF-II is mostly intact (Figure 4). Man-6-P/IGF-II receptor is rapidly recycled to the plasma membrane, both in the presence or absence of the ligand.[150] IGF-I binds with low affinity to the Man-6-P/IGF-II receptor (1 to 2% of IGF-II) and is probably not internalized by this pathway under physiological conditions.[145,146] The endocytosis of phosphomannosylated hydrolases by the Man-6-P/IGF-II receptor has been investigated in great detail with respect to the ligand binding, sugar specificity, kinetics, intracellular transport, and organelle associated (see Section D.4).

In conclusion, endocytosis of IGF-I and IGF-II is mediated by two types of receptors, the IGF-I receptor tyrosine kinase and the Man-6-P/IGF-II receptor, respectively. The two pathways are different with regard to their ligand specificity and turn-over kinetics but, for both receptors, the peptide is almost completely degraded whereas the receptor is recycled to the cell surface. In the last respect, the receptor-mediated endocytosis of IGFs differs from that of EGF and PDGF where the receptors are down-regulated.

4. Nerve Growth Factor

Nerve growth factor (NGF) is a M_r 26,000-Da dimer of two identical subunits held together by noncovalent bonds. It forms a complex with two additional proteins, one with an esteropeptidase activity, probably involved in processing an NGF precursor, and the other with yet unknown function.[151] NGF acts primarily as a neurotrophic factor in the peripheral nervous system on sympathetic and sensory nerve cells in the spinal ganglia. NGF prevents nerve cell death and induces axonic growth during development and regeneration *in vivo* as well as *in vitro*.[151,152] The cellular actions of NGF have, in particular, been studied in cultures of chromaffin cells and their neoplastic counterpart, the pheochromocytoma PC12 cell line. In

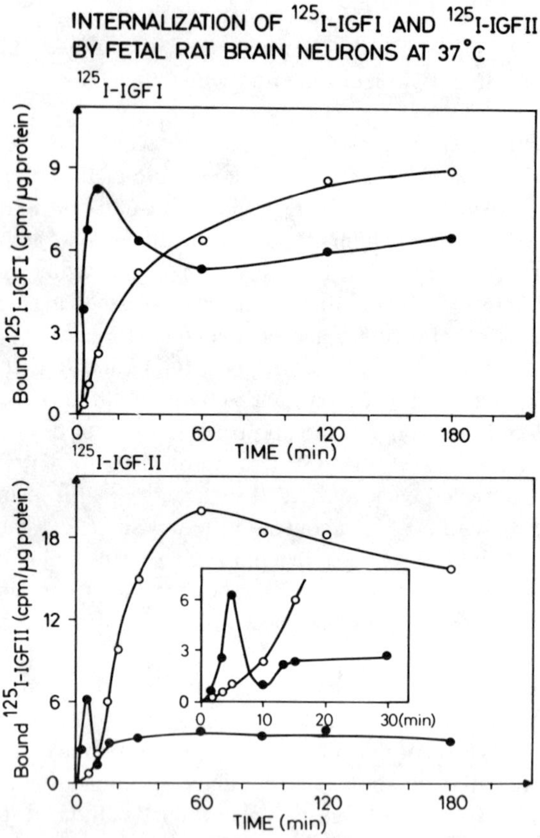

FIGURE 3. Endocytosis of IGF-I and IGF-II in cultured fetal rat brain neurons. Cells were incubated at 37°C for the indicated times with 0.3 n*M* of ^{125}I-labeled IGF-I (A) or IGF-II (B). Surface-bound ^{125}I-activity (o) was released by incubation of the cells for 5 min at 4°C with 0.2 *M* acetic acid and 0.5 *M* NaCl at pH 2.5 followed by extraction of internalized ^{125}I- activity (o) in 0.2 *M* NaOH. The radioactivity was measured, corrected for nonspecific binding, and expressed as cpm per µg cell protein. (Data from Nielsen et al. in Reference 145).

these cells, NGF induces a phenotypic shift resulting in their neuronal differential.[153] Recently, NGF has also been shown to act on neurons in the central nervous system.[151] NGF interacts with two distinct classes of binding sites which are characterized by high and low affinity, respectively.[154] The low-affinity (fast) NGF receptor has been cloned and its amino acid sequence resolved from the cDNA structure as a protein of 425 amino acid residues with a single membrane-spanning domain and a cytoplasmic domain with no ATP-binding site, suggesting that it has no intrinsic kinase activity.[32] It is not clear whether the high- and low-affinity receptors represent different conformations of the same molecule or separate proteins.

Early demonstrations that NGF is taken up by nerve endings of sympathetic or sensory fibers and retrogradely transported to the cell perykarya led to speculations that endocytosed NGF may interact with unknown factors in the cell nucleus and thereby induce its trophic actions.[151] Later, it was shown by electron microscopic autoradiography that NGF was internalized via a receptor-mediated process and transported by an endosomal pathway to lysosomes, whereas the endocytosed NGF was not associated with nuclei.[155,156] This suggests that endocytosis of NGF is involved in degradation of NGF rather than actions of NGF. Furthermore, it was found that exposure to NGF was followed by down-regulation of recep-

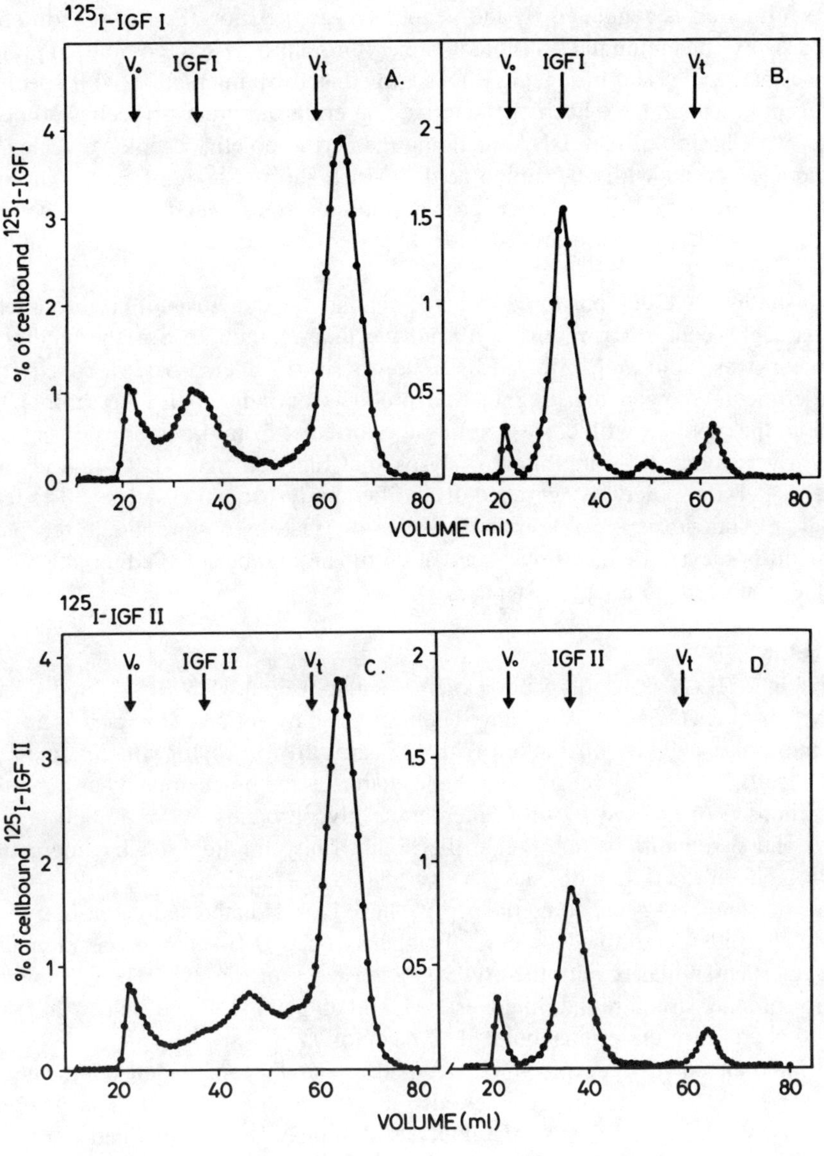

FIGURE 4. Degradation of internalized [^{125}I]IGF-I and [^{125}I]IGF-II by cultured fetal rat brain neurons. Cells were incubated in culture medium for 60 min at 37°C with 0.6 n*M* [^{125}I]IGF-I (A, B) or [^{125}I]IGF-II (C, D). After removal of extracellular peptide by washing, surface-bound IGF-I or IGF-II was removed by incubation at pH 3.5 for 10 min at 4°C. The cells were reincubated in fresh culture medium containing 0.6 n*M* IGF-I or IGF-II. After 3 h at 37°C, the incubation medium was collected and the cells were extracted with 0.1 *M* HCl, 0.1% Triton X-100 and 1 mg/ml albumin. The ^{125}I-activity released to the medium (A, C) or extracted from the cells (B, D) was analyzed by gel filtration on Sephadex G-50 fine and expressed as the percentage of internalized IGF-I or IGF-II. The arrows indicate the void volume (V_o), IGF-I or IGF-II, and the salt volume (V_t). (Data from Nielsen et al. in Reference 145.)

tors at the cell surface, indicating that endocytosed receptors are degraded.[157] Thus, the internalization of NGF and its receptor is similar to that of EGF and PDGF.

5. Erythropoietin

Erythropoietin is a glycoprotein with a molecular weight of approximately 34,000 and is secreted by peritubular cells in the renal cortex and medulla and, to a small extent, by the adult liver. The formation is regulated by the venous oxygen tension. Erythropoietin stimulates erythropoiesis by enhancing the division of late erythroid precursors (committed progenitors) in the bone marrow.[158] The mitogenic effect is mediated by interaction with specific high-affinity receptors which have been characterized on erythroid precursor cells and permanent cell lines.[159,160] Cloning of the cDNA encoding the erythropoietin receptor revealed a single transmembrane protein with 507 amino acids[41] which showed a significant homology with receptors for interleukins[42-45] as well as growth hormone and prolactin.[47,48] It is believed that the erythropoietin receptor is associated with other membrane proteins forming a multisubunit complex.

In vitro studies have demonstrated receptor-mediated endocytosis of erythropoietin in its target cells in the bone marrow, the erythroid precursor cells.[159,160] Surface binding of ^{125}I-erythropoietin was rapid and followed by a slower rise in internalized radioactivity. Pulse chase experiments showed that degradation products of radiolabeled erythropoietin were released into the medium with corresponding loss of label from the interior of the cell. This degradation was reduced by inhibitors of lysosomal function like NH_4Cl and chloroquine, indicating that erythropoietin is degraded in acidified prelysosomal vesicles.[159] The fate of the receptor after endocytosis is not known. It is possible that the erythropoietin receptor is not degraded and is recycled, since down-regulation of the number of binding sites at the cell surface does not occur to a large extent.

6. Interleukin-2

Interleukin-2 (IL-2) is a polypeptide of molecular weight 15,500, secreted by thymus-derived lymphocytes (T-cells) which have been activated by antigen. IL-2 acts as an autocrine growth factor on T-cells by interaction with a specific IL-2 receptor which is expressed on activated T-cells.[161] The IL-2 receptor has been characterized biochemically as a multisubunit protein composed of at least two transmembrane glycoproteins of M_r 55,000 and 75,000, termed α- and β-subunits, which create the high-affinity binding site by interaction with different regions of the IL-2 molecule. The M_r 55,000 α-subunit was originally identified as the Tac peptide and cDNA encoding the peptide have been isolated and the primary sequence of 251 amino acids determined.[161] Recently, the M_r 75,000 β-subunit was cloned and its sequence revealed similarity with the erythropoietin receptor family.[42] There is evidence for a yet more complex structure that involves a M_r 100,00 peptide in addition to the M_r 55,000 and 75,000 peptides in the multisubunit IL-2 receptor.[161]

IL-2 is internalized via a receptor-mediated pathway and recent evidence indicates that the 55,000 and 75,000 subunits are co-internalized.[162–164] Endocytosed IL-2 is degraded and released as trichloroacetic acid-soluble products. Addition of IL-2 to activated T cells expressing the IL-2 receptor subunits leads to a net reduction in high-affinity IL-2 receptor expression and, thus, a movement towards the resting state of the T-cell. This observation suggests that endocytosed receptor subunits are degraded in lysosomes.[161]

7. Transforming Growth Factor-β

Transforming growth factor-β (TGF-β) is a polypeptide which is synthesized by many different cells which also have specific high-affinity receptors for this peptide. TGF-β is a M_r 25,000, disulfide-linked dimer and three forms are known, named TGF-β1, TGF-β2, and TGF-β1,2, which result from homodimeric and heterodimeric combination of the subunits β1 and β2.[165,166] Cell-surface receptors for TGF-βs have only been partially characterized in biochemical studies and include three forms of molecular weight (65,000, 85,000, and 280,000) which bind the different isomers of TGF-β with varying affinity. They appear to have no intrinsic tyrosine kinase activity.[166] TGF-β is multifunctional since it can either stimulate or inhibit cell proliferation and can either stimulate or inhibit cell differentiation. The actual response induced by TGF-β depends on the cell and its state of differentiation.[165] Like other growth factors, TGF-β is internalized by a receptor-mediated pathway. Following endocytosis, TGF-β is degraded in an endosomal compartment. The TGF-β receptor is resistant to down-regulation by its ligand, indicating that endocytosis is followed by recycling to the plasma membrane.[167,168]

C. NEUROPEPTIDES

Neuropeptides constitute a large class of bioactive peptides which act as neurotransmitters in the central and peripheral nervous system. Based on their occurrence and physiological role in vertebrates, the most prominent members of this family are vasoactive intestinal polypeptide (VIP), cholecystokinin, neuropeptide Y, and the tachykinins. Receptors for neuropeptides have been characterized by their ligand binding, molecular size, signal transduction, and cellular effects. Recently, receptors for substance K, substance P, and neuromedin K (members of the tachykinin family) were cloned and the amino acid sequence of the receptors resolved from the cDNA sequences.[14,28,29] Using the Xenopus oocyte expression system, a new cloning strategy was developed. A stomach cDNA library was constructed with a vector that allowed *in vitro* synthesis of mRNAs. A particular clone was identified by electrophysiological cDNA testing for receptor expression following injection of mRNAs synthesized *in vitro* into the oocyte. Because oocytes injected with exogenous mRNAs can express numerous receptors and channels, this new strategy may be applied in the general molecular cloning of these proteins. The substance K receptor has similarity with the rhodopsin-sequence type receptors (the G-protein-coupled family) and thus possesses multiple membrane-spanning domains.[14,22] Receptor-mediated endocytosis of neuropeptides has only received limited attention so far. One example is the internalization of the VIP receptor which has been studied in more detail.

1. Vasoactive Intestinal Polypeptide

Vasoactive intestinal polypeptide (VIP) is a 28-amino-acid peptide which is widely distributed in vertebrate tissues and acts as a neurotransmitter in the peripheral and central nervous systems.[169] The VIP receptor has been described in numerous tissues and cell lines.[170] The VIP receptor is a plasma membrane glycoprotein of 64 kDa which is linked to a G-protein, and interaction with VIP induces a large increase in the intracellular cyclic AMP level.[171] The internalization of VIP has been studied in human adenocarcinoma HT29 cells. After binding to its receptor, VIP is rapidly internalized and then degraded in the lysosome. This internalization leads to a rapid and transient decrease in the cell-surface receptor number ($t_{1/2} = 2$ min) and simultaneously to a homologous desensitization of adenylate cyclase. The decrease in binding capacity is reversible after removal of VIP and the rate of the recovery ($t_{1/2} = 15$ min), as well as its independence of protein synthesis, suggest that receptors are recycled to the plasma membrane. The biochemical data have been confirmed in ultrastructural studies which showed that radiolabeled VIP is located intracellularly in endosomes and lysosomes, where it is degraded.[170] Receptor-mediated degradation of VIP has also been described in porcine liver cells. VIP was unable to elicit a glycogenolytic response, suggesting that the major function of the hepatic VIP receptor is to internalize and degrade VIP in the portal blood.[172]

In conclusion, the interaction of VIP with its receptor leads to internalization and degradation of VIP and a transient desensitization of the adenylate cyclase to VIP, whereafter the unoccupied receptor is recycled to the plasma membrane. The molecular mechanisms of these phenomena are not understood at present. The VIP-induced desensitization of the cell is, however, similar to the desensitization phenomenon described for β-adrenergic receptors.[173,174] Following the binding of β-adrenergic agonists, the receptors become functionally uncoupled from the G-protein-adenylate cyclase complex and therefore are unable to stimulate cyclic

AMP formation. The desensitization involves covalent modification of the receptor such as phosphorylation. The β-adrenergic receptor is phosphorylated on serine residules by a cyclic AMP-independent protein kinase, which leads to uncoupling of the receptor from the stimulatory G-protein.

D. PLASMA PROTEINS

Receptor-mediated endocytosis has been demonstrated for a number of circulating plasma proteins including LDL,[2] transferrin,[175] asialoglycoproteins,[176] polymeric IgA and IgM,[36] α_2-macroglobulin,[177] as well as phosphomannosylated proteins like lysosomal hydrolases.[84,85] As a common feature of the mechanism of endocytosis of plasma proteins, the receptor is rapidly recycled back to the plasma membrane. In the case of LDL, asialoglycoproteins, and α_2-macroglobulin-protease complex, the ligand is degraded in lysosomes. In contrast, transferrin, polymeric IgA and IgM, and phosphomannosylated proteins are not degraded, but transported inside the cell to different compartments including the plasma membrane or the Golgi. The cellular uptake of these proteins serves various functions including cellular metabolism (LDL, iron-transferrin complex), protein removal and breakdown (asialoglycoproteins and α_2-macroglobulin-protease complex), intracellular targeting (phosphomannosylated proteins), and transepithelial transport (polymeric IgA and IgM). A detailed review of the internalization of plasma proteins is outside the scope of this chapter, but it should be emphasized that our concepts of the molecular and cellular mechanisms of receptor-mediated endocytosis have emerged from these systems. Therefore, major aspects will be summarized.

1. Low-Density Lipoproteins

The low-density lipoprotein (LDL) receptor carries cholesterol-bearing lipoproteins into cells where they are degraded and cholesterol is used in biosynthesis of steroids and other lipids.[1,2] The mechanism of receptor-mediated endocytosis of LDL has been resolved in great detail. The LDL receptors are clustered in coated pits on the cell surface and internalized in endosomes.[70,79] In the endosomal vesicle, the receptor-bound LDL is dissociated at the acid pH, sorted to the lysosome, and degraded; whereas the unoccupied receptor is returned to the cell surface in noncoated vesicles.[178] The LDL receptor endocytosis and recycling is a constitutive process which occurs in the absence of ligand.[2] Cloning of cDNA encoding the bovine and human LDL receptor revealed that the mature receptor (without the signal sequence) consists of 839 amino acids divided into 5 distinguishable domains with the following characteristics: ligand-binding, EGF precursor homology, O-linked sugars, hydrophobic plasma membrane-spanning region, and a cytoplasmic tail.[33] It is now clear that the cytoplasmic portion contains the information necessary for concentrating the LDL receptor in coated pits and facilitating its rapid endocytosis. This is based on analysis of naturally occurring mutations in internalization-defective receptors as well as site-directed mutations and deletions *in vitro* leading to decreased endocytosis.[179,180] The mutational analysis of the cytoplasmic tail of the LDL receptor may result in a general model of the structure-function relationship of clathrin-dependent endocytosis of receptors. Other mutations in the LDL receptor gene result in abnormal receptor biosynthesis with no detectable precursor or no processing of the receptor precursor or a normally processed receptor with abnormal binding of LDL.[2]

2. Transferrin

Transferrin receptors are cell-surface proteins which can bind specifically and facilitate entry of iron-transferrin into cells via receptor-mediated endocytosis. The receptor is an integral membrane glycoprotein composed of two identical subunits (molecular weight 95,000) which are linked by a disulfide bridge to form a dimer.[175] Cloning of the human transferrin receptor cDNA revealed a 760-residue polypeptide with a hydrophobic membrane-spanning region located 61 amino acids from the N-terminus.[34] This predicts that the transferrin receptor is oriented in the membrane with a cytoplasmic N-terminus and an extracellular C-terminus like the asialoglycoprotein receptor,[35] but different from other membrane receptors.[2,56] The deduced amino acid sequence shows that it does not contain an N-terminal hydrophobic signal peptide. This is similar to the asialoglycoprotein receptor, suggesting that it may be linked to their membrane orientation with the N-terminus inside the cell.[35,181] Transcription and translation *in vitro* of wild-type and mutant transferrin receptor cDNA suggested that the transmembrane segment functions as a signal peptide.[182]

The pathway of receptor-mediated endocytosis of transferrin is characterized by recycling both the receptor and ligand. On the cell surface, the transferrin receptor is clustered in coated pits and rapidly internalized. When the transferrin/receptor complex reaches the endosome, the two proteins do not dissociate. In contrast to other receptors like those for LDL, asialoglycoproteins, and EGF, the transferrin receptor fails to dissociate its ligand at pH 5.[81,83] However, iron does dissociate from transferrin at the acidic pH in the endosome while the apo-transferrin remains attached to the receptor. The apo-transferrin/receptor complex returns to the cell surface where the ligand is dissociated at neutral pH and the free receptor binds transferrin. The transferrin receptor is degraded slowly with a half-life greater than 30 h even though it enters the cell every 15 to 20 min independently of added ligand.[80] Like other recycling membrane receptors, the cytoplasmic region has been shown to be necessary for endocytosis.[73]

While the transferrin receptor serves a major function in delivering iron-containing transferrin to intracellular iron stores, an alternative role in regulation of cell growth has been suggested.[175] This is based on a number of observations, including that addition of transferrin is necessary for cell growth, that transferrin-like molecules with growth activity are produced by cell lines, and that transferrin receptor-blocking antibodies can inhibit cell growth. The mitogenic action of transferrin is thought to involve induction of surface growth receptors.

3. Asialoglycoproteins

Desialylated serum glycoproteins have drastically reduced survival times in circulation compared to native forms. A specific receptor on hepatocytes mediates their clearance by recognition of galactose residues made terminal by removal of sialic acid.[176] The capture of asialoglycoproteins at the cell surface is followed by rapid endocytosis of receptor-bound ligand and depletion of surface receptors with a $t_{1/2}$ of 0.5 to 1 min. This is followed by a rapid ($t_{1/2}$ = 2 to 4 min) reappearance of surface receptors originating from endocytosed cell-surface receptors.[80] The endocytosis and recycling of asialoglycoprotein receptors depends on binding of the ligand to its receptor and does not occur in the absence of ligand. Thus, basal recycling of receptors seems only to occur to a small extent compared with LDL receptors.[176] A wide range of glycoproteins are known to be endocytosed by the asialoglycoprotein receptor when sialic acid has been removed from their carbohydrate chains. These include regulatory polypeptides (erythropoietin, FSH, and CG), carrier proteins (ceruloplasmin, transferrin, transcobalamin, transcortin, and thyroglobulin), protein complexers (α_2-macroglobulin and haptoglobulin), and others (IgG, orosomucoid, and prothrombin).[176] Internalized asialoglycoproteins are degraded after accumulation in a lysosomal compartment, whereas the receptor is sorted and recycled to the plasma membrane. The recycling of the receptor is an energy-dependent process.[183] Thus, the role of the asialoglycoprotein receptor is to remove and degrade desialated glycoproteins during their physiological turnover.[176.] Cloning of the asialoglycoprotein receptor revealed that it is a single, transmembrane glycoprotein of 291 amino acid residues with the N-terminus on the cytoplasmic side of the membrane. Like the transferrin, it lacks a signal peptide, suggesting that the transmembrane region directs insertion of the receptor across the membrane.

4. Phosphomannosylated Proteins

The targeting of lysosomal enzymes from their site of synthesis in the endoplasmatic reticulum to lysosomes is mediated by Man-6-P receptors in the Golgi which recognize phosphomannosyl residues on the enzymes.[84] Two types of Man-6-P receptors have been biochemically and structurally characterized.[85] They are distinguished by their molecular size and divalent cation requirements. The cation-independent (CI) Man-6-P receptor is a membrane-associated glycoprotein with an apparent molecular weight of 215,000. The cation-dependent (CD) Man-6-P receptor is also a membrane glycoprotein, but it has an apparent molecular weight of 46,000. Both types of Man-6-P receptors have been cloned and sequence analysis indicated that the CI Man-6-P receptor has 2455 amino acid residues of which only 163 constitute the carboxy-terminal cytoplasmic domain.[37,39] The sequence is identical with the IGF-II receptor.[38] The CD Man-6-P receptor is a single chain protein with 251 residues of which 67 are cytoplasmic.[184] The small Man-6-P receptor exists as a noncovalent dimer in the membrane, whereas the large receptor is a monomer. Both receptors function in the sorting of newly synthesized lysosomal enzymes in the Golgi. In addition, the 215,000 CI Man-6-P receptor is also present on the cell surface where it functions in endocytosis of extracellular phosphorylated lysosomal enzymes as well as IGF-II.

The receptor-mediated endocytosis of ligands carrying the Man-6-P recognition marker has been described in great detail.[85] The receptor undergoes rapid endocytosis via clathrin-coated pits and recycling via uncoated endosomes both in the absence and presence of ligand. The lysosomal enzymes are released at the acid pH in the endosome and transported to the Golgi or the lysosome where they are mixed in the pools of endogenous enzymes. The CI Man-6-P receptors are reutilized and can undergo many rounds of ligand delivery. In addition, all CI Man-6-P receptors in the cell are in rapid equilibrium, suggesting that there is only one pool of receptor and that a single CI Man-6-P receptor functions in both the endocytic and biosynthetic pathways.[85] Cloning of the human IGF-II receptor revealed that it is identical to the mammalian CI Man-6-P receptor.[37–39] The identity of the two receptors has been confirmed by biochemical studies which show that this protein can bind and internalize phosphomannosyl residues and IGF-II (a nonglycosylated polypeptide) simultaneously, and that each ligand may influence the binding of the other ligand to the receptor.[141,145,148,185] The biological significance of the presence of Man-6-P and IGF-II binding sites on one membrane receptor is still uncertain. The CI Man-6-P (IGF-II) receptor binds and internalizes IGF-II at the cell surface, resulting in the degradation of the ligand.[145,150] In addition, several reports suggest that IGF-II mediates a response through its own receptor perhaps via coupling to a pertussis toxin-sensitive G-protein.[148] Man-6-P has also been shown to stimulate the division of neuronal precursor cells via the CI Man-6-P receptor.[149] It is also possible that IGF-II may modulate the trafficking of lysosomal enzymes by preventing the binding of enzymes at the cell surface or by altering the intracellular trafficking and distribution of the CI Man-6-P receptor. It may be concluded, however, that the intriguing finding that IGF-II binds to the CI Man-6-P receptor has not yet been explained in functional terms.

REFERENCES

1. **Goldstein, J. L. and Brown, M. S.,** Binding and degradation of low density lipoproteins by cultured human fibroblasts: comparison of cells from a normal subject and from a patient with homozygous familial hypercholesterolemia, *J. Biol. Chem.,* 249, 5153, 1974.
2. **Goldstein, J. L., Brown, M. S., Anderson, R. G. W., Russell, D. W., and Schneider, W. J.,** Receptor-mediated endocytosis: concepts emerging from the LDL receptor system, *Annu. Rev. Cell Biol.,* 1, 1, 1985.
3. **Carpenter, G. and Cohen, S.,** Epidermal growth factor, *Annu. Rev. Biochem.,* 48, 193, 1979.
4. **Gorden, Ph., Carpentier, J.-L., Freychet, P., and Orci, L.,** Internalization of polypeptide hormones. Mechanism, intracellular localization and significance, *Diabetologia,* 18, 263, 1980.
5. **Sonne, O.,** Receptor-mediated endocytosis and degradation of insulin, *Physiol. Rev.,* 68, 1129, 1988.
6. **Carpenter, G.,** Receptors for epidermal growth factor and other polypeptide mitogens, *Annu. Rev. Biochem.,* 56, 881, 1987.
7. **Schlessinger, J.,** The epidermal growth factor receptor as a multifunctional allosteric protein, *Biochemistry,* 27, 3119, 1988.
8. **Mishina, M., Kurosaki, T., Tobimatsu, T., Morimoto, Y., Noda, M., Yamamoti, T., Terao, M., Lindstrom, J., Takahashi, T., Kuno, M., and Numa, S.,** Expression of functional acetylcholine receptor from cloned CDNAs, *Nature (London),* 307, 604, 1984.
9. **Ullrich, A., Coussens, L., Hayflick, J. S., Dull, T. J., Gray, A., Tam, A. W., Lee, J., Yarden, Y., Libermann, T. A., Schlessinger, J., Downward, J., Mayes, E. L. V., Whittle, N., Waterfield, M. D., and Seeburg, P. H.,** Human epidermal growth factor receptor cDNA sequence and aberrant expression of the amplified gene in A 431 epidermoid carcinoma cells, *Nature (London),* 309, 418, 1984.
10. **Coussens, L., Yang-Feng, T. L., Liao, Y.-C., Chen, E., Gray, A., McGrath, J., Seeburg, P. H., Libermann, T. A., Schlessinger, J., Francke, U., Levinson, A., and Ullrich, A.,** Tyrosine kinase receptor with extensive homology to EGF receptor shares chromosomal location with neu oncogene, *Science,* 230, 1132, 1985.
11. **Schechter, A. L., Stern, D. F., Vaidyanathan, L., Decker, S. J., Drebin, J., Greene, M. I., and Weinberg, R. A.,** The neu oncogene: an erb-B-related gene encoding a 185,000-M_r tumour antigen, *Nature (London),* 312, 513, 1984.
12. **Yamamoto, T., Ikawa, S., Akiyama, T., Semba, K., Nomura, N., Miyajima, N., Saito, T., and Toyoshima, K.,** Similarity of protein encoded by the human c-erb-B 2 gene to epidermal growth factor receptor, *Nature (London),* 319, 230, 1986.
13. **Parmentier, M., Libert, F., Maenhaut, C., Lefort, A., Gérard, C., Perret, J., Van Sande, J., Dumont, J. E., and Vassart, G.,** Molecular cloning of the thyrotropin receptor, *Science,* 246, 1620, 1989.
14. **Masu, Y., Nakayama, K., Tamaki, H., Harada, Y., Kuno, M., and Nakanishi, S.,** cDNA cloning of bovine substance-K receptor through oocyte expression system, *Nature (London),* 329, 836, 1987.
15. **Ullrich, A., Bell, J. R., Chen, E. Y., Herrera, R., Petruzzelli, L. M., Dull, T. J., Gray, A., Coussens, L., Liao, Y.-C., Tsubokawa, M., Mason, A., Seeburg, P. H., Grunfeld, C., Rosen, O. M., and Ramachandran, J.,** Human insulin receptor and its relationship to the tyrosine kinase family of oncogenes, *Nature (London),* 313, 756, 1985.
16. **Ebina, Y., Ellis, L., Jarnagin, K., Edery, M., Graf, L., Clauser, E., Ou, J., Masiarz, F., Kan, Y. W., Goldfine, I. D., Roth, R. A., and Rutter, W. J.,** The human insulin receptor cDNA: the structural basis for hormone-activated transmembrane signalling, *Cell,* 40, 747, 1985.
17. **Ullrich, A., Gray, A., Tam, A. W., Yang-Feng, T., Tsubokawa, M., Colins, C., Henzel, W., Le Bon, T., Kathuria, S., Chen, E., Jacobs, S., Francke, U., Ramachandran, J., and Fujita-Yamaguchi, Y.,** Insulin-like growth factor I receptor primary structure: comparison with insulin receptor suggests structural determinants that define functional specificity, *EMBO J.,* 5, 2503, 1986.
18. **Lee, P. L., Johnson, D. E., Cousens, L. S., Fried, V. A., and Williams, L. T.,** Purification and complementary DNA cloning of a receptor for basic fibroblast growth factor, *Science,* 245, 57, 1989.
19. **Yarden, Y., Escobedo, J. A., Kuang, W.-J., Yang-Feng, T. L., Daniel, T. O., Tremble, P. M., Chen, E. Y., Ando, M. E., Harkins, R. N., Francke, U., Fried, V. A., Ullrich, A., and Williams, L. T.,** Structure of the receptor for platelet-derived growth factor helps define a family of closely related growth factor receptors, *Nature (London),* 323, 226, 1986.
20. **Claesson-Welsh, L., Eriksson, A., Westermark, B., and Heldin, C.-H.,** cDNA cloning and expression of the human A type PDGF receptor establishes structural similarity to the B type PDGF receptor, *Proc. Natl. Acad. Sci. U.S.A.,* 86, 4917, 1989.
21. **Sherr, C. J., Rettenmier, C. W., Sacca, R., Roussel, M. F., Look, A. T., and Stanley, E. R.,** The c-fms proto-oncogene product is related to the receptor for the mononuclear phagocyte growth factor, CSF-1, *Cell,* 41, 665, 1985.
22. **Nathans, J. and Hogness, D. S.,** Isolation, sequence analysis, and intron-exon arrangement of the gene encoding bovine rhodopsin, *Cell,* 34, 307, 1983.

23. **Kobilka, B. K., Dixon, R. A. F., Frielle, T., Dohlman, H. G., Bolanowski, M. A., Sigal, I. S., Yang-Feng, T. L., Francke, U., Caron, M. G., and Lefkowitz, R. J.,** cDNA for the human β_2-adrenergic receptor: a protein with multiple membrane-spanning domains and encoded by a gene whose chromosomal location is shared with that of the receptor for platelet-derived growth factor, *Proc. Natl. Acad. Sci. U.S.A.,* 84, 46, 1987.
24. **Kubo, T., Fukuda, K., Mikami, A., Maeda, A., Takahashi, H., Mishina, M., Haga, T., Haga, K., Ichiyama, A., Kangawa, K., Kojima, M., Matsuo, H., Hirose, T., and Numa, S.,** Cloning, sequencing and expression of complementary DNA encoding the muscarinic acetylcholine receptor, *Nature (London),* 323, 411, 1986.
25. **Julius, D., MacDermott, A. B., Axel, R., and Jessell, T. M.,** Molecular characterization of a functional cDNA encoding the serotonin 1c receptor, *Science,* 241, 558, 1988.
26. **Bunzow, J. R., Van Tol, H. H. M., Grandy, D. K., Albert, P., Salon, J., Christie, M., Machida, C. A., Neve, K. A., and Civelli, O.,** Dopanine receptors. Cloning and expression of a rat D_2 dopamine receptor cDNA, *Nature (London),* 336, 783, 1988.
27. **Jackson, T. R., Blair, L. A. C., Marshall, J., Goedert, M., and Hanley, M. R.,** Angiotensin receptor. The mas oncogene encodes an angiotensin receptor, *Nature (London),* 335, 437, 1988.
28. **Yokota, Y., Sasai, Y., Tanaka, K., Fujiwara, T., Tsuchida, K., Shigemoto, Kakizuka, A., Ohkubo, H., and Nakanishi, S.,** Molecular characterization of a function cDNA for rat substance P receptor, *J. Biol. Chem.,* 264, 17649, 1989.
29. **Shigemoto, R., Yokota, Y., Tsuchida, K., and Nakanishi, S.,** Cloning and expression of a rat neuromedin K receptor cDNA, *J. Biol. Chem.,* 265, 1990, in press.
30. **McFarland, K. C., Sprengel, R., Phillips, H. S., Köhler, M., Rosemblit, N., Nikolics, K., Segaloff, D. L., and Seeburg, P. H.,** Lutropin-choriongonadotropin receptor: an unusual member of the G protein-coupled receptor family, *Science,* 245, 494, 1989.
31. **Garbers, D. L.,** Guanylate cyclase, a cell surface receptor, *J. Biol. Chem.,* 264, 9103, 1989.
32. **Radeke, M. J., Misko, T. P., Hsu, C., Herzenberg, L. A., and Shooter, E. M.,** Gene transfer and molecular cloning of the rat nerve growth factor receptor, *Nature (London),* 325, 593, 1987.
33. **Yamamoto, T., Davis, C. G., Brown, M. S., Schneider, W. J., Casey, M. L., Goldstein, J. L., and Russell, D. W.,** The human LDL receptor: a cysteine-rich protein with multiple alu sequences in its mRNA, *Cell,* 39, 27, 1984.
34. **McClelland, A., Kühn, L. C., and Ruddle, F. H.,** The human transferrin receptor gene: genomic organization, and the complete primary structure of the receptor deduced from a cDNA sequence, *Cell,* 39, 267, 1984.
35. **Spiess, M., Schwartz, A. L., and Lodish, H. F.,** Sequence of human asialoglycoprotein receptor cDNA, *J. Biol. Chem.,* 260, 1979, 1985.
36. **Mostov, K. E., Friedlander, M., and Blobel, G.,** The receptor for transepithelial transport of IgA and IgM contains multiple immunoglobulin-like domains, *Nature (London),* 308, 37, 1984.
37. **Lobel, P., Dahns, N. M., and Kornfeld, S.,** Cloning and sequence analysis of the cation-independent mannose-6-phosphate receptor, *J. Biol. Chem.,* 263, 2563, 1988.
38. **Morgan, D. O., Edman, J. C., Standring, D. N., Fried, V. A., Smith, M. C., Roth, R. A., and Rutter, W. J.,** Insulin-like growth factor II receptor as a multifunctional binding protein, *Nature (London),* 329, 301, 1987.
39. **Oshima, A., Nolan, C. M., Kyle, J. W., Grubb, J. H., and Sly, W. S.,** The human cation-independent mannose-6-phosphate receptor. Cloning and sequence of the full-length cDNA and expression of functional receptor in COS cells. *J. Biol. Chem.,* 263, 2553, 1988.
40. **Smith, C. A., Davis, T., Anderson, D., Solam, L., Beckmann, M. P., Jerzy, R., Dower, S. K., Cosman, D., and Goodwin, R. G.,** A receptor for tumor necrosis factor defines an unusual family of cellular and viral proteins, *Science,* 248, 1019, 1990.
41. **D'Andrea, A. D., Lodish, H. F., and Wong, G. G.,** Expression cloning of the murine erythropoietin receptor, *Cell,* 57, 277, 1989a.
42. **Hatakeyama, M., Tsudo, M., Minamoto, S., Kono, T., Doi, T., Miyata, T., Miyasaka, M., and Taniguchi, T.,** Interleukin-2 receptor β chain gene: generation of three receptor forms by cloned human α and β chain cDNA's, *Science,* 244, 551, 1989.
43. **Mosley, B., Beckmann, M. P., March, C. J., Idzerda, R. L., Gimpel, S. D., Van den Bos, T., Friend, D., Alpert, A., Anderson, D., Jackson, J., Wignall, J. M., Smith, C., Gallis, B., Sims, J. E., Urdal, D., Widmer, M. B., Cosman, D., and Park, L. S.,** The murine interleukin-4 receptor: molecular cloning and characterization of secreted and membrane bound forms, *Cell,* 59, 335, 1989.
44. **Yamasaki, K., Taga, T., Hirata, Y., Yawata, H., Kawanishi, Y., Seed, B., Taniguchi, T., Hirano, T., and Kishimoto, T.,** Cloning and expression of the human interleukin-6 (BSF-2/IFNβ) receptor, *Science,* 241, 825, 1988.
45. **Goodwin, R. G., Friend, D., Ziegler, S. F., Jerzy, R., Falk, B. A., Gimpel, S., Cosman, D., Dower, S. K., March, C. J., Namen, A. E., and Park, L. S.,** Cloning of the human and murine interleukin-7 receptors: demonstration of a soluble form and homology to a new receptor superfamily, *Cell,* 60, 941, 1990.

46. **Gearing, D. P., King, J. A., Gough, N. M., and Nicola, N. A.,** Expression cloning of a receptor for human granulocyte-macrophage colony-stimulating factor, *EMBO J.*, 8, 3667, 1989.
47. **Leung, D. W., Spencer, S. A., Cachianes, G., Hammonds, R. G., Collins, C., Henzel, W. J., Barnard, R., Waters, M. J., and Wood, W. I.,** Growth hormone receptor and serum binding protein: purification, cloning and expression, *Nature (London)*, 330, 537, 1987.
48. **Edery, M., Jolicoeur, C., Levi-Meyrueis, C., Dusanter-Fourt, I., Petridou, B., Boutin, J.-M., Lesueur, L., Kelley, P. A., and Djiane, J.,** Identification and sequence analysis of a second form of prolactin receptor by molecular cloning of complementary DNA from rabbit mammary gland, *Proc. Natl. Acad. Sci. U.S.A.*, 86, 950, 1989.
49. **Schofield, P. R., Darlison, M. G., Fujita, N., Burt, D. R., Stephenson, F. A., Rodriquez, H., Rhee, L. M., Ramachandran, J., Reale, V., Glencorse, T. A., Seeburg, P. H., and Barnard, E. A.,** Sequence and functional expression of the GABA receptor shows a ligand-gated receptor super-family, *Nature (London)*, 328, 221, 1987.
50. **Schmieden, V., Grenningloh, Schofield, P. R., and Betz, H.,** Functional expression in Xenopus oocytes of the strychnine binding 48 kd subunit of the glycine receptor, *EMBO J.*, 8, 695, 1989.
51. **Hollmann, M., O'Shea-Greenfield, A., Rogers, S. W., and Heinemann, S.,** Cloning by functional expression of a member of the glutamate receptor family, *Nature (London)*, 342, 643, 1989.
52. **Evans, R. M.,** The steroid and thyroid hormone receptor superfamily, *Science*, 240, 889, 1988.
53. **Yarden, Y. and Ullrich, A.,** Molecular analysis of signal transduction by growth factors, *Biochemistry*, 27, 3113, 1988.
54. **Gilman, A. G.,** G proteins: transducers of receptor-generated signals, *Annu. Rev. Biochem.*, 56, 615, 1987.
55. **Catterall, W. A.,** Molecular properties of voltage-sensitive sodium channels, *Annu. Rev. Biochem.*, 55, 953, 1986.
56. **Yarden, Y. and Ullrich, A.,** Growth factor receptor tyrosine kinases, *Annu. Rev. Biochem.*, 57, 443, 1988.
57. **Williams, L. T.,** Signal transduction by the platelet-derived growth factor receptor, *Science*, 243, 1564, 1989.
58. **Gammeltoft, S. and Van Obberghen, E.,** Protein kinase activity of the insulin receptor, *Biochem. J.*, 235, 1, 1986.
59. **Rosen, O. M.,** After insulin binds, *Science*, 237, 1452, 1987.
60. **Kahn, C. R.,** The molecular mechanism of insulin action, *Annu. Rev. Med.*, 36, 429, 1985.
61. **Berridge, M. J.,** Inositol triphosphate and diacylglycerol: two interacting second messengers, *Annu. Rev. Biochem.*, 56, 159, 1987.
62. **Rowland, E. A., Müller, T. H., Goldstein, M., and Greene, L. A.,** Cell-free detection and characterization of a novel nerve growth factor-activated protein kinase in PC12 cells, *J. Biol. Chem.*, 262, 7504, 1987.
63. **Nicola, N. A.,** Hemopietic cell growth factors and their receptors, *Annu. Rev. Biochem.*, 58, 45, 1989.
64. **Besterman, J. M. and Low, R. B.,** Endocytosis: a review of mechanisms and plasma membrane dynamics, *Biochem. J.*, 210, 1, 1983.
65. **Van Deurs, B., Petersen, O. W., Olsnes, S., and Sandvig, K.,** The ways of endocytosis, *Int. Rev. Cytol.*, 117, 131, 1989.
66. **Mellman, I., Fuchs, R., and Helenius, A.,** Acidification of the endocytic and exocytic pathways, *Annu. Rev. Biochem.*, 55, 663, 1986.
67. **Pastan, I. H. and Willingham, M. C.,** Receptor-mediated endocytosis of hormones in cultured cells, *Annu. Rev. Physiol.*, 43, 239, 1981.
68. **Pearse, B. M. F. and Crowther, R. A.,** Structure and assembly of coated vesicles, *Annu. Rev. Biophys. Chem.*, 16, 49, 1987.
69. **Pearse, B. M. F.,** Coated vesicles from pig brain: purification and biochemical characterization, *J. Mol. Biol.*, 97, 93, 1975.
70. **Anderson, R. G. W., Goldstein, J. L., and Brown, M. S.,** Localization of low density lipoprotein receptors on plasma membrane of normal human fibroblasts and their absence in cells from a familial hypercholesterolemia homozygote, *Proc. Natl. Acad. Sci. U.S.A.*, 73, 2434, 1976.
71. **Goldberg, R. I., Smith, R. M., and Jarett, L.,** Insulin and alpha-2-macroglobulin-methylamine undergo endocytosis by different mechanisms in rat adipocytes. I. Comparison of cell surface events, *J. Cell Physiol.*, 133, 203, 1987.
72. **Pearse, B. M. F.,** Receptors compete for adaptors found in plasma membrane-coated pits, *EMBO J.*, 7, 3331, 1988.
73. **Glickman, J. N., Conibear, and Pearse, B. M. F.,** Specificity of binding of clathrin adaptors to signals on the mannose-6-phosphate/insulin-like growth factor II receptor, *EMBO J.*, 8, 1041, 1989.
74. **Pearse, B. M. F.,** Assembly of the mannose-6-phosphate receptor into reconstituted clathrin coats, *EMBO J.*, 4, 2457, 1985.
75. **Honegger, A. M., Dull, T. J., Felder, S., Van Obberghen, E., Bellot, F., Szapary, D., Schmidt, A., Ullrich, A., and Schlessinger, J.,** Point mutation at the ATP binding site of EGF receptor abolishes protein-tyrosine kinase activity and alters cellular routing, *Cell*, 51, 199, 1987.

76. **Chen, W. S., Lazar, C. S., Lund, K. A., Welsh, J. B., Chang, C.-P., Walton, G. M., Der, C. J., Wiley, H. S., Gill, G. N., and Rosenfield, M. G.,** Functional independence of the epidermal growth factor receptor from a domain required for ligand-induced internalization and calcium regulation, *Cell,* 59, 33, 1989.
77. **Fehlmann, M., Carpentier, J.-L., Van Obberghen, E., Freychet, P., Thamm, P., Saunders, D., Brandenburg, D., and Orci, L.,** Internalized insulin receptors are recycled to the cell surface in rat hepatocytes, *Proc. Natl. Acad. Sci. U.S.A.,* 79, 1187, 1982.
78. **Knutson, V. P., Ronnett, G. V., and Lane, M. D.,** Rapid, reversible internalization of cell surface insulin receptors. Correlation with insulin-induced down-regulation, *J. Biol. Chem.,* 158, 12139, 1983.
79. **Anderson, R. G. W., Brown, M. S., and Goldstein, J. L.,** Role of the coated endocytic vesicle in the uptake of receptor-bound low density lipoprotein in human fibroblasts, *Cell,* 10, 351, 1977.
80. **Chiechanover, Aa., Schwartz, A. L., and Lodish, H. F.,** The asialoglycoprotein receptor internalizes and recycles independently of the transferrin and insulin receptors, *Cell,* 32, 267, 1983.
81. **Dautry-Varsat, A., Chiechanover, Aa., and Lodish, H. F.,** pH and the recycling of transferrin during receptor-mediated endocytosis, *Proc. Natl. Acad. Sci. U.S.A.,* 80, 2258, 1983.
82. **Hanover, J. A., Willingham, M. C., and Pastan, I.,** Kinetics of transit of transferrin and epidermal growth factor through clathrin-coated membranes, *Cell,* 39, 283, 1984.
83. **Klausner, R. D., Ashwell, G., Van Renswoude, J., Harford, J. B., and Bridges, K. R.,** Binding of apotransferrin to K562 cells: explanation of the transferrin cycle, *Proc. Natl. Acad. Sci. U.S.A.,* 80, 2263, 1983.
84. **Von Figura, K. and Hasilik, A.,** Lysosomal enzymes and their receptors, *Annu. Rev. Biochem.,* 55, 167, 1986.
85. **Dahms, N. M., Lobel, P., and Kornfeld, S.,** Mannose 6-phosphate receptors and lysosomal enzyme targeting, *J. Biol. Chem.,* 264, 12115, 1989.
86. **Gammeltoft, S.,** Insulin receptors: binding kinetics and structure-function relationship of insulin, *Physiol. Rev.,* 64, 1321, 1984.
87. **Bergeron, J. J. M., Sikstrom, R., Hand, A. R., and Posner, B. I.,** Binding and uptake of ^{125}I-insulin into rat liver hepatocytes and endothelium. An *in vivo* radioautographic study, *J. Cell Biol.,* 80, 427, 1979.
88. **Carpentier, J.-L., Gorden, P., Freychet, P., Le Cam, A., and Orci, L.,** Lysosomal association of internalized ^{125}I-insulin in isolated rat hepatocytes. Direct demonstration by quantitative electron microscopic autoradiography, *J. Clin. Invest.,* 63, 1249, 1979.
89. **Smith, R. M. and Jarett, L.,** Receptor-mediated endocytosis and intracellular processing of insulin: ultrastructural and biochemical evidence for cell-specific heterogeneity and distinction from nonhormonal ligands, *Lab. Invest.,* 58, 613, 1988.
90. **Posner, B. I., Patel, B. A., Khan, M. N., and Bergeron, J. J. M.,** Effect of chloroquine on the internalization of ^{125}I-insulin into subcellular fractions of rat liver. Evidence for an effect of chloroquine on Golgi elements, *J. Biol. Chem.,* 257, 5789, 1982.
91. **Khan, M. N., Posner, B. I., Kahn, R. J., and Bergeron, J. J. M.,** Internalization of insulin into rat liver Golgi elements. Evidence for vesicle heterogeneity and the path of intracellular processing, *J. Biol. Chem.,* 257, 5969, 1982.
92. **Sonne, O. and Simpson, I. A.,** Internalization of insulin and its receptor in the isolated rat adipose cell. Time-course and insulin concentration dependency, *Biochim. Biophys. Acta,* 804, 404, 1984.
93. **Kahn, C. R. and Baird, K.,** The fate of insulin bound to adipocytes. Evidence for compartmentalization and processing, *J. Biol. Chem.,* 253, 4900, 1978.
94. **Levy, J. R. and Olefsky, J. M.,** The trafficking and processing of insulin and insulin receptors in cultured rat hepatocytes, *Endocrinology,* 121, 1075, 1987.
95. **Marshall, S.,** Degradative processing of internalized insulin in isolated adipocytes, *J. Biol. Chem.,* 260, 13517, 1985.
96. **Terris, S. and Steiner, D. F.,** Binding and degradation of ^{125}I-insulin by rat hepatocytes, *J. Biol. Chem.,* 250, 8389, 1975.
97. **Gammeltoft, S., Kristensen, L. Ø., and Sestoft, L.,** Insulin receptors in isolated rat hepatocytes. Reassessment of binding properties and observations on the inactivation of insulin at 37°C, *J. Biol. Chem.,* 253, 8406, 1978.
98. **Hofmann, C., Marsh, J. W., Miller, B., and Steiner, D. F.,** Cultured hepatoma cells as a model system for studying insulin processing and biologic responsiveness, *Diabetes,* 29, 865, 1980.
99. **Duckworth, W. C. and Kitabchi, A. E.,** Insulin metabolism and degradation, *Endocrinol. Rev.,* 2, 210, 1981.
100. **Affholter, J. A., Fried, V. A., and Roth, R. A.,** Human insulin-degrading enzyme shares structural and functional homologies with *E. coli* protease III, *Science,* 242, 1415, 1988.
101. **Assoian, R. K. and Tager, H. S.,** Peptide intermediates in the cellular metabolism of insulin, *J. Biol. Chem.,* 257, 9078, 1982.
102. **Shii, K. and Roth, R. A.,** Inhibition of insulin degradation by hepatoma cells after microinjection of monoclonal antibodies to a specific cytosolic protease, *Proc. Natl. Acad. Sci. U.S.A.,* 83, 4147, 1986.
103. **Marshall, S.,** Dual pathways for the intracellular processing of insulin, *J. Biol. Chem.,* 260, 13524, 1985.

104. **Marshall, S.,** Kinetics of insulin receptor internalization and recycling in adipocytes. Shunting of receptors to a degradative pathway by inhibitors of recycling, *J. Biol. Chem.*, 260, 4136, 1985.

105. **Banskota, N. K., Carpentier, J.-L., and King, G. L.,** Processing and release of insulin and insulin-like growth factor I by macro- and microvascular endothelial cells, *Endocrinology,* 119, 1904, 1986.

106. **Ronnett, G. V., Knutson, V. P., and Lane, M. D.,** Insulin-induced down-regulation of insulin receptors in 3T3-L1 adipocytes. Altered rate of receptor inactivation, *J. Biol. Chem.*, 257, 4285, 1982.

107. **McClain, D. A., Maegawa, H., Lee, J., Dull, T. J., Ulrich, A., and Olefsky, J. M.,** A mutant insulin receptor with defective tyrosine kinase displays no biologic activity and does not undergo endocytosis, *J. Biol. Chem.*, 262, 14663, 1987.

108. **Wilden, P. A., Backer, J. M., Kahn, C. R., Cahill, D. A., Schroeder, G. J., and White, M. F.,** The insulin receptor with phenylalanine replacing tyrosine-1146 provides evidence for separate signals regulating cellular metabolism and growth, *Proc. Natl. Acad. Sci. U.S.A.*, 87, 3358, 1990.

109. **McClain, D. A., Maegawa, H., Levy, J., Hueckstead, T., Dull, T. J., Lee, J., Ullrich, A., and Olefsky, J. M.,** Properties of a human insulin receptor with a COOH-terminal truncation. I. Insulin binding, autophosphorylation, and endocytosis. *J. Biol. Chem.*, 263, 8904, 1988.

110. **Khan, M. N., Baquiran, G., Brule, C., Burgess, J., Foster, B., Bergeron, J. J. M., and Posner, B. I.,** Internalization and activation of the rat liver insulin receptor kinase *in vivo, J. Biol. Chem.*, 264, 12931, 1989.

111. **Backer, J. M., Kahn, C. R., and White, M. F.,** Tyrosine phsophorylation of the insulin receptor during insulin-stimulated internalization in rat hepatoma cells, *J. Biol. Chem.*, 264, 1694, 1989.

112. **Rodbell, M.,** Glucagon, *Nature (London),* 284, 17, 1980.

113. **Sonne, O. and Gliemann,** The fate of receptor-bound glucagon in rat adipocytes. In *Hormone-Receptors in Digestion and Nutrition,* Rosselin, G., Fromageot, P., and Bonfils S., Eds., Elsevier, Amsterdam, 1979, 319.

114. **Barrazzone, P., Gorden, P., Carpentier, J.-L., Orci, L., Freychet, P., and Canivet, B.,** *J. Clin. Invest.*, 66, 1081, 1980.

115. **Hagopian, W. A. and Tager, S.,** Receptor binding and cell-mediated metabolism of [^{125}I]monoiodoglucagon by isolated canine hepatocytes, *J. Biol. Chem.*, 259, 8986, 1984.

116. **Reichlin, S.,** Somatostatin, *N. Engl. J. Med.*, 309, 1495, 1983.

117. **Schonbrunn, A., Dorflinger, L. J., and Koch, B. C.,** Mechanisms of somatostatin action in pituitary cells, *Adv. Exp. Med. Biol.*, 188, 305, 1985.

118. **Presky, D. H. and Schonbrunn, A.,** Receptor-bound somatostatin and epidermal growth factor are processed differently in GH_4C_1 rat pituitary cells, *J. Cell Biol.*, 102, 878, 1986.

119. **Kiess, W. and Butenandt, O.,** Specific growth hormone receptors on human peripheral mononuclear cells: reexpression, identification, and characterization, *J. Clin. Endocrin. Metabol.*, 60, 740, 1985.

120. **Roupas, P. and Herington, A. C.,** Processing of growth hormone by rat adipocytes in primary culture: differentiation between release of intact hormone and degradative processing, *Endocrinology,* 121, 1521, 1987.

121. **Møldrup, A., Billestrup, N., Thorn, N. A., Lernmark, Å., and Nielsen, J. H.,** Multiple growth hormone-binding proteins are expressed on insulin-producing cells, *Mol. Endocrinol.*, 3, 1173, 1989.

122. **Roupas, P. and Herington, A. C.,** Intracellular processing of growth hormone receptors by adipocytes in primary culture, *Mol. Cell. Endocrinol.*, 57, 93, 1988.

123. **Sanford, J. C. and Batten, B. E.,** Endocytosis of follicle-stimulating hormone by ovarian granulosa cells: analysis of hormone processing and receptor dynamics, *J. Cell. Physiol.*, 138, 154, 1989.

124. **Ahmed, C., Sawyer, H., and Niswender, G.,** Internalization and degradation of human chorionic gonadotropin in ovine luteal cells: kinetic studies, *Endocrinology,* 109, 1380, 1981.

125. **Ascoli, M.,** Lysosomal accumulation of the hormone-receptor complex during receptor-mediated endocytosis of human chorionogonadotropin, *J. Cell Biol.*, 99, 1242, 1984.

126. **Mock, E. J. and Niswender, G. D.,** Differences in the rates of internalization of ^{125}I-human choirionic gonadotropin, luteinizing hormone and epidermal growth factor by ovine leutal cells, *Endocrinology,* 113, 259, 1983.

127. **Mock, E., Papkoff, H., and Niswender, G.,** Internalization of ovine luteinizing hormone/human chorionic gonadotropin recombinants: differential effects of the α and β subunits, *Endocrinology,* 113, 265, 1983.

128. **Robinson, M., Rhodes, J. A., and Albertini, D. F.,** Slow internalization of human chorionic gonadotropin by cultured granulosa cells, *J. Cell. Physiol.*, 117, 43, 1983.

129. **Fletcher, P. and Reichert, L.,** Cellular processing of follicle-stimulating hormone by Sertoli cells in serum-free culture, *Mol. Cell. Endocrinol.*, 34, 39, 1984.

130. **Carpenter, G. and Cohen, S.,** ^{125}-I labeled human epidermal growth factor. Binding, internalization, and degradation in human fibroblasts, *J. Cell Biol.*, 71, 159, 1976.

131. **Shechter, Y., Schlessinger, J., Jacobs, S., Chang, K.-J., and Cuatrecasas, P.,** Fluorescent labeling of hormone receptors in viable cells: preparation and properties of highly fluorescent derivatives of epidermal growth factor and insulin, *Proc. Natl. Acad. Sci. U.S.A.*, 75, 2135, 1978.

132. **Schlessinger, J., Shechter, Y., Willingham, M. C., and Pastan, I.,** Direct visualization of binding, aggregation, and internalization of insulin and epidermal growth factor on living fibroblastic cells, *Proc. Natl. Acad. Sci. U.S.A.,* 75, 2659, 1978.
133. **Schlessinger, J.,** Allosteric regulation of the epidermal growth factor receptor kinase, *J. Cell Biol.,* 103, 2067, 1986.
134. **Felder, S., Miller, K., Moehren, G., Ullrich, A., Schlessinger, J., and Hopkins, C. R.,** Kinase activity controls the sorting of the epidermal growth factor receptor within the multivesicular body, *Cell,* 61, 523, 1990.
135. **Kay, D. G., Lai, W. H., Uchihashi, M., Khan, M. N., Posner, B. I., and Bergeron, J. J. M.,** Epidermal growth factor receptor kinase translocation and activation *in vivo, J. Biol. Chem.,* 261, 8473, 1986.
136. **Heldin, C.-H. and Westermark, B.,** Growth factors: mechanism of action and relation to oncogenes, *Cell,* 37, 9, 1984.
137. **Heldin, C.-H., Bäckström, G., Östman, A., Hammacher, A., Rönnstrand, L., Rubin, K., Nistér, M., and Westermark, B.,** Binding of different dimeric forms of PDGF to human fibroblasts: evidence for two separate receptor types, *EMBO J.,* 7, 1387, 1988.
138. **Nilsson, J., Thyberg, J., Heldin, C.-H., Westermark, B., and Wasterson, Å.,** Surface binding and internalization of platelet-derived growth factor in human fibroblasts, *Proc. Natl. Acad. Sci. U.S.A.,* 80, 5592, 1983.
139. **Rosenfeldt, M. E., Bowen-Pope, D. F., and Ross, R.,** Platelet-derived growth factor: morphologic and biochemical studies of binding, internalization, and degradation, *J. Cell. Physiol.,* 121, 263, 1984.
140. **Heldin, C.-H., Wasteson, Å., and Westermark, B.,** Interaction of platelet-derived growth factor with its fibroblast receptor, *J. Biol. Chem.,* 257, 4216, 1982.
141. **Gammeltoft, S.,** Insulin-like growth factors and insulin: gene expression, receptors and biological actions, in *Peptide Hormones as Prohormones. Processing, Biological Activity and Pharmacology,* Martinez, J., Ed., Ellis Horwood Limited, Chichester, England, 1989, 176.
142. **Czech, M. P.,** Signal transmission by the insulin-like growth factors, *Cell,* 59, 235, 1989.
143. **Gammeltoft, S., Haselbacher, G. K., Humbel, R. E., Fehlmann, M., and Van Obberghen, E.,** Two types of receptor for insulin-like growth factors in mammalian brain, *EMBO J.,* 4, 3407, 1985.
144. **Daughaday, W. H. and Rotwein, P.,** Insulin-like growth factors I and II. Peptide messenger ribonucleic acid and gene structures, serum, and tissue concentrations, *Endocrine Rev.,* 10, 68, 1989.
145. **Nielsen, F. C., Wang, E., and Gammeltoft, S.,** Receptor binding, endocytosis and mitogenesis of insulin-like growth factors I and II in fetal rat brain neurons, *J. Neurochem.,* 56, 12, 1991.
146. **Furlanetto, R. W.,** Receptor-mediated endocytosis and lysosomal processing of insulin-like growth factor I by mitogenically responsive cells, *Endocrinology,* 122, 2044, 1988.
147. **Bar, R. S., Boes, M., and Yorek, M.,** Processing of insulin-like growth factors I and II by capillary and large vessel endothelial cells, *Endocrinology,* 118, 1072, 1986.
148. **Roth, R. A.,** Structure of the receptor for insulin-like growth factor II: the puzzle amplified, *Science,* 239, 1269, 1988.
149. **Nielsen, F. C. and Gammeltoft, S.,** Mannose-6-phosphate stimulates proliferation of neuronal precursor cells, *FEBS Lett.,* 262, 142, 1990.
150. **Oka, Y., Rozek, L. M., and Czech, M. P.,** Direct demonstration of rapid insulin-like growth factor II receptor internalization and recycling in rat adipocytes. Insulin stimulates ^{125}I-insulin-like growth factor II degradation by modulating the IGF-II receptor recycling process, *J. Biol. Chem.,* 260, 9435, 1985.
151. **Levi-Montalcini, R.,** The nerve growth factor 35 years later, *Science,* 237, 1154, 1987.
152. **Yankner, B. A. and Shooter, E. M.,** The biology and mechanism of action of nerve growth factor, *Annu. Rev. Biochem.,* 51, 845, 1982.
153. **Greene, L. A. and Tischler, A. S.,** Establishment of a noradrenergic clonal line of rat adrenal pheochromocytoma cells which responds to nerve growth factor, *Proc. Natl. Acad. Sci. U.S.A.,* 73, 2424, 1976.
154. **Landreth, G. E. and Shooter, E. M.,** Nerve growth factor receptors on PC12 cells: ligand-induced conversion from low- to high-affinity states, *Proc. Natl. Acad. Sci. U.S.A.,* 77, 4751, 1980.
155. **Hougue-Angeletti, R., Stieber, A., and Gonatas, N. K.,** Endocytosis of nerve growth factor by PC12 cells studied by quantitative ultrastructural autoradiography, *Brain Res.,* 241, 145, 1982.
156. **Claude, P., Hawrot, E., Dunis, D. A., and Campenot, R. B.,** Binding, internalization, and retrograde transport of ^{125}I-nerve growth factor in cultured rat sympathetic neurons, *J. Neurosci.,* 2, 431, 1982.
157. **Bernd, P. and Green, L. A.,** Association of ^{125}I-nerve growth factor with PC12 pheochromocytoma cells. Evidence for internalization via high-affinity receptors only and for long-term regulation by nerve growth factor of both high- and low-affinity receptors, *J. Biol. Chem.,* 259, 15509, 1984.
158. **Eckardt, K.-U. and Bauer, C.,** Erythropoietin in health and disease, *Eur. J. Clin. Invest.,* 19, 117, 1989.
159. **Sawyer, S. T., Krantz, S. B., and Goldwasser, E.,** Binding and receptor-mediated endocytosis of erythropoietin in friend virus-infected erythroid cells, *J. Biol. Chem.,* 262, 5554, 1987.
160. **Mufson, R. A. and Gesner, T. G.,** Binding and internalization of recombinant human erythropoietin in murine erythroid precursor cells, *Blood,* 69, 1485, 1987.

161. **Waldmann, T. A.,** The multi-subunit interleukin-2 receptor, *Annu. Rev. Biochem.,* 58, 87, 1989.
162. **Fung, M. R., Ju, G., and Greene, W. C.,** Co-internalization of the p55 and p70 subunits of the high-affinity human interleukin 2 receptor. Evidence for a stable ternary receptor complex, *J. Exp. Med.,* 168, 1923, 1988.
163. **Tanaka, T., Saiki, O., Doi, S., Fuji, M., Sugamura, K., Hara, H., Negoro, S., and Kishimoto, S.,** Novel receptor-mediated internalization of interleukin 2 in B cells, *J. Immunol.,* 140, 866, 1988.
164. **Lee, T.-P. and Mookerjee, B. K.,** Receptor-mediated endocytosis of IL-2 in PHA-stimulated lymphocytes, *Transplant. Proc.,* 21, 100, 1989.
165. **Sporn, M. B., Roberts, A. B., Wakefield, L. M., and Assoian, R. K.,** Transforming growth factor-β: biololgical function and chemical structure, *Science,* 233, 532, 1986.
166. **Massagué, J.,** The TGF-β family of growth and differentiation factors, *Cell,* 49, 437, 1987.
167. **Frolik, C. A., Wakefield, L. M., Smith, D. M., and Sporn, M. B.,** Characterization of a membrane receptor for transforming growth factor-β in normal rat kidney fibroblasts, *J. Biol. Chem.,* 259, 10995, 1984.
168. **Massagué, J. and Kelly, B.,** Internalization of transforming growth factor-β and its receptor in BALB/c 3T3 fibroblasts, *J. Cell. Physiol.,* 128, 216, 1986.
169. **Said, S. I.,** Vasoactive intestinal polypeptide (VIP): current status, *Peptides,* 5, 143, 1984.
170. **Luis, J., Martin, J.-M., El Battari, A., Marvaldi, J., and Pichon, J.,** The vasoactive intestinal peptide (VIP) receptor: recent data and hypothesis, *Biochimie,* 70, 1311, 1988.
171. **Couvineau, A., Amiranoff, B., and Laburthe, M.,** Solubilization of the liver vasoactive intestinal peptide receptor. Hydrodynamic characterization and evidence for an association with a functional GTP regulatory protein, *J. Biol. Chem.,* 261, 14482, 1986.
172. **Gammeltoft, S., Staun-Olsen, P., Ottesen, B., and Fahrenkrug, J.,** VIP receptors in pig liver cells: binding characteristics and role in degradation, *Peptides,* 5, 367, 1984.
173. **Lefkowitz, R. J., Caron, M. G., and Stiles, G. L.,** Mechanisms of membrane-receptor regulation: biochemical, physiological, and clinical insights derived from studies of the adrenergic receptors, *N. Engl. J. Med.,* 310, 1570, 1984.
174. **Benovic, J. L., Bouvier, M., Caron, M. G., and Lefkowitz, R. J.,** Regulation of adenylyl cyclase-coupled β-adrenergic receptors, *Annu. Rev. Cell Biol.,* 4, 405, 1988.
175. **May, W. S., Jr. and Cuatrescasas, P.,** Transferrin receptor: its biological significance, *J. Membrane Biol.,* 88, 205, 1985.
176. **Ashwell, G. and Harford, J.,** Carbohydrate-specific receptors of the liver, *Annu. Rev. Biochem.,* 51, 531, 1982.
177. **Dickson, R. B. R., Schlegel, R., Willingham, M. C., and Pastan, I.,** Binding and internalization of α_2-macroglobulin by cultured fibroblasts. Effects of monovalent ionophores, *Exp. Cell Res.,* 142, 127, 1982.
178. **Davis, C. G., Goldstein, J. L., Südhof, T. C., Anderson, R. G. W., Russell, D. W., and Brown, M. S.,** Acid-dependent ligand dissociation and recycling of LDL receptor mediated by growth factor homology region, *Nature (London),* 326, 760, 1987.
179. **Lehrman, M. A., Goldstein, J. L., Brown, M. S., Russell, D. W., and Schneider, W. J.,** Internalization-defective LDL receptors produced by genes with nonsense and frameshift mutations that truncate the cytoplasmic domain, *Cell,* 41, 735, 1985.
180. **Davis, C. G., van Driel, I. R., Russell, D. W., Brown, M. S., and Goldstein, J. L.,** The low density lipoprotein receptor. Identification of amino acids in cytoplasmic domain required for rapid endocytosis, *J. Biol. Chem.,* 262, 4075, 1987.
181. **Holland, E. C., Leung, J. O., and Drickamer, K.,** Rat liver asialoglycoprotein receptor lacks a cleavable NH_2-terminal signal sequence, *Proc. Natl. Acad. Sci. U.S.A.,* 81, 7338, 1984.
182. **Zerial, M., Melancon, P., Schneider, C., and Garoff, H.,** The transmembrane segment of the human transferrin receptor functions as a signal peptide, *EMBO J.,* 5, 1543, 1986.
183. **Clarke, B. L. and Weigel, P. H.,** Recycling of the asialoglycoprotein receptor in isolated rat hepatocytes. ATP depletion blocks receptor recycling but not a single round of endocytosis, *J. Biol. Chem.,* 260, 128, 1985.
184. **Dahms, N. M., Lobel, P., Breitmeyer, J., Chirgwin, J. M., and Kornfeld, S.,** 46 kDa mannose-6-phosphate receptor: cloning, expression, and homology to the 215 kDa mannose-6-phosphate receptor, *Cell,* 50, 181, 1987.
185. **Kiess, W., Blickenstaff, G. D., Sklar, M. M., Thomas, C. L., Nissley, S. P., and Sahagian, G. G.,** Biochemical evidence that the type II insulin-like growth factor receptor is identical to the cation-independent mannose-6-phosphate receptor, *J. Biol. Chem.,* 263, 9339, 1988.

Chapter 6

DEGRADATION OF CIRCULATING RENIN AND ANGIOTENSIN

Giorgio La Villa and Vicente Arroyo

TABLE OF CONTENTS

I. INTRODUCTION

The renin-angiotensin system plays a major role in the overall regulation of blood pressure, circulatory homeostasis, extracellular fluid volume (including blood volume), sodium and potassium excretion, and the electrolyte composition of the body. Because of the potent vasoconstricting and sodium-retaining effects of angiotensin II, the active component of the renin-angiotensin system, regulation of plasma concentrations of renin and angiotensin II is of major importance for the whole-body homeostasis. This is accomplished by both a complex mechanism which controls renin secretion and a very efficient and rapid removal of circulating renin and angiotensin II. In this chapter, the metabolic fate of circulating renin and angiotensin II will be reviewed after a brief discussion on the biochemistry and physiology of these two compounds. The last section of the chapter will deal with the behavior of the renin-angiotensin system in cirrhosis of the liver, a unique condition in which high plasma levels of renin and angiotensin II occur in the setting of the failure of the main site of renin metabolism.

II. PHYSIOLOGY OF RENIN AND ANGIOTENSIN II

Renin (EC 3.4.99.19) is a highly specific aspartyl proteinase which catalyzes the conversion of the glycoprotein angiotensinogen to the decapeptide angiotensin I (Figure 1). In most species, renin cleaves the Leu^{10}– Leu^{11} bond of angiotensinogen.[1] The amino acid sequence of human angiotensinogen is different at this site (Leu^{10}–Val^{11}) and is cleaved only by renin of humans and subhuman primates.[2] Renin is mainly produced and stored in specialized epithelioid cells of the juxtaglomerular apparatus which is located in the wall of afferent glomerular arterioles, in intimate contact with the vascular pole of the glomerulus and convoluted distal tubule. These cells release renin directly into the bloodstream and into renal lymph.[3] Three main mechanisms control renin release from the kidney:[4-6] (1) the renal baroreceptor mechanism sensitive to changes in renal perfusion pressure, (2) the macula densa mechanism which responds to changes in sodium chloride delivery or transport in the thick ascending limb of Henle,[7] and (3) the renal sympathetic nervous activity which directly stimulates renin release from juxtaglomerular cells via β-1-adrenoreceptors. These three mechanisms are not mutually exclusive but, rather, they probably operate in concert since they are influenced in the same direction by changes in effective circulating blood volume or arterial pressure. In addition, several circulating hormones and locally generated substances participate in the control of renin secretion, including renal prostaglandins, antidiuretic hormone, parathyroid hormone, atrial natriuretic peptide, angiotensin II, acetylcholine, dopamine, histamine, adenosine, platelet-activating factor, and serum potassium concentration.[4,8]

In addition to the kidney, several organs and tissues are known to contain renin, including the adrenal glands,[9,10] heart, brain,[11] pituitary, arterial wall,[12] uterus,[13] testis,[14] and salivary glands.[15] The contribution of these extrarenal sources to the plasma levels of renin, if any, is negligible, since plasma renin concentration falls to low or undetectable levels after bilateral nephrectomy.[16]

Renin shows a high degree of heterogeneity in several animal species, including man. Firstly, renin exists in active and inactive forms. Inactive renin has a higher molecular weight than the active enzyme and can be activated *in vitro* by acidification, cold, or incubation with proteases.[17] It constitutes 60 to 90% of total plasma renin in normal subjects.[18-20] The kidneys are thought to be the principal source of inactive renin,[21] although other sources cannot be excluded since inactive renin has been found in the brain, pituitary, and amniotic fluid.[22,23] The observation that all of the inactive renin in human plasma and kidney is recognized by antibodies specifically directed against fragments of the non-renin portion (prosegment) of human renin precursor strongly suggest that inactive renin is prorenin.[24,25] This has been

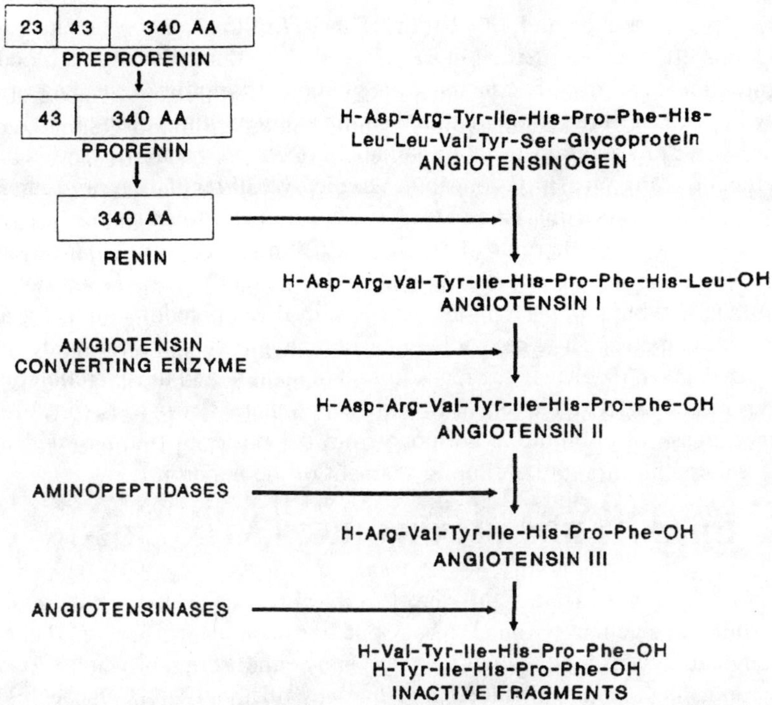

FIGURE 1. Components to the renin-angiotensin system and circulating degradation products of angiotensin II. (AA: amino acids).

recently confirmed by the analysis of amino acid composition of purified inactive renin.[26] The physiological role and metabolic fate of circulating prorenin are still unknown. Multiple forms of active renin have also been reported, which are separable by isoelectric focusing or other techniques.[27-29] This latter phenomenon may be of importance in determining the metabolic fate of circulating renin (see later).

Angiotensin I has no biological activity, but it is immediately transformed to the octapeptide angiotensin II by the action of a specific dipeptidyl carboxypeptidase (angiotensin converting enzyme). Since the largest concentration of this converting enzyme is found in the lung, until recently it was believed that conversion of angiotensin I to angiotensin II was primarily systemic rather than intrarenal. However, subsequent investigations demonstrating that converting enzyme is also present in the juxtaglomerular apparatus, and that significant amounts of angiotensin II are generated locally within the kidney,[30,31] suggest that this biologically active compound is released immediately adjacent to the afferent and efferent glomerular arterioles. It is, therefore, very likely that renal function could be influenced not only by the angiotensin II reaching the kidney via the renal artery but also by that locally generated within or near the juxtaglomerular apparatus. Angiotensin II is among the most active endogenous vasoconstrictors thus far identified. Angiotensin II-induced vasoconstriction is due to a direct interaction with specific receptors located on the vascular smooth muscle cells, leading to an elevation of cytosolic calcium levels and contraction of these cells, to the stimulation of specific receptors in the area postrema of the central nervous system resulting in increased sympathetic nervous activity, and to an enhancement of neurotransmission at the peripheral noradrenergic neuroeffector junction.[32] The renal vasculature is especially sensitive to the vasoconstrictor effect of angiotensin II since a striking reduction of renal blood flow occurs with doses of angiotensin II well below those required to induce a pressor response.[32,33] Angiotensin II also reduces the glomerular filtration rate. This latter effect is related to both the decreased renal perfusion and a reduction of the ultrafiltration coefficient (K_f), due to a direct contracting effect on glomerular mesangial cells.[34] Reductions in renal blood flow tend to be disproportionately greater than those in glomerular filtration rate and are therefore associated with a rise in filtration fraction, indicating a preferential action of angiotensin II on efferent arteriole.[35,36] In addition to its remarkable effects on intrarenal hemodynamics and glomerular function, angiotensin II contributes to the overall regulation of intrarenal sodium handling by several mechanisms. In fact, this substance: (1) influences filtered sodium load because of its actions on glomerular filtration rate, (2) modifies peritubular physical factors governing proximal tubules fluid reabsorption,[37] and (3) directly stimulates sodium reabsorption in the proximal tubule.[38] Furthermore, angiotensin II is a potent stimulus for aldosterone secretion by acting upon specific receptors in the adrenal glomerulosa cells. Aldosterone acts on collecting duct cells to enhance the reabsorption of sodium and the excretion of potassium and hydrogen ions. Finally, angiotensin II influences water homeostasis, since it stimulates thirst and the release of antidiuretic hormone from the posterior pituitary and reduces the delivery of sodium chloride to the diluting segment of the nephron.[39]

III. METABOLISM OF CIRCULATING RENIN

Initial studies on renin metabolism were mainly aimed at evaluating its half-life in the circulation, which was generally found to be about 10 to 20 min.[40–42] The rapid elimination of renin from circulation was not due to its destruction by either serum or whole blood, since the half-life of renin incubated with blood *in vitro* was of about 21 h. Schaechtelin et al.[43] investigated the disappearance rate of renin-like material using isovolemic cross-circulation with nephrectomized indicator rats. After performing bilateral nephrectomy in otherwise untreated rat, endogenous renin disappeared from blood within 1 h. When plasma renin levels were raised prior to the nephrectomy by placing a clip on one renal artery, 2 h were required before renin-like activity was no longer detectable in the indicator rat. Exogenously administered renin had about the same disappearance rate as endogenous renin. The calculated half-life of both endogenous and exogenous renin was about 30 min. The half-life of renin was four times faster in rats depleted of renin by overdosage of mineralocortocoids and salts.[44] Assaykeen et al.[45] reported data on the disappearance rate of exogenously administered renin in nephrectomized dogs. Renin disappearance curves could be resolved into two exponential components, having a half-life of 6 and 60 min, respectively. The rapid component of the curve was considered to be due to the mixing of injected renin in the circulation, while the slower one was thought to indicate true metabolism. Similar results were obtained by Schneider et al.,[46] who investigated the disappearance rate of homologous renin after a 10-min infusion to nephrectomized dogs. Again, a two-component disappearance curve was found, the first component having a half-life of 4 and the second of 45 min. The calculated metabolic clearance rate of renin was 64 to 69 ml/min. Comparable disappearance curves and half-lives for both endogenous and exogenous renin were observed in the possum[47] and the rat.[48] The metabolism of renin in the rat was not modified by treatment with deoxycorticosterone and salt nor by hemorrhage.[48] The disappearance rate of circulating renin was also investigated in humans undergoing bilateral nephrectomy. Again, a two-exponential disappearance curve was observed, with a renin half-life somewhat higher than that previously reported in animals, usually ranging between 1 and 3 h.[19,49,50] The results of all these studies support the contention that plasma renin concentration can change rather rapidly when renal secretion changes in response to physiologic or pathologic stimuli. Accordingly, plasma renin levels are usually a reliable index of renin secretion.[51]

The first attempt to identify the site of renin inactivation was performed by Heacox and associates,[52] who measured the arterio-venous difference for renin across several vascular beds in baseline conditions and after elevation of arterial renin induced by acute salt depletion, which stimulated endogenous renin secretion, or the infusion of exogenous renin. In all experiments, a significant arterial-hepatic venous gradient for renin was observed, the percent hepatic extraction of renin being about 25%. There was no significant arterial-portal venous renin difference, suggesting that the spleen, pancreas, and gastrointestinal tract have no role in renin metabolism. As expected, renin concentration in the renal vein was always higher than the arterial one in baseline conditions and during the stimulation of endogenous renin secretion. Exogenously administered renin usually decreased the renal venous-arterial renin difference, but a net inactivation of renin by the kidney was observed in only 2 out of 12 samples. Furthermore, the hepatic inactivation rate of renin accounted for the calculated metabolic clearance of all infused renin. Based on these results, it was suggested that the liver was the principal organ involved in the metabolism of renin, if not the only. This contention was in agreement with the results of by another study in dogs,[53] showing that the metabolic clearance of renin from peripheral plasma was approximately equal to the simultaneously calculated hepatic clearance of renin. The recognition that the liver plays a major role in renin metabolism led to the hypothesis that a reduction of hepatic blood flow and/or liver function should result in a reduced renin metabolism, accounting, at least partially, for the high levels of renin found in conditions such as salt deprivation, heart failure, and liver cirrhosis.[51–53] Evidence supporting this contention was provided by Schneider et al.[54] who measured hepatic blood flow, hepatic extraction rate, and hepatic clearance of renin in three group of dogs: normal, with low-output heart failure induced by constricting the thoracic inferior vena cava, and with high-output heart failure induced by constructing an aortic-caval fistula. All animals with heart failure developed hyperreninemia, sodium retention and peripheral edema. In normal dogs, the percent hepatic extraction of renin was found to be about 20% and hepatic clearance of renin 104.8 ± 15.5 ml/min. In caval dogs, hepatic extraction of renin increased up to 33%. No changes in the hepatic clearance of renin were therefore observed in these animals despite a marked reduction in hepatic blood flow. In contrast, dogs with the aortic-caval fistula showed a 50% reduction in the hepatic clearance of renin, which was associated with a decreased hepatic extraction rate. A reduced hepatic clearance of renin was also observed in dogs submitted to acute hemorrhage[55] or chronic sodium depletion.[55,56] Other experimental conditions characterized by a reduction in the hepatic metabolism of renin were found to be acute lead poisoning[57] and administration of captopril.[58]

The role of the liver in renin metabolism was also investigated in man. By measuring plasma renin activity in peripheral plasma and the hepatic vein, Christlieb et al.[59] found that the liver extracts about 26% of renin entering this organ. A 29% hepatic renin extraction was reported by Kokot et al.,[60] while somewhat lower values (17 and 16%, respectively) were found by Barnardo[61] and Hesse et al.[62] Since the hepatic clearance of all substances removed by the liver with an extraction ratio lower than 30% is independent of hepatic plasma flow,[63] alterations in liver perfusion should not affect renin metabolism. Indeed, Hesse et al.[62] found no changes in the hepatic clearance of renin in man when hepatic blood flow was reduced by physical exercise and i.v. vasopressin. Barnardo et al.[61] also measured plasma renin activity in samples obtained from the portal vein of 38 patients without liver diseases undergoing abdominal surgery, and found it was higher than that in peripheral plasma in 29 patients, suggesting a splanchnic source of renin. These results, however, are not in agreement with the data by Heacox et al.[52] on dogs, Horky et al.[64] on rats, nor were they later confirmed by Rector and Hossack[65] in man.

As previously mentioned, a significant role of the kidney in renin metabolism was initially excluded, based on the finding that the hepatic clearance of renin approached the metabolic clearance rate.[52–54] However, since the renal venous-arterial gradient for renin was used as a reflection of renin secretion in the calculation of the metabolic clearance rate, the possibility of a dual role of the kidney in renin homeostasis (i.e., inactivation of renin entering this organ via the renal artery and addition of newly released renin to the renal venous blood) could not be ruled out in these studies. On the other hand, evidence for a role of the kidney in renin metabolism was provided by the observation that exogenously administered renin disappeared

TABLE 1
Main Parameters of Plasma Disappearance of Exogenously Administered ^{125}I-Renin in Intact, Conscious Rats

$t_{1/2}$[a] (min)	6.7 ± 0.4
$t_{1/2}$[b] (min)	65.1 ± 5.7
Volume of distribution (ml/kg)	1030 ± 47
MCR[c] (ml/min/kg)	11/4 ± 1.0

Note: Data are expressed as the mean ± SEM.

[a] Half-life for the rapid component of the disappearance curve.
[b] Half-life for the slow component.
[c] MCR: metabolic clearance rate.

From Kim, S., Iwao, H., Nakamura, N., Ikemoto, F., and Yamamoto, K., *Am. J. Physiol.*, 252, E136, 1987.

more rapidly from circulation in intact than in nephrectomized animals.[40,48,66] Renin has a molecular weight of about 40,000 and therefore it is, at least partially, filtered by the glomeruli. Indeed, renin has been isolated from urine of man[66–69] and some animal species.[70–74] In humans, renin excretion is higher in females than in males, shows high inter-individual but low intra-individual variations, and is not dependent on plasma levels of renin, creatinine clearance, protein clearance, and sodium and water excretion.[69] Direct evidence of glomerular filtration of renin has been obtained by micropuncture[73] and stopped-flow techniques.[74] Even if it is not possible to quantitatively assess glomerular filtration of renin due to the unmeasurable contribution of direct secretion of renin into the tubular lumen, urinary excretion of renin apparently accounts for a very small percentage of what might be filtered,[51,68–74] suggesting that renin, as well as other plasma proteins having similar molecular weight, is extensively reabsorbed by renal tubuli. The findings that urinary excretion of renin is higher than normal in patients with renal diseases mainly affecting proximal tubules[68] or following experimentally induced tubular damage[72] are in agreement with this contention. Direct evidence of tubular reabsorption of renin has been provided by stopped-flow techniques.[74]

To directly assess the catabolism of renin in intact animals, the metabolic fate of exogenously administered ^{125}I-labeled renin was investigated in conscious rats. In a first series of experiments, purified ^{125}I-labeled mouse submaxillary renin was found to reach the kidney, where it was filtered, reabsorbed by the proximal tubule, and then degraded within the lysosomes.[75-78] The metabolism of submaxillary renin, however, cannot apply to that of renal renin since submaxillary renin, at difference from renal renin, is not glycosylated[79] and, therefore, cannot be recognized and thus removed by the liver, which possesses specific receptor for the carbohydrate moiety of glycoproteins.[80] A second series of experiments was therefore performed using a highly purified ^{125}I-labeled renal renin.[81-84] The disappearance curve of exogenously administered ^{125}I-renin to intact rats was found to have two exponential components, the half-life of the rapid and slow phases being 6.7 ± 0.4 and 65.1 ± 5.7 min (mean ± SEM), respectively (Table 1). The metabolic clearance rate was 11.4 ± 1.0 ml·min·kg.[81] Rat plasma was unable to degrade renin. In nephrectomized animals, plasma levels of ^{125}I-renin were higher than in the control group at all time points evaluated after the injection and the half-life of the rapid component increased, suggesting that the fast component represented not only distribution of renin but also clearance by the kidney. The half-life of the slow component, however, was similar to that of the control group. In rats submitted to 70% hepatectomy, plasma concentration of ^{125}I-renin was similar to that of control animals in the first 30 min and higher thereafter, up to a maximum of about 4 times at 180 min. Partial hepatectomy did not modify the half-life of the rapid component, but markedly increased that of the slow one. The metabolic clearance rate of renin was reduced by more than 50% in both nephrectomized and hepatectomized rats.[81] Evaluation of ^{125}I-radioactivity in various organs showed that most renin disappeared from circulation within the first minutes, and accumulated in the liver and kidneys. At 5 min after ^{125}I-renin administration, hepatic and renal radioactivity was mainly due to the accumulation of unbroken renin; thereafter, radioactivity decreased in both organs and degradation products having lower molecular weight than the injected ^{125}I-renin became predominant, indicating degradation of ^{125}I-renin. Accumulated ^{125}I-renin per gram tissue was similar in the liver and kidney; however, since the liver mass is much greater than that of the kidneys, the contribution of the liver to the clearance of circulating renin predominated. The accumulation of ^{125}I-renin was less than 1% of the dose injected in other organs such as lungs, heart, spleen, and adrenal glands. No radioactivity was found in bile. Cumulative urinary excretion of ^{125}I-renin was about 2% of the injected dose, indicating a minor role of this route in the removal of circulating renin.[81] Whole-body autoradiography confirmed the prevalent uptake of circulating ^{125}I-renin by the liver and kidneys.[82] These results, obtained in rats given rat renal renin, have been recently confirmed in the marmoset receiving highly purified recombinant human renin.[83] Metabolism of ^{125}I-renin at cellular and subcellular levels was investigated by performing light and electron microscope autoradiography.[84] In the liver, renin was selectively taken up by Kupffer and endothelial cell, where it was transferred to lysosomes to be degraded. In the kidney, renin reached, probably by glomerular filtration, the epithelial cells of proximal convoluted tubules, where it was observed on the brush border and in the apical invaginations, endocytic vacuoles, and lysosomes. No accumulation of ^{125}I-renin was found in the hepatocytes, nor in glomeruli, loop of Henle, distal tubules, or collecting ducts. Comparison with previous results obtained by the same group using submaxillary gland renin[75-78] indicates therefore that glycosylation of renin is essential for its recognition by Kupffer cells and, thus, for the hepatic clearance. In contrast, renal clearance of renin is unaffected by the presence of carbohydrate moieties in the renin molecule.

As previously discussed, the kidney produces and releases into circulation different forms of active renin.[27-29] Recent evidence suggests that each of these different types of renin may have a different metabolic clearance rate and, therefore, that the plasma concentration of renin is the result of the balance between the rate of production and removal of each of these forms.[85-87] The physiological and pathophysiological implications of these findings are unknown at present. However, the possibility should be considered that in various pathophysiological states such as low- and high-sodium intake, diabetes, renovascular hypertension, and heart failure, renal production and secretion of different forms of renin, having different metabolic clearance rates, may play a role in determining plasma renin levels in these conditions.[87]

IV. METABOLISM OF CIRCULATING ANGIOTENSIN II

The removal of angiotensin II from circulation is even faster than that of renin. After intravenous injection of angiotensin II to assay animals, the half-life of the peptide ranged between 30 sec and 4 min.[88-92] A comparable half-life of about 1 min was observed in normal humans receiving radioiodinated angiotensin II.[93] On the other hand, many experiments perfusing angiotensin II into various vascular beds *in vivo* and in isolated organs have shown that 50 to 80% of injected angiotensin II disappeared from circulation during its passage through several vascular beds within one circulation time. Considerable extraction of circulating angiotensin II was found to occur in the liver,[94-99] kidney,[97-102] head,[97] hindquarter[97,98] and heart,[100] among others, while pulmonary circulation was unable to destroy angiotensin II.[97] It is therefore probably right to consider that the half-life of this peptide in plasma should be close to 15 to 20 s, that is, one circulation time.

Using a constant-infusion technique, the metabolic clearance rate of angiotensin II was measured in experimental animals,[90-92,103,104] and humans[93,105,106] and found to be about 50 to 60 ml·min·kg, a figure which approaches the cardiac output, thus confirming that most of the angiotensin II is cleared during one circulation through peripheral tissues. In humans, the metabolic clearance rate of angiotensin II is influenced by sodium intake, being higher in sodium depletion.[105]

Numerous studies have specifically investigated the metabolism of angiotensin II by the kidney. Initial studies, performed by injecting angiotensin II into the renal artery, showed that only 22 to 35% of administered dose could be recovered in the renal vein.[97–102] The renal ability to remove angiotensin II is very wide since the percentage of extraction of the peptide was constant even when the amount injected was 200 times the physiological levels.[101] More recently, the renal degradation rate of angiotensin II was estimated in the dog during continuous arterial infusion on angiotensin II and under conditions of blockade on intrarenal angiotensin II production, in order to eliminate possible errors due to the too rapid variations of angiotensin II concentration occurring after bolus injection of the peptide. Under these conditions, renal degradation of angiotensin II was about 90% at uncontrolled and 86% at reduced renal perfusion pressure, indicating that renal catabolism of angiotensin II is even greater than previously recognized.[107] Mechanisms for renal degradation of angiotensin II involve glomerular filtration followed by hydrolysis at the brush border of the renal proximal tubule and reabsorption of metabolites.[108-111] Urinary excretion of angiotensin II is negligible.[31] Angiotensinase activity has been found in the brush border of renal tubular epithelium.[112] Distal tubules do not significantly contribute to renal metabolism of angiotensin II since, when labeled angiotensin II is directly injected into the luminal fluid of these structures, almost all the injected radioactivity appears in urine as intact angiotensin II.[111] In experimental animals, renal metabolism of angiotensin II is markedly influenced by changes in renal blood flow, renal perfusion pressure, and sodium intake,[102,113] while it is unaffected by modifications in glomerular filtration rate induced by ureteral constriction.[113] The physiological relevance of these observations is unknown.

Degradation of angiotensin II is promoted by several proteolytic enzymes, collectively named angiotensinases. This terminology implies specificity for angiotensin. However, the very multiplicity of these peptidases suggests they are rather nonspecific. A number of proteolytic enzymes are known to possess angiotensinase activity, including trypsin, chymotrypsin, aminopeptidase, and pepsin.[114] Angiotensinase activity has been found in plasma and red cells.[115,116] However, angiotensin metabolism in plasma or whole blood *in vitro* is too slow to account for the very rapid disappearance of the peptide from circulation *in vivo*,[97,101,117] indicating that degradation of angiotensin II in the blood does not appreciably contribute to its disappearance from circulation. This is confirmed by the finding that angiotensin II can be cleared by isolated, Krebs perfused organs.[118]

Degradation of angiotensin II leads to the formation of amino acids and small peptides, including 3-8 hexapeptide and 4-8 pentapeptide derivatives, which may be released into circulation, thus interfering with angiotensin II measurements by cross-reacting with anti-angiotensin II antibodies.[119,120] This phenomenon, however, has little practical relevance, due to the low proportion of these compounds, which constitutes less than 20% of total angiotensin II immunoreactivity. Total angiotensin II determination, therefore, sufficiently reflects the concentration of the active heptaoctapeptide fractions in both arterial and venous plasma.[121]

V. THE RENIN-ANGIOTENSIN SYSTEM IN CIRRHOSIS

The renin-angiotensin system is activated in most cirrhotics with ascites and marked sodium retention (urinary sodium excretion lower than 5 mEq/1) and in all patients with functional renal failure.[122-124] In many of these patients, the plasma levels of renin and aldosterone reach extraordinarily high values. In cirrhotics with ascites and moderate sodium retention, plasma renin activity may be normal or only slightly elevated.[122–124] Plasma renin activity is usually normal in compensated cirrhotics.[123-126] Several lines of evidence indicate that aldosterone plays a major role in the pathogenesis of sodium retention in cirrhosis. Urinary sodium excretion in cirrhotics with ascites correlates inversely with plasma aldosterone concentration.[122,126] Moreover, in most patients, sodium retention can be reversed by blocking the tubular effect of aldosterone with spironolactone.[127] Finally, in rats with experimental cirrhosis, a close chronological relationship has been observed between the stimulation of the renin-angiotensin-aldosterone system and the onset of sodium retention, both features preceding the onset of ascites.[128] The observation that some cirrhotics with ascites show sodium retention in the absence of hyperaldosteronism is generally considered an indication that factors other than aldosterone are involved in the excessive tubular sodium reabsorption occurring in cirrhosis.[129] These latter patients, however, are extremely sensitive to spironolactone,[130] suggesting that they may have an increased tubular sensitivity to aldosterone.[126]

Cirrhosis of the liver is characterized by a marked impairment of hepatic hemodynamics and function. Moreover, in many of these patients, especially when they have ascites, renal perfusion and function are also reduced. Since the liver and kidney play a major role in renin metabolism, the possibility should be considered that hyperreninemia in cirrhosis with ascites might be due to an impaired hepatic catabolism.[51–53] This hypothesis was first tested by Barnardo et al.,[61] who measured plasma renin activity in peripheral and hepatic veins in cirrhotic and control subjects and found that, in cirrhotic patients, at difference from control subjects, hepatic venous renin levels were significantly higher than those in peripheral blood. To explain these findings, portal venous plasma renin activity was also directly measured in a group of patients without liver diseases undergoing abdominal surgery and was found to exceed that in peripheral vein. It was therefore concluded that a splanchnic source of renin exists in humans which, in physiological conditions, is inactivated by the liver. An impaired liver metabolism could therefore contribute to the high plasma renin activity of cirrhosis.[61] A significant reduction in the hepatic extraction of renin in patients with alcoholic cirrhosis was later found by Mitch et al.[131] In this study, however, the hepatic extraction of renin in healthy subjects was about 40%, a very high figure when compared with the 29 to 16% generally observed in normal humans.[59-62] Furthermore, hepatic metabolism of renin in cirrhosis was found to be normal by Wernze et al.[132] and Bosch et al.[133] Finally, a recent investigation by Rector and Hossack[65] failed to find any significant difference between cirrhotic and control subjects with respect to percentage of hepatic extraction, intrinsic hepatic clearance, and hepatic excretion rate of renin. In the same study, a splanchnic source of renin was also excluded by simultaneous measurements of arterial and portal venous plasma renin activity. The increased plasma renin activity of cirrhosis is therefore due solely to increased renin production.

In cirrhotics with ascites, plasma aldosterone concentration closely correlates with the plasma levels of renin, indicating that the main stimulus for aldosterone secretion is angiotensin II. In turn, plasma renin activity inversely correlates with the renal blood flow and glomerular filtration rate.[134] This observation was initially interpreted as an indication that the increased renin release was secondary to an activation of the renal baroreceptor mechanism due to the impaired renal hemodynamics. However, studies in nonazotemic cirrhotics with ascites have clearly demonstrated that the renin-angiotensin system may be markedly activated in patients with normal RBF and GRF.[122] Therefore, alterations in renal perfusion are not major determinants of renin secretion in cirrhotics with ascites.

The introduction of pharmacological agents that interrupt the renin-angiotensin system, such as captopril, an inhibitor of the converting enzyme, and saralasin, a competitive antagonist of angiotensin II, have made it possible to better understand why the renin-angiotensin system is activated in patients with cirrhosis and ascites. These patients have systemic circulatory disturbances characterized by hypervolemia, low arterial pressure, high cardiac output, and low peripheral resistance, which are consistent with a marked arteriolar vasodilation.[135–141] The inhibition of the endogenous renin-angiotensin system in cirrhotic patients with increased plasma levels of renin by the i.v. injection of saralasin (a specific antagonist of angiotensin II) or converting enzyme inhibitors is associated with a further decrease of arterial pressure and peripheral resistance, which may be very intense in patients with marked overactivity of the renin-angiotensin system.[142,143] The systemic circulatory disturbances of these patients would, therefore, be much more intense if endogenous angiotensin II were not acting at the peripheral vasculature to maintain arterial pressure at normal or near normal levels. These data clearly indicate that renin release in cirrhotics with ascites is a homeostatic mechanism to maintain systemic hemodynamics. Arterial hypotension-mediated stimulation of renal and extrarenal baroreceptors, and not a reduced disposal of this peptide, is therefore becoming the most likely mechanism of hyperreninism in these patients.

VI. CONCLUSION

An efficient and rapid degradation of circulating renin and angiotensin II plays a major role in the overall control of the renin-angiotensin system activity. The disappearance rate of circulating renin follows a two-exponential curve, with a fast and a slower component. Renin half-life ranges from 10 to 30 min. Plasma renin concentration, therefore, can change rather rapidly when renal secretion changes in response to physiologic or pathologic stimuli. Renin catabolism mainly occurs in the liver and the kidney, the hepatic contribution being prevalent. Studies performed with ^{125}I-labeled renin indicate that it is selectively taken up in the liver by Kupffer and endothelial cells, where it is transferred to lysosomes to be degraded. Glycosylation of renin is essential for the hepatic clearance. In the kidney, renin reaches, probably by glomerular filtration, the epithelial cells of proximal convoluted tubules, where it is extensively reabsorbed and degraded. Very small amounts of renin are excreted unmodified into the urine. Some recent evidence suggests that the kidney may produce and release different forms of renin, having different metabolic clearance rates. The physiological and pathophysiological implications of these findings, however, are unknown at the present time.

Metabolism of angiotensin II is even faster than that of renin. In fact, almost all angiotensin II is removed from circulation in one circulation time (about 15 s), due to extensive extraction by several vascular beds, including the liver, kidney, hindquarter, heart and others. The pulmonary vascular bed, however, is unable to destroy this compound. Destruction of angiotensin II is due to degradation by several tissue proteolytic enzymes, collectively called angiotensinases, which lead to the formation of amino acids and small peptides, including 3-8 hexapeptide and 4-8 pentapeptide derivatives. These fragments are, at least partially, released into circulation. Blood angiotensinases do not appreciably contribute to angiotensin metabolism.

Present evidence indicates that the severe alterations of hepatic hemodynamics and function occurring in liver cirrhosis do not appreciably impair renin and angiotensin II metabolism.

REFERENCES

1. **Ondetti, M. A. and Cushman, D. W.,** Enzymes of the renin-angiotensin system and their inhibitors, *Annu. Rev. Biochem.,* 51, 283, 1982.
2. **Tewksbury, D. A., Dart, R. A., and Travis, J.,** Studies on the characterization of human angiotensinogen, *Circulation,* 60, II-132, 1970.
3. **Lever, A. F. and Peart, W. S.,** Renin and angiotensin-like activity in renal lymph, *J. Physiol (London),* 160, 548, 1962.
4. **Davis, J. O. and Freeman, R. H.,** Mechanisms regulating renin release, *Physiol. Rev.,* 56, 1, 1976.
5. **Reid, J. A., Morris, B. J., and Ganong, W. F.,** The renin-angiotensin system, *Annu Rev. Physiol.,* 40, 377, 1978.
6. **Schambelan, M. and Stockigt, J. R.,** Pathophysiology of the renin-angiotensin system. in *Hormonal Function and the Kidney,* Brenner, B. M. and Stein, J. H., Eds., Churchill Livingstone, New York, 1979, 1.
7. **Kotchen, T. A., Welch, W. J., Lorenz, J. N., and Ott, C. E.,** Renal tubular chloride and renin release, *J. Lab. Clin. Med.,* 110, 533, 1987.
8. **Hackenthal, E. and Taugner, R.,** Hormonal signals and intracellular messenger for renin secretion, *Mol. Cell Endocrinol.,* 47, 1, 1986.
9. **Hayduc, E., Boucher, R., and Genest J.,** Renin activity content in varous tissues of dogs under different physiopathological states, *Proc. Soc. Exp. Biol. Med.,* 134, 252, 1970.
10. **Naruse, M., Sussman, R., Naruse, K., Jackson, R. V., and Inagami, T.,** Renin exists in human adrenal tissue, *J. Clin. Endocrinol. Met.,* 57, 482, 1983.
11. **Paul, M., Hermann, K., Printz, M., Lang, R. E., Unger, Th., and Ganten, D.,** The brain angiotensin system: subcellular localization and interferences with converting enzyme inhibitors, *J. Hypertension,* 1 (Suppl. 1), 9, 1983.
12. **Swales, J. D., Abramovici, A., Beck, F., Bing, R. F., Loudon, M., and Thurston, H.,** Arterial wall renin, *J. Hypertension,* 1 (Suppl. 1), 17, 1983.
13. **Ferris, T. F., Gorden, P., and Mulrow, P. J.,** Rabbit uterus as a source of renin, *Am. J. Physiol.,* 212, 698, 1967.
14. **Field, L. J., McGowan, R. A., Dickinson, D. P., and Gross, K. W.,** Tissue and gene specificity of mouse renin expression, *Hypertension,* 6, 597, 1984.
15. **Misono, K. S., Chang, J. J., and Inagami, T.,** Amino acid sequence of mouse submaxillary renin, *Proc. Natl. Acad. Sci. U.S.A.,* 79, 4858, 1983.
16. **Osmond, D. H. and Loth, A. Y.,** Effects of pregnancy, oral contraception and binephrectomy on human plasma "prorenin" and its activator(s). *Clin. Exp. Hypertens.,* A4, 2385, 1982.
17. **Sealey, J. E., Atlas, S. A., and Laragh, J. H.,** Prorenin and other large molecular weight forms of renin, *Endocrine Rev.,* 1, 365, 1980.
18. **Hsueh, W., Luetscher, J. A., Carlson, E. J., and Grislis, G.,** Inactive renin of high molecular weight (big renin) in normal human plasma, *Hypertension,* 2, 750, 1980.
19. **Derkx, F. H. M. and Wenting, G. J., Man In't Veld, A. J., Verhoven, R. P., and Schalekamp, M. A. D. H.,** Control of enzymatically inactive renin in man under various pathological conditions: implications for the interpretation of renin measurements in peripheral and renal venous plasma, *Clin. Sci. Mol. Med.,* 54, 529, 1978.
20. **Atlas, S. A., Laragh, J. H., Sealey, J. E., and Hesson, T. E.,** An inactive, prorenin-like substance in human kidney and plasma, *Clin. Sci.,* 59, 29S, 1980.
21. **Hsueh, W. A., Carlson, E. J., and Dzau, V.,** Characterization of inactive renin from human kidney and plasma. Evidence for a renal source of circulating inactive renin, *J. Clin. Invest.,* 71, 506, 1983.
22. **Hirose, S., Naruse, M., Ohtsuki, K., and Inagami, T.,** Totally inactive renin zymogen and different forms of active renin in hog brain tissues, *J. Biol. Chem.,* 256, 5572, 1981.
23. **Lumbers, E. R.,** Activation of renin in human amniotic fluid by low pH, *Enzymologia,* 40, 329, 1971.
24. **Bouhnik, J., Fehrentz, J. A., Galen, F. X., Seyer, R., Evin, G., Castro, B., Menard, J., and Corvol, P.,** Immunologic identification of both plasma and human renal inactive renin as prorenin, *J. Clin. Endocrinol. Met.,* 60, 399, 1985.
25. **Atlas, S. A., Christofalo, P., Hesson, T., Sealey, J. E., and Fritz, L. C.,** Immunological evidence that inactive renin is prorenin, *Biochem. Biophys. Res. Commun.,* 132, 1038, 1985.
26. **Higashimori, K., Mizuno, K., Nakajo, S., Boehm, F. H., Marcotte, P. A., Egan, D. A., Holleman, W. H., Heusser, C., Poisner, A. M., and Inagami, T.,** Pure human inactive renin. Evidence that native inactive renin is prorenin, *J. Biol. Chem.,* 246, 14662, 1989.
27. **Sessler, F. M. and Malvin, R. L.,** Modifications of renin isoelectric heterogeneity in Goldblatt hypertensive rats, *Am. J. Physiol.,* 248, E694, 1985.
28. **Sessler, F. M., Kim, S. H., and Malvin, R. L.,** Changes of renin isoelectric heterogeneity after acute and chronic stimulation of renin secretion, *Proc. Soc. Exp. Biol. Med.,* 182, 258, 1986.

29. **Kim, S., Hosoi, M., Hiruma, M., Ikemoto, F., and Yamamoto, K.,** Heterogeneity of glycosilation of circulating active and inactive renin, *Clin. Exp. Hypertens.,* A10, 1203, 1988.
30. **DiSalvo, J., Peterson, A., Montefusco, C., and Menta, M.,** Intrarenal conversion of angiotensin I to angiotensin II in the dog, *Circ. Res.,* 29, 398, 1971.
31. **Bailie, M. D., Rector, F. C., Jr., and Seldin, D. W.,** Angiotensin II in arterial and renal venous plasma and renal lymph in the dog, *J. Clin. Invest.,* 50, 119, 1971.
32. **Jackson, E. K., Branch, R. A., Margolius, H. S., and Oates, J. A.,** Physiological function of the renal prostaglandin, renin and kallikrein systems, in *The Kidney: Physiology and Pathophysiology,* Seldin, D. W. and Giebisch, G., Eds., Raven Press, New York, 1985, 613.
33. **Hollenberg, N. K., Cheniz, W. R., Adams, D. F., and Williams, G. H.,** Reciprocal influence of salt intake on adrenal glomerulosa and renal vascular response to angiotensin II in normal man, *J. Clin. Invest.,* 54, 34, 1974.
34. **Blantz, R. C., Konnen, K. S., and Tucker, B. J.,** Angiotensin II effects upon the glomerular microcirculation and ultrafiltration coefficient, *J. Clin. Invest.,* 57, 419, 1976.
35. **Frega, N. S., Davalos, M., and Leaf, A.,** Effect of endogenous angiotensin on the efferent glomerular arteriole of rat kidney, *Kidney Int.,* 18, 323, 1980.
36. **Hall, J. E., Guyton, A. C., Jackson, T. E., Coleman, T. G., Lohmeier, T. E., and Trippodo, N. C.,** Control of glomerular filtration rate by circulating angiotensin II, *Am. J. Physiol.,* 241, R190, 1981.
37. **Myers, B. D., Deen, W. M., and Brenner, B. M.,** Effects of norepinephrine and angiotensin II on the determinants of glomerular ultrafiltration and proximal tubule fluid reabsorption in the rat, *Circ. Res.,* 37, 101, 1975.
38. **Schuster, V. L., Kokko, J. P., and Jacobson, H. R.,** Angiotensin II directly stimulates sodium transport in rabbit proximal convoluted tubules, *J. Clin. Invest.,* 73, 507, 1984.
39. **Ballermann, B. J., Levenson, D. J., Brenner, and B. M.,** Renin, angiotensin, kinins, prostaglandins, and leukotrienes, in *The Kidney,* 3rd ed., Brenner, B. M. and Rector, F. C., Eds., W. B. Saunders, Philadelphia, 1986, 281.
40. **Houssay, B. A., Braun-Mendenez, E., and Dexter, L.,** The destruction and elimination of renin, *Ann. Intern. Med.,* 17, 461, 1942.
41. **Ziegler, M. and Gross, F.,** Effect of blood volume changes on renin-like activity in blood, *Proc. Soc. Exp. Biol. Med.,* 116, 774, 1963.
42. **Goldblatt, H., Lamfrom, H., and Haas, E.,** Physiological properties of renin and hypertension, *Am. J. Physiol.,* 175, 75, 1953.
43. **Schaechtelin, G., Regoli, D., and Gross, F.,** Quantitative assay and disappearance rate of circulating renin, *Am. J. Physiol.,* 206, 1361, 1964.
44. **Gross, F., Brunner, H., and Ziegler, M.,** Renin-angiotensin system, aldosterone, and sodium balance, *Rec. Prog. Hormone Res.,* 21, 119, 1965.
45. **Assaykeen, T. A., Otsuda, K., and Ganong, W. F.,** Rate of disappearance of exogenous dog renin from the plasma of nephrectomized dogs, *Proc. Soc. Exp. Biol. Med.,* 127, 306, 1968.
46. **Schneider, E. G., Rostorfer, H. H., and Nash, F. D.,** Distribution volume and metabolic clearance rate of renin in anesthetized nephrectomized dogs, *Am. J. Physiol.,* 215, 1115, 1986.
47. **Reid, I. A. and McDonald, I. R.,** The clearance of renin from the peripheral plasma of the marsupial trichosurus vulpecola, *J. Endocrinol.,* 45, 121, 1969.
48. **Peters-Haefeli, L.,** Rate of inactivation of endogenous and exogenous renin in normal and renin-depleted rats, *Am. J. Physiol.,* 221, 1339, 1971.
49. **Hannon, R. C., Deruyck, R. P., Joossens, J. V., and Amery, A. K.,** Disappearance rate of endogenous renin from the plasma after bilateral nephrectomy in humans, *J. Clin. Endocrinol.,* 29, 1420, 1969.
50. **Michelakis, A. M. and Mizukoshi, H.,** Distribution and disappearance rate of renin in man and dog, *J. Clin. Endocrinol.,* 33, 27, 1971.
51. **Laragh, J. H. and Sealey, J. E.,** The renin-angiotensin-aldosterone hormonal system and regulation of sodium, potassium, and blood pressure homeostasis, in *Handbook of Physiology,* Section 8: Renal Physiology, Orloff, J. and Berliner, R. W., Eds., American Physiologic Society, Washington, D.C., 1973, 831.
52. **Heacox, R., Harvey, A. M., and Vander, A. J.,** Hepatic inactivation of renin, *Circ. Res.,* 21, 149, 1967.
53. **Schneider, E. G., Davis, J. O., Baumber, J. S., and Johnson, J. A.,** The hepatic metabolism of renin and aldosterone: A review with new observation on the hepatic clearance of renin, *Circ. Res.,* 26-27 (Suppl. 1), 175, 1970.
54. **Schneider, E. G., Davis, J. O., Rob, C. A., and Baumber, J. S.,** Hepatic clearance of renin in canine experimental models for low- and high-output heart failure, *Circ. Res.,* 24, 213, 1969.
55. **Johnson, J. A., Davis, J. O., Baumber, J. S., and Schneider, E. G.,** Effects of hemorrhage and chronic sodium depletion on hepatic clearance of renin, *Am. J. Physiol.,* 220, 1677, 1971.
56. **Echtenkamp, S. F., Davis, J. O., Freeman, R. H., Dietz, J. R., and Villarrel, D.,** Splanchnic and renal contribution to circulatory homeostasis is sodium depletion, *Am. J. Physiol.,* 245, H573, 1983.

57. **Goldman, J. M., Vander, A. J., Mouw, D. R., Keiser, J., and Nicholls, M. G.,** Multiple short-term effects of lead on the renin-angiotensin system, *J. Lab. Clin. Med.,* 97, 251, 1981.
58. **Keiser, J. A., Opgenorth, T. J., and Romero, J. C.,** Hepatic clearance of renin after angiotensin blockade, *Am. J. Physiol.,* 248, R23, 1985.
59. **Christlieb, A. R., Couch, N. P., Amsterdam, E. A., Dobrzinsky, S. J., and Hickler, R. B.,** Renin extraction by human liver, *Proc. Soc. Exp. Biol. Med.,* 128, 821, 1968.
60. **Kokot, F., Kuska, J., and Czekala, Z.,** Hepatic inactivation of renin in man, *Enzymol. Biol. Clin.,* 9, 317, 1968.
61. **Barnardo, D. E., Strong, C. G., and Baldus, W. P.,** Failure of the cirrhotic liver to inactivate renin: Evidence for a splanchnic source of renin-like activity, *J. Lab. Clin. Med.,* 74, 495, 1969.
62. **Hesse, B., Andersen, E. D., and Ring-Larsen, H.,** Hepatic elimination of renin in man, *Clin. Sci. Mol. Med.,* 55, 377, 1978.
63. **Wilkinson, G. R. and Shand, D. G.,** A physiological approach to hapatic drug clearance, *Clin. Pharmacol. Ther.,* 18, 377, 1975.
64. **Horky', K., Rojo-Hortega, J. M., Rodriguez, J., and Genest, J.,** Renin uptake and excretion by liver in the rat, *Am. J. Physiol.,* 219, 387, 1970.
65. **Rector, W. G. and Hossack, K. R.,** Splanchnic and hepatic renin metabolism in alcoholic cirrhosis. Lack of evidence of a splanchnic source of production or of a significantly impaired hepatic clearance, *J. Heptaol.,* 7, 93, 1988.
66. **Boyd, G. W.,** The prolonged pressor response to renin in nephrectomized rats, *Circ. Res.,* 45, 396, 1976.
67. **Brown, J. J., Davies, D. L., Lever, A. F., Lloyd, A. M., Robertson, J. I. S., and Tree, M.,** A renin-like enzyme in normal human urine, *Lancet,* 2, 709, 1964.
68. **Pessina, A. C., Hulme, B., Rappelli, A., and Peart, W. S.,** Renin excretion in patients with renal disease, *Circ. Res.,* 27, 891, 1970.
69. **Lambers, E. R. and Skinner, S. L.,** Observations on the origin of renin in human urine, *Circ. Res.,* 24, 689, 1969.
70. **Iwamoto, T., Takaoka, M., Ohyama, M., and Morimoto, S.,** Purification and characterization of rat urinary renin, *Biochem. Int.,* 10, 359, 1985.
71. **Yukimura, T., Miura, K., Matsushima, Y., Ikemoto, F., and Yamamoto, K.,** Urinary excretion of renin and its biochemical properties in dogs, *Hypertension,* 6, 837, 1984.
72. **Rappelli, A. and Peart, W. S.,** Renal excretion of renin in the rat, *Circ. Res.,* 23, 531, 1968.
73. **Leyssac, P. P.,** A micropuncture study of glomerular filtration and tubular reabsorption of endogenous renin in the rat, *Renal Physiol.,* 1, 181, 1978.
74. **Bailie, M. D., Donoso, V. S., and Porter, T. M.,** Urinary excretion of renin in the dog: effect of changes in plasma renin, *J. Pharmacol. Exp. Ther.,* 225, 366, 1983.
75. **Iwao, H., Nakamura, N., Ikemoto F., and Yamamoto, K.,** Whole body autoradiographic distribution of exogenously administered renin in mice, *J. Histochem. Cytochem.,* 31, 776, 1983.
76. **Iwao, H., Nakamura, N., Ikemoto, F., and Yamanoto, K.,** Subcellular localization of exogenously administered renin in mouse kidney, *Jpn. Circ. J.,* 47, 1198, 1983.
77. **Iwao, H., Nakamura, N., Ikemoto, F., Yamamoto, K., Mizuhira, M., Ono, M., and Sugiura, Y.,** Distribution of exogenously administered renin in mouse kidney, *Clin. Exp. Hypertens.,* A4, 2449, 1982.
78. **Nakamura, M., Iwao, H., Ikemoto, F., and Yamamoto, K.,** Renal metabolism in mice of exogenously administered ^{125}I-labeled renin, *J. Hypertens.,* 2, 241, 1984.
79. **Corvol, P., Panther, J. J., Soubrier, F., Menard, J., and Rougeon, F.,** Mouse submaxillary renin: A useful model for the study of renal renin, *J. Hypertens.,* 1 (Suppl. 1), 3, 1979.
80. **Ashwell, G. and Harford, J.,** Carbohydrate-specific receptors of the liver, *Annu. Rev. Biochem.,* 51, 531, 1982.
81. **Kim, S., Iwao, H., Nakamura, N., Ikemoto, F., and Yamamoto, K.,** Fate of circulating renin in conscious rats, *Am. J. Physiol.,* 252, E136, 1987.
82. **Kim, S., Iwao, H., Nakamura, N., Ikemoto, F., and Yamamoto, K.,** Whole body autoradiographic distribution of exogenously administered renal renin in rats, *J. Histochem. Cytochem.,* 35, 549, 1987.
83. **Kim, S., Ikemoto, F., Murakami, K., and Yamamoto, K.,** Fate of recombinant human renin administered exogenously to anesthetized monkeys, *Hypertension,* 12, 317, 1988.
84. **Kim, S., Iwao, H., Nakamura N., Ikemoto, F., Yamamoto, K., Mizuhira, V., and Yokofujita, J.,** Cellular and subcellular distribution of exogenously administered renal renin in rat liver and kidney, *Am. J. Physiol.,* 253, E621, 1987.
85. **Shier, D. N. and Malvin, R. L.,** Differential secretion and removal of multiple renin forms, *Am. J. Physiol.,* 249, R79, 1985.
86. **Sessler, F. M., Jacquez, J. A., and Malvin, R. L.,** Different production and decay rates of six renin forms isolated from rat plasma, *Am. J. Physiol.,* 250, E551, 1986.
87. **Kim, S., Hiruma, M., Ikemoto, F., and Yamanoto, K.,** Importance of glycosilation for hepatic clearance of renal renin, *Am. J. Physiol.,* 255, E642, 1988.

88. **Regoli, D., Riniker, B., and Brunner, H.,** The enzymatic degradation of various angiotensin II drivatives by serum, plasma, or kidney homogenates, *Biochem. Pharmacol.*, 12, 637, 1963.
89. **Hodge, R. L., Lowe, R. D., and Vane, J. R.,** The effects of alterations of blood volume on the concentration of circulating angiotensin in anesthetized dogs, *J. Physiol.*, 185, 613, 1966.
90. **Doyle, A. E., Louis, W. J., Jerums, G., and Osborn, E. C.,** Metabolism and blood levels following infusion of a radioactive analog of angiotensin, *Am. J. Physiol.*, 215, 164, 1968.
91. **Cain, M. D., Catt, J. P., Coghlan, D. A., and Blair-West, J. R.,** Evaluation of angiotensin II in sheep by radioimmunoassay, *Endocrinology,* 86, 955, 1970.
92. **Naden, R. P., Coultrup, S., Arant, B. S., and Rosenfield, C. R.,** Metabolic clearance of angiotensin II in pregnant and nonpregnant sheep, *Am. J. Physiol.*, 249, E49, 1985.
93. **Donato, L., Coli, A., Pasqualini, R., and Duce, T.,** Metabolic clearance rate of radioiodinated angiotensin II in normal men, *Am. J. Physiol.*, 223, 1250, 1972.
94. **Methot, A. L., Meyer, P., Biron, P., Loraine, M. F., Lagrue, G., and Milliez, P.,** Hepatic inactivation of angiotensin, *Nature (London)*, 203, 531, 1964.
95. **Abrams, J. S.,** Blood pressure response to portal infusion of angiotensin II, *Surg. Forum,* 17, 133, 1966.
96. **Kelly, K. A. and Nyhus, L. M.,** Angiotensin and the liver, *Am. J. Physiol.*, 210, 305, 1966.
97. **Hodge, R. L., Ng, K. K. F., and Vane, J. R.,** Disappearance of angiotensin from the circulation of the dog, *Nature, (London),* 215, 138, 1967.
98. **Biron, P., Meyer, P., and Panisset, J.,** Removal of angiotensin from the systemic circulation, *Can. J. Physiol. Pharmacol.*, 46, 175, 1968.
99. **Leary, W. P. P. and Ledingham, J. G. G.,** Renal and hepatic inactivation of angiotensin in rats: influence of sodium balance and renal artery compression, *Clin. Sci.*, 38, 573, 1970.
100. **Osborne, M. J., Angles D'Auriac, C., Meyer, P., and Worcel, M.,** Mechanism of extraction of angiotensin II in coronary and renal circulation, *Life Sci.*, 9, 859, 1970.
101. **Oparil, S. and Bailie, M. D.,** Mechanisms of renal handling of angiotensin II in the dog, *Circ. Res.*, 33, 500, 1973.
102. **Ledingham, J. G. G.,** Removal of angiotensin by isolated perfused organs in rats, *Nature (London)*, 222, 959, 1969.
103. **Fei, D. T. W., Coghlan, J. P., Fernley, R. T., and Scoggins, B. A.,** Blood clearance rate of angiotensin II and its metabolites in sheep: presence of immunoreactive fragments in arterial blood, *Clin. Exp. Pharmacol. Physiol.*, 6, 129, 1979.
104. **Fei, D. T. W., Scoggins, B. A., Treagear, G. W., and Coghlan, J. P.,** Reduction of blood clearance rate of Val^5 angiotensin II by Sar^1, Ile^8 angiotensin II in sheep, *Life Sci.*, 29, 157, 1981.
105. **Johnston, C. I., Mendelsohn, F. A. O., and Doyle, A. E.,** Metabolism of angiotensin II in sodium depletion and hypertension in humans, *Circ. Res.*, 30-31 (Suppl. II), 203, 1972.
106. **Oelkers, W., Dusterdieck, G., and Morton, J. J.,** Arterial angiotensin II and venous immunoreactive material before and after angiotensin infusion in man, *Clin. Sci.*, 43, 209, 1972.
107. **Rosivall, L., Narkates, A. J., Oparil, S., and Navar, L. G.,** *De novo* intrarenal formation of angiotensin II during control and enhanced renin secretion, *Am. J. Physiol.*, 252, F1118, 1987.
108. **Pullmann, T. N., Oparil, S., and Carone, F. A.,** Fate of labeled angiotensin II microinfused into individual nephrons in the rat, *Am. J. Physiol.*, 228, 747, 1975.
109. **Peterson, D. R., Oparil, S., Flouret, G., and Carone, F. A.,** Handling of angiotensin II and oxytocin by renal tubular segments perfused *in vitro, Am. J. Physiol.*, 232, F319, 1977.
110. **Carone, F. A., Peterson, D. R., Oparil, S., and Pullman, T. N.,** Renal tubular transport and catabolism of proteins and peptides, *Kidney Int.*, 16, 271, 1979.
111. **Peterson, D. R., Chrabaszcz, G., Peterson, W. R., and Oparil, S.,** Mechanism for renal tubular handling of angiotensin, *Am. J. Physiol.*, 236, F365, 1979.
112. **Kugler, P.,** Localization of aminopeptidase A (angiotensinase A) in the rat and mouse kidney, *Histochemistry,* 72, 269, 1981.
113. **Bailie, M. D. and Oparil, S.,** Relation of renal hemodynamics to metabolism of angiotensin II by the canine kidney, *Circ. Res.*, 41, 283, 1977.
114. **Pearth, W. S.,** The renin-angiotensin system, *Pharmacol. Rev.*, 17, 143, 1965.
115. **Glenner, G. G., McMilan, P. J., and Folk, L. F.,** A mammalian peptidase specific for the hydrolysis of N-terminal and *L*-glutamyl and aspartic residues, *Nature (London),* 194, 867, 1962.
116. **Khairallah, P. A., Bumpus, F. M., Page, I. H., and Smeby, R. R.,** Angiotensinase with a high degree of specificity in plasma and red cells, *Science,* 140, 672, 1963.
117. **Semple, P. F., Boyd, A. S., Dawes, P. M., and Morton, J. J.,** Angiotensin II and its heptapeptide (2-8), hexapeptide (3-8) and pentapeptide (4-8) metabolites in arterial and venous blood of man, *Circ. Res.*, 39, 671, 1976.
118. **Bakhle, Y. S., Reynard, A. M., and Vane, J. R.,** Metabolism of angiotensin in isolated perfused tissues, *Nature (London),* 222, 956, 1969.

119. **Cain, M. D., Catt, K. J., and Coghlan, J. P.,** Immunoreactive fragments of angiotensin II in blood, *Nature (London),* 223, 617, 1969.
120. **Dietrich, F. M.,** Immunochemical analysis of rabbit antibodies against angiotensin II, *Immunochemistry,* 4, 65, 1967.
121. **Spech, H. J.,** Total immunoreactive angiotensin II, its hepta-octapeptide fraction, and its hexapeptide in patients with liver disease, *Klin. Wschr.,* 56, 399, 1978.
122. **Arroyo, V., Bosch, J., and Mauri, M., et al.,** Renin, aldosterone and renal hemodynamics in cirrhosis with ascites, *Eur. J. Clin. Invest.,* 9, 69, 1979.
123. **Wilkinson, S. P., Smith, K., and Wiliams, R.,** Changes in plasma renin activity in cirrhosis: a reappraisal based on studies in 67 patients and "low renin" cirrhosis, *Hypertension,* 1, 125, 1979.
124. **Sellars, L., Shore, A. C., Mott, V., and Wilkinson, R.,** The renin-angiotensin-aldosterone system in decompensated cirrhosis: its activity in relation to sodium balance, *Quart. J. Med.,* 56, 485, 1985.
125. **Bernardi, M., Trevisani, F., Santini, C., De Palma, R., and Gasbarrini, G.,** Aldosterone related blood volume expansion in cirrhosis before and during the early phase of ascites formation, *Gut,* 24, 761, 1983.
126. **Wilkinson, S. P., Jowett, J. D. H., Slater, J. D. H., Arroyo, V., Moodie, H., and Williams, R.,** Renal sodium retention in cirrhosis: relationship to aldosterone and nephron site, *Clin. Sci.,* 56, 169, 1979.
127. **Perez Ayuso, R. M., Arroyo, V., and Planas, R., et al.,** Randomized comparative study of efficacy of furosemide vs. spironolactone in non-azotemic cirrhosis with ascites. Relationship between the diuretic response and the activity of the renin-angiotensin-aldosterone system, *Gastroenterology,* 84, 961, 1983.
128. **Jimenez, W., Martinez-Pardo, A., and Arroyo, V., et al.,** Temporal relationship between hyperaldosteronism, sodium retention and ascites formation in rats with experimental cirrhosis, Effect of spironolactone, *Hepatology,* 5, 245, 1985.
129. **Epstein, M.,** Renal sodium handling in liver disease, in *The Kidney in Liver Disease,* 3rd ed., Epstein, M., Ed., Williams & Wilkins, Baltimore, 1988, 3.
130. **Bernardi, M., Servadei, U., and Trevisani, F.,** Importance of plasma aldosterone on natriuretic effect of spironolactone in patients with liver cirrhosis and ascites, *Digestion,* 31, 189, 1985.
131. **Mitch, W. E., Whelton, P. K., Cooke, C. R., Walker, W. G., and Maddrey, W. C.,** Plasma levels and hepatic extraction of renin and aldosterone in alcoholic liver disease, *Am. J. Med.,* 66, 804, 1979.
132. **Wernze, H., Seki, A., and Schneider, K. W.,** Hepatische Extraktion und Clearance von Renin bei Lebercirrhosen, *Klin. Wschr.,* 50, 302, 1972.
133. **Bosch, J., Arroyo, V., Mauri, M., Rivera, F., Navarro, F., and Rodés, J.,** Renin secretion rate and hepatic metabolism of renin and aldosterone in cirrhosis of the liver, Proc. 12 Meeting Eur. Association for the Study of the Liver, *Arch. Hellenic Med. Soc.,* 3 (Suppl.), 18, 1977.
134. **Perez Ayuso, R. M., Arroyo, V., Camps, J., et al.,** Renal kallikrein excretion in cirrhotics with ascites: relationship to renal hemodynamics, *Hepatology,* 4, 247, 1984.
135. **Bosch, J. Arroyo, V., Betriu, A., et al.,** Hepatic hemodynamics and the renin-angiotensin-aldosterone system in cirrhosis, *Gastroenterology,* 78, 92, 1980.
136. **Kotelanski, B., Groszmann, R. J., and Cohn, J. N.,** Circulation times in the splanchnic and hepatic beds in alcoholic liver disease, *Gastroenterology,* 63, 102, 1972.
137. **Nicholls, K. M., Shapiro, M. D., and Van Putten, V. J., et al.,** Elevated plasma norepinephrine concentrations in decompensated cirrhosis. Association with increased secretion rate, normal clearance rate and suppressibility by central blood volume expansion, *Circ. Res.,* 56, 457, 1985.
138. **Bichet, D. G., Groves, B., and Schrier, R. W.,** Mechanism of improvement of water and sodium excretion by immersion in decompensated cirrhotic patients, *Kidney Int.,* 24, 788, 1983.
139. **Kontos, H. H., Shapiro, W., Mauck, H. P., and Patterson, J. L.,** General and regional circulatory alterations in cirrhosis of the liver, *Am. J. Med.,* 37, 526, 1964.
140. **Lieberman, F. L. and Reynolds, T. B.,** Plasma volume in cirrhosis of the liver: its relation to portal hypertension, ascites and renal failure, *J. Clin. Invest.,* 46, 1297, 1967.
141. **Schrier, R. W., Arroyo, V., Bernardi, M., Epstein, M., Henriksen, J. H., and Rodés, J.,** Peripheral arterial vasodilation hypothesis: a proposal for the initiation of renal sodium and water retention in cirrhosis, *Hepatology,* 8, 1151, 1988.
142. **Schroeder, E. T., Anderson, G. H., Goldman, S. H., and Streeten, D. H. P.,** Effect of blockade of angiotensin II on blood pressure, renin and aldosterone in cirrhosis, *Kidney Int.,* 9, 511, 1976.
143. **Pariente, E. A., Bataille, C., Bercoff, E., and Lebrec, D.,** Acute effects of captopril on systemic and renal hemodynamics and on renal function in cirrhotic patients with ascites, *Gastroenterology,* 88, 1255, 1985.

Chapter 7

PROCESSING AND CLEARANCE OF ATRIAL NATRIURETIC FACTORS (ANF)

Alexander L. Gerbes and Angelika M. Vollmar

TABLE OF CONTENTS

I. INTRODUCTION

Atrial natriuretic factor (ANF) is the first well-defined natriuretic hormone. The best characterized pharmacological and partly physiological actions of ANF are natriuresis, diuresis, and smooth muscle relaxation (for review, see References 1 to 12). Recently, evidence has been provided that ANF might be involved in various other biological actions apart from its role in volume homeostasis such as immune or reproductive functions.[13–17]

ANF is synthesized in the human heart as a 151 amino acid preprohormone and stored as a 126 amino acid prohormone. Upon appropriate stimulation, such as atrial stretch, the prohormone is cleaved into an N-terminal and C-terminal fragment. The latter has been identified as the circulating bioactive ANF 99-126.[18,19] A body of investigations has concentrated on the mechanisms of ANF release and on its biological actions (for review, see References 1 to 14). Thus far, there is much less information on the processing of ANF prohormone as well as on the clearance of bioactive ANF. However, these seem to be topics of considerable interest, since plasma levels and biological activity of ANF are significantly influenced by these processes. This review article will concentrate on the mechanisms of ANF prohormone processing and of degradation and clearance of bioactive ANF, including pathophysiological aspects and possible therapeutic implications. The still poorly understood metabolism of ANF in the central nervous system will not be addressed in this chapter.

II. PROCESSING OF ATRIAL NATRIURETIC PROHORMONE

A. THE HEART

It is well established that the heart atria represent the major site of synthesis for ANF (for review, see References 20 and 21). Atrial muscle cells possess well-developed structures necessary for synthesizing, packaging, and storing this peptide. Initially, a common precursor termed prepro-ANF, which comprises between 149 and 153 amino acids depending on the species (151 amino acids in human), is being synthesized (see Figure 1). This prepro ANF molecule contains a signal peptide sequence at its amino terminus which seems to provide the signal for the co-translational transport of the peptide across the endoplasmatic reticulum. During this transport, the signal sequence is removed and the resulting 126-amino acid peptide is stored in secretory granules. In contrast to other prohormone forms of secreted proteins (e.g., proinsulin, prepromelanocortin, and proenkephalin) which are being cleaved to the mature form during vesicular packaging,[22–24] the ANF prohormone is stored in the atrial granules as the unprocessed prohormone.[25] The only 28-amino acid comprising carboxyterminus ANF(99-126) (see Figure 2) represents the biologically active hormone and is released together with the N-terminal ANF(1-98) into the circulation.[19,26–30] The site and exact mechanism of this proteolytic conversion of ANF(1-126) to ANF(99-126) is not yet known. Two possible mechanisms have to be considered. First, the cleavage occurs during secretional processing; this concept is supported by *in vitro* experiments showing that ANF(99-126) is released from isolated perfused hearts[27,31] and by the fact that plasma is not capable of performing the cleavage.[32] Secondly, data showing that isolated cultured myocytes release the intact pro-ANF provide evidence that the cleavage does occur extracellular of myocytes.[33,34] However, cleavage does not seem to take place in the circulation, because the release of ANF(99-126) from the isolated perfused rat heart seems not to be dependent on serum related enzymes.[32] In this regard, one must be aware of the fact that processing of ANF(1-126) from isolated cultured cells must not necessarily correspond to the *in vivo* conditions. Taken together, the cleavage of pro-ANF may take place during secretion, as well as shortly after secretion from the myocytes.[35–38]

In this context, the knowledge of the enzyme responsible for selectively cleaving the prohormone and, moreover, of its cellular or subcellular localization would be helpful to

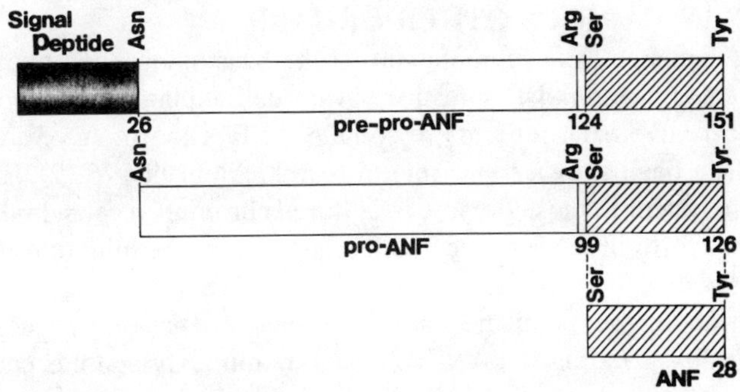

FIGURE 1. Schematic illustration of various ANF forms: synthesized prepro-ANF is stored as ANF(1-126), the prohormone. Upon release and cleavage, ANF(99-126), the circulating C-terminal fragment, exerts biological activities.

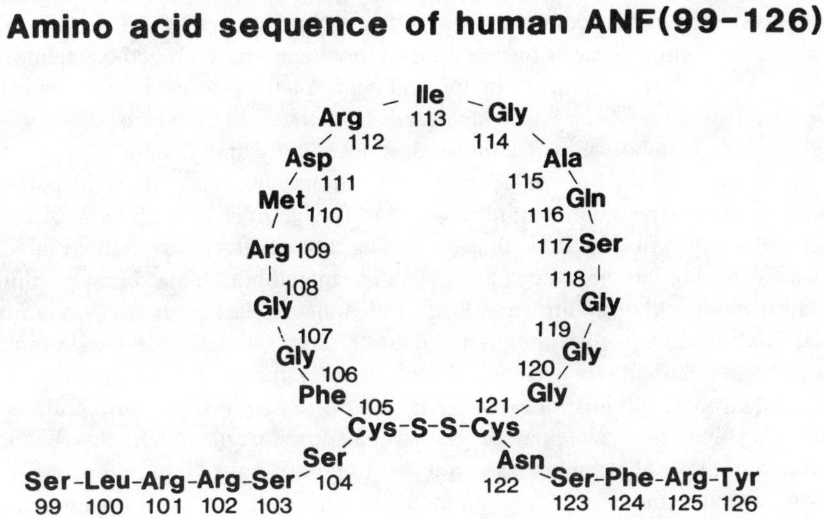

FIGURE 2. Amino acid sequence of the biologically active ANF(99-126).

elucidate this question. Recently, a serylprotease has been isolated from atrial tissue which selectively cleaves the Arg 98-Ser 99 bond of pro-ANF, resulting in ANF(99-126) and ANF(1-98).[39,40] The exact location of this peptide is still controversial. Several possibilities are being proposed:

1. Location of a serine protease with ANF (99-126) substrate specificity was described in the atrial granules containing ANF(1-126).[40] This finding can hardly be reconciled with the absence of ANF(99-126) in these granules.[25]
2. The microsomal fraction,[39] rather than the granule fraction of atrial cells, containing this enzyme activity would be in line with our understanding that pro-ANF is most likely cleaved during exocytotic secretion from its storage granules. Myocytes, but also adjacent atrial mesenchymal cells, may possess such an extragranularly located enzyme.[36]
3. Alternatively, this enzyme might be situated extracellularly;[36] however, this hypothesis has not been backed up by experimental findings thus far.

B. PROCESSING IN VARIOUS OTHER ORGANS

Over the years, evidence has accumulated that the heart is not the only peripheral site of ANF synthesis.[40a] Lung,[41] adrenals,[42] intestine,[17] various lymphoid organs,[15,43] and female reproductive organs[16,44] have been reported to contain ANF prohormone. So far, only the lung and the adrenals have been shown to be able to secrete ANF(99-126).[45,46] This suggests that maturation processes occur in these organs. In cultured chromaffin cells, both the prohormone as well as the mature form of ANF are co-secreted,[45] whereas the lung releases only ANF(99-126).[46]

No information about the exact mechanism of these maturation processes is available, however, a contribution of extraatrial ANF to the circulating ANF pool is being discussed.[41,46]

III. CLEARANCE OF BIOACTIVE ANF

A. ORGAN CLEARANCE

Plasma levels of bioactive ANF are determined by both release and clearance. While there is a rich body of information about ANF release, the number of studies on the clearance of this peptide is rather limited.[47–63] The ability of various organs to clear ANF from the circulation can be described by the extraction ratio or by the organ clearance. The extraction ratio is calculated as arterial minus venous blood concentration over arterial blood concentration (%). Multiplication of the extraction ratio with the organ plasma flow results in the organ clearance (liter/minute). Several aspects and possible pitfalls have to be taken into consideration regarding studies on ANF metabolism: since the studies necessarily apply venous and even arterial catheterization, for ethical reasons, many investigations have been done in patients with hypertension or congestive heart failure rather than in healthy subjects. Some of the studies did not determine organ plasma flow, thus restricting their results to extraction ratios. Extraction ratios as well as organ clearance have been determined both under basal conditions and following intravenous ANF administration. For clearance determinations of pharmacologically active ANF doses, specific attention should be paid to the rate of infusion in order to achieve steady-state conditions. Problems may arise by inappropriate intervals of plasma sampling. Furthermore, administration of ANF may decrease blood pressure,[56] affect cardiac output and organ blood flow, and increase glomerular filtration rate; this might result in under- or overestimation of ANF clearance as compared to physiological conditions.

Furthermore, some studies are based on the assumption that there is no pulmonary extraction of ANF. However, there is now evidence, obtained by pulmonary wedge or pulmonary venous sampling,[47,48] that the lung is an important site for ANF extraction.

With these reservations in mind, it seems that there is no major organ specificity for ANF extraction. Lung, liver, kidney, intestine, and lower limb show similar extraction ratios within the respective studies. Values reported, however, differ considerably between various studies (Table 1).[47–51,53,54,56,58–61,63] Extraction ratios were reported to be 0 to 33% for lung, 22 to 75% for liver and intestine, 23 to 67% for kidney, and 15 to 61% for the lower limb. Taken together, it seems that there is a general arterio-venous organ extraction ratio for ANF of about 35%.

Since there are marked differences of organ blood flow, obviously total ANF clearance differs considerably between various organs. Few investigations have determined organ blood flow together with ANF extraction under physiological or steady-state conditions.[47,49,51,56] About 50% of the total body clearance of ANF seems to be furnished by the lung.[47] The renal[47,49,51] (around 15%) and splanchnic-hepatic[47,49,56] (about 20%) vascular beds also contribute significantly to ANF clearance.

Several mechanisms could account for the above-mentioned organ clearance of ANF: receptor-mediated uptake, enzymatic degradation, or organ-specific excretion (urine or bile). However, there are only a few reports of concentrations of ANF in bile[47] or urine.[57,62] The following sections in this chapter will cover enzymatic degradation of ANF, as well as ANF clearance receptors.

TABLE 1
Clearance of ANF by Various Organs

Author/ year	Species	Healthy	(n)	Arterial sampling site	ANF arterio-venous extraction ratio (%) Kidney	Liver/ intestine	Lung	Lower limb	Ref.
Crozier, 1986	Man	No	4—8	Pulmonary/ aortic	50	45	<10	50	58
Henriksen, 1986	Man	Yes	8—14	Femoral	54	28	—	40	54
Schütten, 1987	Man	?	8—14	Pulmonary/ femoral	54	28	0	40	53
Gines, 1988	Man	?	11	Pulmonary	—	71	—	61	63
Vierhapper, 1988	Man	Yes	6	Femoral	—	75	—	—	56
Woods, 1988	Dog	Yes	5	Aortic	55	—	—	—	51
Bates, 1989	Dog	Yes	14	Pulmonary	—	—	32	—	48
Hollister, 1989	Man	No	8—39	Pulmonary/ aortic	35	30	24	38	47
	Dog	Yes	4—7	Pulmonary/ aortic	42	36	19	—	
Krieter, 1989[a]	Rat	Yes	4—7	Femoral	55—61	22—44	—	43—54	49
Kurosawa, 1989	Man	No	6—56	Pulmonary/ femoral/coeliac/ renal	23	24	12	15	59
Masuda, 1989	Dog	Yes	6	Pulmonary/ renal/ hepatic	34	28	33	—	50
Sudhir, 1989	Man	Yes	6	Brachial	67	—	—	—	61
Henriksen, 1990	Man	Yes	6	Femoral	—	53	—	—	60

[a] Data obtained during ANF infusion.

Note: Arteriovenous extraction ratios are calculated as (arterial minus venous) over arterial ANF plasma concentrations and given in percent (%). Sites of sampling for arterial blood are shown in the shown in the table; ideally the arterial plasma concentrations should be determined in the respective organ-supplying arteries. —: not determined.

1. Enzymatic Degradation

Since the kidney is a major target organ for ANF, interest has focused on this organ, not only as a site for receptor mediated clearance of this peptide (see Section III.A.2), but also as a tissue which may be able to enzymatically degrade circulating ANF(99-126).

Peptidase degradation of ANF is considered to play an important role among possible clearance mechanisms[64,65] (see also Chapter 4). The brush border of the renal proximal tubule is known to be very rich in peptidases. These enzymes are membrane bound and their active sites face the lumen of the tubule.[66] Incubating ANF(99-126) with these kidney microvilli results in a major degradation product cleaved between cysteine and phenylalanine (Cys 105-Phe 106).[64,65,67–69] Since the intact ring structure of ANF(99-126) is essential for its biological activity,[70,71] ANF is effectively inactivated by this cleavage process.

The properties of the ANF degrading activity have been investigated by several groups. The responsible substance is a glycoprotein[64,67,69] with a molecular weight of approximately 94 kDa, sensitive to metalloendoprotease inhibitors such as phosphoramidon and thiorphan. In view of these findings, the major proteolytic enzyme responsible for the ANF cleavage seems to be the endoprotease EC 3.4.24.11.[63,67,69] EC 3.4.24.11 has been reported also to possess enkephalin degrading activity.[72] However, there may be various forms of EC 3.4.24.11 differing in cleavage site specificity and tissue distribution.[67,73-77] Thus, one may speculate that the kidney possesses a specific type of endoprotease EC 3.4.24.11.

Apart from the major cleavage site within the loop structure, secondary attack by aminopeptidases has been hypothesized.[64] These enzymes are not initially responsible for degradation because specific inhibitors like captopril or amastatin have no significant effect.[64] After the major cleavage process at the Cys-Phe bond of the ring structure, more complex degradation processes seem to follow, producing smaller ANF fragments.[68] The major sites of renal ANF degradation by the endopeptidase seem to be the proximal tubule and the glomerulus.[78]

Prompted by these observations, efforts have been made to develop ANF analogues which are resistant to this enzyme, as well as specific enzyme inhibitors which could be therapeutically employed (see Section III.C).

Recent reports indicate ANF degrading activity by the vasculature[79] and the adrenal capsule.[73] No information is available about the enzymatic degradation activity of other organs reported to contribute to ANF clearance (see Section III.A).

2. Clearance Receptors

Specific binding sites for ANF have been identified on various cells such as endothelial cells, vascular smooth muscle cells, inner medullary collecting duct cells, and in various organs such as lung, kidney, adrenal gland, liver, and intestine.[80–82] Several observations demonstrate that there are at least two different types of ANF receptors with different molecular size and different biological properties.[83–88] The development of truncated ANF analogues which selectively bind to one receptor type such as des[(18-22)]-rANF(4-23)-NH_2,[89,90] recently has allowed for better quantitative discrimination of both receptors.

The current concept of both receptors is as follows: the 130-kDa receptor exhibits a higher affinity for circulating biologically active ANF, is coupled to cyclic guanylate cyclase, and mediates the biological effects. Thus, it has been called B- (biological) or R_1-receptor. The other receptor type shows a molecular weight of 65 kDa, exhibits a lower affinity to biologically active ANF, and binds truncated ANF analogues; it has thus been called C- (clearance) or R_2-receptor.[91–94] C-receptors might mediate suppression of adenylate cyclase activity by ANF.[95,96] Interestingly, it has been shown that the C-receptor may regulate the biologic activity of endogenously released ANF,[94] and there is evidence that this receptor type indeed has a clearance function.[91]

In anesthetized rats, administration of an ANF analogue selectively binding to C-receptors induced marked dose-dependent increases of ANF(99-126) plasma concentration and decreases of its volume of distribution and metabolic clearance rate. Furthermore, there is evidence that a major part of ANF mertabolism is due to an internalization following the binding of circulating ANF to the C-receptor and subsequent intralysomal hydrolysis.[91] The regulation of the C-receptor is poorly understood as yet. There is evidence that maneuvers decreasing intravascular volume decrease ANF plasma levels and, in turn, increase total ANF receptor density.[97–99] Surprisingly, this increase of ANF receptor density is coupled to a reduced biological response to ANF. The observation that C-receptors, but not B-receptors, are augmented following dehydration[100] could explain this seeming contradiction and indicates that regulation of C-receptor density might be involved in volume homeostasis. There are few data on the quantitative relationship between B- and C-receptors in various tissues. In the kidney, in particular on glomeruli, the majority of ANF receptors seems to be of the clearance type.[100,101] The same has been observed on cultured bovine aorta endothelial cells,[83] on adrenal zona glomerulosa cells,[101] and on lung membranes.[88]

These recent observations are compatible with the concept of a functional importance for the C-receptors, which are distinctly different from the B-receptors, in the clearance of ANF — a rather unique mechanism in the metabolism of peptides.

B. PATHOPHYSIOLOGY

Elevated plasma concentrations of ANF have been found in various pathophysiological states, such as congestive heart failure, renal failure, and in some patients with liver disease (see References 102 to 105). An increased release of ANF has been suggested in these diseases.[63,106–109] However, elevated plasma concentrations might also be caused by decreased clearance. To date, little is known about clearance in pathophysiological states. Recently, it has been reported that the fractional extraction of various organs is unchanged in patients with liver cirrhosis.[54] In an experimental model, it has been shown that the renal clearance of ANF varies considerably, depending upon glomerular filtration rate;[51] thus, augmented ANF plasma concentrations in renal diseases might be partly due to reduced ANF elimination. The proportion of B- and C-receptors for ANF in aorta, adrenal, and kidney seems unchanged in experimental congestive heart failure.[101] In an experimental model of cirrhosis, C-receptor density on glomeruli was increased, whereas B-receptors tended to be decreased;[110] however, these animals were not investigated for ANF clearance. Clearly, there is a need for further investigations on whether and how metabolic clearance and degradation of ANF is affected by pathophysiological states.

C. THERAPEUTIC IMPLICATIONS

Beneficial effects of ANF application have been reported in congestive heart failure, hypertension, renal failure, chronic obstructive lung disease, and liver disease.[111–118] Augmented circulating levels of ANF may be achieved by exogenous administration of ANF. Due to the peptide character of the substance, ANF can be administered only intravenously. The rapid metabolic clearance[119–121] makes repeated bolus injections or continuous infusions necessary. Alternatively, circulating ANF concentrations might be elevated by blocking its clearance. This might be achieved by administering truncated analogues, binding to clearance receptors only and, thus, augmenting the concentration of ANF reaching the B-receptors.[91,94,122]

Furthermore, inhibitors of the ANF degrading peptidase might show the same results. Indeed, several investigations employing endopeptidase EC 3.4.24.11 inhibitors[122–132] showed that the circulating ANF immunoreactivity was elevated and this was accompanied by an augmented biological activity, such as increased urinary volume and sodium excretion. Interestingly, these chemicals seem to be especially useful in pathophysiological conditions. They have been shown to enhance the biological activities (natriuretic and depressor responses) of ANF in experimental models of hypertension[124,130–132] and volume expansion.[128] Further investigations will evaluate the potential of this rather elegant therapeutic approach.

IV. SUMMARY

Atrial natriuretic factor, the first well-defined natriuretic hormone is synthesized in the human heart as 151 amino acid (AA) preprohormone and stored as 126 AA prohormone in atrial granules. Upon appropriate stimulation, the prohormone is cleaved into a 98 AA N-terminal fragment and a 28 AA C-terminal fragment, the biological active ANF(99-126), both circulating in plasma. The cleavage occurs most likely either during or shortly after the release from the myocyte granules. The respective enzyme, a serylprotease, is discussed as being bound to plasma membranes of myocytes or of atrial mesenchymal cells. There is increasing evidence for ANF synthesis and processing in peripheral organs other than the heart.

Circulating ANF(99-126) is cleared by various organs such as lung, liver, intestine, kidney, and upper and lower limbs. Reported arterio-venous extraction ratios vary greatly, but do not differ much between organs, the average extraction ratio being bout 35%. Due to marked differences of organ blood flow, the contribution of various organs to total body ANF clearance differs considerably.

Major mechanisms for ANF clearance are uptake by clearance receptors and degradation by an endoprotease EC 3.4.24.11. Clearance receptors, distinctly different from the receptors mediating the biological actions of ANF, have been demonstrated in various organs. Characterization of the ANF degrading enzyme activity has been performed in kidney tissue. Whether and how pathophysiological states affect ANF clearance is still poorly understood. Inhibition of clearance by ANF analogues binding to clearance receptors and by inhibitors of degrading peptidase can increase the biological action of circulating ANF. This may prove to be a therapeutic approach in diseases with smooth muscle contraction or volume overload.

ACKNOWLEDGMENTS

Jens H. Henriksen, M.D., Ph.D., University of Copenhagen, Denmark, and William R. Samson, Ph.D., University of Missouri–Columbia, are thanked for critical reading of the manuscript. M. Hummel is thanked for preparation of the manuscript. Alexander L. Gerbes, M.D., has been supported by the Deutsche Ferschungsgemeinschaft (Ge 576).

REFERENCES

1. **Anderson, J. V. and Bloom, S. R.,** Atrial natriuretic peptide: what is all the excitement about?, *J. Endocrinol.,* 110, 7, 1986.
2. **Arendt, R. M. and Gerbes, A. L.,** Atrialer natriuretischer Faktor. Die endokrine Funktion des Herzens, *Dtsch. Med. Wschr.,* 111, 1849, 1986.
3. **Ballermann, B. J. and Brenner, B. M.,** Biologically active atrial peptides, *J. Clin. Invest.,* 76, 2041, 1985.
4. **Cantin, M. and Genest, J.,** The heart as an endocrine gland, *Sci. Am.,* 254, 76, 1986.
5. **De Bold, A. J.,** Atrial natriuretic factor: an overview, *Fed. Proc.,* 45, 2081, 1986.
6. **Espiner, E. A. and Richards, A. M.,** Atrial natriuretic peptide. An important factor in sodium and blood pressure regulation, *Lancet,* 707, 1989.
7. **Inagami, T.,** Atrial natriuretic factor, *J. Biol. Chem.,* 264, 3043, 1989.
8. **Lang, R. E., Unger, T., and Ganten, D.,** Atrial natriuretic peptide: a new factor in blood pressure control, *J. Hypertens.,* 5, 255, 1987.
9. **Laragh, J. H. and Atlas, S. A.,** Atrial natriuretic hormone: a regulator of blood pressure and volume homeostasis, *Kidney Intern.,* 34, S64, 1988.
10. **Needleman, P. J. and Greenwald, J. E.,** Atriopeptin: a cardiac hormone intimately involved in fluid, electrolyte, and blood-pressure homeostasis, *N. Engl. J. Med.,* 314, 828, 1986.

11. **Thibault, G., Garcia, R., Gutkowska, J., Genest, J., and Cantin, M.,** Atrial natriuretic factor. A newly discovered hormone with significant clinical implications, *Drugs,* 31, 369, 1986.
12. **Trippodo, N. C.,** An update on the physiology of atrial natriuretic factor, *Hypertension,* 10 (Suppl.I), 122, 1987.
13. **Vollmar, A. M.,** Atrial natriuretic peptide in peripheral organs other than the heart, *Klin. Wochenschr.,* 68, 699, 1990.
14. **Gutkowska, J. and Nemer, M.,** Structure, expression and function of atrial natriuretic factor in extraatrial tissues, *Endocrine Rev.,* 10, 519, 1989.
15. **Vollmar, A. M. and Schulz, R.,** Atrial natriuretic peptide is synthesized in the human thymus, *Endocrinology,* 126, 2277, 1990.
16. **Vollmar, A. M., Mytzka, C., Arendt, R. M., and Schulz, R.,** Atrial natriuretic peptide in bovine corpus luteum, *Endocrinology,* 123, 762, 1988.
17. **Vollmar, A. M., Friedrich, A., Sinowatz, F., and Schulz, R.,** Presence of atrial natriuretic peptide like material in guinea pig intestine, *Peptides,* 9, 965, 1988.

17a. **Nemer, M., Chamberlend, M., Sirois, M., Argertin, S., Drouin, J., Dixon, R. A., Fivin, R., and Condra, J. A.,** Gene structure of human cardiac hormone precursor, pro-natriodilatin, *Nature,* 392, 654, 1984.

18. **Sugiyama, M., Fukume, H., Gammer, R. T., Misono, K. S., Yabe, Y., Morisawa, Y., and Inagami, T.,** Synthesis of atrial natriuretic peptide and studies on structural factors in tissue specificity, *Biochem. Biophys. Res. Commun.,* 123, 338, 1984.
19. **Schwartz, D., Geller, D. M., Manning, P. T., Siegel, N. R., Fok, K. R., Smith, C. E., and Needleman, P.,** Ser-Leu-Arg-Arg-atriopeptin III: the major circulating form of atrial peptide, *Science,* 229, 397, 1985.
20. **Cantin, M. and Genest, J.,** The heart and the atrial natriuretic factor, *Endocrinol. Rev.,* 6, 107, 1985.
21. **De Bold, A. J.,** Atrial natriuretic factor: a hormone produced by the heart, *Science,* 230, 767, 1985.
22. **Moore, H. P. H., Walker, M. D., Lee, F., and Kelly, R. B.,** Expressing a human proinsulin cDNA in a mouse ACTH-secreting cell. Intracellular storage, proteolytic processing, and secretion on stimulation, *Cell,* 35, 531, 1983.
23. **Moore, H. P. H. and Kelly, R. B.,** Secretory protein targeting in a pituitary cell line: differential transport of foreign secretory proteins to distinct secretory pathways, *J. Cell Biol.,* 101, 1773, 1985.
24. **Comb, M., Liston, D., Martin, M., Rosen, H., and Herbert, E.,** Expression of the human enkephalin gene in mouse pituitary cells: accurate and efficient mRNA production and proteolytic processing, *EMBO J.,* 4, 3115, 1985.
25. **Thibault, G., Garcia, R., Gutkowska, J., Bilodeau, J., Lazure, C., Seidah, N. G., Chretien, M., Genest, J., and Cantin, M.,** The propeptide Asn1-Tyr126 is the storage form of rat atrial natriuretic factor, *Biochem. J.,* 241, 265, 1987.
26. **Thibault, G., Lazure, C., Schiffrin, E. L., Gutkowska, J., Cartier, L., Garcia, R., Seidah, N. G., Chretien, M., Genest, J., and Cantin, M.,** Identification of a biologically active circulating form of rat atrial natriuretic factor, *Biochem. Biophys. Res. Commun.,* 130, 981, 1987.
27. **Thibault, G., Garcia, R., Gukowska, J., Lazure, C., Seidah, N. G., Chrétien, M., Genest, J., and Cantin, M.,** Identification of the released form of atrial natriuretic factor by the perfused rat heart, *Proc. Soc. Exp. Biol. Med.,* 182, 137, 1986.
28. **Lang, E., Tholken, H., Ganten, D., Luft, F. C., Ruskoaho, H., and Unger, T.,** Atrial natriuretic factor — a circulating hormone stimulated by volume loading, *Nature (London),* 314, 264, 1985.
29. **Gerbes, A. L. and Vollmar, A. M.,** Water immersion increases the concentration of the immunoreactive N-terminal fragment of proatrial natriuretic factor in human plasma, *Biochem. Biophys. Res. Commun.,* 156, 228, 1988.
30. **Sundsfjord, J. A., Thibault, G., Larochelle, P., and Cantin, M.,** Identification and plasma concentrations of the N-terminal fragment of proatrial natriuretic factor in man, *J. Clin. Endocrinol. Metab.,* 66, 605, 1988.
31. **Trippodo, N. C., Chai, R. D., MacPhee, A. A., and Cole, F. E.,** Atrial natriuretic factor: atrial conversion of high to low molecular weight form, *Biochem. Biophys. Res. Commun.,* 119, 282, 1984.
32. **Gibson, T. R., Shields, P. P., and Glembotski, C. C.,** The conversion of atrial natriuretic peptide (ANP)-(1-126) to ANP-(99-126) by rat serum: contribution to ANP cleavage in isolated perfused rat hearts, *Endocrinology,* 120, 764, 1987.
33. **Bloch, K. D., Scott, J. A., Zisfein, J. B., Fallon, J. T., Seidman, C. E., Matsueda, G. R., Margolis, M. N., Homcy, C. J., Graham, R. M., and Seidman, J. G.,** Biosynthesis and secretion of proatrial natriuretic factor by cultured rat cardiocytes, *Science,* 230, 1168, 1985.
34. **Glembotski, C. C. and Gibson, T. R.,** Molecular forms of immunoreactive atrial natriuretic peptide released from cultured rat atrial myocytes, *Biochem. Biophys. Res. Commun.,* 132, 1008, 1985.
35. **Wildey, G. M., Fischman, A. J., Fallon, J. T., Matsueda, G. R., Zisfein, J. B., Preibisch, G., Seipke, G., Homcy, C. J., and Graham, R. M.,** Cellular processing of pro-atrial natriuretic factor (pro-ANF): studies using an antiserum that selectively binds ANF(99-126) after its cleavage from pro-ANF, *Endocrinology,* 123, 2054, 1988.

36. **Michener, M. L., Gierse, J. K., Seetharam, R., Fok, K. F., Olins, P. O., Mai, M. S., and Needleman, P.,** Proteolytic processing of atriopeptin prohormone, *Mol. Pharmacol.*, 30, 552, 1986.
37. **Ito, T., Toki, Y., Siegel, N., Gierse, J. K., and Needleman, P.,** Manipulation of stretch-induced atriopeptin prohormone release and processing in the perfused rat heart, *Proc. Natl. Acad. Sci. U.S.A.*, 85, 8365, 1988.
38. **Thibault, G., Haile-Meskel, H., Wrobel-Konrad, E., Ballak, M., Garcia, R., Genest, J., and Cantin, M.,** Processing of the atrial natriuretic factor propeptide by atrial cardiocytes as revealed by immunocryoultramicrotomy, *Endocrinology*, 124, 3109, 1989.
39. **Imada, T., Takayanagi, R., and Inagami, T.,** Atrioactivase, a specific peptidase in bovine atria for the processing of proatrial natriuretic factor, *J. Biol. Chem.*, 263, 9515, 1988.
40. **Wypij, D. M. and Harris, R. B.,** Atrial granules contain an amino-terminal processing enzyme of atrial natriuretic factor, *J. Biol. Chem.*, 263, 7079, 1988.

40a. **Gardner, D. G., Deschepper, C. F., Ganong, W. F., Hane, S., Fides, J., Baxter, J. D., and Lewicki, J.,** Extra-atrial expression of the gene for atrial natriuretic factor, *Proc. Natl. Acad. Sci. U.S.A.*, 83, 6697, 1986.

41. **Gutkowska, J., Nemer, M., Sole, M. J., Drouin, J., and Sirois, P.,** The lung is an important source of atrial natriuretic factor in experimental cardiomyopathy, *J. Clin. Invest.*, 83, 1500, 1989.
42. **Ong, H., Lazure, C., Nguyen, T. T., McNicoll, N., Seidah, N., Chretien, M., and DeLean, A.,** Bovine adrenal chromaffin granules are a site of synthesis of atrial natriuretic factor, *Biochem. Biophys. Res. Commun.*, 147, 957, 1987.
43. **Vollmar, A. M., Friedrich, A., and Schulz, R.,** Immunoreactive atrial natriuretic peptide in the guinea pig spleen, *Life Sci.*, 45, 1293, 1989.
44. **Kim, S. H., Cho, W. K., Seul, K. H., Ryn, H., and Koh, G. Y.,** Presence of immunoreactive atrial natriuretic peptide in follicular fluid, ovary and ovarian perfusates, *Life Sci.*, 45, 1581, 1989.
45. **Nguyen, T. T., Ong, H., and DeLean, A.,** Secretion and biosynthesis of atrial natriuretic factor from altered chromaffin cells, *FEBS Lett.*, 231, 393, 1988.
46. **Gutkowska, J., Cantin, M., Genest, J., and Sirois, P.,** Release of immunoreactive atrial natriuretic factor from the isolated rat lung, *FEBS Lett.*, 214, 17, 1987.
47. **Hollister, A. S., Rodeheffer, R. J., White, F. J., Potts, J. R., Imada, T., and Inagami, T.,** Clearance of atrial natriuretic factor by lung, liver and kidney in human subjects and the dog, *J. Clin. Invest.*, 83, 623, 1989.
48. **Bates, E. R., McGillem, M. J., Mancini, J., and Grekin, R. J.,** Pulmonary extraction of immunoreactive atrial natriuretic factor in dogs, *Am. J. Cardiol.*, 63, 372, 1989.
49. **Krieter, P. A. and Trapani, A. J.,** Metabolism of atrial natriuretic peptide. Extraction by organs in the rat, *Drug Metab. Dispos.*, 17, 14, 1989.
50. **Masuda, T., Shichiri, M., and Marumo, F.,** Degradation of atrial natriuretic peptide in dogs, *Acta Endocrinol.*, 120, 170, 1989.
51. **Woods, R. L.,** Contribution of the kidney to metabolic clearance of atrial natriuretic peptide, *Am. J. Physiol.*, 255, E934, 1988.
52. **Yukimura, T., Ito, K., Miura, K., and Yamamoto, K.,** Plasma disappearance and urinary excretion of alpha-human atrial natriuretic polypeptide in anaesthetized dogs, *J. Hypertens.*, 4 (Suppl. 6), S532, 1986.
53. **Schütten, H. J., Henriksen, J. H., and Warberg, J.,** Organ extraction of atrial natriuretic peptide (ANP) in man. Significance of sampling site, *Clin. Physiol.*, 7, 125, 1987.
54. **Henriksen, J. H., Schütten, H. J., Bendtsen, F., and Warberg, J.,** Circulating atrial natriuretic peptide (ANP) and central blood volume (CBV) in cirrhosis, *Liver*, 6, 361, 1986.
55. **Tang, J., Webber, R. J., Chang, D., Chang, J. K., Kiang, J., and Wei, E. T.,** Depressor and natriuretic activities of several atrial peptides, *Regul. Pept.*, 9, 53, 1984.
56. **Vierhapper, H., Gasic, S., Nowotny, P., and Waldhäusl, W.,** Splanchnic disposal of atrial natriuretic peptide in humans, *Metabolism*, 37, 973, 1988.
57. **Marumo, F., Sakamoto, H., Ando, K., Ishigami, T., and Kawakami, M.,** A highly sensitive radioimmunoassay of atrial natriuretic peptide (ANP) in human plasma and urine, *Biochem. Biophys. Res. Commun.*, 137, 231, 1986.
58. **Crozier, I. G., Nicholls, M. G., Ikram, H., Espiner, E. A., Yandle, T. G., and Jans, S.,** Atrial natriuretic peptide in humans. Production and clearance by various tissues, *Hypertension*, 8 (Suppl. II), 11, 1986.
59. **Kurosawa, T., Hiramatsu, S., Katoh, Y., Kikawada, R., and Marumo, F.,** No specific organ removes atrial natriuretic peptide from the human plasma, *Jpn. Circ. J.*, 53, 779, 1989.
60. **Henriksen, J. H., Bendtsen, F., and Gerbes, A. L.,** Splanchnic removal of human alpha atrial natriuretic peptide (human ANF 99-126) in man: enhancement after food intake, *Metabolism*, 38, 553, 1990.
61. **Sudhir, K., Friberg, P., Meredith, I. T., Woods, R. L., Esler, M. D., and Jennings, G. L.,** Cardiac secretion and renal clearance of atrial natriuretic peptide in normal man: effect of salt restriction, *Clin. Sci.*, 77, 605, 1989.
62. **Ando, K., Umetani, N., Kurosawa, T., Takeda, S., Katoh, Y., and Marumo, F.,** Atrial natriuretic peptide in human urine, *Klin. Wochenschr.*, 66, 768, 1988.

63. **Ginès, P., Jimenéz, W., Arroyo, V., Navasa, M., López, C., Tító, L., Serra, A., Bosch, J., Sanz, G., Rivera, F., and Rodés, J.,** Atrial natriuretic factor in cirrhosis with ascites: plasma levels, cardiac release and splanchnic extraction, *Hepatology,* 8, 636, 1988.
64. **Stephenson, S. L. and Kenny, A. J.,** The hydrolysis of α-human atrial natriuretic peptide by pig kidney microvillar membranes is initiated by endopeptidase 24.11, *Biochem. J.,* 243, 183, 1987.
65. **Olins, G. M., Spear, K. L., Siegel, N. R., Reinhard, E. J., and Zurcher-Neely, H. A.,** Atrial peptide inactivation by rabbit-kidney brush-border membranes, *Eur. J. Biochem.,* 170, 431, 1987.
66. **Kenny, A. J. and Maroux, S.,** Topology of microvillar membrane hydrolases of kidney and intestine, *Physiol. Rev.,* 62, 91, 1982.
67. **Bertrand, P. and Doble, A.,** Degradation of atrial natriuretic peptides by an enzyme in rat kidney resembling neutral endopeptidase 24.11, *Biochem. Pharmacol.,* 37, 3817, 1988.
68. **Koehn, J. A., Norman, B. N., Jones, L., LeSueur, L., Sakane, Y., and Ghai, R. D.,** Degradation of atrial natriuretic factor by kidney cortex membranes. Isolation and characteristics of the primary proteolytic products, *J. Biol. Chem.,* 262, 11623, 1987.
69. **Sonnenberg, J. L., Sakane, Y., Jeng, A. Y., Koehn, J. A., Ansell, J. A., Wennogle, L. P., and Ghai, R. D.,** Identification of protease 3.4.24.11 as the major atrial natriuretic factor degrading enzyme in the rat kidney, *Peptides,* 9, 173, 1988.
70. **Misono, K. S., Fukumi, H., Grammer, R. T., and Inagami, T.,** Rat atrial natriuretic factor: complete amino acid sequence and disulfide linkage essential for biological activity, *Biochem. Biophys. Res. Commun.,* 119, 524, 1984.
71. **Chartier, L., Schiffrin, E., and Thibault, G.,** Effect of atrial natriuretic factor (ANF)-related peptides on aldosterone secretion by adrenal glomerulosa cells: critical role of the intramolecular disulphide bond, *Biochem. Biophys. Res. Commun.,* 122, 171, 1984.
72. **Fulcher, I. S., Matsas, R., Turner, J., and Kenny, A. J.,** Kidney neutral endopeptidase and the hydrolysis of enkephaline by synaptic membranes show similar sensitivity to inhibitors, *Biochem. J.,* 203, 519, 1982.
73. **Schiebinger, R. J., Pratt, J. H., and Kem, D. C.,** The adrenal capsule alters the response of zona glomerulosa cells to atrial natriuretic peptide, *Endocrinology,* 123, 492, 1988.
74. **Lieorens, C. and Schwartz, J. C.,** Enkephalinase activity in rat peripheral organs, *Eur. J. Pharmacol.,* 69, 113, 1981.
75. **Matsas, R., Fulcher, I. S., Kenny, A. J., and Turner, A. J.,** Substance P and [leu]enkephalin are hydrolysed by an enzyme in pig caudate synaptic membranes that is identical with the endopeptidase of kidney microvilli, *Proc. Natl. Acad. Sci. U.S.A.,* 80, 3111, 1983.
76. **Matsas, R., Kenny, A. J., and Turner, A. J.,** The metabolism of neuropeptides, *Biochem. J.,* 223, 433, 1984.
77. **Matsas, R., Turner, A. J., and Kenny, A. J.,** Endopeptidase 24.11 and aminopeptidase activity in brain synaptic membranes are jointly responsible for the hydrolysis of cholecystokinin octapeptide (CCK-8), *FEBS Lett.,* 175, 124, 1984.
78. **Shima, M., Seino, Y., Torikai, S., and Imai, M.,** Intrarenal localization of degradation of atrial natriuretic peptide in isolated glomeruli and cortical nephron segments, *Life Sci.,* 43, 357, 1988.
79. **Tamburini, P. P., Koehn, J. A., Gilligan, J. P., Charles, D., Palmesino, R. A., Sharif, R., McMartin, C., Erion, M. D., and Miller, M. J. S.,** Rat vascular tissue contains a neutral endopeptidase capable of degrading atrial natriuretic peptide, *J. Pharm. Exp. Ther.,* 251, 956, 1989.
80. **Mantyh, C. R., Kruger, L., Brecha, N. C., and Mantyh, P. W.,** Localization of specific binding sites for atrial natriuretic factor in peripheral tissues of the guinea pig, rat, and human, *Hypertension,* 8, 712, 1986.
81. **Bianchi, C., Gutkowska, J., Thibault, G., Garcia, R., Genest, J., and Cantin, M.,** Radioautographic localization of ^{125}I-atrial natriuretic factor (ANF) in rat tissues, *Histochemistry,* 82, 441, 1985.
82. **Stewart, R. E., Swithers, S. E., Plunkett, L. M., and McCarthy, R.,** ANF receptors: distribution and regulation in central and peripheral tissues, *Neurosci. Biobehav. Rev.,* 12, 151, 1988.
83. **Leitman, D. C., Andresen, J. W., Kuno, T., Kamisaki, Y., Chang, J. K., and Murad, F.,** Identification of multiple binding sites for atrial natriuretic factor by affinity cross-linking in cultured endothelial cells, *J. Biol. Chem.,* 261, 11650, 1986.
84. **Koseki, C., Hayashi, Y., Ohnuma, N., and Imai, M.,** Difference in molecular size of receptors for α-atrial natriuretic polypeptide among the kidney, aorta and adrenal gland as identified by direct UV-photoaffinity labeling, *Biochem. Biophys. Res. Commun.,* 136, 200, 1986.
85. **Pandey, K. and Inagami, T.,** Kidney tubular epithelium cells and vascular smooth muscle cells contain different types of atrial natriuretic factor receptors, *J. Hypertens.,* 6 (Suppl. 4), S292, 1988.
86. **Féthière, J., Meloche, S., Nguyen, T. T., Ong, H., and De Léan, A.,** Distinct properties of atrial natriuretic factor receptor subpopulations in epithelial and fibroblast cell lines, *Mol. Pharmacol.,* 35, 584, 1989.
87. **Takayanagi, R., Snajdar, R. M., Imada, T., Tamura, M., Pandey, M., Misono, K. S., and Inagami, T.,** Purification and characterization of two types of atrial natriuretic factor receptors from bovine adrenal cortex: guanylate cyclase-linked and cyclase-free receptors, *Biochem. Biophys. Res. Commun.,* 144, 244, 1987.

88. **Porter, J. G., Wang, Y., Schwartz, K., Arfsten, A., Loffredo, A., Spratt, K., Schenk, D. B., Fuller, F., Scarborough, R. M., and Lewicki, J. A.,** Characterization of the atrial natriuretic peptide clearance receptor using a vaccinia virus expression vector, *J. Biol. Chem.*, 263, 18827, 1988.
89. **Scarborough, R. M., Schenk, D. B., McEnroe, G. A., Arfsten, A., Kang, L. L., Schwartz, K., and Lewicki, J. A.,** Truncated atrial natriuretic peptide analogs, *J. Biol. Chem.*, 261, 12960, 1986.
90. **Olins, G. M., Patton, D. R., Bovy, P. R., and Mehta, P. P.,** A linear analog of atrial natriuretic peptide (ANP) discriminates guanylate cyclase-coupled ANP receptors from non-coupled receptors, *J. Biol. Chem.*, 263, 10989, 1988.
91. **Almeida, F. A., Suzuki, M., Scarborough, R. M., Lewicki, J. A., and Maack, T.,** Clearance function of type C receptors of atrial natriuretic factor in rats, *Am. J. Physiol.*, 256, R469, 1989.
92. **Chinkers, M., Garbers, D. L., Chang, M. S., Lowe, D. G., Chin, H., Goeddel, D. V., and Schulz, S.,** A membrane form of guanylate cyclase is an atrial natriuretic peptide receptor, *Nature (London)*, 338, 78, 1989.
93. **De Léan, A., Vinay, P., and Cantin, M.,** Distribution of atrial natriuretic factor receptors in dog kidney fractions, *FEBS Lett.*, 193, 239, 1985.
94. **Maack, T., Suzuki, M., Almeida, F. A., Nussenzveig, D., Scarborough, R. M., McEnroe, G. A., and Lewicky, J. A.,** Physiological role of silent receptors of atrial natriuretic factor, *Science*, 238, 675, 1987.
95. **Anand-Srivastava, M. B., Srivastava, A. K., and Cantin, M.,** Pertussis toxin attenuates atrial natriuretic factor-mediated inhibition of adenylate cyclase, *J. Biol. Chem.*, 262, 4931, 1987.
96. **Anand-Srivastava, M. B., Gutkowska, J., and Cantin, M.,** Atrial natriuretic factor receptors of ANF-R2 type are coupled to adenylate cyclase system in rat platelets, *J. Biol. Chem.*, 265, 8566, 1990.
97. **Ballermann, B. J., Hoover, R. L., Karnovsky, M. J., and Brenner, B. M.,** Physiologic regulation of atrial natriuretic peptide receptors in rat renal glomeruli, *J. Clin. Invest.*, 76, 2049, 1985.
98. **Cachofeiro, V., Schiffrin, E. L., Cantin, M., and Garcia, R.,** Glomerular and vascular atrial natriuretic factor receptors in adrenal-ectomized rats, *Am. J. Physiol.*, 256, R1250, 1989.
99. **Hinko, A., Thibonnier, M., and Rapp, J. P.,** Binding characteristics of atrial natriuretic factor and the production of cyclic GMP in kidneys of DAHL salt-sensitive and salt-resistant rats, *Biochem. Biophys. Res. Commun.*, 144, 1076, 1987.
100. **Kollenda, M. C., Vollmar, A. M., McEnroe, G. A., and Gerbes, A. L.,** Dehydration increases the density of C-receptors for ANF on rat glomerular membranes, *Am. J. Physiol.*, 258, R1084, 1990.
101. **Bianchi, C., Thibault, G., Wrobel-Konrad, E., De Lean, A., Genest, J., and Cantin, M.,** Atrial natriuretic factor binding sites in experimental congestive heart failure, *Am. J. Physiol.*, 257, F515, 1989.
102. **Gerbes, A. L., Arendt, R. M., and Paumgartner, G.,** Atrial natriuretic factor — possible implications in liver disease, *J. Hepatol.*, 5, 123, 1987.
103. **Arendt, R. M., Gerbes, A. L., Ritter, D., Stangl, E., Bach, P., and Zähringer, J.,** Atrial natriuretic factor in plasma of patients with arterial hypertension, heart failure or cirrhosis of the liver, *J. Hypertens.*, 4 (Suppl. 2), S131, 1986.
104. **Predel, H. G. Bäcker, A., Kipnowski, J., Düsing, R., and Kramer, H. J.,** Relationship of plasma concentrations of human atrial natriuretic peptide to renal function and blood pressure in patients with progressive chronic renal failure, *Klin. Wochenschr.*, 65 (Suppl. 8), 127, 1987.
105. **Raine, A. E. G., Erne, P., Bürgisser, E., Müller, F. B., Bolli, P., Burkart, F., and Bühler, F. R.,** Atrial natriuretic peptide and atrial pressure in patients with congestive heart failure, *N. Engl. J. Med.*, 315, 533, 1986.
106. **Cantin, M., Thibault, G., Ding, J., Gutkowska, J., Garcia, R., Jasmin, G., Hamet, P., and Genest, J.,** ANF in experimental congestive heart failure, *Am. J. Pathol.*, 130, 552, 1988.
107. **Hasegawa, K., Matsushita, Y., Inoue, T., Morii, H., Ishibashi, M., and Yamaji, T.,** Plasma levels of atrial natriuretic peptide in patients with chronic renal failure, *J. Clin. Endocrinol. Metab.*, 63, 819, 1986.
108. **Lang, R. E., Dietz, R., Merkel, A., Unger, T., Ruskooaho, H., and Ganten, D.,** Plasma atrial natriuretic peptide values in cardiac disease, *J. Hypertens.*, 4 (Suppl. 2), S119, 1986.
109. **Drexler, H., Hänze, J., Finckh, M., Lu, W., Just, H., and Lang, R. E.,** Atrial natriuretic peptide in a rat model of cardiac failure, *Circulation*, 79, 620, 1989.
110. **Gerbes, A. L., Kollenda, M. C., Vollmar, A. M., Reichen, J., Vakil, N., and Scarborough, R. M.,** Density of glomerular binding sites for atrial natriuretic factor (ANF) is altered in rats with cirrhosis and ascites, *Hepatology*, 10, 588, 1989.
111. **Schafferhans, K., Heidbreder, E., Sperber, S., Dämmrich, J., and Heidland, A.,** Atrial natriuretic peptide in gentamicin-induced acute renal failure, *Kidney Intern.*, 34 (Suppl. 25), S101, 1988.
112. **Gerbes, A. L., Arendt, R. M. Stangl, E., Gülberg, V., Sauerbruch, T., Jüngst, D., and Paumgartner, G.,** Characterization of the ANF system in patients with cirrhosis of the liver, in *Functional Morphology of the Endocrine Heart*, Forssmann, W. G., Scheuermann, D. W., and Alt, J., Eds., Steinkopff, Darmstadt, 1988, 229.
113. **Laffi, G., Pinzani, M., Meacci, E., LaVilla, G., Renzi, D., Baldi, E., Cominelli, F., Marra, F., and Gentilini, P.,** Renal hemodynamic and natriuretic effects of human atrial natriuretic factor infusion in cirrhosis with ascites, *Gastroenterology*, 96, 167, 1989.

114. **Adnot, S., Andrivet, P., Chabrier, P. E., Piquet, J., Plas, P., Braquet, P., Roudot-Thoraval, F., and Brun-Buisson, C.,** Atrial natriuretic factor in chronic obstructive lung disease with pulmonary hypertension, *J. Clin. Invest.,* 83, 986, 1989.
115. **Richards, A. M., Nicholls, M. G., Espiner, E. A., Ikram, H., Yandle, T. G., Joyce, S. L., and Cullens, M. M.,** Effects of α-human atrial natriuretic peptide in essential hypertension, *Hypertension,* 7, 812, 1985.
116. **Cody, R. J., Atlas, S. A., Laragh, J. H., Kubo, S. H., Covit, A. B., Ryman, K. S., Shaknovich, A., Pondolfino, K., Clark, M., Camargo, M. J. F., and Scarborough, R. M.,** Atrial natriuretic factor in normal subjects and heart failure patients, *J. Clin. Invest.,* 78, 1362, 1986.
117. **Saito, Y., Nakao, K., Nishimura, K., Sugawara, A., Pkumura, K., Obata, K., Sonoda, R., Ban, T., Yasue, H., and Imura, H.,** Clinical application of atrial natriuretic polypeptide in patients with congestive heart failure: beneficial effects on left ventricular function, *Circulation,* 76, 115, 1987.
118. **Riegger, G. A. J., Kromer, E. P., and Kochsiek, K.,** Human atrial natriuretic peptide: plasma levels, hemodynamic, hormonal, and renal effects in patients with severe congestive heart failure, *J. Cardiovasc. Pharmacol.,* 8, 1107, 1986.
119. **Nakao, K., Sugawara, A., Morii, N., Sakamoto, M., Yamada, T., Itoh, H., Shiono, S., Saita, Y., Nishimura, K., Ban, T., Kangawa, K., Matsuo, H., and Imura, H.,** The pharmacokinetics of α-human atrial natriuretic polypeptide in healthy subjects, *Eur. J. Clin. Pharmacol.,* 31, 101, 1986.
120. **Luft, F. C., Lang, R. E., Aronoff, G. R., Ruskoaho, H., Toth, M., Ganten, D., Sterzel, R. B., and Unger, T.,** Atriopeptin III kinetics and pharmacodynamics in normal and anephric rats, *J. Pharmacol. Exp. Ther.,* 236, 416, 1986.
121. **Murthy, K. K., Thibault, G., Schiffrin, E. L., Garcia, R., Chartier, L, Gutklowska, J., Genest, J., and Cantin, M.,** Disappearance of atrial natriuretic factor from circulation in the rat, *Peptides,* 7, 241, 1986.
122. **Koepke, J. P., Tyller, L. D., Trapani, A. J., Bovy, P. R., Spear, K. L., Olins, G. M., and Blaine, E. H.,** Interaction of non-guanylate cyclase linked atriopeptin receptor ligand and endopeptidase inhibitor in conscious rats, *J. Pharmacol. Exp. Ther.,* 249, 172, 1989.
123. **Olins, G. M., Krieter, P. A., Trapani, A. J., Spear, K. L., and Bovy, P. R.,** Specific inhibitors of endopeptidase 24.11 inhibit the metabolism of atrial natriuretic peptides *in vitro* and *in vivo, Mol. Cell. Endocrinol.,* 61, 201, 1988.
124. **Sybertz, E. J., Chiu, P. J. S., Vemulapalli, S., Pitts, B., Foster, J., Watkins, R. W., Barnett, A., and Haslanger, M. F.,** SCH 39370, a neutral metallo-endopeptidase inhibitor, potentiates biological responses to atrial natriuretic factor and lowers blood pressure in desoxycorticosterone acetate-sodium hypertensive rats, *J. Pharmacol. Exp. Ther.,* 250, 624, 1989.
125. **Gros, C., Souque, A., Schwartz, J. C., Duchier, J., Cournot, A., Baumer, P., and Lecomte, J. M.,** Protection of atrial natriuretic factor against degradation: diuretic and natriuretic responses after *in vivo* inhibition of enkephalinase (EC 3.4.24.11) by acetorphan, *Proc. Natl. Acad. Sci. U.S.A.,* 86, 7580, 1989.
126. **Schwartz, J. C., Gres, C., Lecomte, J. M., and Bralet, J.,** Enkephalinase (ec 3.4.24.11) inhibitors: protection of endogenous ANF against inactivation and potential therapeutic applications, *Life Sci.,* 47, 1279, 1990.
127. **Northridge, D. B., Alabaster, C. T., Connell, J. M. C., Dilly, S. G., Lever, A. F., Jardine, A. G., Barclay, P. L., Dargie, H. J., Findlay, I. N., and Samuels, G. M. R.,** Effects of VK 69578: a novel atriopeptidase inhibitor, *Lancet,* ii, 591, 1989.
128. **Lafferty, H. M., Gunnin, G. M., Silva, P., Zimmermann, M. B., Brenner, B. M., and Anderson, S.,** Enkephalinase inhibition increases plasma atrial natriuretic peptide levels, glomerular filtration rate and urinary sodium excretion in rats with reduced renal mass, *Circ. Res.,* 65, 640, 1989.
129. **Trapani, A. J., Smits, G. J., McGraw, D. E., Spear, K. L., Koepke, J. P., Olins, G. M., and Blaine, E. H.,** Thiorphan, an inhibitor of endopeptidase 24.11, potentiates the natriuretic activity of atrial natriuretic peptide, *J. Cardiovasc. Pharmacol.,* 14, 419, 1989.
130. **Webb, R. L., Yasay, G. D., McMartin, C., McNeal, R. B., and Zimmermann, M. B.,** Degradation of atrial natriuretic peptide: pharmacologic effects of portease EC 24.11 inhibition, *J. Cardiovasc. Pharmacol.,* 14, 285, 1989.
131. **Seymour, A. A., Swerdel, J. N., Fennell, S. A., Druckman, S. P., Neubeck, R., and Delaney, N. G.,** Potentiation of the depressor responses to atrial natriuretic peptides in conscious SHR by an inhibitor of neutral endopeptidase, *J. Cardiovasc. Pharmacol.,* 14, 194, 1989.
132. **Sybertz, E. J., Chiu, P. J. S., Vemulapalli, S., Watkins, R., and Haslanger, M. F.,** Atrial natriuretic factor-potentiating and antihypertensive activity of SCH-3426. An orally active neutral metalloendopeptidase inhibition, *Hypertension,* 15, 152, 1990.

Chapter 8

DEGRADATION OF GASTRIC PEPTIDES: METABOLISM OF THE GASTRINS

Linda Hilsted and Jens F. Rehfeld

TABLE OF CONTENTS

I. INTRODUCTION

The stomach is a sophisticated organ. In addition to its well-known mechanical and chemical digestive functions, the stomach is also a highly differentiated neuroendocrine organ. Both the antral and fundic mucosa contain a large number of endocrine and paracrine cells that synthesize an array of hormonal peptides. Moreover, the stomach is also equipped with a network of peptidergic neurons that release peptides as transmitters.

The classical gastric hormone is gastrin. Gastrin was discovered in 1905[1] only 3 years after secretin, the hormone which founded endocrinology. In accordance with its name (a contraction of the words, gastric secretin), gastrin was, during most of this century, considered the only gastric hormone. During the 1970s and 1980s, however, intensive peptide research has revealed that the stomach also synthesizes substantial amounts of newer hormonal peptides such as somatostatin, gastrin releasing peptide (GRP), glucagon, and glucagon-like peptides (GLP-1 and GLP-2). Furthermore, the stomach contains trace amounts of neuropeptides like the enkephalins, vasoactive intestinal polypeptide (VIP), neurotensin, and adrenocorticotropic hormone (ACTH). Finally, the gastric mucosa contains peptidergic neurons and endocrine cells whose products still remain to be identified.

In this chapter, we have, for a number of reasons, deliberately focused only on gastrin. First, the other gastric hormones, such as somatostatin, the glucagons, and GRP, are to a major extent synthesized and released to blood from extragastric organs. Consequently, they are not primarily gastric peptides. Second, gastric somatostatin, GRP, and several of other the neuropeptides function mainly as local paracrine and neurotransmitter substances about whose degradation in the stomach very little is known. Finally, the gastrin system of peptides is, in itself, so comprehensive due to the multiple hormonal forms in circulation that it amply deserves a chapter of its own. Accordingly, presentation of the structures of the heterogenous gastrin system may be a pertinent start.

II. STRUCTURE AND SYNTHESIS OF THE GASTRINS

Gastrin is the principal hormonal regulator of gastric acid secretion. Edkins discovered gastrin in 1905 when he found that injection of extracts of antral mucosa stimulated the secretion of gastric juice.[1] For decades, however, it was debated whether the active antral factor might be histamine contaminating the extracts, or whether the antrum contained a protein or peptide hormone of its own. In 1964, the identity of gastrin was established when Gregory and Tracy isolated peptidal gastrin from the porcine antral mucosa. Gastrin was the first gut hormone for which the structure was determined.[2–4]

Gregory and Tracy isolated two heptadecapeptides (gastrin-17) that differed only by sulfation of the tyrosin-12 residue.[3] Both the N- and C-terminal residues were blocked, the N-terminus being pyroglutamic acid and the C-terminal phenylalanine being carboxyamidated. Moreover, a characteristic pentaglutamyl sequence was found in the middle of the peptide. Subsequent sequence determinations of human, feline, canine, bovine, and rat gastrin-17 have revealed only one or two amino acid substitutions[5–7] (see also Figure 1). Structure-function studies have demonstrated an unusually well-defined active site; that is, the C-terminal Trp-Met-Asp-Phe-NH_2 sequence where the carboxyamide moiety is absolutely decisive for biological activity as found for many α-amidated regulatory peptides.[8,9] Since the isolation of gastrin-17, other gastrin forms have been discovered, including Component-I, gastrin-34, gastrin-14, and gastrin-6. They all possess the active site.[10–14]

Recently, gastrin cDNA for several mammalian species has been cloned and sequenced, and the amino acid sequence of pig, human, rat, cow, and canine preprogastrin consequently deduced.[15–19] The preprogastrins consist of 101 or 104 amino acids and share the same overall structure: an N-terminal signal sequence, followed by a prosequence, gastrin-34, a proteolytic cleavage/amidation signal (Gly-Arg-Arg), and a hexa- or nonapeptide at the extreme C-

PRIMARY STRUCTURE OF HEPTADECAPEPTIDE GASTRIN IN DIFFERENT MAMMALS

												(SO_3H)					
MAN	pGlu	Gly	Pro	Trp	Leu	Glu	Glu	Glu	Glu	Glu	Ala	Tyr	Gly	Trp	Met	Asp	$PheNH_2$
PIG					Met												
DOG					Met					Ala							
CAT										Ala							
RAT		Arg		Pro	Met												
COW					Val					Ala							
SHEEP					Val					Ala							
GOAT					Val					Ala							

FIGURE 1. Structure of the principal molecular form of gastrin, the heptadecapeptide amide, from differenmt mammalian species. The common active site of the C-terminus is framed. The sulfate group in brackets indicates that approximately half of the mammalian gastrins are sulfated.

terminus of the precursor. The multiple bioactive (i.e., α-amidated) forms of gastrin occur by the differentiated post-translational processing of progastrin as illustrated in Figure 2.

The gastrins are mainly produced in the endocrine G-cells.[20,21] These cells are most abundant in the pyloric glands of the antrum, but also occur to a lesser extent in the proximal duodenum, distally from where their density decreases abruptly. In the antrum, gastrin-17 peptides constitute more than 90% of the amidated forms, and gastrin-34 comprises 5 to 10%.[22] In the circulation, the component distribution is quite different. Gastrin-34 accounts for more than half of the amidated gastrins in normal subjects, a pattern also found after meal stimulation of gastrin release.[23,24]

As shown by the absence of gastrin in plasma after antroduodenectomy,[25,26] the gastrins in circulation originate almost exclusively from antroduodenal G-cells. The gastrin gene is, however, also expressed in the pancreas, the pituitary, in vagal and hypothalamic neurons, as well as in spermatogenic cells.[27–32] The size of the gastrin synthesis in these different cell types is, in normal adults, so moderate and the processing of progastrin so different[28,32,33] that their contribution to the bioactive, carboxyamidated gastrins in normal plasma is undetectable. Instead, this widespread extraintestinal synthesis of gastrin seems to serve local regulatory purposes.

III. DEGRADATION OF THE GASTRINS

A. METABOLIC CLEARANCE RATE AND HALF-LIFE

In 1968, the first useful radioimmunoassay for gastrin and, by the way for a gut hormone also, was developed.[34] The sensitivity and specificity inherent in the immunological determination permitted studies on the metabolism of gastrin. Two techniques were used, either bolus injections or continuous infusion (see Chapter 2). If the latter was employed, the plasma concentrations after infusion were converted to the natural logarithm and regression of ln plasma gastrin concentrations vs. time computed to yield the slope (k) from which the half-life ($t_{1/2}$) was determined by dividing into ln 2. If the bolus technique was used, the half-life was evaluated from the slope of the curve after completion of the distribution phase.

The metabolic clearance rate (MCR) was calculated according to the plateau principle, assuming a steady state:

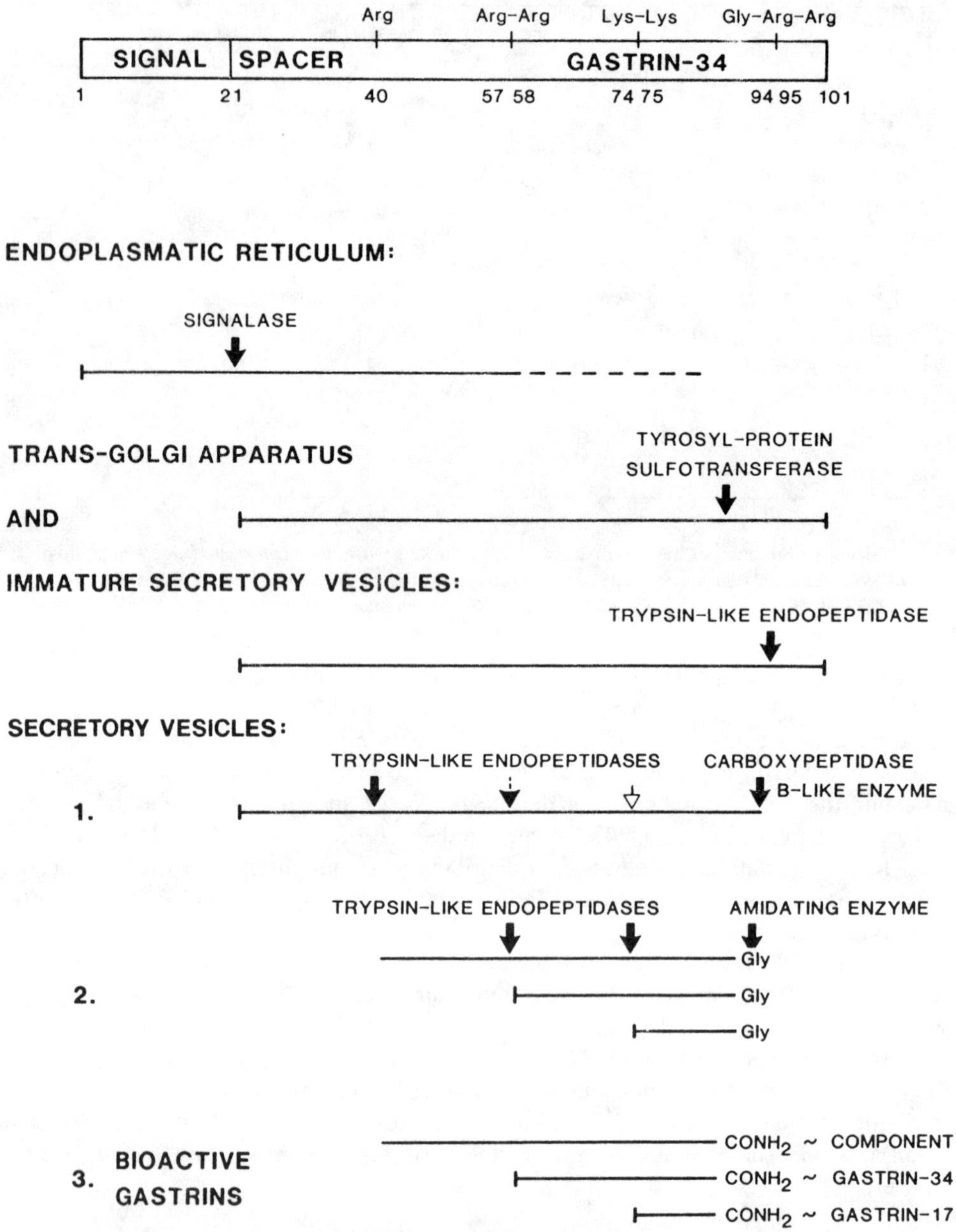

FIGURE 2. The posttranslational processing of human preprogastrin in antral cells.

$$\text{MCR} = \frac{\text{infusion dose}}{\text{increment of plasma concentration}}$$

Alternatively, MCR was obtained by plotting steady-state plasma concentrations against infusion doses and calculating the reciprocal of the slope, which represents clearance. Assuming a one-compartment system, distribution volume (V_d) was calculated from the equation

$$\text{MCR} = V_d \cdot \frac{\ln 2}{t_{1/2}}$$

Numerous studies have been performed using the methods described above. Early experiments were mainly performed in dogs using human or porcine nonsulfated gastrin-17, and it was found that exogenous gastrin-17 disappears rapidly from the circulation with a half-life of 2 to 4 min.[35–39] In some experiments where continuous infusion was used, half-life was slightly longer (i.e., 6 to 10 min.[40,41] The MCR in dogs was found to be 40 to 50 ml/kg/min, and the distribution volume 15 to 28% of the body weight.[35,37–39]

Subsequent studies have demonstrated a longer half-life of gastrin-17 in man; that is, between 5 and 11 min.[38,42–46] Clearance rates of 9 to 20 ml/kg/min are correspondingly lower than in the dog.[38,43,44,46] Distribution volume per kilogram body weight was similar to that found in dog.[38,44–46] Differences in pharmacokinetic parameters have also been observed between other species, MCR being 135 ml/kg/min in the cat and 60 to 100 ml/kg/min in pigs.[39,47] It has been shown that the concentration of gastrin required to elicit a given response varies little between species. Thus, animals of smaller size require greater exogenous doses to obtain a given response due to the higher disposal rate.[38]

The infusion doses used in these studies have covered a wide range and plasma concentrations have varied accordingly, from physiological to supraphysiological levels. Plasma concentrations, however, have been found to increase in proportion to the dose used. Therefore, in most cases disappearance curves have been found to be monoexponential, that is, kinetics are first order. In a few studies, regression analysis has indicated a better fit by a model with at least two compartments.[41,43]

Pharmacokinetic parameters have also been established for other gastrin components besides gastrin-17. Gastrin-34 has a half-life of 9 to 15 min in dogs, 8 min in cats, and 39 to 44 min in man; that is, five times longer than that of gastrin-17.[36,37,44,45,48] This difference may explain why the ratio of gastrin-34 to gastrin-17 is five times higher in peripheral plasma than in antral tissue. Half-lives, distribution volumes, and clearance rates of gastrin-14 and gastrin-17 are similar.[39] All these studies have employed nonsulfated gastrins. However, comparison of disappearance rates of sulfated and nonsulfated gastrin has been performed. Sulfation of tyrosine-12 in gastrin-17 has been reported by one group to infer a fourfold longer half-life in man;[49] but these results have now been questioned in a recent study in which sulfated and nonsulfated gastrin-17 showed similar disappearance rates.[50] Tyrosine-O-sulfation of a gastrin-6 analogue has, however, shown a fivefold lower clearance rate in cat,[51] but thus far gastrin-6 has not been demonstrated to occur in normal plasma.

B. SITES OF DEGRADATION

1. Renal Elimination

The sites of gastrin removal have been searched for extensively. The distribution volume for exogenous gastrin-17 is close to the total extracellular volume, and the short half-life could indicate enzymatic degradation. In dogs, the MCR is close to resting cardiac output, thus exceeding splanchnic and renal plasma flow and suggesting that gastrin is also degraded in other vascular beds. It has been suggested that the kidneys play a major role in gastrin breakdown in the dog. The canine kidneys remove 40% of exogenous human gastrin-17 with a linear relationship between arterial concentrations and amount of gastrin removed across the kidney, and nephrectomy causes an elevation of plasma gastrin levels.[52] Endogenous gastrin is also removed to the same degree in the canine kidneys.[53] Renal extraction of gastrin is almost similar to glomerular filtration and very small amounts of extracted gastrin appear in the urine, suggesting that the dog kidneys are a major site of gastrin degradation. Possibly, the mechanism whereby the kidney removes gastrin from the circulation includes filtering at the glomerulus and reabsorption and subsequent degradation in the proximal tubules. Enzymatic activity capable of cleaving phenylalanine-amide from gastrin analogues has been demonstrated in rat kidneys.[54,55] Fragments formed by such cleavage would be biologically inactive and would not be measured by the conventional C-terminal directed radioimmunoassays.

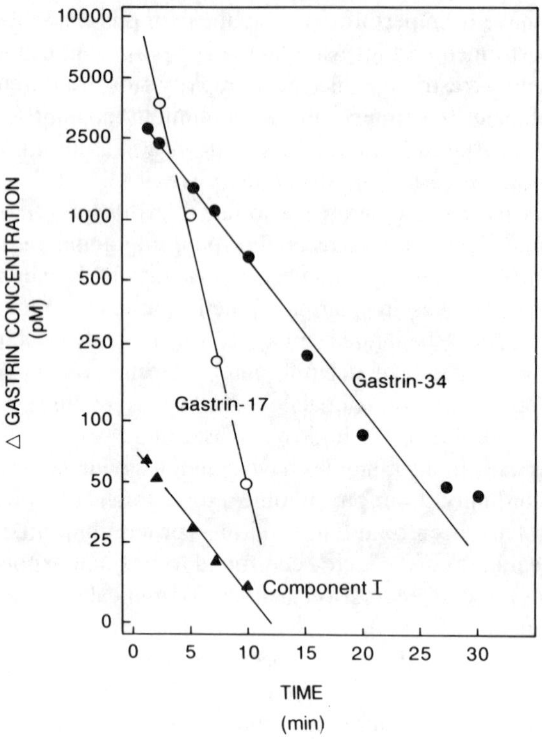

FIGURE 3. Degradation of the gastrins as measured in peripheral venous blood of cats after intravenous bolus injection of feline component I (▲—▲), gastrin-34 (●—●), and gastrin-17 (○—○). The gastrin components were isolated from extracts of feline antral mucosa. For further details, see reference 48.

In man, the role of the kidneys in the elimination of gastrins is less evident. Acute and chronic renal failure is associated with hypergastrinemia, the degree of which reflects the degree of renal dysfunction.[56,57] Hypergastrinemia is also found in bilaterally nephrectomized patients.[58] However, in fasting as well as meal-stimulated subjects, there are no significant differences between arterial and renal venous gastrin concentrations, suggesting that the kidney is not a major site of gastrin degradation in man.[59,60] The observed renal extraction ratio would — using a distribution volume of 12% of the body weight — correspond to a half-life of 441 min. Whether the hypergastrinemia observed in renal failure can be ascribed to an increased gastrin synthesis and/or release has not been demonstrated. However, hyposecretion of gastric acid in uremic patients has been reported.[61] Consequently, hypergastrinemia might be secondary by means of negative feedback; that is, the inhibited acid secretion leads to an increased synthesis and release of gastrin.[62] In hypergastrinemia associated with Zollinger-Ellison's syndrome or pernicious anemia, the urine contains gastrins in significant concentrations that increase with increasing gastrin concentrations in plasma.[63] Apparently the renal threshold is 100 to 200 pmol/l plasma, a level which is not reached in normal subjects.

2. Hepatic Elimination

After release from antral and duodenal mucosa, gastrin enters the portal vein and passes through the liver before reaching the target cells. Consequently, the liver could play an important role in regulating the systemic blood concentration and the relative distribution of gastrin components. Several studies have been carried out to ascertain whether gastrin is indeed metabolized in the liver, but the results have been conflicting. Early studies used measurement of gastrin by bioassay. Systemic infusion or portal infusion/hepatic perfusion in dogs of natural porcine gastrin-17 elicited the same gastric acid secretion.[64,65] Others found that portal injection of synthetic human gastric in the dog caused a 40% lower Heidenhain pouch acid output than did injection in the caval vein.[66] Apparently, shorter fragments are much more susceptible to hepatic elimination. Thus, fragments with 8 or fewer amino acids were more than 90% inactivated by hepatic transit in the dog.[67] In this study, elimination of human gastrin-17 was 30% which, in turn, was not significantly different from the arteriovenous difference in immunoreactive gastrin previously observed in most capillary beds during infusion of human gastrin-17.[68] In a human study injection of human gastrin-17 in the portal vein resulted in a 40% loss of gastric acid output.[69]

Radioimmunochemical measurements have been equally contradictory. After portal infusion of supraphysiological doses of heterologous gastrin in dogs, some hepatic elimination was observed.[66] During stimulation of endogenous gastrin release, a higher concentration of gastrin was measured in portal than in hepatic venous blood and this difference was ascribed to dilution with hepatic arterial blood.[70] Correspondingly, when flow measurements were incorporated in calculation of the mass of endogenous gastrin entering and leaving the liver per minute, the results on dogs showed no significant gastrin removal upon hepatic transit.[71]

In man, gastrin concentrations were not significantly different in portal vs. peripheral blood and the composition of gastrin-17 and gastrin-34 was similar.[60,72]

In the pig, about half of the endogenous gastrin in portal venous blood was eliminated in the liver, before and after meal stimulation, and the elimination rate was proportional to the gastrin concentration in portal venous blood, indicating first-order kinetics within the physiological range of gastrin concentrations.[73] In contrast, no net loss or gain of endogenous gastrin was observed across the porcine liver in fasting animals, neither did supraphysiological doses of synthetic human gastrin-17 induce any hepatic elimination.[47]

These diverging results may be explained by several factors: different surgical preparations (ligation of hepatic arteries), addition of gastrin to systemic vs. portal circulation, variations in doses of exogenous gastrin, and different preparations of gastrin used for different species. Furthermore, although virtually all assays applied in the studies mentioned are C-terminally directed, modified gastrins may not be measured equally by different assay systems. However, it would seem — at least in man — that the liver does not play a substantial role in the elimination of endogenous gastrin.

C. MECHANISM OF ELIMINATION

In some studies, measurements on uptake of exogenous or endogenous gastrin have demonstrated a similar extraction (20 to 40%) over several vascular beds including kidney, liver, leg, and head.[68,74] In man, extraction was found to be below 10% across these vascular beds and lungs.[60] It has been suggested that extraction of gastrin occurs on transit of all capillary beds, either by proteases in the vascular endothelium, elsewhere in tissues, or in blood[68,74] (see Chapter 4).

Recent studies on the metabolism of gastrin have applied sequence-specific radioimmunoassays directed against different sequences of the gastrin molecule. Moreover, chromatographic characterization has been further developed to include separation of peptides according to size, charge, and hydrophobicity.

In the dog, i.v. infusion of canine gastrin-17 gave rise to circulating C-terminal immunoreactive fragments which had a longer half-life than the intact gastrin-17 molecule. N-terminal fragments were not found in systemic circulation. It was suggested that enzymes responsible for formation of the fragments could be associated with capillary endothelial cells and might well be associated with degradation of gastrin.[75] During infusion of human gastrin-17 in man, 15% of C-terminal immunoreactivity was attributable to gastrin-14-like material and more than 25% of circulating N-terminal immunoreactivity as attributable to two N-terminal fragments, one of which corresponds to the N-terminal tridecapeptide of gastrin-17, a biologically inactive fragment which is found in antral mucosa.[76,77] After the infusion was terminated, these fragments constituted a progressively larger fraction of total gastrin immunoreactivity. *In vitro,* similar fragments were formed although the rate of conversion was much lower.[76]

When sulfated human gastrin-17 was infused in man, the same fragments were formed; however, a short N-terminal form constituted a smaller part of N-terminal immunoreactivity, suggesting the sulfate group prevents conversion to this fragment.[49]

In porcine antral mucosa, gastrin-17 accounts for over 90% of C-terminal immunoreactivity but only 40% in antral venous blood, the remaining C-terminal immunoreactivity is accounted for by shorter C-terminal fragments.[78] Pig plasma, like human plasma, contains enzymes capable of this conversion. However, loss of immunoreactivity observed when plasma is incubated with gastrin-17 is far too slow to account for the loss observed *in vivo.*[76,78] Recent studies have focused on the membrane associated metalloendopeptidase 24.11 that hydrolyzes bonds involving the amino group of hydrophobic amino acid residues. Endopeptidase 24.11 was first isolated from the brush border in kidney and later from brain, intestine, and stomach of several species and it is found widely distributed throughout the body.[79,80] It has been implicated in the metabolism of several bioactive peptides and may be responsible for the previously described gastrin-degrading activity in homogenates of liver and kidney.[55,81] Potential sites for hydrolysis in gastrin-17 include Pro_3-Trp_4, Ala_{11}-Tyr_{12}, Gly_{13}-Trp_{14}, Asp_{16}-Phe_{17}. Cleavage at Ala_{11}-Tyr_{12} would cause the formation of 1-11 gastrin-17 and gastrin-6, while cleavage at Gly_{13}-Trp_{14} would liberate 1-13 gastrin-17 and gastrin-4, and these fragments are known to occur naturally. Cleavage at Asp_{16}-Phe_{17} would abolish biological activity. *In vitro,* endopeptidase 24.11 hydrolyzes porcine gastrin-17 and the hydrolysis is completely inhibited by the specific inhibitor phosphoramidon. *In vivo,* close arterial infusion of phosphoramidon to the pig stomach reduces the conversion of gastrin-17 to shorter C-terminal fragments, suggesting that endopeptidase 24.11 may participate in hydrolysis of gastrin-17 during secretion.[82] Incubation of human nonsulfated gastrin-17 with endopeptidase 24.11 from porcine stomach resulted in cleavage at the bonds described above. Hydrolysis of the Asp_{16}-Phe_{17} bond was a major cleavage site as none of the identified metabolites contained the bioactive tetrapeptide amide, Trp_{14} -Phe_{17}.[83] Further studies including amino acid analysis have demonstrated that cleavage *in vitro* occurs most rapidly at Asp_{16}-Phe_{17} and Gly_{13}-Trp_{14}, and the fragments observed after *in vitro* incubation are also formed during infusion of human gastrin-17 in human volunteers.[84] The half-lives of the 1-11 and 1-13 fragments of gastrin-17 are longer than that of gastrin-17 itself, but it is not known if they are converted in the circulation to smaller fragments such as the 1-4 and 5-17 fragments as found *in vitro.*[84] The degradation of gastrin-17 *in vitro* by endopeptidase 24.11 is slow ($t_{1/2}$ = 67 min) compared to elimination *in vivo*. However, endopeptidase 24.11 is found throughout the body and it has been suggested that local concentrations are sufficiently high to sustain the degradation rate observed *in vivo.*[85] The *in vitro* degradation pattern of sulfated gastrin-17 is similar to that of the nonsulfated form, but the half-life has been reported to be threefold higher.[85] One possible explanation could be that sulfation of Tyr_{12} induces modifications of tertiary structure, giving localized protection against cleavage.[85]

Although the recent methodological advances have permitted new insight into potential degradation mechanisms, these have not been fully elucidated. Possibly, a combination of enzymes are involved in the degradation of gastrin, with endopeptidase 24.11 participating in the early steps.

Enzymes capable of deamidating gastrin-4 have been found widespread in rat tissues and peptidase activity cleaving between Met_{15}-Asp_{16} has been demonstrated in rat kidney. The failure of phosphoramidon to inhibit the degradation of gastrin completely also suggests that other enzymes are involved. Localization of endopeptidase 24.11 by immunocytochemistry has not been accomplished, but will yield further information regarding its physiological role in the degradation of gastrin.

IV. SUMMARY

Since the discovery in 1905 of the principal hormonal regulator of gastric acid secretion, gastrin, several biologically active forms have been described; that is, the gastrin system is heterogeneous. Pharmacokinetic parameters have been established for the various bioactive forms and vary not only between the gastrins, but also between species. Species variations are also found in the sites of elimination. While the liver and the kidneys apparently play a major role in the degradation of gastrins in some species, the localization of gastrin-degrading activity is less evident in man. It has been suggested that degradation of gastrins may occur across all vascular beds, and the neutral endopeptidase 24.11, which cleaves peptide bonds involving the amino group of hydrophobic amino acids, has been implicated. *In vitro* cleavage yields fragments corresponding to those observed *in vivo* after infusion of gastrins. However, the half-life of gastrins observed *in vitro* cannot account for the *in vivo* elimination. Thus, although gastrin is one of the oldest and most extensively studied gastrointestinal hormones, mechanisms and sites of elimination are still only partially elucidated.

ABBREVIATIONS

ACTH:	Adrenocorticotropic hormone
cDNA:	Complementary deoxyribonucleic acid
GLP:	Glucagon-like peptide
GRP:	Gastrin releasing peptide
MCR:	Metabolic clearance rate
$t_{1/2}$:	Half-life
V_d:	Distribution volume
VIP:	Vasoactive intestinal polypeptide

REFERENCES

1. **Edkins, J. S.,** On the chemical mechanism of gastric secretion, *Proc. R. Soc. (London),* 76, 376, 1905.
2. **Gregory, R. A. and Tracy, H. J.,** The constitution and properties of two gastrins extracted from hog antral mucosa, *Gut,* 5, 103, 1964.
3. **Gregory, H., Hardy, P. M., Jones, D. S., Kenner, G. W., and Sheppard, R. C.,** Structure of gastrin, *Nature (London)*, 204, 931, 1964.
4. **Anderson, J. C., Barton, M. A., Gregory, R. A., Hardy, P. M., Kenner, G. W., MacLeod, J. K., Preston, J., and Sheppard, R. C.,** Synthesis of gastrin, *Nature (London)*, 204, 933, 1964.
5. **Bentley, P. H., Kenner, G. W., and Sheppard, R. C.,** Structures of human gastrins I and II, *Nature (London)*, 209, 583, 1966.
6. **Kenner, G. W. and Sheppard, R. C.,** Gastrins of various species, in *Frontiers in Gastrointestinal Hormone Research*, Andersoon, S., Ed., Almquist and Wiksell, Stockholm, 1973.
7. **Reeve, J. R., Jr., Dimaline, R., Shively, J. E., Hawke, D., Chew, P., and Walsh, J. H.,** Unique amino terminal structure of rat little gastrin, *Peptides*, 2, 453, 1981.
8. **Morley, J. S., Tracy, H. J., and Gregory, R. A.,** Structure function relationships in the active C-terminal tetrapeptide sequence of gastrin, *Nature (London)*, 207, 1356, 1965.
9. **McGuigan, J. E. and Thomas, H. F.,** Physiological and immunological studies with desamidogastrin, *Gastroenterology*, 62, 553, 1972.

10. **Rehfeld, J. F.**, Three components of gastrin in human serum, *Biochim. Biophys. Acta*, 285, 364, 1972.
11. **Greogry, R. A. and Tracy, H. J.**, Isolation of two big gastrins from Zollinger-Ellison tumour tissues, *Lancet*, ii, 797, 1972.
12. **Rehfeld, J. F. and Stadil, F.**, Gel filtration studies on immunoreactive gastrin in serum from Zollinger-Ellison patients, *Gut*, 14, 369, 1973.
13. **Greogry, R. A. and Tracy, H. J.**, Isolation of two minigastrins from Zollinger-Ellison tumour tissue, *Gut*, 15, 683, 1974.
14. **Gregory, R. A., Dockray, G. J., Reeve, J. R., Jr., Shively, J. E., and Miller, C.**, Isolation from porcine antral mucosa of a hexapeptide corresponding to the C-terminal sequence of gastrin, *Peptides*, 4, 319, 1983.
15. **Yoo, O. J., Powell, C. T., and Agarwal, K. L.**, Molecular cloning and nucleotide sequence of full-length cDNA coding for porcine gastrin, *Proc. Natl. Acad. Sci. U.S.A.*, 79, 1049, 1982.
16. **Boel, E., Vuust, J., Norris, K., Norris, F., Wind, A., Rehfeld, J. F., and Marcker, K. A.**, Molecular cloning of human gastrin cDNA: evidence for evolution of gastrin by gene duplication, *Proc. Natl. Acad. Sci. U.S.A.*, 80, 2866, 1983.
17. **Fuller, P. J., Stone, D. L., and Brand, S. J.**, Molecular cloning and sequencing of a rat preprogastrin complementary deoxyribonucleic acid, *Mol. Endocrinol.*, 1, 306, 1987.
18. **Lund, T., Olsen, J., and Rehfeld, J. F.**, Cloning and sequencing of the bovine gastrin gene, *Mol. Endocrinol.*, 3, 1585, 1989.
19. **Kang, S. C., Agarwal, K. L., and Yoo, O. J.**, Molecular cloning and sequence analysis of cDNA coding for canine gastrin, *Biochem. Int.*, 18, 631, 1989.
20. **Solcia, E., Vassallo, G., and Capella, C.**, Studies on the G cells of the pyloric mucosa, the propable site of gastrin secretion, *Gut*, 10, 379, 1969.
21. **Pearse, A. G. E. and Bussolati, G.**, The identification of gastrin cells as G-cells, *Virchows Arch. (Path. Anat.)*, 355, 99, 1972.
22. **Malmstrøm, J., Stadil, F., and Rehfeld, J. F.**, Gastrins in tissue: concentration and component pattern in gastric, duodenal, and jejunal mucosa of normal human subjects and patients with duodenal ulcer, *Gastroenterology*, 70, 697, 1976.
23. **Yalow, R. S. and Berson, S. A.**, Size and charge distinctions between endogenous human plasma gastrin in peripheral blood and heptadecapeptide gastrins, *Gastroenterology*, 58, 609, 1970.
24. **Rehfeld, J. F., Stadil, F., and Vikelsøe, J.**, Immunoreactive gastrin components in human serum, *Gut*, 15, 102, 1974.
25. **Rehfeld, J. F., Stadil, F., Baden, H., and Fischermann, K.**, The enteral insulin-stimulation after Whipple's operation, *Diabetologia*, 11, 207, 1975.
26. **Creutzfeldt, W., Ebert, R., Arnold, R., Frerichs, H., and Brown, J. C.**, Gastric inhibitory polypeptide (GIP), gastrin and insulin: response to a test-meal in coeliac disease and after duodeno-pancreatectomy, *Diabetologia*, 12, 279, 1976.
27. **Larsson, L.-I., Rehfeld, J. F., Sundler, F., and Håkanson, R.**, Pancreatic gastrin in foetal and neonatal rats, *Nature, (London)*, 262, 609, 1976.
28. **Bardram, L., Hilsted, L., and Rehfeld, J. F.**, Progastrin expression in the mammalian pancreas, *Proc. Natl. Acad. Sci. U.S.A.*, 87, 298, 1990.
29. **Rehfeld, J. F.**, Localisation of gastrins to neuro- and adeno-hypophysis, *Nature (London)*, 271, 771, 1978.
30. **Uvnäs-Wallensten, K., Rehfeld, J. F., Larsson, L.-I., and Uvnäs, B.**, Heptadecapeptide gastrin in the vagal nerve, *Proc. Natl. Acad. Sci. U.S.A.*, 74, 5707, 1977.
31. **Rehfeld, J. F., Hansen, H. F., Larsson, L.-I., Stengaard-Pedersen, K., and Thorn, N. A.**, Gastrin and cholecystokinin in pituitary neurons, *Proc. Natl. Acad. Sci. U.S.A.*, 81, 1902, 1984.
32. Schalling, M., Persson, H., Pelto-Huikko, M., Ødum, L., Ekman, P., Hökfelt, T., and Rehfeld, J. F., Expression of gastrin in human spermatogenic cells, *Mol. Endocrinol.*, in press, 1990.
33. **Rehfeld, J. F.**, Accumulation of non-amidated preprogastrin and preprocholecystokinin products in the porcine anterior pituitary, *J. Biol. Chem.*, 261, 5841, 1986.
34. **McGuigan, J. E.**, Immunochemical studies with synthetic human gastrin, *Gastroenterology*, 54, 1005, 1968.
35. **Schrumpf, E. and Semb, L. S.**, The metab.olic clearance rate and half-life of synthetic human gastrin in dogs, *Scand J. Gastroenterol.*, 8, 203, 1973.
36. **Straus, E. and Yalow, R. S.**, Studies on the distribution and degradation of heptadecapeptide, big, and big big gastrin, *Gastroenterology*, 66, 936, 1974.
37. **Walsh, J. H., Debas, H. T., and Grossman, M. I.**, Pure human big gastrin: immunochemical properties, disappearance half time and acid-stimulating action in dogs, *J. Clin. Invest.*, 54, 477, 1974.
38. **Boniface, J., Picone, D., Schebalin, M., Zfass, A. M., and Makhlouf, G. M.**, Clearance rate, half-life, and secretory potency of human gastrin-17-I in different species, *Gastroenterology*, 71, 291, 1976.
39. **Carter, D. C., Taylor, J. L., Elashoff, J., and Grossman, M. I.**, Reappraisal of the secretory potency and disappearance rate of pure human minigastrin, *Gut*, 20, 705, 1979.
40. **McGuigan, J. E., Isaza, J., and Landor, J. H.**, Relationships of gastrin dose, serum gastrin, and acid secretion, *Gastroenterology*, 61, 659, 1971.

41. **Reeder, D. D., Jackson, B. M., Brandt, E. N. Jr., and Thompson, J. C.,** Rate and pattern of disappearance of exogenous gastrin in dogs, *Am. J. Physiol.*, 222, 1571, 1972.
42. **Ganguli, P. C., Elder, J. B., Smith, I. S., and Hunter, W. M.,** The half-life ($t_{1/2}$) of synthetic human gastrin I in man, *Br. J. Surg.*, 57, 848, 1970.
43. **Schrumpf, E., Semb, L. S., and Vold, H.,** Metabolic clearance and disappearance rates of synthetic human gastrin in man, *Scand. J. Gastroenterol.*, 8, 731, 1973.
44. **Walsh, J. H., Isenberg, J. I., Ansfield, J., and Maxwell, V.,** Clearance and acid-stimulating action of human big and little gastrins in duodenal ulcer subjects, *J. Clin. Invest.*, 57, 1125, 1976.
45. **Eysselein, V. E., Maxwell, V., Reedy, T., Wünsch, E., and Walsh, J. H.,** Similar acid stimulatory potencies of synthetic human big and little gastrins in man, *J. Clin. Invest.*, 73, 1284, 1984.
46. **Blair, A. J., III, Richardson, C. T., Vasko, M., Walsh, J. H., and Feldman, M.,** Comparison of acid secretory responsiveness to gastrin heptadecapeptide and of gastrin heptadecapeptide pharmacokinetics in jeodenal ulcer patients and normal subjects, *J. Clin. Invest.*, 78, 779, 1986.
47. **Vinik, A. J., Hickman, R., and Grant, B. J.,** Endogenous and exogenous heptadecapeptide gastrin transport across the liver, *S. A. Med. J.*, 53, 759, 1978.
48. **Rehfeld, J. F. and Uvnäs-Wallensten, K.,** Gastrins in cat and dog: evidence for a biosynthetic relationship between the large molecular forms of gastrin and heptadecapeptide gastrin, *J. Physiol.*, 283, 379, 1978.
49. **Pauwels, S., Dockray, G. J., and Walker, R.,** Comparison of the metabolism of sulfated and unsulfated heptadecapeptide gastrin in humans, *Gastroenterology*, 92, 1220, 1987.
50. **Cantor, P., Petersen, M. B., Christiansen, J., and Rehfeld, J. F.,** Does sulfation of gastrin influence acid secretion in man?, *Scand. J. Gastroenterol.*, in press, 1990.
51. **Hirst, B. H., Holland, J., Marr, A. P., Parsons, M. E., and Sanders, D. J.,** Comparison of effect of peptide length and sulphation on acid secretory potency of gastrin in the cat *in vivo* and *in vitro*, *J. Physiol.*, 357, 441, 1984.
52. **Clendinnen, B. G., Davidson, W. D., Reeder, D. D., Jackson, B. M., and Thompson, J. C.,** Renal uptake and excretion of gastrin in the dog, *Surg. Gynecol. Obstet.*, 132, 1039, 1971.
53. **Davidson, W. D., Springberg, P. D., and Falkinburg, N. R.,** Renal extraction and excretion of endogenous gastrin in the dog, *Gastroenterology*, 64, 955, 1973.
54. **Glass, J. D., Schwartz, I. L., and Walter, R.,** Enzymatic inactivation of peptide hormones possessing a C-terminal amide group, *Proc. Natl. Acad. Sci. U.S.A.*, 63, 1426, 1969.
55. **Laster, L. and Walsh, J. H.,** Enzymatic degradation of C-terminal tetrapeptide amide of gastrin by mammalian tissue extracts, *Fed. Proc.*, 27, 1328, 1968.
56. **Korman, M. G., Laver, M. C., and Hansky, J.,** Hypergastrinaemia in chronic renal failure, *Br. Med. J.*, 1, 209, 1972.
57. **Dent, R. I., Hirsch, H., James, J. H., and Fischer, J. E.,** Hypergastrinemia in patients with acute renal failure, *Surg. Forum*, 23, 312, 1972.
58. **Maxwell, J. C., Moore, J. G., Dixon, J., and Stevens, L. E.,** Gastrin levels in anephric patients, *Surg. Forum*, 22, 305, 1971.
59. **Schjönsby, H. and Willassen, Y.,** Renal extraction of endogenous gastrin in patients with normal renal function, *Scand. J. Gastroenterol.*, 12, 205, 1977.
60. **Petersen, B., Henriksen, J. H., Rehfeld, J. F., and Christiansen, J.,** Removal of endogenous gastrin in man, *Scand. J. Gastroenterol.*, 16, 727, 1981.
61. **Ala-Kaila, K., Kataja, M., Keyriläinen, O., and Pasternack, A.,** The gastric secretion profile of patients with chronic renal failure, *Ann. Clin. Res.*, 11, 107, 1981.
62. **Ala-Kaila, K., Kekki, M., Paronen, I., and Paakkala, T.,** Serum gastrin in chronic renal failure: its relation to acid secretion, G-cell density and upper gastrointestinal findings, *Scand. J. Gastroenterol.*, 24, 939, 1989.
63. **Schwartz, T. W., Saksø, P., and Rehfeld, J. F.,** Immunochemical studies on gastrins in the urine, *Clin. Chim. Acta*, 89, 381, 1978.
64. **Gillespie, I. E. and Grossman, M. I.,** Gastric secretion of acid in response to portal and systemic venous injection of gastrin, *Gastroenterology*, 43, 189, 1962.
65. **Lick, R. F., Welsch, K. H., Hart, W., Brückner, W., Balser, D., and Gürtner, Th.,** Zur sekretorischen Funktion des Magens nach Injektion von Histamin, Gastrin, und syntetischen Tetrapeptid in den grossen Kreislauf und in die Pfortader, *Z. Gastroentrologie*, 5, 7, 1967.
66. **Thompson, J. C., Reeder, D. D., Davidson, W. D., Charters, A. C., Brückner, W. L., Lemmi, C. A. E., and Miller, J. H.,** Effect of hepatic transit of gastrin, pentagastrin, and histamine measured by gastric secretion and by assay of hepatic vein blood, *Ann. Surg.*, 170, 493, 1969.
67. **Strunz, U. T., Thompson, M. R., Elashoff, J., and Grossman, M. I.,** Hepatic inactivation of gastrins of various chain lengths in dogs, *Gastroenterology*, 74, 550, 1978.
68. **Strunz, U. T., Walsh, J. H., and Grossman, M. I.,** Removal of gastrin by various organs in dogs, *Gastroenterology*, 74, 32, 1978.
69. **Beger, H. G., Meves, M., Witte, C., and Kraas, E.,** The effect of the liver on the gastric secretion stimulated with gastrin II and gastrin-like substances in human, *Acta Hepato. Gastroenterol.*, 19, 14, 1972.

70. **McGuigan, J. E., Jaffe, B. M., and Newton, W. T.,** Immunochemical m easurements of endogenous gastrin release, *Gastroenterology,* 59, 499, 1970.
71. **Reeder, D. D., Brandt, E. N., Jr., Watson, L. C., Hjelmquist, U. B. E., and Thompson, J. C.,** Pre- and posthepatic measurements of mass of endogenous gastrin, *Surgery*, 72, 34, 1972.
72. **Dencker, H., Håkanson, R., Liedberg, G., Norryd, C., Oscarson, J., Rehfeld, J. F., and Stadil, F.,** Gastrin in portal and peripheral venous blood after feeding in man, *Gut*, 14, 856, 1973.
73. **Christiansen, L. A., Keiding, S., and Winkler, K.,** Hepatic elimination of endogenous gastrin in pigs, *Scand. J. Gastroenterol.*, 17, 113, 1982.
74. **Thompson, J. C., Llanos, O. L., Teichmann, R. K., Schafmayer, A., and Rayford, P. L.,** Catabolism of gastrin and secretin, *World J. Surg.*, 3, 469, 1979.
75. **Dockray, G. J., Gregory, R. A., Tracy, H. J., and Zhu, W.-Y.,** Postsecretory processing of heptadecapeptide gastrin: conversion to C-terminal immunoreactive fragments in the circulation of the dog, *Gastroenterology*, 83, 224, 1982.
76. **Pauwels, S., Dockray, G. J., Walker, R., and Marcus, S.,** Metabolism of heptadecapeptide gastrins in humans studied by regionspecific antisera, *J. Clin. Invest.*, 75, 2006, 1985.
77. **Gregory, R. A.,** The gastrointestinal hormones: a review of recent advances, *J. Physiol.*, 241, 1, 1974.
78. **Power, D. M., Bunnett, N., and Dimaline, R.,** Chromatographic and immunochemical studies on postsecretory processing of gastrin in the pig, *Am. J. Physiol.*, 251, G300, 1986.
79. **Kerr, M. A. and Kenny, A. J.,** The purification and specificity of a neutral endopeptidase from rabbit kidney brush border, *Biochem. J.,* 137, 477, 1974.
80. **Turner, A. J., Hryszko, J., and Bunnett, N. W.,** Endopeptidase 24.11 (enkephalinase) from porcine fundic muscle: purification and characterization, *Regul. Pept.*, 13, 72, 1985.
81. **Gafford, J. T., Skidgel, R. A., Erdös, E. G., and Hersh, L. B.,** Human kidney "enkephalinase", a neutral metalloendopeptidase that cleaves active peptides, *Biochemistry*, 22, 3265, 1983.
82. **Power, D. M., Bunnett, N., Turner, A. J., and Dimaline, R.,** Degradation of endogenous heptadecapeptide gastrin by endopeptidase 24.11 in the pig, *Am. J. Physiol.*, 253, G33, 1987.
83. **Bunnett, N. W., Debas, H. T., Turner, A. J., Kobayashi, R., and Walsh, J. H.,** Metabolism of gastrin and cholecystokinin by endopeptidase 24.11 from the pig stomach, *Am. J. Physiol.*, 255, G676, 1988.
84. **Deschodt-Lanckman, M., Pauwels, S., Nadjovski, T., Dimaline, R., and Dockray, G. J.,** *In vitro* and *in vivo* degradation of human gastrin by endopeptidase 24.11, *Gastroenterology*, 94, 712, 1988.
85. **Pauwels, S., Najdovski, T., Dimaline, R., Lee, C. M., and Deschodt-Lanckman, M.,** Degradationof human gastrin and CCK by endopeptidase 24.11: differential behaviour of the sulphated and unsulphated peptides, *Biochim. Biophys. Acta*, 996, 82, 1989.

Chapter 9

DEGRADATION OF CIRCULATING INTESTINAL PEPTIDES: METABOLISM OF SUBSTANCE P AND VASOACTIVE INTESTINAL POLYPEPTIDE

Peter Schmidt and Jan Fahrenkrug

TABLE OF CONTENTS

I. INTRODUCTION

The gastrointestinal tract is a complex organ system under autonomic nervous control. The nerves supplying the gastrointestinal tract are either coming to the gut (extrinsic nerves) or they are intrinsic, constituting the enteric nervous system. Besides the classical neurotransmitters acetylcholine and norepinephrine, a variety of other substances has been explored as possible neurotransmitters in the gastrointestinal functions. Since the mid-1970s, an increasing number of biologically active peptides have been shown to occur in autonomic nerves of the gut, and many of these peptides have potent effects on motility, secretion, and circulation. The neurons and nerve fibers containing substance P and vasoactive intestinal polypeptide (VIP) comprise major and widespread populations of the peptide-containing systems in the gut, and their role as neurotransmitters seems to be established. The present chapter will deal with the degradation of these two vasoactive peptides which are released from the gastrointestinal tract.

II. DEGRADATION OF SUBSTANCE P

A. STRUCTURE, SYNTHESIS, AND FUNCTION OF SUBSTANCE P

Substance P was isolated from equine brain and intestine in 1931 and was found to be a potent hypotensive and spasmogenic agent.[1] Forty years later, the amino acid sequence of substance P was established.[2] It is now known that substance P belongs to a family of peptides called tachykinins, which includes both mammalian and nonmammalian peptides (see Table 1). They share a common C-terminal sequence (Phe-X-gly-leu-met-NH_2) and have similar biological effects such as smooth muscle contraction, saliva secretion, and lowering of the blood pressure.[3] The biological effect thus seems to be linked to the C-terminal amino acids of the peptide sequence. From cDNA studies, the structures of two substance P precursors, α- and β-prepro-tachykinin were established in 1983 (Figure 1).[4] While the α-prepro-tachykinin contains only substance P in its sequence, the β-prepro-tachykinin also contains neurokinin-A and the extended form of neurokinin-A, designated neuropeptide K.[5] Substance P and other tachykinins are located in neurons, mainly in the central nervous system and the gastrointestinal tract. In the intestine, substance P-containing fibers seem to supply myenteric and submucosal ganglia, the circular smooth muscle, and the mucosa. Most fibers seem to be intrinsic of origin.[6] The endocrine mucosa cells have been reported to contain substance P immunoreactivity, at least in some species.[7–9]

The main function of substance P seems to be a role as sensory afferent neurotransmitter. The peptide circulates in a concentration in the low picomolar range, which probably reflects spillover of locally acting peptide.[10–12] The majority of substance P in the circulation is most likely derived from the gastrointestinal tract. Thus, exvisceration or ligation of all blood vessels from the intestine in cats leads, within 15 min, to a decrease in circulating substance P to 29% and 44%, respectively.[13] An endocrine role for substance P is rather unlikely since the circulating levels are below those producing biological effects. The plasma concentrations of neurokinin-A and neuropeptide K in man are at the same level as the substance P concentration.[14] In patients with carcinoid tumors however, plasma concentrations of neurokinin-A and neuropeptide K were demonstrated to be constantly elevated, while the increase in substance P was more variable.[15,16] Whether these peptides are involved in the symptoms developed by some of these patients, including flushing, hypotension, and asthma, remains to be clarified, but infusion of substance P in man can mimic these symptoms and peak concentrations of neurokinin-A and neuropeptide K seem to coincide with carcinoid flushing episodes.[14,17]

B. METABOLIC CLEARANCE RATE AND HALF-LIFE OF SUBSTANCE P

The pharmacokinetics of substance P have been examined in pigs, where the peptide was

TABLE 1
Primary Sequence of Mammalian Tachykinins

Substance P
Arg-Pro-Lys-Pro-Gln-Gln-Phe-Phe-Gly-Leu-Met-NH_2
Neurokinin A
His-Lys-Thr-Asp-Ser-Phe-Val-Gly-Leu-Met-NH_2
Neurokinin B
Asp-Met-His-Asp-Phe-Phe-Val-Gly-Leu-Met-NH_2
Neuropeptide K
Asp-Ala-Asp-Ser-Ser-Ile-Glu-Lys-Gln-Val-Ala-Leu-Leu-Lys-Ala-Leu-Tyr-Gly-His-Gly-Gln-Leu-Ser-His-Lys-Arg-His-Lys-Thr-Asp-Ser-Phe-Val-Gly-Leu-Met-NH_2

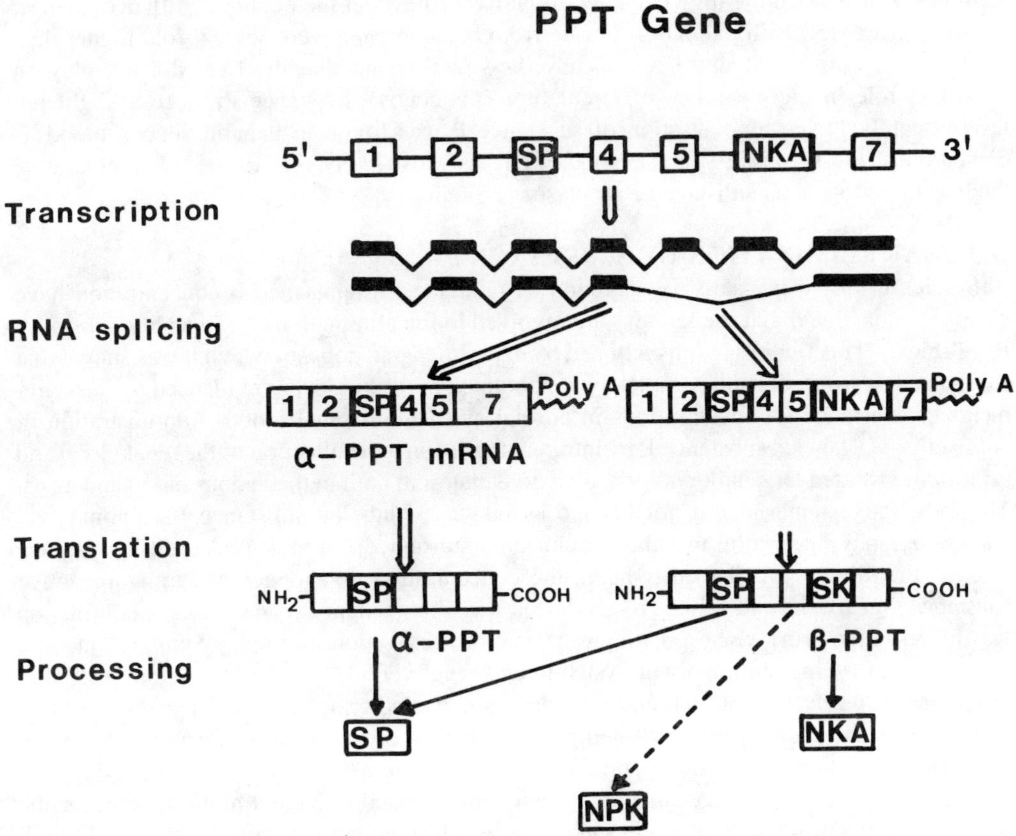

FIGURE 1. Schematic representation of alternative RNA splicing pathways in the expression of the preprotachykinin (PPT) gene. In the structure of the gene, numbered boxes and those indicated as SP and NKA stand for the corresponding exons and the exons encoding substance P (exon 3) and neurokinin A (exon 6), respectively. Note that the translational initiation and termination sites are present in the mRNA sequences encoded by exon 2 and exon 7, respectively. NPK stands for neuropeptide K.

infused to plateau level and blood samples were repeatedly collected after stopping the infusion. In these experiments, the half-life of substance P was found to be 3.0 min; the metabolic clearance rate (MCR) and the apparent distribution space were estimated to be 39 ml/kg/min and 168 ml/kg, respectively.[18] The short half-life of substance P in the circulation correlates well with studies in man where the symptoms, in response to the substance P infusion, subside 1 to 2 min after cessation of the peptide administration.[17] By incubation of substance P with plasma *in vitro,* it was found that the exogenous substance P was more rapidly degraded than endogenous substance P, and this degradation was temperature dependent.[19–22] Evidence has been provided that the endogenous substance P is more stable in plasma due to protein binding and protection of the peptide.[23]

Information on degradation of the other members of the tachykinin family is limited, but it has been shown *in vitro* that substance P is more rapidly degraded by tissue homogenates than neurokinin-A.[24]

C. HEPATIC ELIMINATION OF SUBSTANCE P

The contribution of the liver to the elimination of substance P has been studied in rats where the peptide was infused into the portal vein, or into the jugular vein and the effect on salivary secretion (substance P bioassay) was compared. The results indicated that 84% of substance P entering the liver was eliminated[20] (Table 2). Similar studies in the dog using bioassay for substance P supported the high capacity of the liver to extract the peptide.[25,26] In accordance, the substance P levels in patients suffering from hepatic coma were several-fold higher than the levels in normal subjects.[11] One study, however, found that the liver did not play an important role in the clearance of circulating endogenous substance P, since a slight but insignificantly lower concentration of substance P was found in hepatic venous blood as compared to the systemic venous blood.[12] This is most likely due to the relatively stable binding of endogenous substance P to plasma proteins.

D. RENAL ELIMINATION OF SUBSTANCE P

Studies in rats using salivary secretion as bioassay for substance P concentration have clearly demonstrated that the kidneys are involved in the elimination of exogenous substance P[20] (Table 2). This finding is substantiated by experiments in dogs, in which it was shown that substance P was rapidly and completely metabolized in the kidney, as evidenced by measurements of substance P concentrations in arterial and venous renal blood. Administration of radioactive ^{125}I-labeled substance P by intrarenal infusion and collection of the renal blood and urine demonstrated large amounts of radioactive material, both in the venous blood and urine. The radioactive material did not behave as intact ^{125}I-labeled substance P on thin layer chromatography, indicating that the administered substance P was degraded during passage through the kidney. On the other hand, the concentration of endogenous immunoreactive substance P in arterial and venous renal plasma was almost identical and only a small fraction was found in the urine, corresponding to a fractional excretion of 6.6%.[27] Similar data were found in man by measurements of systemic and renal venous plasma concentrations.[12] The difference between the results found with endogenous and exogenous substance P could be explained by the previously mentioned protein binding of the peptide and protection from glomerular filtration and tubular peptidases.

In vitro rat kidney homogenates have been demonstrated to have a high capacity for the inactivation of substance P[28] and pig kidney microvilli membranes from the proximal tubuli could also degrade substance P effectively.[29]

E. OTHER SITES FOR SUBSTANCE P ELIMINATION

The contribution of the lungs to the metabolism of circulating substance P has, in rats and dogs, been demonstrated to be small or negligible.[20,30] The role of the rat brain in the elimination of substance P has been studied by comparing the saliva secretion obtained by injection of substance P into either the internal carotid artery or the jugular vein. Since the biological response was approximately equal after the two routes of administration, it was concluded that no or limited elimination of substance P takes place in the brain.[20] Substance P, however, seems to be eliminated in the hind limb of the rat, since injection of substance P into the femoral vein gave a salivary secretion that was six times as large as that obtained by injection into the iliac artery[20] (Table 2). As substance P is degraded in circulating plasma,

TABLE 2
Salivary Response (Bioassay for Substance P) to the Infusion of 1 μg SP/kg/min into Various Vascular Beds, as Measured during the 5-min Infusion Period

	Ligation of coeliac artery, superior and inferior mesenteric artery	Ligation of renal vessels	Site of infusion of SP	mg saliva per 5 min ($x \pm s_x$; n = 6)	% of "control"	P-value	Difference between 1 and 2 represents elimination of SP mainly in
A1	–	–	Jugular vein	306 ± 38	100	<0.0001	Liver
A2	–	–	Portal vein	35 ± 17	11		
B1	+	+	Jugular vein	292 ± 32	100	<0.05	Kidney
B2	+	+	Jugular vein	214 ± 37	73		
C1	+	+	Femoral vein	294 ± 58	100	<0.002	Hind leg
C2	+	+	Iliac artery	49 ± 5	17		

From Lembeck, F., Holzer, P., Schweditsch, M., and Gamse, R., *Naunyn-Schmiedebergs Arch. Pharmacol.*, 305, 9, 1978.

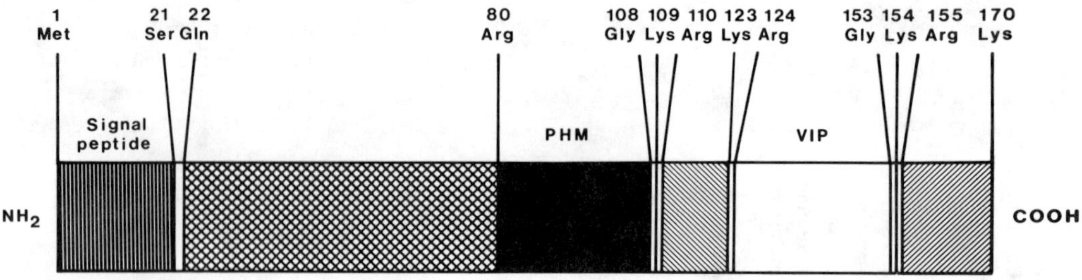

FIGURE 2. Schematic representation of structure of human prepro-VIP/PHM. Putative signal peptide is from 1 to 21, PHM is from 81 to 107, and VIP is from 125 to 152. Amino acid residues at post-translational processing site are shown.

differences in transit times between injection on venous and arterial side of the circulation may also, at least in part, explain the difference in salivary response.

F. MECHANISMS FOR DEGRADATION OF SUBSTANCE P

In vitro substance P is degraded by cultured endothelial cells, tissue homogenates, and, as previously mentioned, in plasma.[24,31,32] Proteolytic enzymes seem to be important since the degradation in plasma is temperature-dependent and could be prevented or reduced by various enzyme inhibitors in plasma and tissue homogenates.[20,32,33]

These *in vitro* studies suggest that endopeptidase 24.22 and angiotensin-converting enzyme could be involved in the elimination of substance P. Both enzymes are widely distributed in the body bound to plasma membranes, and endopeptidase 24.11 is capable of cleaving all tachykinins, whereas angiotensin-containing enzyme can only cleave substance P.[34–37]

From human brain, a specific substance P endopeptidase has been isolated; but to what extent this enzyme is involved in degradation of circulating substance P is not known.[38]

III. DEGRADATION OF VIP

A. STRUCTURE, SYNTHESIS, AND FUNCTION OF VIP

Vasoactive intestinal polypeptide (VIP) is a 28 amino acid peptide which, in 1970, was isolated from extracts of porcine intestine and recognized for its potent vasodilatory effects.[39,40] The peptide is a highly basic, single-chain, linear polypeptide, which terminates in a carboxyamide group.[41,42] The primary structure of VIP is closely related to secretin, a peptide with N-terminal histidine and C-terminal isoleucine amide (PHI) and, to a lesser extent, to a pancreatic glucagon, gastric inhibitory polypeptide, and growth hormone releasing factor. There is evidence that VIP may have a short helical component in the structure encompassing amino acid residues 15 to 21.[43] VIP of mammalian origin has a remarkably constant structure, since the sequence of bovine, human, and rat VIP has shown to be identical to the porcine peptide.[44–46] Guinea pig VIP, however, differs in four amino acid residues from VIP in other mammalian species.[47] VIP is synthesized from a 170 amino acid precursor, the structure of which was established by means of nucleotide sequence analysis of cDNA complementary to the mRNA coding for VIP[48,49] (Figure 2). This precursor contains, in addition to VIP, another biologically active peptide in its sequence called PHM [peptide with N-terminal histidine and C-terminal methionin amide (human form)] or PHI [peptide with N-terminal histidine and C-terminal isoleucine amide (rat and porcine form)]. Although the deduced structure of the common precursor implies production and release of VIP and PHI/PHM on an equimolar basis, there is evidence that the two peptides are differentially processed from the common precursor.[50]

VIP is present in neurons, with widespread occurrence in both the central and the peripheral nervous system. VIP is synthesized by a ribosomal mechanism in the neuronal cell bodies and the newly formed VIP, which is derived from the precursor by proteolytic cleavage, is then exported along the axones or dendrites to the terminal stores from which VIP is released.[51] The VIP-containing neurons and fibers are mainly intrinsic or post-ganglionic parasympathetic located in the gut, the respiratory and urogenital tracts, exocrine and endocrine glands, certain autonomic nerves and ganglia, and in the central nervous system.[52] The VIP-containing fibers and terminals seem to innervate vascular and nonvascular smooth muscle cells as well as glands.

According to the occurrence of the VIP-containing nerves, the peptide is a potent vasodilator in most vascular beds. The peptide relaxes the smooth musculature in the digestive, respiratory, and urogenital tracts and, in these systems, VIP has been established as a noncholinergic, nonadrenergic inhibitory neurotransmitter.[53] In addition, VIP stimulates secretion from various endocrine and exocrine glands. Finally, VIP has some metabolic effects which most likely reflect the structural resemblance to other peptides, such as glucagon. Partial sequences of VIP display limited biological effects; the activity increases with increasing chain length, and the whole molecule is needed for full bioactivity.[54,55] The effect of VIP is elicited by its interaction with a specific membrane-bound receptor. How the binding of VIP to its receptor is coupled to the biological response is only partially understood but, in most cells, VIP stimulates adenylate cyclase activity and formation of cAMP.[56]

VIP is released during nervous activation and the circulating VIP probably reflects neuronal overflow from activated VIP-containing neurons. The majority of the circulating VIP in peripheral venous blood probably originates from neurons in the gut, where the bulk of VIP is located. The circulating levels of VIP measured by radioimmunoassay is below 20 pmol/l.[57,58] Elevated levels, however, can be observed in conditions with impaired renal or liver function such as uremia and liver cirrhosis.[58,59] Grossly elevated plasma levels of VIP are observed in patients with VIP-secreting tumors associated with the watery diarrhea kalemia syndrome.[60]

B. METABOLIC CLEARANCE RATE AND HALF-LIFE OF VIP

Studies on the pharmacokinetics of VIP in man have shown that the half-life of the peptide is very short (i.e., 1 to 2 min) as shown in Figure 3.[61,62] A slightly longer half-life was found in dogs.[63] Results on the MCR and distribution volume in man are conflicting. Thus, one study reported a metabolic clearance rate of 9 ml/kg/min and a distribution volume of 14 ml/kg, while other investigations found the MCR and distribution volume to be 42 ml/kg/min and 135 ml/kg, respectively.[61,62] The reason for this discrepancy is not obvious, but is noteworthy that a distribution volume of 14 ml/kg is less than the plasma volume in man. It should be emphasized, however, that substances with degradation in circulating plasma may not fulfill the basic assumptions of kinetic concepts like distribution volume and MCR (see Chapter 2).

C. HEPATIC ELIMINATION OF VIP

As previously mentioned, the majority of circulating VIP originates from neurons in the gastrointestinal tract. Consequently, the concentrations of VIP are found to be two- to threefold higher in the portal blood than in the systemic circulation.[64]

The role of the liver in the metabolism of splanchnically released VIP is not fully clarified and the literature is rather conflicting. In dogs, it was shown by simultaneous determination of VIP in plasma from the portal and inferior caval vein that, under basal conditions, the liver extracted 42% of the endogenous VIP passing through the organ.[64] However, in another study in the same species, systematically administered VIP did not result in significantly higher plasma concentrations as compared to the same dose of VIP given into the portal vein.[63] It was concluded that the liver did not play a role in the elimination of VIP. This interpretation is, however, debatable since the amount of peptide used resulted in plasma concentrations far above the physiological levels.

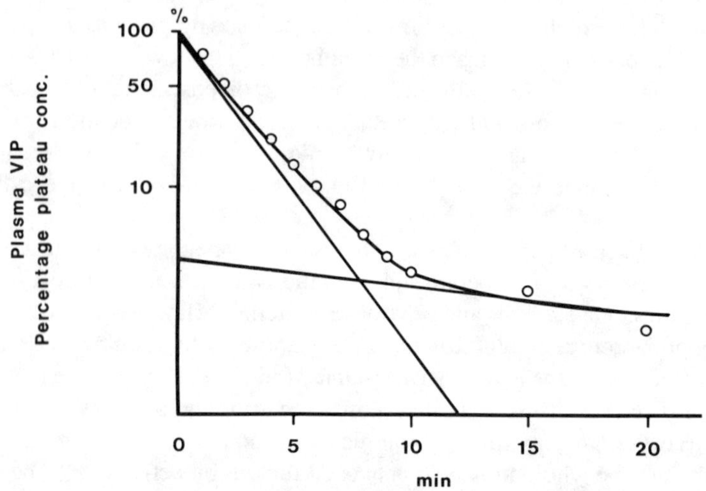

FIGURE 3. Elimination of VIP in plasma. VIP concentration plotted semilogarithmically as per cent of level obtained by steady-state infusion vs. time in minutes. The curve consists of two exponential functions with a $t_{1/2}$ of 2 min and 21 min, respectively. (From Holm-Bentzen, M., Christiansen, J., Petersen, B., Fahrenkrug, J., Schultz, A., and Kirkegaard, P., *Scand. J. Gastroenterol.*, 16, 429, 1981. With permission.)

Evidence for a concentration-dependent and substrate-activated VIP degradation in the liver has been shown in pig studies. By measuring the VIP concentration in portal and right atrial blood, it was observed that when the VIP concentration was below 30 pmol/l, 6.6% of the peptide was removed during its passage through the liver while 29% was extracted when the VIP concentration was above this level.[65] On the other hand, in isolated, perfused rat liver, VIP added in low concentrations is rapidly degraded whereas a fall in clearance of VIP in the perfusate was found at higher peptide concentrations.[66] Studies *in vitro* support a hepatic elimination of VIP since it was found that the liver cells, primarily the nonparenchymal cells, degraded the peptide.[67]

Results on the plasma concentrations of VIP in patients with impaired liver function (i.e., cirrhosis of the liver) are conflicting. One group of investigators was unable to show elevated VIP levels in cirrhotic patients, while other reports found significantly higher concentrations of VIP in this condition as compared to normal controls.[57,59,68] In the largest study comprising 136 patients with cirrhosis and 112 normal subjects, the VIP concentrations were 9.4 pmol/l and 6.2 pmol/l, respectively[59] (p-value < 0.01).

The elevated VIP concentrations could be explained by reduced liver and/or kidney function or porto-caval shunting. However, even in patients with cirrhosis, the liver extracted 43% of the endogenous VIP passing through, a finding which supports a role for the organ in the elimination of the peptide.[59]

D. RENAL ELIMINATION OF VIP

There is also evidence in man that the kidney plays a role in the elimination of VIP. Thus, in normal subjects, the kidney was found to extract 23% of the circulating endogenous VIP passing through the organ, while substantially no renal extraction of VIP was observed in patients with uremia. Accordingly, plasma levels of VIP in the circulation are elevated in patients with impaired renal function. A positive correlation between serum creatinine and peripheral venous VIP was found in patients with elevated serum creatinine concentrations.[58]

E. OTHER SITES OF VIP ELIMINATION

Besides renal and hepatic elimination of VIP, it is likely that the lungs play a role in the clearance of circulating VIP. Thus, a pulmonary extraction of administered peptide has been demonstrated in experiments on pigs.[69]

In man, no arterio-venous differences in VIP concentrations were found in the brain or the lower limb, findings which exclude major roles of these areas in the elimination of VIP.[58]

F. MECHANISMS FOR DEGRADATION OF VIP

In vitro VIP is degraded by tissue homogenates from various organs.[67,70–73] Proteolytic enzymes seem to be involved, since studies on homogenates from rat ileum revealed that VIP degradation was temperature dependent and could be reduced or annulled by various enzyme inhibitors.[73] In studies on porcine enterocytes, it was demonstrated that VIP, shortly after its receptor binding, loses the N-terminal histidine residue. This process involves proteolytic cleavage by an aminopeptidase, most likely aminopeptidase N.[72] Based on experiments with specific inhibitors of proteolytic enzymes, it was suggested that the degradation of VIP also involved endopeptidase 24.11.[72]

Other results have been found in studies on the mechanism of VIP elimination in the liver. VIP was, after the initial binding to membrane receptors on hepatocytes and Kupffer cells, found to be internalized and degraded by lysosomal enzymes.[71]

IV. SUMMARY

Substance P infused into the circulation is rapidly eliminated with a half-life of 3.0 min. The liver, kidneys, and lungs have high capacity if degrading exogenous substance P. Degradation of **endogenous substance P** is not related to a particular organ, but probably takes place in all vascular beds mediated by enzymes bound to endothelial cells and enzymes in plasma.

Infused VIP is rapidly cleared from the circulation with a half-life of 2.0 min. The liver and the kidneys play important roles in the elimination of **endogenous VIP**, whereas the lung also has a high capacity for extraction of **exogenous VIP**. Mechanisms in the VIP degradation could include enzymatic cleavage and endocytosis.

ABBREVIATIONS

cAMP	Cyclic adenosine monophosphate
cDNA	Complementary deoxyribonucleic acid
PHI	Peptide with N-terminal histidine and C-terminal isoleucine amide
PHM	Peptide with N-terminal histidine and C-terminal methionine amide
mRNA	Messenger ribonucleic acid
VIP	Vasoactive intestinal polypeptide

REFERENCES

1. **von Euler, U. S. and Gaddum, J. H.,** An unidentified depressor substance in certain tissue extracts, *J. Physiol. (London),* 72, 74, 1931.
2. **Chang, M. M., Leeman, S. E., and Niall, H. D.,** Amino acid sequence of substance P, *Nature (London),* New Biol., 232, 86, 1971.
3. **Pernow, B.**, Substance P, *Pharmacol. Rev.,* 35, 86, 1983.
4. **Nawa, H., Hirose, T., Takashima, H., Inayama, S., and Nakanishi, S.,** Nucleotide sequences of cloned cDNA for two types of bovine brain substance P precursor, *Nature (London),* 306, 32, 1983.

5. **Tatemoto, K., Lundberg, J. M., Jörvall, H., and Mutt, V.**, Neuropeptide K: isolation, structure and biological activities of a novel brain tachykinin, *Biochem. Biophys. Res. Commun.,* 128, 947, 1985.
6. **Furness, J. B. and Marcello, C.,** *The Enteric Nervous System,* Churchill Livingstone, New York, 1987, chap. 4.
7. **Nilsson, G., Larsson, L.-I., Håkanson, R., Brodin, E., Pernow, B., and Sundler, F.,** Localization of substance P-like immunoreactivity in mouse gut, *Histochemistry,* 43, 97, 1975.
8. **Pearse, A. G. E. and Polak, J. M.,** Immunocytochemical localization of substance P in mammalian intestine, *Histochemistry,* 41, 373, 1975.
9. **Sjölund, K., Sanden, G., Håkanson, R., and Sundler, F.,** Endocrine cells in human intestine: an immunocytochemical study, *Gastroenterology,* 85, 1120, 1983.
10. **Emson, P. C., Gilbert, R. F. T., Martensen, H., and Nobin, A.,** Elevated concentrations of substance P and 5-HT in plasma in patients with carcinoid tumors, *Cancer,* 54, 715, 1984.
11. **Hörtnagl, H., Singer, E. A., Lenz, K., Kleinberger, G., and Lochs, H.,** Substance P is markedly increased in plasma of patients with hepatic coma, *Lancet,* 1, 480, 1984.
12. **Henriksen, J. H., Schaffalitzky de Muckadell, O. B., and Bülow, J. B.,** Does liver-intestine significantly degrade circulating endogenous substance P in man?, *Scand. J. Gastroenterol.,* 21, 300, 1986.
13. **Gamse, R., Mroz, E., Leeman, S., and Lembeck, F.,** The intestine as source of immunoreactive substance P in plasma of the cat, *Naynyn-Schmiedebergs Arch. Pharmacol.,* 305, 17, 1978.
14. **Norheim, I., Theodorsson-Norheim, E., Brodin, E., and Öberg, K.,** Tachykinins in carcinoid tumors: their use as a tumor marker and possible role in the carcinoid flush, *J. Clin. Endocrinol. Metab.,* 63, 605, 1986.
15. **Theodorsson-Norheim, E., Norheim, I., Öberg, K., Brodin, E., Lundberg, J. M., Tatemoto, K., and Lindgren, P. G.,** Neuropeptide K: a major tachykinin in plasma and tumor tissue from carcinoid patients, *Biochem. Biophys. Res. Commun.,* 131, 77, 1985.
16. **Norheim, I., Theodorsson-Norheim, E., Brodin, E., Öberg, K., Lundquist, G., and Rosell, S.,** Antisera raised against eledoisin and kassinin detect elevated levels of immunoreactive material in plasma and tumor tissue from patient with carcinoid tumors, *Peptides,* 9, 245, 1984.
17. **Duner, H. and Pernow, B.,** Circulatory studies on substance P in man, *Acta Physiol. Scand.,* 49, 261, 1960.
18. **MacGregor, G. P., Blank, M. A., Manolas, K., Farmer, H., and Bloom, S. R.,** Investigation of the pharmacokinetics of substance P in the pig by radioimmunoassay, in *Proc. Int. Symp. Substance P,* Skrabanek, P. and Powell, D., Eds., Boole Press, Dublin, 1983, 137.
19. **Skrabanek, P., Cannon, D., Kirranem, J., Legge, D., and Powell, D.,** Circulatory immunoreactive substance P in man, *Ir. J. Med. Sci.,* 145, 399, 1976.
20. **Lembeck, F., Holzer, P., Schweditsch, M., and Gamse, R.,** Elimination of substance P from the circulation of the rats and its inhibition by bacitracin, *Naunyn-Schmiedebergs Arch. Pharmacol.,* 305, 9, 1978.
21. **Berger, H., Fencher, K., Albrect, E., and Niedrich, H.,** Substance P: *in vitro* inactivation by rat brain fractions and human plasma, *Biochem. Pharmacol.,* 28, 3173, 1979.
22. **Couture, R. and Regoli, D.,** Inactivation of substance P and its C-terminal fragments in plasma and its inhibition by captopril, *Can. J. Physiol. Pharmacol.,* 59, 621, 1981.
23. **Cannon, D., Skrabenek, P., and Powell, D.,** Difference in behavior between synthetic and endogenous substance P in human plasma, *Naunyn-Schmiedebergs Arch Pharmacol.,* 307, 251, 1979.
24. **Holzer-Petsche, U., Saria, A., Ruptiz, M., Schuligoi, R., and Lembeck, F.,** Degradation of substance P, neurokinin A, and calcitonin gene-related peptide in rat plasma *in vivo, Regul. Pept.,* 22, 89, 1988.
25. **Hallberg, D. and Pernow, B.,** Effect of substance P on various vascular beds in the dog, *Acta Physiol. Scand.,* 93, 277, 1975.
26. **Melchiorri, P., Tonelli, F., and Negri, L.,** Comparative circulatory effects of substance P, eledoisin, and physalemin in the dog, in *Nobel Symp. 37 Substance P,* von Euler, U.S. and Pernow, B., Eds., Raven Press, New York, 1977, 311.
27. **Campbell, W. B. and Ward, P. E.,** Renal metabolism of substance P in the dog, *Life Sci.,* 24, 1995, 1979.
28. **Ward, P. E. and Johnsson, A. R.,** Renal inactivation of substance P in the rat, *Biochem. J.,* 171, 143, 1977.
29. **Stephenson, S. L. and Kenny, A. J.,** Metabolism of neuropeptides, *Biochem. J.,* 241, 237, 1987.
30. **Boileau, J., Campeau, L., and Biron, P.,** Pulmonary fate of histamine, isoproterenol, physalaemin and substance P, *Can. J. Physiol. Pharmacol.,* 48, 681, 1970.
31. **Johnson, A. R. and Erdos, E. G.,** Inactivation of substance P by cultured human endothelial cells, in *Nobel Symp. 37 Substance P,* von Euler, U.S. and Pernow, B., Eds., Raven Press, New York, 1977, 253.
32. **Orloff, M. S., Turner, A. J., and Bunnett, N. W.,** Catabolism of substance P and neurotensin in the rat stomach wall is susceptible to inhibitors of angiotensin converting enzyme, *Regul. Pept.,* 14, 21, 1986.
33. **Iwamoto, I., Ueki, I., and Nadel, J. A.,** Effect of neutral endopeptidase inhibitors on ^{3}H-substance P binding in rat ileum, *Neuropeptides,* 11, 185, 1988.
34. **Erdos, E. G.,** Kininases, in *Handbook of Experimental Pharmacology,* Vol. 25, Suppl., Erdos, E. G., Ed., Springer Verlag, Heidelberg, 1979, 427.
35. **Gee, N. S., Matsas, R., and Kenny, A. J.,** A monoclonal antibody to kidney endopeptidase 24.11, *Biochem. J.,* 214, 377, 1983.

36. **Hersh, L. B.,** Degradation of enkephalins: the search for an enkephalinase, *Mol. Cell Biochem.*, 47, 35, 1982.
37. **Hooper, N. M., Kenny, A. J., and Turner, A. J.,** The metabolism of neuropeptides, *Biochem. J.*, 231, 357, 1985.
38. **Lee, C., Sandberg, B. E. B., Hanley, M. R., and Iversen, L. L.,** Purification and characterization of a membrane-bound substance P-degrading enzyme from human brain, *Eur. J. Biochem.*, 114, 315, 1981.
39. **Said, S. I. and Mutt, V.,** Polypeptide with broad biological activity: isolation from small intestine, *Science,* 169, 1217, 1970.
40. **Said, S. I. and Mutt, V.,** Isolation from porcine-intestinal wall of a vasoactive octacosapeptide related to secretin and to glucagon, *Eur. J. Biochem.*, 28, 199, 1972.
41. **Mutt, V. and Said, S. I.,** Structure of the porcine vasoactive intestinal octacosapeptide. The amino acid sequence. Use of kallikrein in its determination, *Eur. J. Biochem.*, 42, 581, 1974.
42. **Bodanszky, M., Klausner, U. S., Lin, C. Y., Mutt, V., and Said, S. I.,** Synthesis of the vasoactive intestinal peptide (VIP), *J. Am. Chem. Soc.*, 96, 4973, 1974.
43. **Bodanszky, M., Bodanszky, A., Klausner, U. S., and Said, S. I.,** A preferred conformation in the vasoactive intestinal peptide (VIP). Molecular architecture of gastrointestinal hormones, *Biorg. Chem.*, 3, 133, 1974.
44. **Carlquist, M., Mutt, V., and Jörnvall, H.,** Isolation and characterization of bovine vasoactive intestinal peptide (VIP), *FEBS Lett.*, 108, 457, 1979.
45. **Carlquist, M., McDonald, T. J., Go, V. L. W., Bataille, D., Johansson, C., and Mutt, V.,** Isolation and amino acid composition of human vasoactive intestinal polypeptide (VIP), *Horm. Metab. Res.*, 14, 28, 1982.
46. **Dimaline, R., Reeve, J. R., Shively, J. E., and Hawke, D.,** Isolation and characterization of rat vasoactive intestinal peptide, *Peptides,* 5, 183, 1984.
47. **Du, B.-H., Eng, J., Hulmes, J. D., Chang, M., Pan, Y.-C. E., and Yalow, R. S.,** Guinea pig has a unique mammalian VIP, *Biochem. Biophys. Res. Commun.*, 128, 1093, 1983.
48. **Itoh, N., Obata, K., Yanaihara, N., and Okamoto, H.,** Human preprovasoactive intestinal polypeptide contains a novel PHI-27-like peptide, PHM-27, *Nature (London),* 304, 547, 1983.
49. **Nishizawa, M., Hayakawa, Y., Yanaihara, M., and Okamoto, H.,** Nucleotide sequence divergence and functional constraint in VIP precursor mRNA evolution between human and rat, *FEBS Lett.*, 183, 55, 1985.
50. **Fahrenkrug, J., Bek, T., Lundberg, J. M., and Hökfelt, T.,** VIP and PHI in cat neurons: colocalization but variable tissue content possibly due to differential processing, *Regul. Pept.*, 12, 21, 1985.
51. **Gilbert, R. F. T., Emson, P. C., Fahrenkrug, J., Lee, C. M., Penman, E., and Wass, J.,** Axonal transport of neuropeptides in the cervical vagus nerve of the rat, *J. Neurochem.*, 34, 108, 1980.
52. **Hökfelt, T., Schultzberg, M., Lundberg, J. M., Fuxe, K., Mutt, V., Fahrenkrug, J., and Said, S. I.,** Distribution of vasoactive intestinal polypeptide in the central and peripheral nervous systems as revealed by immunocytochemistry, in *Vasoactive Intestinal Polypeptides,* Said, S. I., Ed., Raven Press, new York, 1982, 65.
53. **Fahrenkrug, J.,** VIP and autonomic neurotransmission, *Pharmacol. Ther.*, 41, 515, 1989.
54. **Yanaihara, N., Sakagami, M., Sato, H., Yamamoto, K., Hashimoto, T., Yanaihara, T., Ito, Z., Yamaguchi, K., and Abe, K.,** Immunological aspects of secretin, substance P and VIP, *Gastroenterology,* 72, 803, 1979.
55. **Fahrenkrug, J.,** Vasoactive intestinal polypeptide: measurement, distribution and putative neurotransmitter function, *Digestion,* 19, 149, 1979.
56. **Amiranoff, B. and Rosselin, G.,** VIP receptors and control of cyclic AMP production, in *Vasoactive Intestinal Polypeptides,* Said, S. I., Ed., Raven Press, New york, 1982, 307.
57. **Elias, E., Mitchell, S. J., and Bloom, S. R.,** Vasoactive intestinal peptide in cirrhosis, *Lancet,* 2, 1312, 1975.
58. **Henriksen, J. H., Staun-Olsen, P., Mortensen, N. B., and Fahrenkrug, J.,** Circulating endogenous vasoactive intestinal polypeptide (VIP) in patients with uraemia and liver cirrhosis, *Eur. J. Clin. Invest.*, 16, 211, 1986.
59. **Henriksen, J. H., Staun-Olsen, P., Fahrenkrug, J., and Ring-Larsen, H.,** Vasoactive intestinal polypeptide (VIP) in cirrhosis: arteriovenous extraction in different vascular beds, *Scand. J. Gastroenterol.*, 15, 787, 1980.
60. **Bloom, S. R. and Polak, J. M.,** VIPomas, in *Vasoactive Intestinal Polypeptides,* Said, S. I., Ed., Raven Press, New York, 1982, 457.
61. **Domschke, S., Domschke, W., Bloom, S. R., Mitznegg, P., Mitchell, S. J., Lux, G., and Strunz, U.,** Vasoactive intestinal peptide in man: pharmacokinetic, metabolic and circulatory effects, *Gut,* 19, 1049, 1978.
62. **Holm-Bentzen, M., Christiansen, J., Petersen, B., Fahrenkrug, J., Schultz, A., and Kirkegaard, P.,** Infusion of vasoactive intestinal polypeptide in man: pharmacokinetics and effect on gastric acid secretion, *Scand. J. Gastroenterol.*, 16, 429, 1981.
63. **Strunz, U. T., Walsh, J. H., Bloom, S. R., Thompson, M. R., and Grossmann, M. I.,** Lack of hepatic inactivation of canine vasoactive intestinal peptide, *Gastroenterology,* 73, 768, 1977.
64. **Ebeid, A. M., Escourrou, J., Soeters, P. B., Murray, P., and Fisher, J. E.,** Hepatic inactivation of vasoactive intestinal peptide in man and dog, *Ann. Surg.*, 188, 28, 1978.
65. **Hansen, C. P. Boesby, S., and Kirkegaard, P.,** Hepatic metabolism of vasoactive intestinal polypeptide (VIP) during liver transplantation in pig, unpublished data, 1990.

66. **Brook, C. W., Sewell, R. B., Shulkes, A., and Smallwood, R. A.,** Hepatic metabolism of vasoactive intestinal polypeptide (VIP) in the rat, *Regul. Pet.,* 20, 311, 1988.
67. **Gardner, D. F., Kilberg, M. S., Wolfe, M. M., McGuigan, J. E., and Misbin, R. I.,** Preferential binding of vasoactive intestinal peptide to hepatic nonparenchymal cells, *Am. J. Physiol.,* 248, G663, 1985.
68. **Said, S. I., Faloona, G. R., Deon, H., Unger, R. H., and Siegel, S. R.,** Vasoactive intestinal polypeptide: elevated levels in patients with hepatic cirrhosis, *Clin. Res.,* 22, 367, 1974.
69. **Hvidsten, D., Lindsetmo, R. O., Rokke, O., Jenssen, T. G., Burhol, P. G., and Revhaug, A.,** Clearance of vasoactive intestinal polypeptide (VIP) in the porcine pulmonary circulation, *Regul. Pept.,* 26, 267, 1989.
70. **Keltz, T. N., Staus, E., and Yalow, R. S.,** Degradation of vasoactive intestinal polypeptide by tissue homogenates, *Biochem. Biophys. Res. Commun.,* 92, 669, 1980.
71. **Misbin, R. I., Wolfe, M. M., Morris, P., Buynitzky, J., and McGuigan, J. E.,** Uptake of vasoactive intestinal peptide by rat liver, *Am. J. Physiol.,* 243, G103, 1982.
72. **Nau, R., Ballmann, M., and Conlon, J. M.,** Binding of vasoactive intestinal polypeptide to dispersed enterocytes results in rapid removal of the NH_2-terminal histidyl residue, *Mol. Cell Endocrinol.,* 52, 97, 1987.
73. **Chakder, S. K. and Zeitlin, I. J.,** Degradation of vasoactive intestinal polypeptide (VIP) by enzymes from rat gastrointestinal tissue, *J. Physiol.,* 392, 96P, 1987.

Chapter 10

DEGRADATION OF GLUCAGONS

Jens Juul Holst

TABLE OF CONTENTS

I. INTRODUCTION

Glucagon is a peptide hormone of 29 amino acids, secreted from the A-cells of the islets of Langerhans. The available evidence suggests that glucagon is an essential hormone with important effects on glucose and amino acid metabolism.[1] Thus, glucagon is responsible for — via its effect on hepatic glucose production — the maintenance of the glucose levels in the basal state[2] and after protein-rich meals,[3] where it accurately adjusts the glucose production according to the needs of the body, in the face of increasing insulin concentrations.[4] Glucagon is also essential in the defense against hypoglycemia.[4] It regulates the rate of urea synthesis[5] and is at least partly responsible for the regulation of amino acid turnover.[6] States of glucagon deficiency are extremely rare, but are associated with disturbed amino acid metabolism and a tendency to hypoglycemia that may be fatal.[7] Hyperglucagonemia, which may be found in cases of glucagon-producing tumors,[8] is associated with glucose intolerance and excessively low levels of circulating amino acids and, as a consequence, necrolysis of the skin and general wasting.

In this chapter, the degradation of glucagon will be reviewed. Glucagon, however, is but one of the members of a family of glucagon-like or -related peptides that are all products of glucagon gene expression. Although the information about the other members of the family is much more scanty, their degradation will also be discussed.

II. STRUCTURE AND SYNTHESIS OF GLUCAGON AND RELATED PEPTIDES

The single, mammalian glucagon gene[9] is expressed in two cell types, namely the A-cells of the islets of Langerhans and the L-cells of the intestinal mucosa.[10] In some mammals (but not in humans), the gene is also expressed in gastric so-called A-like cells.[11] In both locations, the mRNAs resulting from gene transcription are identical, though the resulting (identical) precursor, proglucagon, is processed differently[12] (see Figure 1). In the pancreas, the products are glucagon itself [occupying residues 33 to 61 in proglucagon (henceforth referred to as PG)]; the glicentin-related pancreatic polypeptide, GRPP (PG 1-30),[13] and the so-called major proglucagon fragment (MPGF) that occupies the entire C-terminal part of proglucagon (PG 72-158).[14,15] In the intestinal mucosa, the major products are (1) glicentin (PG 1-69)[16] which may be cleaved to release GRPP (PG 1-30) plus the remaining fragment of glicentin: oxyntomodulin (PG 33-69);[17,18] (2) glucagon-like peptide-1 (GLP-1, PG 78-107amide);[19] (3) intervening peptide 2 (PG 111-123amide);[20] and (4) glucagon-like peptide-2 (GLP-2, PG 126-158).[20,21]

In both the pancreas and the intestine, small amounts of proglucagon 1-61 may also be formed.[22] As will be discussed below, this peptide may be particularly prominent in renal failure.

The functions of glucagon have been briefly outlined above, but nothing is known about the biological activities, if any, of the remaining *pancreatic* products of proglucagon. Among the *intestinal* products, glicentin has been reported to inhibit gastric acid secretion in rats[23] and has been associated with mucosal growth in intestinal adaptation.[24] The fragment, oxyntomodulin, is glucagon-like in its activities and competes with glucagon for binding to the hepatic glucagon receptor, but with lower affinity.[25] It also stimulates insulin secretion[25] and inhibits gastric acid secretion[26,27] (like glucagon), the latter apparently with higher potency than glucagon. The glucagon-like peptide-1 is the most potent insulin-stimulating hormone known.[28,30] In addition, it inhibits glucagon secretion.[29] When released, it will thus greatly increase the insulin/glucagon ratio in the portal venous plasma which, in turn, strongly inhibits the hepatic glucose production.[2] GLP-1 also and very potently inhibits gastric acid secretion[31] and upper gastrointestinal motility, and may thus function as an enterogastron. Nothing is known about the biological activities of the remaining intestinal products of proglucagon.[12]

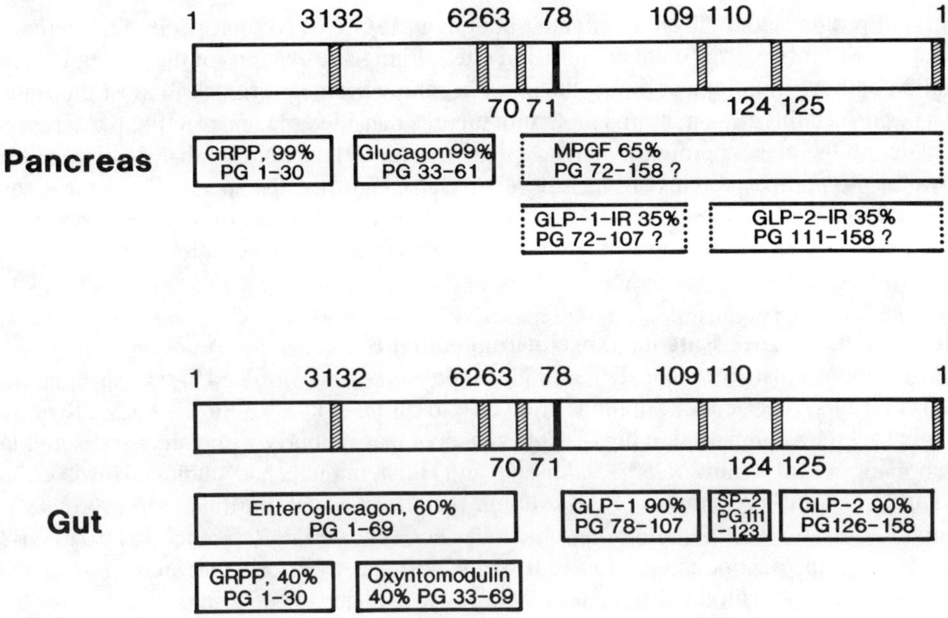

FIGURE 1. Products of proglucagon gene expression in the pancreas and gut. Proglucagon is indicated schematically at the top. The numbering refers to the individual amino acid residues. Hatched vertical bars indicate dibasic cleavage sites; the heavy vertical line indicates a monobasic cleavage site. The per cent values indicate the extent to which proglucagon is processed to this particular component. The abbreviations PG 1-30 refer to the position in proglucagon (PG) of that particular fragment. GRPP: Glicentin related pancreatic peptide; MPGF: Major proglucagon fragment; GLP-1-IR: Glucagon-like peptide-1-immunoreactivity; GLP-2-IR: Glucagon-like peptide-2-immunoreactivity; and SP-2: spacer-peptide-2.

The secretion of the pancreatic proglucagon products is inversely correlated to the glucose concentration of the blood perfusing the islets.[32,33] Further stimuli are amino acids[34] that powerfully increase glucagon secretion and activity in both divisions of the autonomic nervous system.[35] The pancreatic proglucagon products are thus released from the pancreas during activation of the sympatho-adrenal system (during stress), during prolonged fasting, and in response to meals with a high protein/carbohydrate ratio of its contents.

The intestinal products are mainly released in response to luminal stimulation of the mucosa of the intestine.[10,36] Particularly actively absorbed carbohydrates, but also lipids, stimulate their secretion. Most meals, therefore, will cause the release of these peptides.

III. DEGRADATION OF GLUCAGON

A. OVERALL METABOLIC CLEARANCE

In agreement with its function in the minute-to-minute regulation of blood glucose, glucagon is eliminated rapidly from the circulation. Important differences between species seem to exist, however. In most studies, the metabolism has been studied after primed infusions of exogenous glucagon, with subsequent calculation of metabolic clearance rate (MCR) on the basis of radioimmunological determination of the increment caused by the infusion.[37] In most studies, the levels of endogenous glucagon have been assumed to remain constant during the infusion. In a single study, the endogenous glucagon levels were depressed by simultaneous infusion of somatostatin.[38] The validity of the data thus depends on the accuracy of the assay and the completeness of the recovery of glucagon added to blood. In experiments with high infusion rates of exogenous glucagon, errors deriving from co-determination of nonglucagon immunoreactive substances (i.e., the other glucagon-containing products of proglucagon: glicentin, oxyntomodulin, and PG 1-61) or nonspecific interference in the assay will be small. With lower infusion rates aiming at elevations of the glucagon concentration within physiological limits, the assay accuracy becomes critical. In most glucagon assays performed directly on plasma or serum, there is considerable nonspecific interference, mainly due to the plasma proteins.[39] Thus, data obtained with techniques that do not include removal of the plasma proteins are not entirely reliable. Further, the specificity of the assay with regard to the other immunoreactive components must be characterized, or the assay must be combined with chromatography that will allow separation and quantitation of the different components.

1. Metabolic Clearance Rate in Experimental Animals

The metabolic clearance rate (MCR) for glucagon has been determined in rats, sheep, pigs, and dogs. In rats, reported values range from 32[40] to 55 ml × kg^{-1} × min^{-1}.[23] An MCR of 38 ml × min^{-1} × kg^{-1} was reported in pigs.[41] This value compares poorly with values measured in this laboratory (7 ml × min^{-1} × kg^{-1}; Baldissera and Holst, unpublished studies). However, in such studies, careful preparation of the solution for infusion with purified carrier protein is essential. Omission of carrier addition (which was not indicated in Reference 41) may cause major losses of the peptide to be infused. It seems possible, therefore, that an overestimated infusion rate may explain the discrepancy. The MCR in sheep was 5.2 ml × kg^{-1} × min^{-1};[42] in dogs, values between 9.1[43] and 12.5 ml × kg^{-1} × min^{-1} [38] have been reported.

2. Metabolic Clearance Rate in Humans

Alford et al.[44] reported a MCR of 9 ml × kg^{-1} × min^{-1} in normal subjects (and similar values in diabetics) and, in a later study,[45] 13 ml × kg^{-1} × min^{-1} (and similar values in cirrhotics). Sherwin et al.[46] reported 503 ml/m^2 × min, corresponding to approximately 13 ml × kg^{-1} × min^{-1} (and similar values in cirrhotics), and Burman et al.[47] reported similar values (but surprisingly increases in response to fasting — similar increases could be induced by elevated levels of thyroid hormones). From data reported by this laboratory,[48] a value of approximately 10 ml × kg^{-1} × min^{-1} could be calculated.

There is, thus, general agreement regarding values around 10 ml × kg^{-1} × min^{-1} for the metabolic clearance of glucagon in man, dogs, pigs, and sheep. Rats apparently distinguish themselves by having a much higher rate of elimination of glucagon.

3. Half-Life of Glucagon in the Circulation

In several of the studies cited above, the disappearance rate of glucagon was measured after a plateau elevation of the plasma concentration had been established by means of infusion of the peptide. The reported values were all in the range of 5 to 6 min.[38,44] Nonlinear disappearance kinetics have been reported,[37,41] however, suggesting that simple, single-compartment analysis cannot rightfully be applied to glucagon's disappearance kinetics.

B. ORGAN CONTRIBUTION TO THE METABOLIC CLEARANCE OF GLUCAGON

The elimination of glucagon from the circulation could be primarily due to degradation of glucagon in the blood, and subsequently to binding and degradation of the peptide in tissues supplied by the arterial blood.

1. Degradation in Blood or Plasma

According to studies by Hendriks and Beenraad[49] and Märki,[50] glucagon is not degraded in blood or plasma. However, several authors[51] recommend addition to blood samples of protease inhibitors (e.g., aprotinin, benzamidine, or EDTA) which may be essential for the protection of radioactively labeled glucagon used in the radioimmunoassays for glucagon.

2. Degradation in the Limbs

There is general agreement that arterio-venous gradients across the extremities do not exist.[52,96] Thus, there seems be no "nonspecific" loss of glucagon as it passes the capillary beds in general.

3. Degradation in the Liver

As already discussed, the major target for glucagon is the liver, and the hepatocytes are densely equipped with receptors for glucagon.[53] This raises the question of whether or not glucagon is internalized in the liver cells subsequent to binding to the liver cells. Indeed, such internalization has been reported for isolated rat hepatocytes.[54] However, this finding may be at variance with other results (see below) and, on the whole, the issue of internalization of hormone receptor complexes for receptors that stimulate adenylate cyclase remains controversial.[55] In addition, it is extremely difficult to demonstrate that the hepatocellular uptake of radioactivity (which is the technique usually emloyed for the demonstration of internalization) represents uptake of intact, labeled glucagon. It now appears that glucagon may also cause a phospholipase C-induced hydrolysis of inositol phospholipids with formation of inositol trisphoshate and diacylglycerol in liver cells.[56] In other systems, such receptors have been associated with internalization. On the other hand, a receptor-linked, glucagon-specific protease has recently been identified in dog liver cell membranes,[55] suggesting that receptor-mediated cleavage of the peptide may contribute to the termination of the hormonal stimulation of the cell.

It is technically difficult to analyze the degradation of glucagon by the liver *in vitro*. Isolated hepatocytes do degrade glucagon, but the degrading activity (which can be partly recovered in the culture medium) does not seem to reflect the degradation that occurs *in vivo*.[57,58] Attempts have been made to study glucagon degradation in an isolated, perfused liver preparation.[59] However, it was not excluded (with the recirculation technique employed) that a similar degradation could not have been measured in any other perfused tissue preparation. In addition, contrary to the conclusion of the authors, it appears from the data presented that the hepatic extraction of glucagon was not impressive; about 5% of the glucagon presented to the liver appeared to be extracted.

In order to approach the question *in vivo*, several workers have measured the ratio between the glucagon concentrations in portal and peripheral venous plasma. Generally, it has been assumed that a ratio above 1.0 would indicate hepatic extraction of the peptide. This, however, is not the case. In steady state, the net splanchnic glucagon production equals the product of the metabolic clearance rate (MCR) and the systemic plasma concentration of glucagon. Using an MCR of 9 ml × kg^{-1}, a hematocrit of 45%, a cardiac output of 5 1/min, and a portal flow of 25% of cardiac output for a man of 70 kg, and assuming *no* splanchnic degradation, we have

$$C_{port} \cdot 0.69\ 1/\text{min} - C_{art} \cdot 0.69\ 1/\text{min} = C_{art} \cdot 0.63\ 1/\text{min}$$

$$C_{port}/C_{art} = 1.32/0.69 = 1.9$$

where C_{port} is the portal venous concentration and C_{art} the arterial plasma concentration. Thus, only values above 1.9 would be indicative of hepatic extraction. In fact, in normal man (during a normal meal), we found a portal/peripheral ratio of 1.7.[60] A similar value was found by Jaspan and co-workers about 10 years later.[61] In anesthetized subjects, Blackard et al.[62] reported a ratio of 1.7 and Felig et al.[63] found a ratio of 1.1 to 1.5. Values around 1.5 were measured by Greco et al.[64] In rats, however, Jaspan et al.[65] found ratios of 2.8 and 5.2, the latter ratio pertaining to glucagon alone, after removal of nonspecific cross-reacting material by gel filtration. These results strongly suggest that the liver contributes little to the degradation of glucagon in humans, but extensively so in rats.

The hepatic extraction of glucagon has been determined directly in several experiments performed in dogs. In a study by Röjdmark et al.[43] in anaesthetized dogs during infusion of glucagon (and insulin), the hepatic extraction was 5%. During augmented secretion of endogenous glucagon (by arginine), an extraction of 6 ± 2% was found (recalculated from Table 3 in Reference 66). Augmentation of glucagon secretion by insulin-induced hypoglycemia did not change the extraction.[67] In a subsequent study, however, ingestion of a meat meal was associated with an increased extraction of glucagon[68] by unexplained mechanisms that were not reflected in an increased hepatic glucose production. In a final, careful study,[69] the same authors studied the influence of anaesthesia on glucagon extraction both in the basal state, during infusions of arginine and cholecystokinin, and during insulin-glucagon infusions. However, similarly low and constant values for glucagon extraction were found. In fact, the most interesting result of these extensive studies is the marked difference between hepatic extraction of glucagon and insulin, the latter of which is extensive and varies characteristically with the experimental conditions. In a study in which the hepatic extraction of glucagon was measured in dogs in which both the molecular heterogeneity of glucagon in plasma and the contribution from the arterial blood to the supply of glucagon to the liver were taken into account, extraction values of 19% (for exogenous glucagon) were found.[38] In another study from the same period, Jaspan and co-workers calculated an extraction of 34% for glucagon, identified by chromatography and stimulated by arginine and cholecystokinin.[70] In this study, however, the authors found hepatic venous and arterial levels of glucagon that were almost identical, an observation that is difficult to reconcile with the fact that — also in dogs (see below) — the kidneys seem to be of utmost importance for the clearance of glucagon. Thus, it seems that dogs, as well as humans, exhibit a rather low hepatic extraction of glucagon. In this laboratory, we studied the hepatic extraction in conscious pigs in which catheters had been placed in an artery, a hepatic vein, and the portal vein (Christensen, Christiansen, and Holst, unpublished studies). We specifically measured both pancreatic glucagon (comprising glucagon itself and a small contribution of glucagon 1-61) and total glucagon (comprising glicentin, oxyntomodulin, and both of the pancreatic forms). At the operation where the catheters were inserted, the arterial supply to the liver was ligated, and all ligaments to the liver that might contain additional blood vessels were cut. Thus, the difference between portal and hepatic venous plasma concentrations directly reflects the hepatic extraction. The concentrations were followed for 150 min during which pigs were given a mixed meal. The passage fraction for total glucagon remained close to 1.0, whereas the passage fraction for pancreatic glucagon varied around 0.8 (see Figure 2). Thus, in this species, the hepatic extraction of glucagon would be about 20%.

In conclusion, the hepatic extraction of glucagon seems to vary between species. It is pronounced in rats, about 10 to 20% in dogs and pigs, and probably somewhat lower in man. The rather low values do not contrast with the fact that the liver is the main target organ for glucagon and does not argue against receptor-mediated degradation of glucagon, since the liver cell glucagon receptors clearly exhibit the spare-receptor phenomenon — that is, only a minor fraction of receptors need to be occupied in order to elicit maximal biological effect.[53] Whatever the extent of hepatic degradation in humans, one would expect, if the liver contributes at all, that hepatocellular damage could influence glucagon levels. Indeed, in advanced cirrhosis of the liver, most workers have described increased glucagon levels.[71] However, the cause of the hyperglucagonemia has not been established. Sherwin and co-workers[46] found that the MCR was normal in the cirrhotics and therefore concluded that the hyperglucagonemia was due to an increased secretion. Dudley et al.[45] concluded that hyperglucagonemia depended on porto-systemic shunting rather than impaired hepatocellular function. In our own

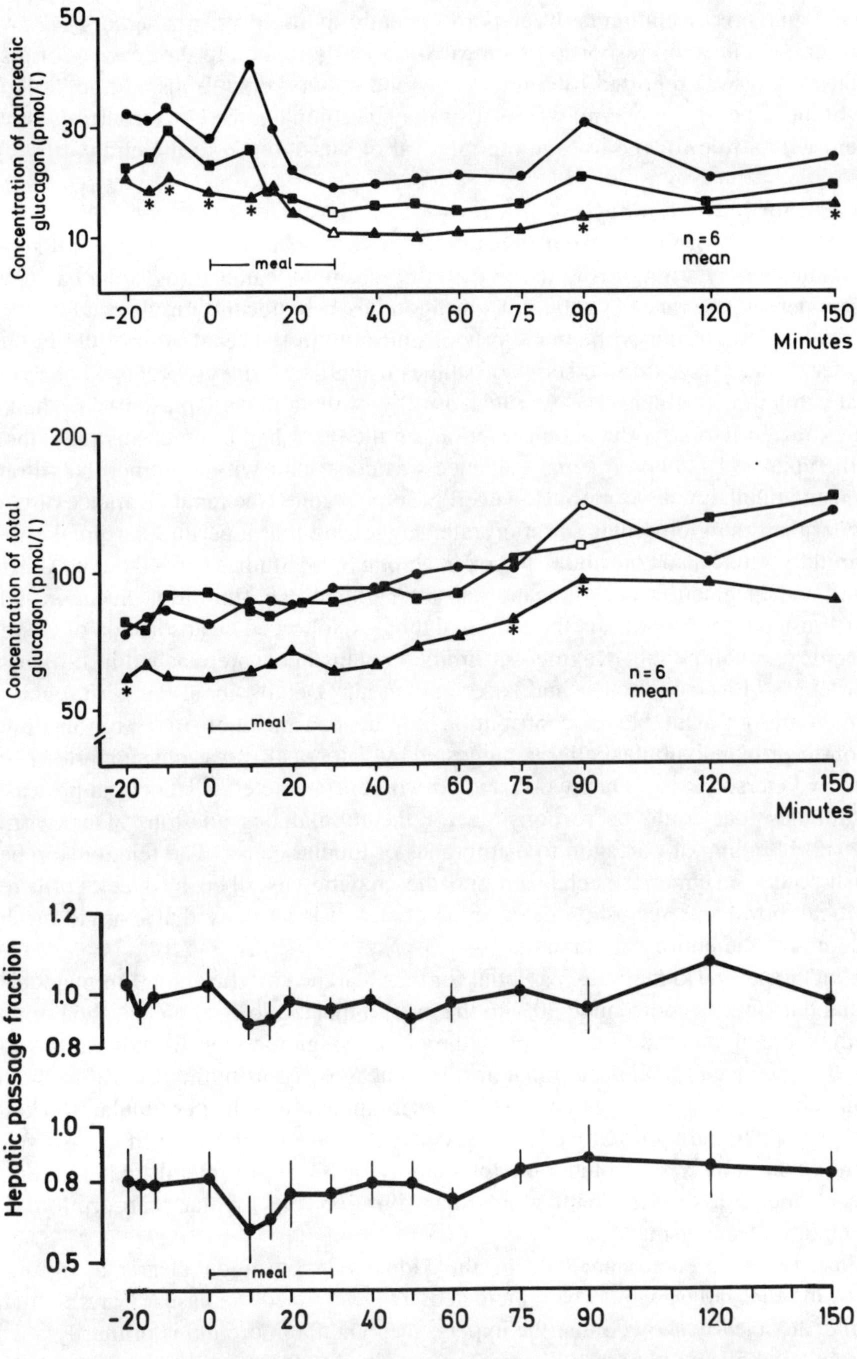

FIGURE 2. Hepatic extraction of pancreatic glucagon and total glucagon (pancreatic + gut) in conscious pigs. Filled circles indicate concentrations in portal venous plasma; filled squares indicate concentrations in hepatic venous plasma; and triangles indicate arterial values. The pigs were given a mixed meal as indicated. The hepatic passage fraction (concentration in hepatic vein divided by portal vein concentration) is shown in the lower panel, total glucagon above pancreatic glucagon. Mean ± SEM.

study, we found no correlation between portosystemic shunting and glucagon levels, whereas the exaggerated glucagon response to a meal was clearly related to the severity of disease.[72] A similar finding was reported by Smith-Laing and co-workers.[73] Thus, although the pathogenesis of the hyperglucagonemia of cirrhosis may be complex, the elevated levels are at least consistent with a role for the liver in the removal of glucagon from the circulation.

4. Degradation in the Kidneys

Whereas the role of the liver remains somewhat controversial, there is general agreement that the kidneys play a major role in the overall glucagon clearance (as is the case for other small peptides).[74] The renal handling of glucagon has been recently reviewed.[75]

In 1972, Assan[76] reported higher levels of glucagon in the renal artery than in the renal vein. Lefebvre and Luyckx, in a series of studies using dog kidneys, precisely characterized the renal extraction of glucagon.[75] As much as 40% of the glucagon presented to the kidneys may be extracted. Urinary glucagon excretion, on the other hand, represents only about 2 to 4% of the uptake. The experimental evidence was consistent with a glomerular filtration of glucagon and tubular reabsorption. However, in some studies, the renal clearance of glucagon was higher than the glomerular filtration rate, suggesting that a peritubular uptake from the renal capillaries may also contribute to the renal uptake.[77,78] Studies carried out by Carone and Peterson[79] and Peterson et al.[80] indicate that, after glomerular filtration, glucagon is hydrolyzed by brush border enzymes in the proximal tubules, followed by absorption of the released amino acids or small peptide fragments. Similar conclusions were reached in further studies by Talor et al.,[81] Peterson et al.,[82] and Nakamura et al.[83] Details about the peritubular uptake mechanism are not available. A contribution to glucagon metabolism by basolateral membranes of the proximal tubular cells, as suggested by Talor et al.,[81] was not supported by results obtained by Peterson et al.[82] Duckworth and co-workers characterized a neutral protease from kidney tubules that could be responsible for the tubular degradation.[84] Duckworth also characterized binding of glucagon to membranes of tubular cells.[85] The relationship between the brush-border enzymatic mechanism and the enzyme described by Duckworth remains hypothetical, however. According to Peterson et al.,[82] it is unlikely that a single mechanism can account for the entire degradation.

Also, in the rat, the kidneys are essential for the clearance of glucagon. Emmanouel et al.[40] found that the kidneys contributed 30% to the overall metabolic clearance of the hormone. In rats with bilateral ureteral ligation (and therefore no glomerular filtration), the kidneys continued to extract 11.5% of the renal arterial glucagon, contributing by 9% to the overall metabolic clearance, suggesting that these values might apply to the peritubular uptake. In our own studies, we found approximately equal concentrations of glucagon in plasma from rats that were either subjected to bilateral nephrectomy or bilateral ureteral ligation, suggesting that the clearance associated with glomerular filtration and tubular reabsorption was the predominating mechanism.[86]

Because of the predominant role of the kidneys in glucagon clearance, a decreased clearance in renal failure would be anticipated. This is indeed the case.[87] Careful studies by Sherwin et al.[88] clearly showed that the hyperglucagonemia in uremia is primarily a result of a decreased catabolism rather than hypersecretion. The MCR in uremics was reduced by 58% as compared to controls, whereas the estimated basal systemic delivery rate of glucagon was not significantly different from controls. Similar results were obtained in rats.[89] In further chromatographic studies Kuku et al.[90] and Emmanouel et al.[89] showed that, although the concentration of glucagon was indeed elevated as a consequence of renal failure, the proportion of the total glucagon-like immunoreactivity that was due to the large-molecular-weight pancreatic glucagon (at that time called "glucagon^{9000}") increased more than that due to glucagon. Finally, it was deduced from studies of the gel filtration pattern in rats that were bilaterally nephrectomized or had a ureteral ligation that the renal metabolism of glucagon depended on a preserved glomerular filtration, whereas the uptake of glucagon^{9000} persisted after ureteral ligation, thus suggesting peritubular uptake for this component in particular.

In later studies, Baldissera and Holst unambiguously identified the elevated glucagon^{9000} component in uremic pigs as proglucagon 1-61.[22] Apparently this component,which is present in rather low concentrations in normal subjects,[91] is metabolized almost exclusviely in the kidneys, as opposed to glucagon which is also metabolized elsewhere. In agreement with this, Jaspan et al.[65] found for this component a portal/peripheral ratio of 1.5 (in nephrectomized rats). As already discussed, such a value suggests a minor role, if any, for the liver in its removal.

IV. DEGRADATION OF THE OTHER GLUCAGON-RELATED PEPTIDES

A. PANCREATIC FORMS

As discussed above, the pancreatic forms comprise the glicentin-related pancreatic peptide (GRPP, PG 1-30), glucagon itself, and the major proglucagon fragment PG 72-158. Nothing is known about the metabolism of GRPP. Glucagon clearance has been discussed above, and the information about "glucagon^{9000}" (PG 1-61), which may also result from expression in the pancreas of the glucagon gene, has also been reviewed. The major proglucagon fragment (PG 72-158) has been identified in plasma,[92] but nothing is known about its metabolism. Its concentration increases after stimulation of the pancreatic glucagon cells with arginine and to a similar degree as the increase of glucagon secretion.[93] This could indicate that the MCRs of the two components were similar.

B. INTESTINAL FORMS

The intestinal forms, glicentin (PG 1-69) and oxyntomodulin (PG 33-69), have been known for several years and some information is therefore available.

Because of the limited amount of glicentin isolated, very few infusion studies have been made. In one study, glicentin was infused into rats. Judging from the infusion rate and the steady-state plasma concentrations obtained, the MCR of glicentin in rats can be calculated to be 12 ml $\times$ kg^{-1} $\times$ min^{-1}.[23] In the same study, the MCR for glucagon was calculated to be 60 ml $\times$ kg^{-1} $\times$ min^{-1}, suggesting that glicentin is metabolized at a much lower rate than glucagon (in rats). This is in agreement with the observation that glicentin is the predominating component of glucagon-related peptides in plasma in normal man in the fasting state.[91] Buchanan et al.[59] studied the clearance of glucagon-like immunoreactivity in an intestinal extract by isolated perfused rat liver and found a half-life of 27 min (as compared to 19 min for glucagon). However, as discussed above, it is difficult to transfer these results to *in vivo* metabolism. Tanaka et al.[94] measured the half-life of glucagon-like immunoreactivity after injection of (porcine) intestinal extracts (containing mainly glicentin) into dogs and found a half-life of 16 min, as compared to 5 min for that of glucagon. Koranyi et al.[95] infused a purified intestinal extract into piglets and found an increase in the plasma concentration of pancreatic glucagon, suggesting that glicentin had been transformed by some enzymatic mechanism into glucagon. There is reason to believe, however, that the extract used contained large amounts of oxyntomodulin, and recent studies by Baldissera and co-workers[25] have shown that oxyntomodulin can stimulate the secretion of pancreatic glucagon in the pig pancreas. The increase noted by Koranyi et al.[95] may therefore represent an increased secretion of glucagon. Roviera et al.[96] studied glucose-induced glicentin secretion in control and nephrectomized dogs. There was a similar increase in glicentin in both groups. Furthermore, in isolated perfused rat kidneys, this component was not removed, suggesting that the kidneys are not essential for the removal of glicentin. In our studies of the hepatic extraction of immunoreactive glucagon in pigs, we found a passage fraction close to 1.0 for "total glucagon", the majority of which is composed of glicentin (Figure 2). This suggest that the liver is not very important for elimination of glicentin.

The metabolic clearance of oxyntomodulin (PG 33-69) has been investigated in humans.[26,27] At infusion rates resulting in plasma concentrations of 300 pmol/l the MCR was 5 ml $\times$ kg^{-1} $\times$ min^{-1}, and the half-life in plasma 12 min. At an infusion rate resulting in a plateau of 200 pmol/l the MCR was calculated to be 9 ml $\times$ kg^{-1} $\times$ min^{-1} and the half-life 8 min.

The MCR of the glucagon-like peptide-1 has been studied in humans. Scholdager et al.[31] infused both full-length GLP-1(PG 72-107amide), originally believed to be a product of proglucagon processing,[9] and truncated GLP-1 (PG 78-107amide), later demonstrated to be the naturally occurring peptide.[19] For PG 72-107amide, the MCR values between 2.2 and 2.6 ml $\times$ kg^{-1} $\times$ min^{-1} and a half-life of 17 min were found; whereas for the natural peptide, the MCR was 13 ml $\times$ kg^{-1} $\times$ min^{-1}, and the half-life 11 min, levels that are comparable to those of glucagon. Similar values were reported by Kreyman et al.[30] After infusion of ^{125}I-labeled truncated GLP-1 into rats, most of the radioactivity was recovered in the kidneys after a very few minutes, suggesting that the kidneys could be involved in the clearance of GLP-1 (Buhl, Seier-Poulsen, and Holst, unpublished studies). Obviously, further studies are needed concerning the metabolism of this interesting new intestinal hormone.

Thus far, nothing is known about the metabolism of the remaining two products of intestinal proglucagon: intervening peptide-2 and GLP-2.

V. CONCLUSION

The glucagon family of peptides comprises four pancreatic members: GRPP (proglucagon 1-30), glucagon (PG-33-61), PG 1-61, and the major proglucagon fragment PG 72-158; and five intestinal members: glicentin (PG 1-69), its fragment, oxyntomodulin (PG 33-69), glucagon-like peptide-1 (PG 78-107 amide), intervening peptide-2 (PG 111-123amide), and glucagon-like peptide-2 (PG 126-158).

Nothing is known about the metabolism of GRPP. Glucagon is cleared rapidly from the circulation with an MCR of about 10 ml $\times$ kg^{-1} $\times$ min^{-1} in man, 20 ml $\times$ kg^{-1} $\times$ min^{-1} in dogs and pigs, and up to 60 ml $\times$ kg^{-1} $\times$ min^{-1} in rats. The kidneys are responsible for most of the elimination, particularly in man,[97] and glomerular filtration, hydrolysis, and uptake in the proximal tubules are the underlying mechanisms. The liver may also contribute, particularly in rats. The component PG 1-61, which is a minor component in normal plasma, it is primarily metabolized in the kidneys by a mechanism that presumably also includes peritubular uptake. In uremia, this component is particularly elevated. Very little is known about the metabolism of MPGF.

Glicentin is eliminated at a much slower rate than glucagon in rats. Neither the liver nor the kidneys extract large amounts of the glicentin presented to them, but this may simply reflect the rather slow MCR of this component. Oxyntomodulin and GLP-1 are metabolized in man at rates that are comparable to that of glucagon, but little is known about the metabolism of intervening peptide-2 and GLP-2.

REFERENCES

1. **Lefebvre, P. J., Ed.**, *Glucagon I and II. Handbook of Experimental Pharmacology,* Vol. 66/ I and II. Springer-Verlag, New York, 1983, 1-535 and 1-700.
2. **Chiasson, J.-L. and Cherrington, A. D.**, Glucagon and liver glucose output *in vivo*, in *Glucagon I, Handbook of Experimental Pharmacology,*Vol. 66/I, Lefebvre, P. J., Ed., Springer-Verlag, New York, 1983, 361.

3. **Unger, R. H., Ohneda,A., Aguilar-Parada, E., and Eisentraut, A. M.,** The role of aminogenic glucagon secretion in blood glucose homeostasis, *J. Clin. Invest.*, 48, 810, 1969.
4. **Gerich, J. E.**, Glucagon as a counterregulatory hormone, in *Glucagon II, Handbook of Experimental Pharmacology,* Vol. 66/II, Lefebvre, P. J., Ed., Springer-Verlag, New York, 1983, 274.
5. **Almdal, T. P., Holst, J. J., and Vilstrup, H.**, Glucagon immunoneutralization in diabetic rats normalizes urea synthesis and decreases nitrogen wasting, *Diabetologia,* 33, A222, 1990.
6. **Cahill, G. F., Jr., Aoki, T. T., and Smith, R. J.**, Glucagon and amino acid metabolism, in *Glucagon I, Handbook of Experimental Pharmacology,* Vol. 66/I, Lefebvre, P. J., Ed., Springer-Verlag, New York, 1983, 399-417.
7. **Boden, G.**, Glucagon deficiency, in *Glucagon II, Handbook of Experimental Pharmacology*, Vol. 66/II, Lefebvre, P. J., Ed., Springer-Verlag, New York, 1983, 399.
8. **Holst, J. J.**, Glucagon-producing tumors, in *Hormone-Producing Tumors of the Gastrointestinal Tract,* Cohen, S. and Soloway, R. D., Eds., Churchill Livingstone, New York, 1985, 57.
9. **Bell, G. I., Sanchez-Pescador, R., Laybourn, P. J., and Narajian, R. C.**, Exon duplication and divergence in the human preproglucagon gene, *Nature (London),* 304, 368, 1983.
10. **Holst, J. J.,** Enteroglucagon, *Advances in Metabolic Disorders*, Vol. 11, Mutt, V., Ed., Academic Press, New York, 1988, 393.
11. **Larsson, L.-I., Holst, J. J., Håkanson, R., and Sundler, F.,** Distribution and properties of glucagon immunoreactivity in the digestive tract of various mammals: an immunohistochemical and immunochemical study, *Histochemistry,* 44, 281, 1975.
12. **Holst, J. J., Ørskov, C., Olsen, J., Buhl, T., Schjoldager, B., and Kofod, H.**, Secretion and effects of the naturally occurring products of proglucagon, *Biomed. Res.*, 9 (Suppl. 3), 181, 1988.
13. **Thim, L. and Moody, A. J.,** Purification and chemical characterization of a glicentin-related pancreatic peptide (proglucagon fragment) from porcine pancreas, *Biochim. Biophys. Acta,* 703, 134, 1982.
14. **Patzelt, C. and Schiltz, E.,** Conversion of proglucagon in pancreatic alpha cells: the major endproducts are glucagon and a single peptide, the major proglucagon fragment that contains two glucagon-like sequences, *Proc. Natl. Acad. Sci. U.S.A.*, 81, 5007, 1984.
15. **Ørskov, C., Holst, J. J., Knuhtsen, S., Baldissera, F. G. A., Poulsen, S. S., and Nielsen, O. V.,** Glucagon-like peptides GLP-1 and GLP-2, predicted products of the glucagon gene, are secreted separately from the pig small intestine, but not pancreas, *Endocrinology,* 119, 1467, 1986.
16. **Thim, L. and Moody, A. J.,** The primary structure of porcine glicentin (proglucagon), *Regul. Pept.*, 2, 139, 1981.
17. **Holst, J. J.,** Evidence that peak II GLI or enteroglucagon is identical to the C-terminal sequence (residues 33-69) of glicentin, *Biochem. J.*, 201, 381, 1982.
18. **Bataille, D., Coudray, A. M., and Carlquist, M.,** Isolation of glucagon-37 (bioactive enteroglucagon/oxyntomodulin) from porcine jejuno-ileum, *FEBS Lett.,* 146, 77, 1982.
19. **Ørskov, C., Bersani, M., Johnsen, A. H., Højrup P., and Holst, J. J.,** Complete sequences of glucagon-like peptide-1 from human and pig small intestine, *J. Biol. Chem.*, 264, 12826, 1989.
20. **Buhl, T., Thim, L., Kofod, H., Ørskov, C., Harling, H., and Holst, J. J.,** Naturally occurring products of proglucagon 111-160 in the porcine and human small intestine, *J. Biol. Chem.*, 263, 8621, 1988.
21. **Ørskov, C., Buhl, T., Rabenhøj, L., Kofod, H., and Holst, J. J.,** Carboxypeptidase-B-like processing of the C-terminus of the glucagon-like peptide-2 in pig and human small intestine, *FEBS Lett.*, 247, 193, 1989.
22. **Baldissera, F. G. A. and Holst, J. J.,** Glicentin 1-61 probably represents a major fraction of glucagon-related peptides in plasma of anaesthetized uraemic pigs, *Diabetologia*, 29, 462, 1986.
23. **Kirkegaard, P., Moody, A. J., Holst, J. J., Loud, F. B., Skov Olsen, P., and Christiansen, J.,** Glicentin inhibits gastric acid secretion in the rat, *Nature (London),* 297, 156, 1982.
24. **Bloom, S. R.,** Gut hormones in adaptation, *Gut,* 28 (Suppl. 28), 31, 1987.
25. **Baldissera, F. G. A., Holst, J. J., Knuhtsen, S., Hilsted, L., and Nielsen, O. V.,** Oxyntomodulin (glicentin-33-69): pharmacokinetics, binding to liver cell membranes, effects on isolated perfused pig pancreas, and secretion from isolated perfused lower small intestine of pigs, *Regul. Pept.*, 21, 151, 1988.
26. **Schjoldager, B. T. G., Baldissera, F. G. A., Mortensen, P. E., Holst, J. J., and Christiansen, J.,** Oxyntomodulin: a potential hormone from the distal gut. Pharmacokinetics and effects on gastric acid and insulin secretion in man, *Eur. J. Clin. Invest.*, 18, 499, 1988.
27. **Schjoldager, B., Mortensen, P. E., Myhre, J., Christiansen, J., and Holst, J. J.,** Oxyntomodulin from distal gut: role in regulation of gastric and pancreatic functions, *Dig. Dis. Sci.*, 34, 1411, 1989.
28. **Holst, J. J., Ørskov, C., Schwartz, T. W., and Nielsen, O. V.,** Truncated glucagon-like peptide-1, an insulin-releasing hormone from the distal gut, *FEBS Lett.*, 211, 169, 1987.
29. **Ørskov, C., Holst, J. J., and Nielsen, O. V.,** Effect of truncated glucagon-like peptide-1 (proglucagon-(78-107)amide) on endocrine secretion from pig pancreas, antrum, and nonantral stomach, *Endocrinology*, 123, 2009, 1988.
30. **Kreyman, B., Williams, G., Ghatei, M. A., and Bloom, S. R.,** Glucagon-like peptide-1 7-36: a physiological incretin in man, *Lancet*, 2, 1300, 1987.

31. **Schjoldager, B., Mortensen, P. E., Christiansen, J., Ørskov, C., and Holst, J. J.,** GLP-1 (glucagon-like peptide 1) and truncated GLP-1, fragments of human proglucagon, inhibit gastric acid secretion in man, *Dig. Dis. Sci.,* 35, 703, 1989.
32. **Moody, A. J., Holst, J. J., Thim, L., and Jensen, S. L.,** Relationship of glicentin to proglucagon and glucagon in the porcine pancreas, *Nature (London),* 289, 514, 1981.
33. **Gerich, J. E.,** Glucose in the control of glucagon secretion, in *Glucagon II, Handbook of Experimental Pharmacology,* Vol. 66/II, Lefebvre, P. J., Ed., Springer-Verlag, New York, 1983, 3.
34. **Assan, R., Marre, M., and Gormley, M.,** The amino acid-induced secretion of glucagon, in Glucagon II, Handbook of Experimental Pharmacology, Vol. 66/II, Lefebvre, P. J., Ed., Springer- Verlag, New York, 1983, 19.
35. **Holst, J. J., Schwartz, T. W., Knuhsen, S., Jensen, S. L., and Nielsen, O. V.,** Autonomic nervous control of pancreatic endocrine secretion from the isolated perfused pig pancreas, *J. Autonom. Nerv. Syst.,* 17, 71, 1986.
36. **Holst, J. J.,** Gut glucagon, enteroglucagon, gut glucagon-like immunoreactivity, glicentin — current status, *Gastroenterology,* 84, 1602, 1983.
37. **Polonsky, K. S., Jaspan, J. B., and Rubenstein, A. H.,** The metabolic clearance rate of glucagon, in *Glucagon II, Handbook of Experimental Pharmacology,* Vol. 66/II, Lefebvre, P. J., Ed., Springer-Verlag, New York, 1983, 353.
38. **Jaspan, J. B., Polonsky, K. S., Lewis, M., Pensler, J., Pugh, W., Moossa, A. R., and Rubenstein, A. H.,** Hepatic metabolism of glucagon in the dog: contribution of the liver to overall metabolic disposal of glucagon, *Am. J. Physiol.,* 240, E233, 1981.
39. **von Schenck, H. and Grubb, A. O.,** Interference of immunoglobulins in two glucagon radioimmunoassays, *Clin. Chem.,* 28, 1103, 1982.
40. **Emmanouel, D. S., Jaspan, J. B., Rubenstein, A. H., Huen, A. H.-J., Fink, E., and Katz, A. I.,** Glucagon metabolism in the rat. Contribution of the kidney to the metabolic clearance rate of the hormone, *J. Clin. Invest.,* 62, 6, 1978.
41. **Van Hoorn, W. A., Vinik, A. I., and Van Hoorn-Hickman, R.,** The metabolic clearance rate of endogenous immunoreactive glucagon and exogenous glucagon in pancreatectomized and sham operated pigs, *Endocrinology,* 103, 1084, 1978.
42. **Brockman, R. P.,** Rates of removal of exogenous and endogenous glucagon from the plasma of sheep *in vivo, Aust. J. Biol. Sci.,* 33, 531, 1980.
43. **Röjdmark, S., Bloom, G., Chou, M. C. Y., and Field, J. B.,** Hepatic extraction of exogenous insulin and glucagon in the dogs, *Endocrinology,* 102, 806, 1978.
44. **Alford, F. P., Bloom, S. R., and Nabarro, J. D. N.,** Glucagon metabolism in man. Studies on the metabolic clearance rate and the plasma acute disappearance time of glucagon in normal and diabetic subjects, *J. Clin. Endocrinol. Metab.,* 42, 830, 1976.
45. **Dudley, F. J., Alford, F. P., Chisholm, D. J., and Findlay, D. M.,** Effect of portasystemic venous shunt surgery on hyperglucagonemia in cirrhosis: paired studies of the pre- and post-shunt subjects, *Gut,* 20, 817, 1979.
46. **Sherwin, R. S., Fisher, M., Bessoff, J., Snyder, N., Hendler, R., Conn, H. O., and Felig, P.,** Hyperglucagonemia in cirrhosis: altered secretion and sensitivity to glucagon, *Gastroenterology,* 74, 1224, 1978.
47. **Burman, K. D., Smallridge, R. C., Jones, L., Ramos, E. A., O'Brian, J. T., Wright, F. D., and Wartofsky, L.,** Glucagon kinetics in fasting: physiological elevations in serum 3′5′3′-triiodothyronine increases the metabolic clearance rate of glucagon, *J. Clin. Endocrinol. Metab.,* 51, 1158, 1980.
48. **Holst, J. J., Guldberg Madsen, O., Knop, J., and Schmidt, A.,** The effect of intraportal and peripheral infusions of glucagon on insulin and glucose concentrations and glucose tolerance in normal man, *Diabetologia,* 13, 487, 1977.
49. **Hendriks, T. and Benraad, T. J.,** On the stability of immunoreactive glucagon in plasma samples, *Diabetologia,* 20, 553, 1981.
50. **Märki, F.,** Stability of endogenous immunoreactive glucagon (IRG) in animal blood, *Horm. Metab. Res.,* 15, 307, 1983.
51. **Ensinck, J. W.,** Immunoassays for glucagon, in *Glucagon I, Handbook of Experimental Pharmacology,* Vol. 66/I, Lefebvre, P. J., Ed., Springer-Verlag, New York, 1983, 203.
52. **Krarup, T., Nürnberg, B., Mikkelsen, K., Astrup, A., Madsbad, S., Henriksen, J., and Holst, J. J.,** How arterialized is blood sampled from a heated superficial hand vein?, *Diabetes,* 39, 88A, 1990.
53. **Rodbell, M.,** The action of glucagon at its receptor: regulation of adenylate cyclase, in *Glucagon I, Handbook of Experimental Pharmacology,* Vol. 66/I, Lefebvre, P. J., Ed., Springer-Verlag, New York, 1983, 263.
54. **Barazzone, P., Gorden, P., Carpentier, J.-L., Orci, L., Freychet, P., and Canivet, B.,** Binding, internalization, and lysosomal association of ^{125}-I-glucagon in isolated rat hepatocytes, *J. Clin. Invest.,* 66, 1081, 1980.
55. **Sheetz, M. J. and Tager, H. S.,** Characterization of a glucagon receptor-linked protease from canine hepatic plasma membranes, *J. Biol. Chem.,* 263, 19210, 1988.

56. **Wakelam, J. O. M., Murphy, G. J., Hruby, V. J., and Houslay, M. D.,** Activation of two signal-transduction systems in hepatocytes by glucagon, *Nature (London)*, 323, 68, 1986.
57. **Thieden, H. I. D., Holst, J. J., Dich, J., Moody, A. J., and Sundby, F.,** Effect of highly purified porcine gut glucagon-like immunoreactivity (glicentin) on glucose relase from isolated hepatocytes, *Biochim. Biophys. Acta,* 675, 163, 1981.
58. **Ishida, T. and Field, J. B.,** Hepatic handling of glucagon, in *Glucagon II, Handbook of Experimental Pharmacology,* Vol. 66/II, Lefebvre, P. J., Ed., Springer-Verlag, New York, 1983, 361.
59. **Buchanan, K. D., Solomon, S. S., Vance, J. E., Porter, H. P., and Williams, R. H.,** Glucagon clearance by the isolated perfused rat liver, *Proc. Soc. Exp. Biol. Med.*, 128, 620, 1968.
60. **Dencker, H., Hedner, P., Holst, J., and Tranberg, K.-G.,** Pancreatic glucagon responses to an ordinary meal, *Scand. J. Gastroenterol.*, 10, 471, 1975.
61. **Jaspan, J. B., Ruddick, J., and Rayfield, E.,** Transhepatic glucagon gradients in man: evidence for glucagon extraction by human liver, *J. Clin. Endocrinol. Metab.*, 58, 287, 1984.
62. **Blackard, W. G., Nelson, N. C., and Andrews, S. S.,** Portal and peripheral vein immunoreactive glucagon concentrations after arginine or glucose infusions, *Diabetes*, 23, 199, 1974.
63. **Felig, P., Gusberg, R., Hendler, R., Gump, F. E., and Kinney, J. M.,** Concentrations of glucagon and the insulin: glucagon ratio in the portal and peripheral circulation, *Proc. Soc. Exp. Biol. Med.,* 147, 88, 1974.
64. **Greco, A. V., Crucitti, F., Ghirlanda, G., Manna, R., Altomonte, L., Rebuzzi, A. G., and Bertoli, A.,** Insulin and glucagon concentrations in portal and peripheral veins in patients with hepatic cirrhosis, *Diabetologia*, 17, 23, 1979.
65. **Jaspan, J. B., Huen, A. H.-J., Morley, C. G., Moossa, A. R., and Rubenstein, A. H.,** The role of the liver in glucagon metabolism, *J. Clin. Invest.*, 60, 421, 1977.
66. **Röjdmark, S., Bloom, G., Chou, M. C. Y., Jaspan, J. B., and Field, J. B.,** Hepatic insulin and glucagon extraction after their augmented secretion in dogs, *Am. J. Physiol.*, 235, E88, 1978.
67. **Ishida, T., Chou, M. C. Y., Lewis, R. M., Hartley, C. G., Entman, M., and Field, J. B.,** The effect of tolbutamide on hepatic extraction of insulin and glucagon and hepatic glucose output in anesthetized dogs, *Endocrinology,* 109, 443, 1981.
68. **Ishida, T., Chou, J., Lewis, C. J., Entman, M., and Field, J. B.,** The effect of ingestion of meat on hepatic extraction of insulin and glucagon and hepatic glucose output in conscious dogs, *Metabolism*, 32, 558, 1983.
69. **Ishida, T., Lewis, R. M., Hartley, C. J., Entman, M. L., and Field, J. B.,** Comparison of hepatic extraction of insulin and glucagon in conscious and anesthetized dogs, *Endocrinology,* 112, 1098, 1983.
70. **Jaspan, J. B., Polonsky, K. S., Röjdmark, S., Field, J. B., and Rubenstein, A. H.,** Hepatic extraction of plasma immunoreactive glucagon components. Predilection for 3500-dalton glucagon metabolism by the liver, *Diabetes*, 30, 767, 1981.
71. **Marco, J.,** Glucagon in cirrhosis of the liver, in *Glucagon II, Handbook of Experimental Pharmacology,* Vol. 66/II, Lefebvre, P. J., Ed., Springer-Verlag, New York, 1983, 491.
72. **Holst, J. J., Burchardt, F., and Kühl, C.,** Pancreatic glucoregulatory hormones in cirrhosis of the liver: portal vein concentrations during intravenous glucose tolerance tests and in response to a meal, *Diab. Metab.,* 6, 117, 1980.
73. **Smith-Laing, G., Ørskov, H., Gore, M. B. R., and Sherlock, S.,** Hyperglucagonemia in cirrhosis. Relationship to hepatocellular damage, *Diabetologia*, 19, 103, 1980.
74. **Carone, F. A., Petersen, D. R., and Flouret, G.,** Renal tubular processing of small peptide hormones, *J. Lab. Clin. Med.*, 100, 1, 1982.
75. **Lefebvre, P. J. and Luyckx, A. S.,** The renal handling of glucagon, in *Glucagon II, Handbook of Experimental Pharmacology,* Vol. 66/II, Lefebvre, P. J., Ed., Springer-Verlag, New York, 1983, 389.
76. **Assan, R.,** *In vivo* metabolism of glucagon, in *Glucagon: Molecular Physiology, Clinical and Therapeutic Implications,* Lefebvre, P. J. and Unger, R. H., Eds., Pergamon, Oxford, New York, 1972, 45.
77. **Bastl, C., Finkelstein, F. O., Sherwin, R., Hendler, R., Felig, P., and Hayslett, J. P.,** Renal extraction of glucagon in rats with normal and reduced renal function, *Am. J. Physiol.*, 233, F67, 1977.
78. **Lefebvre, P. J., Luyckx, A. S., and Nizet, A. H.,** Independence of glucagon and insulin handling by isolated perfused dog kidney, *Diabetologia*, 12, 359, 1976.
79. **Carone, F. A. and Peterson, D. R.,** Hydrolysis and transport of small peptides by the proximal tubule, *Am. J. Physiol.*, 238, F151, 1980.
80. **Peterson, D. R., Carone, F. A., Oparil, S., and Christensen, E. I.,** Differences between renal tubular processing glucagon and insulin, *Am. J. Physiol.*, 242, F112, 1982.
81. **Talor, Z., Emmanouel, D. S., and Katz, A. I.,** Glucagon degradation by luminal and basolateral rabbit tubular membranes, *Am. J. Physiol.*, 244, F297, 1983.
82. **Peterson, D. R., Green, E. A., Oiparil, S., and Hjelle, J. T.,** Transport and hydrolysis of glucagon in the proximal nephron, *Am. J. Physiol.*, 251, F460, 1986.
83. **Nakamura, R., Hayashi, M., Emmanouel, D. S., and Katz, A. I.,** Sites of insulin and glucagon metabolism in the rabbit nephron, *Am. J. Physiol.*, 250, F144, 1986.

84. **Duckworth, W. C.,** Insulin and glucagon degradation by the kidney. II. Characterization of the mechanisms at neutral pH, *Biochim. Biophys. Acta,* 437, 531, 1976.
85. **Duckworth, W. C.,** Insulin and glucagon binding and degradation by kidney cell membranes, *Endocrinology,* 102, 1766, 1978.
86. **El Munshid, H. A., Holst, J. J., Lundquist, I., Lundquist, G., Schaffalitzky de Muckadell, O. B., and Håkanson, R.,** Role of the kidneys in elimination of glucagon, insulin, secretin and somatostatin in the rat, *Acta Physiol. Scand.,* 117, 27, 1983.
87. **Jaspan, J. B., Polonsky, K. S., and Rubenstein, A. H.,** Glucagon and renal insufficiency, in *Glucagon II, Handbook of Experimental Pharmacology,* Vol. 66/II, Lefebvre, P. J., Ed., Springer-Verlag, New York, 1983, 477.
88. **Sherwin, R. S., Bastl, C., Finkelstein, F. O., Fisher, M., Black, H., Hendler, R., and Felig, P.,** Influence of uremia and hemodialysis on the turnover and metabolic effects of glucagon, *J. Clin. Invest.,* 57, 722, 1976.
89. **Emmanouel, D. S., Jaspan, J. B., Kuku, S. F., Rubenstein, A., H., and Katz, A. I.,** Pathogenesis and characterization of hyperglucagonemia in the uremic rat, *J. Clin. Invest.,* 58, 1266, 1976.
90. **Kuku, S. F., Jaspan, J. B., Emmanouel, D. S., Zeidler, A., Katz, A. I., and Rubenstein, A. H.,** Heterogeneity of glucagon. Circulating components in normal subjects and in patients with renal failure, *J. Clin. Invest.,* 58, 742, 1976.
91. **Holst, J. J.,** Molecular heterogeneity of glucagon in normal subjects and in patients with glucagon-producing tumours, *Diabtetologia,* 24, 359, 1983.
92. **Ørskov, C., Holst, J. J., Seier-Poulsen, S., and Kirkegaard, P.,** Pancreatic and intestinal processing of proglucagon in man, *Diabetologia,* 30, 874, 1987.
93. **Hirota, M., Shimizu, I., Ohboshi, C., Nishino, T., and Shima, K.,** A large molecular form of glucagon-like peptide-1 (GLP-1) immunoreactivity is co-released with glucagon from pancreas by arginine in normal subjects, *Clin. Chim. Acta,* 167, 293, 1987.
94. **Tanaka, R., Matsuyama, T., Shima, K., Sawazaki, N., Tarui, S., and Kumahara, Y.,** Half-life of gastrointestinal glucagon-like immunoreactive materials, *Endocrinol. Jpn.,* 26, 59, 1979.
95. **Koranyi, L., Peterfy, F., Szabo, J., Török, A., Guoth, M., and Tamas, G. Y., Jr.,** Evidence for transformation of glucagon-like immunoreactivity of gut into pancreatic glucagon *in vivo, Diabetes,* 30, 792, 1981.
96. **Rovira. A., Lopez-Novoa, J. M., Ghiglione, M., Zubiaur, M., Pascual, J. M., and Valverde, I.,** Lack of gut-GLI removal by the kidney, *Diabetologia,* 19, 311, 1980.
97. **Petersen, K. F. and Henriksen, J. H.,** Endogenous glucagon in different vascular beds: The role of the kidney in degradation of circulating glucagon in man, *Diab. Nutr. Metab.,* 2, 119, 1989.

Chapter 11

DEGRADATION OF PANCREATIC POLYPEPTIDE — A PEPTIDE HORMONE WITH A GLOBULAR STRUCTURE IN AQUEOUS SOLUTION

Mairead M. T. O'Hare and Thue W. Schwartz

TABLE OF CONTENTS

I. INTRODUCTION

Pancreatic polypeptide (PP), a 36 amino acid peptide, was initially discovered as a contaminant of insulin preparations,[1,2] rather than by a distinct biological action. After extensive investigation, the peptide is now known to inhibit pancreatic secretion of bicarbonate and protein and gall bladder contraction.[3-5] PP is found predominantly in the pancreas, in distinct PP cells which are located both in the islets and throughout the exocrine tissue.[6,7] It belongs to a family of regulatory peptides, each with 36 amino acid residues and a C-terminal amidated tyrosine residue. The other two members which have been characterized to date are neuropeptide Y (NPY),[8] first isolated from porcine brain, and peptide YY (PYY)[9] isolated from porcine intestine. Although the peptides are only 45 to 70% homologous in their primary structure (Figure 1), they share a common 3-D structure, known as the PP-fold. Accordingly, the peptide family has been named the PP-fold family.

II. STRUCTURE

The structure of PP and related peptides has been extensively characterized by X-ray crystallography and circular dichroism.[10-12] The moelcule is compact, extremely stable, and the structure is maintained even in dilute aqueous solution. The folded molecule consists of two antiparallel helices: an N-terminal polyproline helix and an amphiphilic alpha helix, connected by a beta turn, as shown in Figure 2. The helices are maintained in cose proximity by tightly packed, interdigitating hydrophobic side chains. The C-terminal hexapeptide is relatively mobile and the C-terminal amide group, in particular, is crucial for the biological effects of the peptide and possibly also for receptor binding.[13,14] Specific receptors have been described for PP on PC 12 cells, a rat phaeochromocytoma cell line, and on the vascular surface of canine small intestinal mucosa.[15,16]

III. BIOSYNTHESIS

PP, like many other regulatory peptides, is synthesized as a larger precursor form, which is proteolytically processed to the bioactive form.[17-19] The structure of the precursor for PP and the processing sequences have been studied in detail and are similar to those observed for the structurally related peptide, NPY (Figure 3). The PP sequence constitutes approximately the N-terminal half of the precursor (Figure 1) and it is separated from the C-terminal extension peptide by three residues, Gly-Lys-Arg, the combined cleavage, and amidation site. The C-peptide is further processed at a monobasic cleavage site to yield pancreatic icosapeptide and a heptapeptide, both of unknown function.[20] The primary sequence of the precursor for PP has been determined in a number of species (Figure 1). The PP sequence itself is highly conserved among different species, whereas the C-peptide is poorly conserved in comparison.[21]

IV. MOLECULAR SPECIES IN PLASMA

Four molecular components of PP immunoreactivity have been described in plasma using gel filtration chromatography. Two molecular species of 9 to10 and 4.2 kDa correspond to pro-PP and PP^{1-36}, respectively, and increase following food ingestion and insulin-induced hypoglycemia.[22-25] Of the two additional components, the apparent PP immunoreactivity, which elutes in the void volume region, most likely represents nonspecific binding to large molecular proteins since this material cannot be removed by immunoabsorption[25] but can be generated following tryptic cleavage of PP^{1-36}.[24] The fourth component, which corresponds to a potential C-terminal fragment of PP, has been more difficult to identify. It can be extracted from tumors[26] and transfected neuroendocrine cell lines expressing PP[27] and, in these instances, it is detectable only using C-terminal specific PP antisera (e.g., Ab 734 and Ab 221,

```
Pro-peptide:    [        PP-FOLD PEPTIDE        ]  GKR  [          C-peptide          ]

Mammalian PP-like PP-fold peptides:

                    5    10   15   20   25   30   35            5    10   15   20   25   30
                ----+----+----+----+----+----+----+-      ----+----+----+----+----+----+
Human      PP   APLEPVYPGDNATPEQMAQYAADLRRYINMLTRPRY GKR HKEDTLAFSGWGSPHAAVPRELSPLDL.
Canine     PP   ----------D------------E------------- --- DRGEMRDIL--------A----MDE.
Feline     PP   ----------------------E-------------      DRGE--DILE-------A------M-V.
Ovine      PP   -S---E----------------E-------------      D--G--D-L-C----S----(   ?   )
Bovine     PP   -----E----------------E-------------
Porcine    PP   ----------D-----------E-------------
Guinea pig PP   ----------D--Q--------EM------------ --S AE--A-GLPV-RNS---A-GGSHRHPPAGL/
                                                         /PAAKGGTGVSGSPPKPWDCLPCRAHSLPSNS.
Rat        PP   -----M----Y--H--R---ETQ------T------ --- DEDTAGLPGRNLP-CTSLLVG-M-CAAARS.
Bullfrog   PP   --S--HH---Q---D-L---YS--YQ--TFI----F

Avian PP-like PP-fold peptides:

                    5    10   15   20   25   30   35
                ----+----+----+----+----+----+----+-
Turkey     PP   GPSQPTYPGDDAPVEDLIRFYNDLQQYLNVVTRHRY
Goose      PP   ---------N-------?---DN----RL--F----
Alligator  PP   T-L--K----G-------Q--------------P-F

NPY-like PP-fold peptides:

                    5    10   15   20   25   30   35            5    10   15   20   25   30
                ----+----+----+----+----+----+----+-      ----+----+----+----+----+----+-
Human      NPY  YPSKPDNPGEDAPAEDMARYYSALRHYINLITRQRY GKR SSPETLISDLLMRESTENVPRTRLEDPAMW.
Rat        NPY  ------------------------------------ --- -----------------A--------S--.
Rabbit     NPY  ------------------------------------
Guinea pig NPY  ------------------------------------
Porcine    NPY  ----------------L-------------------

Human      PYY  --I--EA-----SP-ELN---AS----L--V-----
Porcine    PYY  --A--EA-----SP-ELS---AS----L--V-----
Rat        PYY  --A--EA-----SP-ELS---AS----L--V----- --- GV-AA-F-K--FTDDS--L-FRSRPEGVDQW.
```

FIGURE 1. Primary amino acid sequences (one-letter code) of PP-fold peptides and their biosynthetic precursors characterized to date. The GKR (G: glycine, K: lysine, R: arginine) residues between the peptide and the C-peptide indicate the combined cleavage and amidation site of the precursor. All the PP-fold peptides are amidated at the C-terminus in their final, active form. Amino Acid one-letter code: A, alanine; C, cysteine; D, aspartic acid; E, glutamic acid; F, phenylalanine; G, glycine,;H, histidine; I; isoleucine; K, lysine; L, leucine; M, methionine; N, asparagine; P, proline; Q, glutamine; R, arginine; S, serine; T, threonine; V, valine; W, tryptophan; and Y, tyrosine.

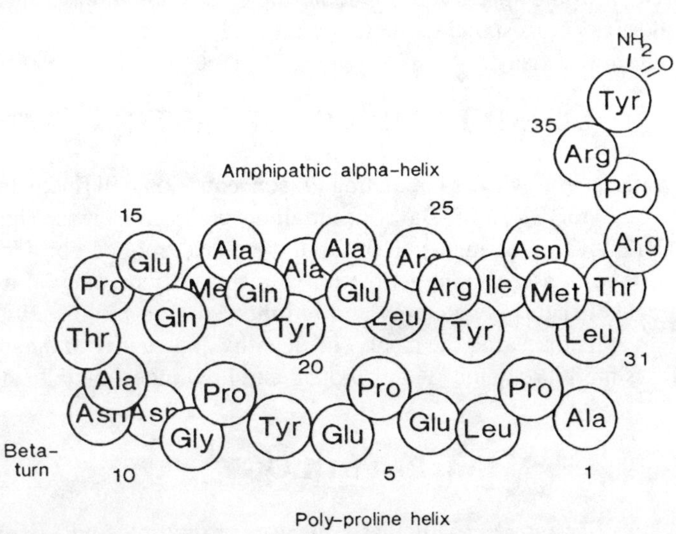

FIGURE 2. Schematic representation of the tertiary structure of PP based on the X-ray structure of avian PP, showing the polyproline helix, the beta turn, and the amphipathic alpha helix.

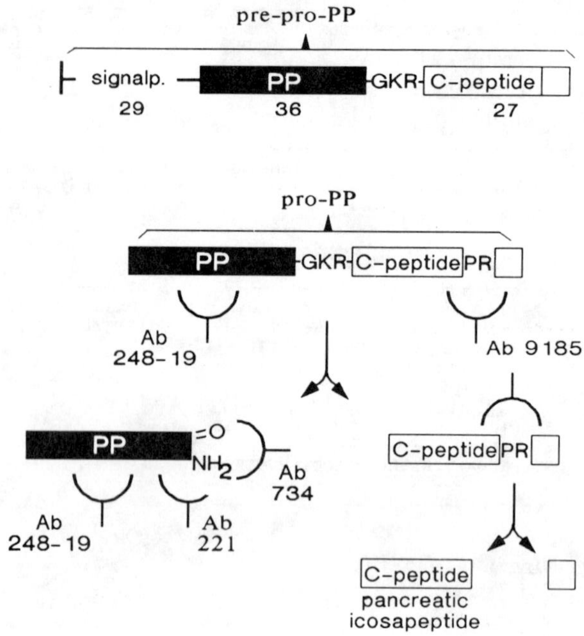

FIGURE 3. Schematic representation of the biosynthetic processing of the primary translation products of the PP gene. The number of amino acids constituting the signal peptide (signalp.), PP, and the C-terminal extension peptide (C-peptide) are indicated. The combined processing and amidation site of the precursor is shown in the amino acid one-letter code (G: glycine, K: lysine, and R: arginine). The specificities of the most commonly used PP antisera are also shown.

(see Figure 3). Although a similar size of immunoreactive species has been described in plasma following gel filtration, the antiserum used in most of these studies (Ab 248-19, Figure 3) does not cross-react with the C-Terminal fragment extracted from tissue, presumably because this antiserum requires the intact 1-36 sequence and it is not C-terminal specific (see Figure 3). Therefore, it would appear likely that the apparent, small molecular weight material found in plasma does not correspond to the tissue-derived fragment and may be due to some nonspecific binding in the assay.

V. REGULATION OF SECRETION

In the fasting state, an individual's circulating PP concentration will fluctuate considerably, with a periodicity of 90 to 120 min.[28] The concentrations will also increase significantly with increasing age.[29,30] Following the ingestion of food, there is a rapid and prolonged rise in PP concentrations.[31] The response is biphasic; the initial response, maximum after 10 to 20 min (cephalic phase) is abolished by vagotomy and is followed by a prolonged gastrointestinal phase lasting for several hours after the meal. The quantity and quality of the meal affects the PP response and it is more pronounced after protein and lipid ingestion than after a carbohydrate-rich meal.[32]

VI. PATHOLOGY

Patients with endocrine tumors (particularly in cases of multiple endocrine adenomatosis and in the watery diarrhea hypokalemia achlorhydria syndrome) frequently have elevated circulating concentrations of PP and the tumors can contain large numbers of PP-producing cells.[26,33-36] Although no clinical features have been attributed to PP hypersecretion, the measurement of plasma PP concentrations can be used to monitor the patients response to therapy. Autonomous secretion of PP by tumor cells can be demonstrted by the Atropine Suppression Test, which has the advantage that it only depends on the suppression of the PP secretion from the normal PP cells.[37,38] Elevated PP concentrations have also been reported in patients with type I and type II diabetes mellitus,[29] chronic renal failure,[39-41] and duodenal ulcer[28] and can be related to the clinical severity of the disease.

Patients with pancreatic exocrine disease (e.g., pancreatitis and cystic fibrosis) tend to be PP deficient and demonstrate impaired responses to food and insulin-induced hypolgycaemia.[42-44] The deficiency tends to be related to the degree of pancreatic damage. Also, the impaired PP secretion during stimulation is used as a sensitive indicator of autonomic neuropathy, for example, in patients with diabetes mellitus.[45]

VII. METABOLIC CLEARANCE

The half-time for disappearance of PP is 5 to 10 min in the basal state, depending on the species examined. This has been studied both by infusion of exogenous PP and by following the disappearance rate after cessation of electrical nerve stimulation in pigs.[46-49] This rapid clearance of PP suggests that PP is constantly secreted and degraded, as for example, during a meal response, which can last for 4 to 5 h. The finding of elevated circulating concentrations of PP in patients with even modest renal failure indicates that the kidneys may be involved in the removal of PP from the circulation. A number of studies have been undertaken to investigate the clearance of PP and to identify the site of degradation. An infusion of the bovine peptide in humans revealed a metabolic clearance rate of 5.1 ml/kg·min[46] and similar results have been reported by others, in humans and dogs.[40,47] In humans, a metabolic clearance rate of 8 to 11 ml/kg·min in the basal state has been calculated on the basis of hepaticosystemic differences in PP concentrations without infusion of exogenous peptide.[40] A higher rate of clearance of porcine PP, 25 ml/kg·min in dogs has been reported by Taylor and co-workers.[48] There appears to be no single site of degradation of PP and many organs are probably involved, including the kidneys but apparently not the liver.[40]

VIII. INFLUENCE OF STRUCTURE ON STABILITY

The relatively short half-life of PP *in vivo* is interesting in view of the fact that the peptide is known to be extremely stable. For example, there is no significant loss of immunoreactivity if the peptide is repeatedly frozen and thawed or is present in whole blood for 24 h at room temperature.[32] This stability is conceivably due to the special 3-D structure of PP (Figure 2). The stability of the peptide can be assessed by comparing the enzymatic cleavage rates of the intact 1-36 peptide and the 1-26 peptide, which does not hold the stable, globular structure in solution. In these studies, we have used two enzymes, endoproteinase Asp-N and endoproteinase Glu-C. $PP^{1-36 \text{ and}}$ PP^{1-26} have identical cleavage sites with respect to primary sequence for both endoproteinase Asp-N and endoproteinase Glu-C. Nevertheless, in PP^{1-26}, these sites are totally cleaved by endoproteinase Asp-N within half an hour, whereas PP^{1-36} is almost totally intact at this time, and after 20 h of digestion, 15% of the PP^{1-36} is still intact (Figure 4). With endoproteinase Glu-C, it also takes half an hour to degrade PP^{1-26}, whereas 3 h are required to cleave the same sites in PP^{1-36}. These data, together with physiochemical studies including circular dichroism, indicate that essentially all PP molecules in aqueous solution are folded in a stable, globular structure which protects against enzymatic degradation and is important for receptor binding of the molecule.[21]

IX. CONCLUSIONS

PP is a pancreatic hormone with a well-defined, 3-D structure consisting of two antiparallel

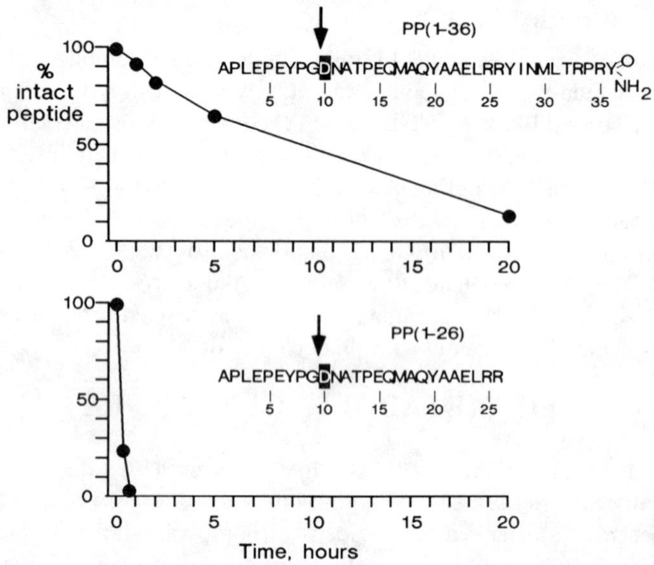

FIGURE 4. Proteolytic stability of intact, folded PP compared to the unfolded (?), long N-terminal fragment of $PP^{1\text{-}26}$. The digestion of $PP^{1\text{-}36}$ and $PP^{1\text{-}26}$ with endoproteinase Asp-N was performed in 0.01 *M* Tris/HCl buffer, pH 7.5, at room temperature. The products were separated by HPLC and identified by amino acid analysis. The identical cleavage site in both peptides is indicated by an arrow.

helices; a polyproline helix and an amphiphilic alpha helix. This secondary structure is maintained in aqueous solution and is most likely responsible for the high degree of stability of the peptide. *In vivo*, circulating PP concentrations vary dramatically, especially following food ingestion. Although PP is efficiently cleared from the circulation, the site of degradation remains obscure.

ABBREVIATIONS

PP	Pancreatic polypeptide
pro-PP	Propancreatic polypeptide
NPY	Neuropeptide Y
PYY	Peptide YY
min	Minute
ml	Milliliter
kg	Kilogram

REFERENCES

1. **Chance, R. E., Moon, N. E., and Johnson, M. G.,** Human pancreatic-polypeptide (HPP) and bovine pancreatic-polypeptide (BPP), in *Methods of Hormone Radioimmunoassay,* Jaffe, B. M. and Behlman, H. R., Eds., Academic, New York, 1979, 657.
2. **Kimmel, J. R., Hayden, L. J., and Pollock, H. G.,** Isolation and characterization of a new pancreatic polypeptide hormone, *J. Biol. Chem.*, 250, 9369, 1975.

3. **Lin, T.-M., Evans, D. C., Chance, R. E., and Spray, G. F.,** Bovine pancreatic peptide: action on gastric and pancreatic secretion in dogs, *Am. J. Physiol.*, 232, E311, 1977.
4. **Greenberg, G. R., McCloy, R. F., Chadwick, V. S., Adrian, T. E., Baron, J. H., and Bloom, S. R.,** Effect of bovine pancreatic polypeptide on basal pancreatic and biliary outputs in man, *Dig. Dis. Sci.*, 24, 11, 1979.
5. **Greenberg, G. R., McCloy, R. F., Adrian, T. E., Chadwick, V. S., Baron, J. H., and Bloom, S. R.,** Inhibition of pancreas and gallbladder by pancreatic polypeptide, *Lancet,* 2, 1280, 1978.
6. **Larsson, L.-I., Sundler, F., and Håkanson, R.,** Pancreatic-polypeptide, a postulated hormone: identification of its cellular storage site by light and electron microscopic immunocytochemistry, *Diabetologia*, 12, 211, 1976.
7. **Bergstrom, B. H., Loo, S., Hirsch, H. J., Schutzengel, D., and Gabbay, K. H.,** Ultrastructural localization of pancreatic-polypeptide in human pancreas, *J. Clin. Endocrinol. Metab.*, 44, 795, 1977.
8. **Tatemoto, K.,** Neuropeptide Y: complete amino acid sequence of the brain peptide, *Proc. Natl. Acad. Sci. U.S.A.,* 79, 5485, 1982.
9. **Tatemoto, K.,** Isolation and characterization of peptide YY (PYY), a candidate gut hormone that inhibits pancreatic exocrine secretion, *Proc. Natl. Acad. Sci. U.S.A.*, 79, 2514, 1982.
10. **Glover, I., Haneef, I., P:itts, J. E., Wood, S. P., Moss, D., Tickle, I. J., and Blundell, T.,** Conformational flexibility in a small globular hormone: X-ray analysis of avian pancreatic polypeptide at 0.98-Å resolution, *Biopolymers*, 22, 293, 1983.
11. **Glover, I., Barlow, D. J., Pitts, J. E., Wood, S. P., Tickle, I. J., Blundell, T., Tatemoto, K., Kimmel, J. R., Wollmer, A., Strassburger, W., and Zhang, Y. S.,** Conformational studies on the pancreatic polypeptide hormone family, *Eur. J. Biochem.*, 142, 379, 1985.
12. **Krstenansky, J. L. and Buck, S. H.,** The synthesis, physical characterization and receptor binding affinity of neuropeptide Y (NPY), *Neuropeptides*, 10, 77, 1987.
13. **Chance, R. E., Cieszkowski, M., Jaworek, J., Konturek, S. J., Swierczek, J., and Tasler, J.,** Effect of pancreatic polypeptide and its C-terminal hexapeptide on meal and secretin induced pancreatic secretion in dogs, *J. Physiol.*, 314, 1, 1981.
14. **Lin, T.-M.,** Pancreatic polypeptide: isolation, chemistry, and biological function, in *Gastrointestinal Hormones,* Jerzy Glass G. B., Ed., Raven Press, New York, 1980, chap. 11.
15. **Schwartz, T. W., Sheikh, S. P., and O'Hare, M. M. T.,** Receptors on phaeochromocytoma cells for two members of the PP-fold family — NPY and PP, *FEBS Lett.*, 225, 209, 1987.
16. **Gilbert, W. R., Frank, B. H., Gavin J. R., III, and Gingerich, R. L.,** Characterization of specific pancreatic polypeptide receptors on basolateral membranes of the canine small intestine, *Proc. Natl. Acad. Sci. U.S.A.*, 85, 4745, 1988.
17. **Schwartz, T. W., Gingerich, R. L., and Tager, H. S.,** Biosynthesis of pancreatic polypeptide, *J. Biol. Chem.*, 255, 11494, 1980.
18. **Boel, E., Schwartz, T. W., Norris, K. E., and Fiil, N. P.,** A cDNA encoding a small common precursor for human pancreatic polypeptide and pancreatic icosapeptide, *EMBO J.*, 3, 909, 1984.
19. **Leiter, A. B., Keutmann, H. T., and Goodman, R. H.,** Structure of a precursor to human pancreatic polypeptide, *J. Biol. Chem.*, 259, 14702, 1984.
20. **Schwartz, T. W.,** Cellular peptide processing after a single arginyl residue — studies on the common precursor for pancreatic polypeptide and pancreatic icosapeptide, *J. Biol. Chem.*, 262, 5093, 1987.
21. **Schwartz, T. W., Fuhlendorff, J., Langeland, N., Thøgersen, H., Jøgensen, J. Ch., and Sheikh, S. P.,** Y_1 and Y_2 receptors for NPY — the evolution of PP-fold peptides and their receptors, in *Neuropeptide Y, XIV Nobel Symposium,* Mutt, V., Hökfelt, T., Fuxe, K., and Lundberg, J., Eds., Raven Press, New York, 1989, 143.
22. **Villanueva, M. L., Hedo, J. A., and Marco, J.,** Heterogeneity of pancreatic polypeptide immunoreactivity in human plasma, *FEBS Lett.*, 80, 99, 1977.
23. **Ishida, T., Chou, M. C. Y., Chou, J., Lewis, R. M., Hartley, C. J., Entman, M., Kimmel, J. R., and Field, J. B.,** Absence of hepatic extraction of pancreatic polypeptide in conscious dogs, *Am. J. Physiol.*, 245, E401, 1983.
24. **Garrote, F. J., Rovira, A., Casado, S., Hernando, L., Castrillo, J. M., and Valverde, I.,** Immunoreactive pancreatic polypeptide components in plasma from normal subjects and patients with chronic renal failure in basal and postprandial conditions, *Metabolism*, 33, 244, 1984.
25. **Schwartz, T. W.,** Pancreatic polypeptide: a hormone under vagal control, *Gastroenterology*, 85, 1411, 1983.
26. **O'Hare, M. M. T., Huda, I., Sloan, J. M., Kennedy, T. L., and Buchanan, K. D.,** Characterization of immunoreactive forms of pancreatic polypeptide in islet cell tumors using antisera with different regional specificities, *Cancer*, 55, 1895, 1985.
27. **Schwartz, T. W., Wulff, B. S., Fuhlendorff, J., Johansen, T. E., Pedersen, M. S., Vogel, C. K., Tolstoy, S., and O'Hare, M. M. T.,** Signal epitopes of PP-fold peptides and their precursors, in *Neuropeptides and their Receptors, Alfred Benzon Symposium 29*, Schwartz, T. W., Hilsted, L., and Rehfeld, J. F., Eds., Munksgaard, Copenhagen, 1990.

28. **Schwartz, T. W., Stenquist, B., Olbe, L., and Stadil, F.,** Synchronous oscillations in the basal secretion of pancreatic-polypeptide and gastric acid, *Gastroenterology,* 76, 14, 1979.
29. **Floyd, J. C., Jr., Fajans, S. S., Pek, S., and Chance, R. E.,** A newly recognized pancreatic polypeptide; plasma levels in health and disease, *Rec. Prog. Hormone Res.*, 33, 519, 1977.
30. **O'Hare, M. M. T., Daly, J. G., and Buchanan, K. D.,** Radioimmunoassay for pancreatic polypeptide, and its age-related changes in concentration, *Clin. Chem.*, 29, 1923, 1983.
31. **Schwartz, T. W., Rehfeld, J. F., Stadil, F., Larsson, L.-I., Chance, R. E., and Moon, M.,** Pancreatic-polypeptide response to food in duodenal ulcer patients before and after vagotomy, *Lancet,* 1, 1102, 1976.
32. **Adrian, T. E., Bloom, S. R., Besterman, H. S., and Bryant, M. G.,** PP-physiology and pathology, in *Gut Hormones,* Bloom, S. R., Ed., Churchill Livingstone, Edinburgh, 1978, chap. 40.
33. **Larsson, L.-I., Schwartz, T. W., Lundquist, G., Chance, R. E., Sundler, F., Rehfeld, J. F., Grimelius, L., Fahrenkrug, J., and Schaffalitzky de Muckadell, O. B.,** Occurrence of human pancreatic polypeptide in pancreatic endocrine tumors, *Am. J. Pathol.*, 85, 675, 1976.
34. **Schwartz, T. W.,** Pancreatic polypeptide and endocrine tumors of the pancreas, *Scand. J. Gastroenterol.*, 14, 93, 1979.
35. **Friesen, S. R., Kimmel, J. R., and Tomita, T.,** Pancreatic polypeptide as screening marker for pancreatic polypeptide apudomas in multiple endocrinopathies, *Am. J. Surg.*, 139, 61, 1980.
36. **Bloom, S. R., Adrian, T. E., and Polak, J. M.,** Pancreatic polypeptide from pancreatic endocrine tumors, *Lancet,* 2, 1026, 1980.
37. **Schwartz, T. W.,** Atropine suppression test for pancreatic polypeptide, *Lancet*, 2, 42, 1978.
38. **Adrian, T. E., Uttenthal, L. O., Williams, S. J., and Bloom, S. R.,** Secretion of pancreatic polypeptide in patients with pancreatic endocrine tumors, *N. Engl. J. Med.,* 315, 287, 1986.
39. **Hällgren, R., Lunqvist, G., and Chance, R. E.,** Serum levels of human pancreatic polypeptide in renal disease, *Scand. J. Gastroenterol.*, 12, 923, 197.
40. **Henriksen, J. H., Schwartz, T. W., and Bülow, J. B.,** Endogenous pancreatic polypeptide in different vascular beds: relationship to release and degradation subjects with normal and decrased kidney function, *Metabolism*, 35, 542, 1986.
41. **Boden, G., Master, R. W., Owen, O. E., and Rudnick, M. R.,** Human pancreatic polypeptide in chronic renal failure and cirrhosis of the liver: role of kidneys and liver in pancreatic polypeptide metabolism, *J. Clin. Endocrinol. Metab*., 51, 573, 1980.
42. **Sive, A., Vinik, A. I., Van Tonder, S., and Lund, A.,** Impaired pancreatic polypeptide secretion in chronic pancreatitis, *J. Clin. Endocrinol. Metab*., 47, 556, 1978.
43. **Adrian, T. E., Besterman, H. S., Mallinson, C. N., Garalotis, C., and Bloom, S. R.,** Impaired pancreatic polypeptide release in chronic pancreatitis with steatorrhoea, *Gut*, 20, 98, 1979.
44. **Adrian, T. E., McKiernan, J., Johnstone, D. I., et al.,** Hormonal abnormalities of the pancreas and gut in cystic fibrosis, *Gastroenterology,* 79, 460, 1980.
45. **Krarup, T., Schwartz, T. W., Hilsted, J., Madsbad, S., Verlaege, O., and Sestoft, L.,** Impaired response of pancreatic polypeptide to hypoglycemia: an early sign of autonomic neuropathy in diabetics, *Br. Med. J.*, 15, 1544, 1979.
46. **Adrian, T. E., Greenberg, G. R., Besterman, H. S., and Bloom, S. R.,** Pharmacokinetics of pancreatic polypeptide in man, *Gut*, 19, 907, 1978.
47. **Hagopian, W., Lever, E. G., Cohen, D., Emmanouel, D., Polonsky, K. S., Pugh, W., Moossa, A., and Jaspan, J. B.,** Predominance of renal and absence of hepatic metabolism of pancreatic polypeptide in the dog, *Am. J. Physiol*., 245, E171, 1983.
48. **Taylor, I. L., Solomon, T. E., Walsh, J. H., and Grossman, M. I.,** Pancreatic polypeptide. Metabolism and effect on pancreatic secretion in dogs, *Gastroenterology,* 76, 524, 1979.
49. **Schwartz, T. W., Holst, J. J., Fahrenkrug, J., Jensen, S. L., Nielsen, O. V., Rehfeld, J. F., Schaffalitzky de Muckadell, O. B., and Stadil, F.,** Cholinergic, vagal regulation of pancreatic-polypeptide secretion, *J. Clin. Invest*., 61, 781, 1978.

Chapter 12

ENZYMATIC INACTIVATION OF NEUROPEPTIDES

Fred Nyberg and Lars Terenius

TABLE OF CONTENTS

I. THE PEPTIDERGIC NEURON

Peptides form a large group of signal substances in the nervous system. For instance, there are at least 20 major peptides with opioid activity. All peptides derive from protein precursors, so-called *prohormones*, which are synthesized in the endoplasmic reticulum and packed into secretory vesicles in the Golgi apparatus. The peptides are liberated from the prohormones by the successive cleavage by several so-called *processing enzymes*. The neuropeptides and their precursors are moved to the nerve terminals by fast axonal transport. Despite the enormous variety at the neuropeptide level, the prohormones are very similar in their general structure. The active peptides are flanked by pairs of basic amino acids which serve as processing signals for a variety of enzymes. The prohormones can be grouped in families with certain homologies (Table 1). Several prohormones are pluripotent and able to release several peptides with similar (such as proenkephalin, which has no less than seven enkephalin sequences) or different (such as preprotachykinins, which are precursors for substance P and neurokinin A) activities. The spectrum of neuropeptides which may be released from a prohormone is not absolute and may be modulated, depending on the rate of secretion or enzymatic processing.

The role of the neuropeptides in the nervous system is not so easily defined. Conceivably, they could serve as neurotransmitters, tuning or modifying the actions of a "classic" neurotransmitter, or they may have a neurohormonal effect influencing excitability or other membrane properties, have a metabotrophic effect, or affect neuronal growth and differentiation. It is possible that any one of the neuropeptides could serve in several capacities depending on the anatomic site, the state of development or the general neuronal activity. The action of neuropeptides on neuronal cells have certain characteristics which distinguish from that of classic transmitters; action is slower in onset and of longer duration, and the effect may be cumulative over longer periods of time. An obvious explanation for the long duration of effects is the slower receptor kinetics, the peptides binding very avidly to their receptors with K_D values in the 0.5 to 5 n*M* range (see Chapter 5). It has been shown to be very difficult to prove whether or not a particular neuropeptide is a neurotransmitter at a given synaps. The usual criteria for a neurotransmitter are the following:

1. Synthesis occurs in the neuron
2. Release from the neuron upon proper stimulation can be demonstrated
3. Pharmacologic actions of the transmitters, when introduced in the area of the synaps, are the same as the effects of stimulation of the nerve
4. Inactivation of the transmitter occurs in the synaptic area

Thus far, the only known mechanism for inactivation of neuropeptides is metabolic fragmentation. It may therefore seem so much more surprising that relatively little is known about enzymes which terminate peptide actions in the synaptic area. This, in turn, is due to the lack of suitable enzyme inhibitors and knowledge about the exact localization of candidate enzymes.

This chapter attempts to illustrate some issues central to the definition of enzymes that are important for the termination of the action of neuropeptides. One instructive example is provided by the studies of the degradation of enkephalins. Another example has the focus on one enzyme, angiotensin converting enzyme (ACE), which has a distribution in the central nervous system (CNS) suggesting that it has a more general role in peptide degradation than offered by its name (see Chapter 4). These and other examples will consider the question of specificity in terms of enzyme action or enzyme localization. The focus of this chapter is on the degradation of opioid peptides and tachykinins.

TABLE 1
Representative Neuropeptide Superfamilies

Cholecystokinin/gastrin family
Progastrin
Procholecystokinin
Opioid peptide family
Prodynorphin
Proenkephalin
Proopiomelanocortin
Tachykinins
Proneurokinin-B
Proneuromedin-B
Protachykinin-α
Protachykinin-β

II. ATTEMPTS TO DEFINE ROUTES OF INACTIVATION OF NEUROPEPTIDES

A. ENKEPHALIN DEGRADATION

The discovery of the enkephalins as potent morphino-mimetics, with a simple structure Tyr-Gly-Gly-Phe-X (where X is Leu or Met), generated unprecedented research activities on peptides. It was soon realized that the pentapeptide sequence was a minimum for significant opioid activity and that the enkephalins were very easily degraded in biological fluids or tissues. Several hundred analogues have been assayed for opioid activity. Replacement of Gly^2 with D-Ala drastically improved the metabolic stability and pharmacologic potency. The prototype drug, D-Ala^2, Met^5-enkephalin amide[1] and the even more stable D-Ala^2, D-Leu^5-enkephalin amide, DADLE,[2] are examples of enkephalin analogs with high metabolic stability. The major aminopeptidase cleaving the enkephalins seems to be aminopeptidase N (also referred to as aminopeptidase M). This enzyme is distributed not only in the CNS but also in kidney and intestine. The enzyme has been cloned and found to belong to the zinc-metallohydrolase superfamily (cf. References 3 to 5 and Chapter 4). It is selectively blocked by amastatin or bestatin[6,7] and these inhibitors are routinely used to protect the release of N-terminal Tyr^1 from the enkephalins. Aminopeptidase N is a membrane-bound enzyme with its active site on the outer cell surface. This enzyme could therefore be involved in the degradation of extracellular enkephalins. Since the localization of aminopeptidase N is imperfectly known, it is not known whether it could be responsible for regulation of enkephalin concentration in the synaptic area.

Another aminopeptidase sensitive to inhibition by puromycin catalyzes enkephalin degradation. This enzyme is either cytosolic or associated to membrane with its active site probably facing the cytosol.[8] This is,therefore, not a candidate enzyme for degradation of extracellular enkephalins.

The neutral endopeptidase 24.11 was originally proposed as a selective enkephalin-degrading enzyme and termed enkephalinase. It is now known to be a rather general endopeptidase with the active site on the outer cell surface.[9] It occurs at high concentrations in certain areas of the CNS. Quite generally, the assumption is made that peptide release or turnover can be measured by the changes in total concentration in brain (or other tissues). However, tissue concentrations are a complex function of peptide biosynthesis, intracellular degradation, and release. Since the releasable pool is small, this approach is likely to be inaccurate. Endopeptidase 24.11 cleaves enkephalins at the Gly^3-Phe^4 bond and has much lower affinity for longer peptides such as dynorphins or beta-endorphin.[10] The N-terminal fragment, Tyr-Gly-Gly could therefore be an extraneuronal marker of enkephalin release. Schwartz and co-workers have devised a radioimmunologic procedure to measure this fragment with reasonable sensitivity.[11,12] They could show that electric stimulation analgesia is accompanied by an increase in the concentration of the fragment in cerebroventricular perfusate.[13]

Several inhibitors to endopeptidase 24.11 are known, including phosphoramidone,[14] a bacterial product, and synthetic compounds such as thiorphan.[15] Inhibition of endopeptidase 24.11 activity by intracerebroventricular thiorphan or the more specific inhibitor, retro-thiorphan, induces analgesia in mice.[15,16] Another enkephalinase inhibitor, kelatorphan, which also inhibits aminopeptidase N is an even more potent analgesic,[17] suggesting that both the endopeptidase and aminopeptidase routes play significant roles in enkephalin degradation.

B. ANGIOTENSIN CONVERTING ENZYME (ACE)

ACE is a transmembrane peptidase with its active site on the external surface of the cell membrane[18-20] (see Chapters 4 and 6). It is widely distributed in the brain and occupies territories in the basal ganglia outside the limited distribution of angiotensin neurons. There is, for instance, a major ACE pathway from the striatum to the substantia nigra which is parallel to pathways with substance P (protachykinin), dynorphin, and GABA. It is not known, however, whether ACE is actually co-localized with any of these substances. Based on its effect on angiotensin I, ACE has been described as a dipeptidyl carboxypeptidase. However, testing of other neuropeptides as substrates has led to an expanding view of ACE as a tripeptide carboxypeptidase (against substance P) and even a tripeptide aminopeptidase (LHRH).[21,22]

Studies of ACE have been greatly facilitated by the available inhibitors, some of which have clinical application (like captopril and enalapril).[23,24] Captopril has been labeled to high specific activity and used for autoradiographic mapping of ACE pathways.[25] Captopril has also been used as a tool to identify the physiological roles of ACE.

Although ACE in the brain has somewhat different physicochemical properties than the pulmonary enzyme,[26] it is probably not an isozyme but differs in the extent of glycosylation.[27] Besides its actions in angiotensin pathways, its functional role in the CNS is not known. If captopril is given intraventricularly to rats, there is an increase in substance P levels in the substantia nigra,[28] suggesting that substance P degradation is affected. Other and still preliminary studies indicate that captopril given into the substantia nigra of the rat induces contralateral rotation and a parallel inhibition of dopamine release. The substantia nigra is the terminal area for both a substance P and a dynorphin pathway. Although captopril could block substance P degradation, ACE also very potently inactivates Leu-enkephalin-Arg^6 by cleavage of the Phe^4-Leu^5 bond.[29] This enkephalyl peptide is a major peptide in dynorphinergic pathways.[30] Interestingly, Leu-enkephalin-Arg^6, given into the substantia nigra, produces contralateral rotation *and* inhibition of dopamine release in the striatum, whereas substance P produces contralateral rotation and stimulation of dopamine release. Thus, these observations suggest that ACE has a more important role to degrade peptides in the dynorphin pathway rather than substance P (Figure 1).

In summary, ACE has a cellular localization and a localization to discrete pathways which are compatible with a role as a neuropeptide-inactivating peptidase. The pharmacologic actions of captopril given centrally suggest that ACE actually serves in this capacity.

III. ENZYMES WITH UNKNOWN LOCALIZATION, EXTRACTED FROM BRAIN

The number of genes expressed in the brain is 10 to 20 times higher than in the liver or

Tyr - Gly - Gly - Phe ↓ **- Leu- Arg**

Arg - Pro - Lys - Pro - Gln - Gln - Phe ↓ **- Phe** ↓ **- Gly** ↓ **- Leu - Met** - NH_2

FIGURE 1. Possible substrates for fragmentation by angiotensin-converting enzyme (ACE) in the striato-nigral pathway, the opioid peptide Leu-enkephalin-Arg[6] from prodynorphin, and substance P from protachykinin.

kidney. Several of these may be uniquely expressed in the brain. Just considering the number of total proteins and the likelihood that a large number of these and/or their fragments have signal function, there has to be a large number of peptidases expressed in the brain which have a "housekeeping" function. Peptidases, inactivating neuropeptides in the synaptic area and terminating their function, may constitute a minority. Enzyme characterization has to follow the conventional way via purification, isolation, structure determination, and design of specific inhibitors. Enzyme characterization and design of inhibitors is very tedious. The enzyme candidates for making these efforts therefore have to be very clearly defined. There is probably no way to tell *a priori* whether a particular enzyme is a neuropeptide inactivation enzyme or simply a nonspecific housekeeping protease. Criteria for a neuropeptidase could include substrate specificity, generation of neuropeptide fragments which can be isolated from tissue, and subcellular distribution to the so-called synaptosomes, artificial organelles formed during homogenization of brain and containing elements from the synapses. If bound to a membrane, it should also be an ectoenzyme with the active site on the cell surfaces. The following list is not intended to be exhaustive. Examples have been selected when there is some reason to believe that the enzyme may be an inactivating or transforming neuropeptidase.

A. AMINOPEPTIDASES

The importance of aminopeptidases in the inactivation of the enkephalins has been demonstrated by Bauer and co-workers.[31] This study has been followed up by many others and several enzymes capable of releasing the N-terminal tyrosine of the enkephalin pentapeptides have been described.[32-36] Two brain aminopeptidases which potently hydrolyze the enkephalins have been partially characterized. One of these was recently purified from human brain.[34] It was found to hydrolyze only the Tyr^1-Gly^2 bond of enkephalins and their C-terminally extended forms. The enkephalin pentapeptides were hydrolyzed at a higher rate. Larger peptides containing the enkephalin N-terminus such as beta-endorphin or dynorphins are not hydrolyzed. The enzyme has a molecular weight of about 100 kDa and shows optimum activity at pH 7.5. It is inhibited by puromycin, bacitracin, and mercuric acetate, and activated by dithiothreitol, suggesting that thiol groups are essential for activity. Inhibition by divalent metal ions such as Zn^{2+}, Cu^{2+} and Ni^{2+} was also described. Another brain enkephalin degrading aminopeptidase was identified in a monkey cerebral membrane fraction.[35] It has a molecular weight of 53 kDa and a pH optimum in the neutral region (pH 6.5). It hydrolyzes Leu-enkephalin with a K_M value of around 240 μM. The enzyme is potently inhibited by bestatin, but shows low sensitivity to puromycin and actinomycin. Both the human and monkey enzymes are distinct from aminopeptidase N. Aminopeptidases releasing N-terminal tyrosine from the enkephalins have also been found in the cerebrospinal fluid. Hazato et al.[36] described a bestatin-sensitive enkephalin degrading aminopeptidase in crude CSF. We have recently recovered and characterized an amastatin- and bestatin-sensitive enzyme capable of hydrolyzing the Tyr^1-Gly bond of enkephalin.[37] Except for a lower molecular weight, this enzyme exhibits properties similar to those of aminopeptidase N.

Dipeptidyl- and tripeptidyl-aminopeptidases are also known to degrade the enkephalins and their C-terminal extensions. Enkephalinase B which hydrolyzes the Gly^2-Gly^3 bond of the pentapeptide is one of the best characterized dipeptidyl aminopeptidases. It was recently purified to homogeneity from rat brain membranes.[38] It has a molecular weight of 82 kDa and degrades Leu-enkephalin and Met-enkephalin with K_M values of 53 and 50 μM, respectively. The enzyme is inactivated by metal chelators such as EDTA and *o*-phenanthroline, but is not affected by bestatin, amastatin, captopril, or phosphoramidon. Enkephalinase B also hydrolyzes C-terminal enkephalin extensions, including Leu-enkephalin-Arg^6, $dynorphin^{1-13}$, and alpha-neoendorphin. The enzyme has not been found to act on other biologically active peptides. A similar enzyme but of lower molecular weight (approximately 51 kDa) has been purified from porcine brain.[39] However, this latter enzyme also releases the N-terminal dipeptide from cholecystokinin (CCK)-4 and CCK-5. The enkephalins and their C-terminally extended forms may also be hydrolyzed by tripeptidyl peptidase II,[40] an enzyme present in human erythrocytes[41] and also in the brain.[31] The enzyme hydrolyzes the opioid peptides at the Gly^3-Phe^4 bond, thus releasing the N-terminal tripeptide Tyr-Gly-Gly from the substrate. The importance of tripeptidyl peptidase II for the inactivation of the enkephalins is not known.

B. POST-PROLINE CLEAVING ENZYMES

Prolyl endopeptidases are of potential importance in the degradation of several neuropeptides. Substance P has been one much-studied substrate which is cleaved at its Pro^2-Lys^3 or Pro^4-Gln^5 bond. Post-proline dipeptidyl aminopeptidase (PPDAP) successively releases the dipeptides Arg-Pro and Lys-Pro from substance P. PPDAP is a serine protease and has been identified in several peripheral tissues (as dipeptidyl peptidase IV), but is also present in the brain. Kato and co-workers reported two forms of the enzyme present in porcine brain.[42] They also found a similar activity in the cerebrospinal fluid.[43] PPDAP activity is present in the brain, both in a membrane-bound form and in a soluble form, with over 50% of total peptidase activity in guinea-pig brain being associated with the soluble fraction and the remaining activity distributed throughout the particulate fractions.[44] The synaptosomal membrane PPDAP has been solubilized and recovered for further characterization.[44] It differs in properties from the soluble brain enzyme with regard to its inhibition profile. Recently, two separate membrane-bound enzymes that liberate dipeptides of the type X-Pro were purified to homogeneity from rat brain.[45] The isolated enzymes had similar catalytic properties and were of similar molecular weight (approximately 110 kDa). They were both inhibited by typical serine protease inhibitors like diisopropyl fluorophosphate (DFP) or *p*-methylphenylsulphonyl fluoride but not by the aminopeptidase inhibitor, bestatin. The substance P (1-4) fragment was a good substrate, whereas the parent molecule was degraded at a lower rate. Both enzymes have been classified as dipeptidyl peptidase II (EC 3.4.14.2) species.

Another post-proline cleaving enzyme acting on substance P has been described by several investigators (e.g., see References 46 and 47). This is an endopeptidase which was first described as a thyroliberin-inactivating enzyme.[47] The enzyme has a molecular weight of 50 kDa and is classified as a thiol-dependent endopeptidase (EC 3.4.21.26). It cleaves substance P at its Pro^4-Gln^5 bond, thus generating the C-terminal substance P (5-11) fragment. This fragment has been identified in human CSF,[48] which suggests that the enzyme may be present in neuronal areas containing substance P. In fact, Kato and co-workers found that following kainic acid lesion in rat caudate-putamen, the prolyl endopeptidase activity was decreased.[49] This decrease was accompanied by changes in substance P content, indicating that the enzyme may be involved in specific neuronal functions in this particular area.

C. ENDOPEPTIDASES

In addition to the post-proline enzyme, there are several other endopeptidases capable of hydrolyzing substance P. Benuck and Marks reported a cytosolic enzyme from rat brain which cleaves the peptide at the Gln^6-Phe^7, Phe^7-Phe^{8,} and Gly^9-Leu^{10} bonds.[50] A membrane-bound enzyme cleaving substance P at its Gln^6-Phe^7, Phe^7-Phe^{8,} and Phe^8-Gly^9 bonds was isolated from human brain.[51] We have identified a substance P (1-7) and (1-8) generating endopeptidase in human CSF.[52] Except for the absence of formation of substance P (1-6), the CSF enzyme shows molecular and catalytic properties in good agreement with those of the enzyme in human brain.[51] The CSF enzyme has a molecular weight of 43 kDa and is sensitive to metal chelators such as EDTA and EGTA. A membrane-bound substance P degrading endopeptidase has also been purified from rat brain.[54] This enzyme, with a molecular weight of 70 kDa, was optimally active at pH 7.5. It was strongly inhibited by metal chelators and bacitracin, but was not affected by serine protease inhibitors or by captopril and phosphoramidon. Substance P is preferentially hydrolyzed at the Gln^6-Phe^7 bond, but fragmentation at other bonds (e.g., Gln^5-Gln^6, Phe^7-Phe^{8,} and Gly^9-Leu^{10}) was also observed. A similar enzyme also cleaving substance P at the Gln^6-Phe^7 bond was recently isolated from a solubilized membrane fraction of rat brain.[55] Except for a lower molecular weight (58 kDa) the latter enzyme showed almost identical properties to those of the 70-kDa rat brain enzyme. Among other peptidases which may act on substance P with endopeptidase activity are angiotensin converting enzyme (EC 3.4.15.1) and neutral endopeptidase (EC 3.4.24.11). Both these enzymes are discussed in detail in Section II.

A number of endopeptidases acting on the opioid peptides have been described during the last decade. While the major route for enkephalin inactivation seems to follow the action of aminopeptidases, peptides with extended sequences appear to be inactivated in several conversion steps involving both endo- and exopeptidases. An important step in the conversion of the dynorphins and alpha-neoendorphin is the hydrolysis of peptide bonds in the vicinity of their dibasic stretches (Arg-Arg or Arg-Lys). Several peptidases with affinity for the paired basic residues of the dynorphins have been described.[56-60] However, most products generated by these enzymes retain opioid activity, if however of a different receptor profile than of the parent peptide. An endopeptidase in human CSF releasing the N-terminal fragment, Leu-enkephalin-Arg^6 from dynorphin A and dynorphin B as well as alpha-neoendorphin has been described.[56] The enzyme has a molecular weight of around 40 kDa and is classified as a serine protease. In our attempts to find the origin of the CSF enzyme, we also have searched for similar activities in the CNS. This led us to identify at least three different endopeptidases hydrolyzing the dynorphins at their dibasic stretches[57] in human spinal cord. One of these endoproteinases is similar to the CSF enzyme with regard to its activity optimum at neutral pH and specificity for all three prodynorphin-derived opioid peptides. However, it differs with regard to molecular size. All peptidases preferentially release Leu-enkephalin-Arg^6 from the substrate peptide; however, hydrolysis with the spinal cord enzymes also gave Leu-enkephalin and Leu-enkephalin-Arg^6-Arg^7 as products.

A second enzyme from human spinal cord and specific for dynorphin B shows close similarity to one isolated from bovine spinal cord.[58] The third enzyme recovered from human spinal cord has an activity optimum at pH 5.6. It releases Leu-enkephalin and Leu-enkephalin-Arg^6 in equimolar concentrations from both dynorphin A and dynorphin B. Recent studies in our laboratory have also revealed the presence of dynorphin converting enzymes in human choroid plexus.[59] Both a soluble and membrane-bound enzyme were purified and characterized. Their estimated molecular weight values (58 and 55 kDa, respectively) differ from those of the CSF and spinal cord enzymes. They hydrolyze the Arg^6-Arg^7 and Arg^6-Lys^7 bonds of the two dynorphins and alpha-neoendorphin at approximately similar rates. Brain endopeptidases capable of converting the dynorphins to Leu-enkephalin have also been described by others. For instance, Chu and Orlowski described a soluble metalloendopeptidase in rat brain which releases Leu-enkephalin from dynorphin A(1-8), as well as from alpha- and betaneoendorphin.[61] This enzyme, however, also degrades nonopioid peptides such as bradykinin, neurotensin, and angiotensin and prefers a hydrophobic amino acid in the P_1, P_2, and P_3 positions. In this capacity, it is an inactivation enzyme.

An endoprotease capable of releasing Leu-enkephalin and Met-enkephalin from small enkephalin-containing peptides named endo-oligopeptidase A has been purified from the cytosol fraction of bovine brain homogenates.[62] The enzyme has been characterized as a thiol endopeptidase. Like the metalloendopeptidase,[61] it also hydrolyzes and inactivates bradykinin and neurotensin.

D. CARBOXYPEPTIDASES

Inactivation of the enkephalins by removal of a single amino acid from their C-terminal end seems to be of comparatively little importance. However, the dipeptidyl carboxypeptidase activities of ACE and neutral endopeptidase are important, as discussed in Section II. As also discussed earlier, substance P is degraded by these enzymes.

Many other neuropeptides, particularly those which have an alpha-amidated C-terminus, appear to be relatively stable against carboxypeptidase degradation.

IV. PROTEOLYTIC ENZYMES IN CEREBROSPINAL FLUID (CSF)

The classic model of neurotransmission is the directed release of a signal substance in a synaps and its inactivation in the synaptic area. However, the transmitter may also spread beyond the synaps or even be released from nerve terminals that do not form synapses. This more diffuse mode of communication with a signal substance has been called "volume transmission".[63] Here, the role of the signal substance approaches that of a hormone. In all probability, classic and volume transmission are the two extreme cases and, in the actual reality, a given signal substance in a particular synaps may act along both mechanisms. Neuropeptides are likely candidates for diffuse signal communication by volume transmission. If so, one might anticipate that the distribution of inactivating enzymes is rather broad and not necessarily restricted to the terminal area. Available data on enzyme distributions within the CNS seem compatible with a rather general distribution.

Another mechanism for peptide inactivation is by enzymes secreted into the CSF. The CSF compartment is engaged in the elimination of bioactive substances in general, including peptides. A relatively large number of peptidases have been identified in the CSF (Table 2). Several of these enzymes occur in substantial quantities, which has allowed at least a partial characterization. In those cases where the corresponding enzyme has been isolated from brain tissue, the CSF enzyme has a smaller apparent molecular weight. This is particularly obvious with ACE which, in its native form, is about 165 kDa, but in the CSF occurs in several fully active species of about 150, 80, and 40 kDa, respectively.[65] ACE is probably released by proteolysis from the membranes where it is incorporated. Proteolysis can apparently occur at several sites of the molecule. A dynorphin converting enzyme of about 43 kDa is very abundant in CSF[56] and distinctly smaller than the enzymes with corresponding activity extracted from CNS tissue.[57] The tissular enzymes are apparently cytosolic within a vesicular compartment, probably synaptic vesicles (unpublished). Here, a possible mechanism of secretion is via exocytosis, which would release the neuropeptide as well. The simultaneous secretion of neurotransmitter and enzyme protein has been observed outside the peptide field. Thus, in the substantia nigra, the transmitter (dopamine) and an enzyme (acetylcholinesterase) are apparently co-released.[66] It is intriguing that acetylcholinesterase also has peptidase activity.[67]

Protease and peptidase activity in the CNS may serve an important scavenger role in the normal peptide homeostasis. It can be anticipated that barrier defects or cerebral bleedings

TABLE 2
Enzymes in Human Cerebrospinal Fluid (CSF) which Inactivate Neuropeptides

Enzyme	Substrate	Ref.
Aminopeptidases	Leu- and Met-enkephalin δ-sleep-inducing peptide	36, 37
Post-proline cleaving enzyme	Substance P	42, 64
Substance P endopeptidase	Substance P	52
Dynorphin-converting enzyme	Dynorphin B α-neoendorphin	56
Endopeptidase 24.11	Leu- and Met-enkephalin	64
Angiotensin-converting cnzyme	Substance P Leu-enkephalin-Arg^6	29 65

could severely interfere with this function and, for instance, contribute to the complications of subarachnoidal hemorrhage, including cerebral vasospasm.

V. INACTIVATING PEPTIDASES AS PHARMACOLOGIC TARGETS

Specific inhibitors are crucial in establishing that a particular enzyme is important for the inactivation of neuropeptides. Progress in this area has been substantial, with the development of inhibitors specific for ACE and endopeptidase 24.11. The medicinal chemistry of peptidase inhibitors has made considerable progress and rational design has recently led to a new ACE inhibitor, enalapril, which has been found to have clinical utility besides the traditional inhibitor, captopril.[24] Inhibitors to endopeptidase 24.11 include thiorphan, which is relatively specific, and retro-thiorphan, which also has activity against aminopeptidases.[15,16] Endopeptidase 24.11 is an ubiquitous enzyme with particularly high concentrations in the kidney brush border membranes. However, when given centrally, thiorphan apparently acts against enkephalin degradation since it produces naloxone-reversible analgesia. Chronic administration of thiorphan produces an opiate-type dependence, as evidenced by the precipitation of withdrawal signs following naloxone injection.[68]

Since several enzymes are not particularly substrate-specific, several nonspecific inhibitors have been used primarily to investigate the general importance of peptidase inactivation. Thus, the aminopeptidase inhibitors bestatin and amastatin have been widely used to protect endogenous or exogenous peptides from inactivation.[69] Similarly, leupeptin has been used to inhibit endopeptidase activity. All these agents are themselves peptides with chemical modifications in their termini. They are also naturally occurring substances.

Although it has been shown that peptidase inactivation of neuropeptides is very efficient due to the ubiquitous distribution of highly potent enzymes, it has also been possible to demonstrate a pharmacologic effect by the administration of recombinant endopeptidase 24.11, which reduced the effects of a neurogenically induced inflammation. This effect was probably due to a decrease in substance P.[70]

VI. CONCLUSIONS

A number of enzymes which inactivate neuropeptides can be isolated from CNS tissue. In all likelihood, most of these enzymes are of housekeeping character, contributing to the general protein/peptide catabolism within the CNS compartment. However, there are also inactivating enzymes with a particular, defined distribution, such as ACE. The transmitter role of many neuropeptides is unclear and, of the criteria required and usually missing, is their regulated inactivation. Identification of inactivating enzymes in such a strict sense is therefore required for the assignment of a transmitter role to a particular peptide. With the possible exception of enkephalin inactivation, the routes of inactivation of most neuropeptides are entirely unclear.

Inactivation enzymes are also highly relevant targets for inhibitor development. Such inhibitors could serve to define the functional importance of neuropeptides and also become candidates for drug development, as illustrated for the ACE and endopeptidase 24.11 (enkephalinase) inhibitors.

VII. ACKNOWLEDGMENTS

This work has been supported by the Swedish Medical Research Council and the Swedish Board of Technological Development.

REFERENCES

1. **Pert, C. B., Pert, A., Chang, J. K., and Fong, B. T. W.,** D-Ala^2-Met-enkephalinamide: a potent, long-lasting synthetic penta-peptide analgesic, *Science*, 194, 330, 1976.
2. **Wei, E., Tseng, L. F., Loh, H. H., and Li, C. H.,** Comparison of the behavioral effects of β-endorphin and enkephalin analogs, *Life Sci.*, 21, 321, 1977.
3. **Malfroy, B., Kado-Fong, H., Gros, C., Giros, B., Schwartz, J.-C., and Hellmiss, R.,** Molecular cloning and amino acid sequence of rat kidney aminopeptidase M: a member of a superfamily of zinc-metallohydrolases, *Biochem. Biophys. Res. Commun.*, 161, 236, 1989.
4. **Olsen, J., Cowell, G. M., Konigshofer, E., Danielsen, M., Moller, J., Laustsen, L., Hansen, O. C., Welinder, K. G., Engberg, J., Hunziker, W., Spiess, M., Sjostrom, H., and Norén, O.,** Complete amino acid sequence of human intestinal aminopeptidase N as deduced from cloned cDNA, *FEBS Lett.*, 238, 307, 1988.
5. **Watt, V. M. and Yip, C. C.,** Amino acid sequence deduced from a rat kidney cDNA suggests it encodes the Zn-peptidase aminopeptidase N, *J. Biol. Chem.*, 264, 5480, 1989.
6. **Rich, D. H., Moon, B. J., and Harbeson, S.,** Inhibition of aminopeptidases by amastatin and bestatin derivatives. Effect of inhibitor structure on slow-binding processes, *J. Med. Chem.*, 27, 417, 1984.
7. **Giros, B., Gros, C., Solhonne, B., and Schwartz, J. C.,** Characterization of aminopeptidases responsible for inactivating endogenous Met^5 enkephalin in brain slices using peptidase inhibitors and anti-aminopeptidase M antibodies, *Mol. Pharmacol.*, 29, 281, 1986.
8. **Dyer, H., Slaughter, A., Orth, K., Moomaw, R., and Hersh, B.,** Comparison of the soluble and membrane-bound forms of the puromycin-sensitive enkephalin-degrading aminopeptidases from rat, *J. Neurochem.*, 54, 547, 1990.
9. **Fulcher, I. S., Matsas, R., Turner, A. J., and Kenny, A. J.,** Kidney neutral endopeptidase and the hydrolysis of enkephalin by synaptic membranes show similar sensitivity to inhibitors, *Biochem. J.*, 203, 519, 1982.
10. **Matsas, R., Kenny, A. J., and Turner, A. J.,** The metabolism of neuropeptides. The hydrolysis of peptides, including enkephalins, tachykinins and their analogues, by endopeptidase-24.11, *Biochem. J.*, 223, 433, 1984.
11. **Llorens-Cortes, C., Gros, C., and Schwartz, J.-C.,** Study of endogenous Try-Gly-Gly, a putative enkephalin metabolite, in mouse brain: validation of a radioimmunoassay, localization and effects of peptidase inhibitors, *Eur. J. Pharmacol.*, 119, 183, 1985.
12. **Llorens-Cortes, C., Schwartz, J.-C., and Gros, C.,** Detection of the tripeptide Tyr-Gly-Gly, a putative enkephalin metabolite in brain, using a sensitive radioimmunoassay, *FEBS Lett.*, 189, 325, 1985.
13. **Llorens-Cortes, C., Gros, C., and Schwartz, J.-C.,** Steady-state level and turnover rate of the tripeptide Tyr-Gly-Gly as indexes of striatal enkephalin release *in vivo* and their reduction during pentobarbital anesthesia, *Proc. Natl. Acad. Sci. U.S.A.*, 83, 6226, 1986.
14. **Suda, H., Aoyagi, T., Takeuchi, T., and Umezawa, H.,** A thermolysin inhibitor produced by actinomycetes: phosphoramidon, *J. Antibiot.*, 26, 621, 1973.

15. **Roques, B. P., Fournié-Zaluski, M. C., Soroca, E., Lecomte, J. M., Malfroy, B., Llorens, C., and Schwartz, J.-C.,** The enkephalinase inhibitor thiorphan shows antinociceptive activity in mice, *Nature (London)*, 288, 286, 1980.
16. **Roques, B. P., Lucas-Soroca, E., Chaillet, P., Costentin, C., and Fournié-Zaluski, M. C.,** Complete differentiation between enkephalinase and angiotensin-converting enzyme inhibition by retro-thiorphan, *Proc. Natl. Acad. Sci. U.S.A.*, 50, 3178, 1983.
17. **Fournié-Zaluksi, M.-C., Chaillet, P., Bouboutou, R., Coulaud, A., Cherot, P., Waksman, G., Costentin, J., and Roques, B. P.,** Analgesic effects of kelartophan, a new highly potent inhibitor of multiple enkephalin degrading enzymes, *Eur. J. Pharmacol.*, 102, 525, 1984.
18. **Johnson, A. R., Skidgel, R. A., Gafford, J. T., and Erdös, E. G.,** Enzymes in placental microvilli: angiotensin I converting enzyme, angiotensinase A, carboxypeptidase, and neutral endopeptidase ("enkephalinase"), *Peptides*, 5, 789, 1984.
19. **Erdos, E. G. and Skidgel, R. A.,** Structure and functions of human angiotensin I converting enzyme (kininase II), *Biochem. Soc. Trans.*, 13, 42, 1985.
20. **Erdos, E. G. and Skidgel, R. A.,** The angiotensin I-converting enzyme, *Lab. Invest.*, 56, 345, 1987.
21. **Yokosawa, H., Endo, S., Ogura, Y., and Ishii, S.-I.,** A new feature of angiotensin-converting enzyme in the brain: hydrolysis of substance P, *Biochem. Biophys. Res. Commun.*, 116, 735, 1983.
22. **Skidgel, R. A. and Erdos, E. G.,** Novel activity of human angiotensin I converting enzyme: release of the NH_2- and COOH-terminal tripeptides from the luteinizing hormone-releasing hormone, *Proc. Natl. Acad. Sci. U.S.A.*, 82, 1025, 1985.
23. **Strittmatter, S. and Snyder, S.,** Characterization of angiotensin converting enzyme by ^3H captopril binding, *Mol. Pharmacol.*, 29, 142, 1986.
24. **Unger, T., Ganten, D., and Lang, R. E.,** Converting enzyme inhibitors: antihypertensive drugs with unexpected mechanisms, *Trends Pharm. Sci.*, 12, 514, 1983.
25. **Strittmatter, S. M., Thiele, E., Kapiloff, M., and Snyder, S.,** A rat brain isoenzyme of angiotensin-converting enzyme, *J. Biol. Chem.*, 260, 9825, 1985.
26. **Rix, E., Ganten, D., Schull, B., Unger, Th., and Taughńeor, R.,** Converting-enzyme in choroid plexus, brain and kidney: immunohistochemical and biochemical studies in rats, *Neurosci. Lett.*, 22, 125, 1981.
27. **Hooper, N. M. and Turner, A. J.,** Isolation of two differently glycosylated forms of peptidyl-dipeptidase A ("angiotensin converting enzyme") from pig brain: a reevaluation of their role in neuropeptide metabolism, *Biochem. J.*, 241, 625, 1987.
28. **Hanson, G. R. and Lowenberg, W.,** Elevation of substance P-like immunoreactivity in rat central nervous system by protease inhibitors, *J. Neurochem.*, 35, 1370, 1980.
29. **Lantz, I. and Terenius, L.,** High enkephalyl peptide degradation due to angiotensin converting enzyme-like activity in human CSF, *FEBS Lett.*, 193, 31, 1985.
30. **Christensson-Nylander, I., Herrera-Marschitz, M., Staines, S., Hökfelt, T., Terenius, L., Ungerstedt, U., Cuello, C., Oertel, W. H., and Goldstein, M.,** Striato-nigral dynorphin and substance P pathways in the rat. I. Biochemical and immunohistochemical studies, *Exp. Brain Res.*, 64, 169, 1986.
31. **Horsthemke, D., Hamprecht, B., and Bauer, K.,** Heterogenous distribution of enkephalin-degrading peptidases between neuronal and glial cells, *Biochem. Biophys. Res. Commun.*, 115, 423, 1983.
32. **Churchill, L., Bausback, H. H., Gerritsen, M. E., and Ward, P. E.,** Metabolism of opioid peptides by cerebral microvascular aminopeptidase M, *Biochim. Biophys. Acta,* 923, 35, 1987.
33. **Malfroy, B., Kado-Fony, H., Gros, C., Giros, B., Schwartz, J.-C., and Hellmiss, R.,** *Biochem. Biophys. Res. Commun.*, 161, 236, 1989.
34. **Durate, G. I., Souza, F. G., Bruno, J. A., Camargo, A. C., and Carvalho, K. M.,** Isolation and characterization of human brain aminopeptidase which hydrolyze enkephalin-containing peptides, *Neuropeptides*, 12, 67, 1988.
35. **Shimamura, M., Hazato, T., and Iwaguchi, T.,** A new aminopeptidase in monkey cerebral membrane fraction: hydrolysis of enkephalin, *Brain Res.*, 445, 350, 1988.
36. **Hazato, T., Shimamura, M., Komuro, T., and Katayama, T.,** Partial purification of two distinct enkephalin degrading aminopeptidases from human cerebrospinal fluid, *Biochem. Int.*, 10, 813, 1985.
37. **Nyberg, F., Thörnwall, M., and Hetta, J.,** Aminopeptidase in human cerebrospinal fluid which degrades delta-sleep-inducing peptide (DSIP), *Biochem. Biophys. Res. Commun.*, 167, 1256, 1990.
38. **Inaoka, Y. and Tamaoki, H.,** Purification and characterization of enkephalinase B from rat brain, *Biochem. Biophys. Acta*, 925, 27, 1987.
39. **Cherot, P., Fournie-Zaluski, M. C., and Laval, J.,** Purification and characterization of an enkephalin-degrading dipeptidyl-aminopeptidase from porcine brain, *Biochemistry,* 25, 8184, 1986.
40. **Nyberg, F., Bålöw, R., Tomkinson, B., and Zetterquist, Ö.,** Degradation of Leu- and Met-enkephalin and their C-terminal extensions by tripeptidyl peptidase II, in *Protides of the Biological Fluids,* Peeters, W., Ed., Pergamon Press, Oxford, Vol. 35, 193, 1987.
41. **Bålöw, R. M., Tomkinson, B., Ragnarsson, U., and Zetterquist, Ö.,** Purification, substrate specificity and classification of tripeptidyl peptidase II, *J. Biol. Chem.*, 261, 2409, 1986.

42. **Kato, T., Hama, T., and Nagatsu, T.,** Separation of two dipeptidylaminopeptidases in the human brain, *J. Neurochem.*, 34, 602, 1980.
43. **Kato, T., Fumita, K., Iwase, K., Nagutsu, T., and Sakakibara, S.,** Comparison of X-prolyl dipeptidylaminopeptidase activity in human cerebrospinal fluid and that in serum, *Experientia*, 35, 20, 1979.
44. **O'Connor, B. and O'Quinn, G.,** Post-proline dipeptidylaminopeptidase from synaptosomal membranes of guinea-pig brain, *Eur. J. Biochem.*, 154, 329, 1986.
45. **Mentlein, R. and Struckhoff, G.,** Purification of two dipeptidyl aminopeptidases II from rat brain and their action on proline-containing peptides, *J. Neurochem.*, 52, 1284, 1989.
46. **Orlowski, M., Wilke, E., Pierce, S., and Wilk, S.,** Purification and properties of a prolyl endopeptidase from rabbit brain, *J. Neurochem.*, 33, 461, 1979.
47. **Knisatschek, H. and Bauer, K.,** Characterization of "thyroliberin-deamidating enzyme" as a post-proline-cleaving enzyme, *J. Biol. Chem.*, 254, 1093, 1979.
48. **Rimon, R., Le Grevès, P., Nyberg, F., Heikkilä, L., Salema, H., and Terenius, L.,** Elevation of substance P-like peptides in the CSF of psychiatric patients, *Biol. Psychiatry*, 19, 509, 1984.
49. **Koshiya, K., Kato, T., Tanaka, R., and Kato, T.,** Brain peptidases: their possible neuronal and glial localization, *Brain Res.*, 324, 261, 1984.
50. **Benuck, M. and Marks, N.,** Enzymatic inactivation of substance P by a partially purified enzyme from rat brain, *Biochem. Biophys. Res. Commun.*, 65, 153, 1975.
51. **Lee, C. M., Sandberg, B. E. B., Hanley, M. R., and Iversen, L. L.,** Purification and characterization of a membrane-bound substance P degrading enzyme from human brain, *Eur. J. Biochem.*, 114, 315, 1981.
52. **Nyberg, F., Le Grevès, P., Sundqvist, C., and Terenius, L.,** Characterization of substance P (1-7) and (1-8) generating enzyme in human cerebrospinal fluid, *Biochem. Biophys. Res. Commun.*, 125, 244, 1984.
53. **Stewart, J. M., Hall, M. E., Harkins, J., Frederickson, R. C. A., Terenius, L., Hökfelt, T., and Krivoy, W.,** A fragment of substance P with specific central activity: SP (1-7), *Peptides*, 3, 851, 1982.
54. **Bergmann, A. and Bauer, K.,** A membrane bound substance P degrading endopeptidase from rat brain, *Natl. Inst. Drug Abuse Res. Monogr.*, Ser. 75, 283, 1986.
55. **Endo, S., Yokosawa, H., and Ishii, S.,** Purification and characterization of a substance P-degrading endopeptidase from rat brain, *J. Biochem. (Tokyo)*, 104, 999, 1988.
56. **Nyberg, F., Nordström, K., and Terenius, L.,** Endopeptidase in human cerebrospinal fluid which cleaves proenkephalin B opioid peptides at consecutive basic amino acids, *Biochem. Biophys. Res. Commun.*, 131, 1069, 1985.
57. **Nyberg, F. and Silberring, J.,** Conversion of the dynorphins to Leu-enkephalin in human spinal cord, in *The International Narcotic Research Conference* (INRC) 89, Quirion, C., Jhamandas, K., and Gianoulakis, C., Eds., Alan R. Liss, New York, 261, 1990.
58. **Silberring, J. and Nyberg, F.,** A novel bovine spinal cord endoprotease with high specificity for dynorphin B, *J. Biol. Chem.*, 264, 11082, 1989.
59. **Nyberg, F., Kankaanranta, S., Brostedt, P., and Silberring, J.,** Purification and characterization from human choroid plexus of endoproteases which hydrolyse prodynorphin derived opioid peptides at paired basic amino acid residues, *Brain Res.*, in press, 1990.
60. **Chromlich, J. A., Seidah, N. G., and Chretien, M. N.,** Isolation and characterization of four proteases from porcine pituitary intermediate lobe. Relation to the maturation enzyme of prohormones, *Neuropeptides*, 5, 493, 1985.
61. **Chu, T. G. and Orlowski, M.,** Soluble metalloendopeptidase from rat brain: action on enkephalin-containing peptides and other bioactive peptides, *Endocrinology*, 116, 1418, 1985.
62. **Camargo, A. C., Oliveira, E. B., Toffoletto, O., Metters, K. M., and Rossier, J.,** Brain endo-oligopeptidase A, a putative enkephalin converting enzyme, *J. Neurochem.*, 48, 1258, 1987.
63. **Terenius, L. and Nyberg, F.,** Peptidases and proteases in cerebrospinal fluid, significance in volume transmission, in *Volume Transmission in the Brain: New Aspects in Electrical and Chemical Communication*, Fuxe, K., Ed., Raven Press, New York, in press, 1990.
64. **Nyberg, F.,** Peptidases in human cerebrospinal fluid, *Adv. Biosci.*, 65, 65, 1987.
65. **Lantz, I., Nyberg, F., and Terenius, L.,** Molecular heterogeneity and angiotensin-converting enzyme in human cerebrospinal fluid, manuscript submitted, 1990.
66. **Kreutzberg, G. W.,** The significance of dendritic release of transmitter and protein in the substantia nigra, *Neurochem. Int.*, 7, 913, 1985.
67. **Lockridge, O.,** Substance P hydrolysis by human serum cholinesterase, *J. Neurochem.*, 39, 106, 1982.
68. **Bean, A. J. and Vaught, J. L.,** Physical dependence by chronic intracerebroventricular infusion of D-Arg kyotorphin or thiorphan to rats, *Eur. J. Pharmacol.*, 105, 333, 1984.
69. **Tholrsett, E. D. and Matthew, J. W.,** Inhibition of zinc peptidases that hydrolyse neuropeptides, in *Neuropeptides and Their Peptidases,* Turner, A. J., Ed., Ellis Horwood, Chichester, England, 1987, 229.
70. **Malfroy, B., Ligett, D., Mc Gabe, J., Baughman, R., Kado-Fong, H., Mulholland, K., Bridenbaugh, R., and Anderson, J.,** Administration of recombinant enkephalinase (neutral endopeptidase) prevents capsaicin-induced miosis in the rabbit, *J. Pharmacol. Exp. Ther.*, 252, 462, 1989.

Chapter 13

DEGRADATION OF NEUROHYPOPHYSIAL PEPTIDES

Peter Bie, Jens L. Andersen, and Lars J. Andersen

TABLE OF CONTENTS

I. INTRODUCTION

V. du Vigneaud was awarded the 1955 Nobel Chemistry prize "for his work on biochemically important sulfur compounds, especially for the first synthesis of a polypeptide hormone". The hormone synthesized was oxytocin, one of the two classical posterior pituitary hormones. The composition of both hormones was determined through the work of du Vigneaud and coworkers. Synthesis of the other one, vasopressin, was delayed because of the difficulty in handling the amino acid arginine. Arginine is one of the two amino acids distinguishing arginine vasopressin from oxytocin (Figure 1). Vasopressin is the antidiuretic substance in all mammals except the pig and other members of the suborder Suina; in these species, the arginine (position 8) is replaced by lysine.

In the body, oxytocin and vasopressin are synthesized in the magnocellular neurons (Figure 2) of the supraoptic and paraventricular nuclei in the hypothalamus. To a lesser extent, magnocellular neurosecretory cell bodies are found elsewhere in the hypothalamus. Virtually all axons from the neurosecretory neurons in the supraoptic nucleus are via the median eminence and neural stalk projected to the neurohypophysis where they terminate as bulbous enlargements along the capillaries. Some axons from the paraventricular nucleus terminate in the median eminence or project to areas outside the hypothalamo-neurohypophysial system. Hormones, particularly vasopressin, released from the latter may function as a synaptic transmitter.

Oxytocin and vasopressin are synthesized as parts of separate, cystine-rich precursor molecules of approximately 20,000 Da. The precursor molecules consist of a signal peptide, the nonapeptide hormone, a larger peptide called neurophysin, and, for vasopressin but not oxytocin, a glycoprotein. Cleavage and modifications take place during transportation down the axon to yield free nonapeptide and neurophysin and — in the case of vasopressin — glycopeptide. Secretion occurs when action potentials move down the axon and cause depolarization of the nerve terminal. This results in a flux of Ca^{2+} into the cell and, by still unknown mechanisms, the neurosecretory granules fuse with the cell membrane and release their content to the interstitium and the blood of the capillaries.

Under physiological conditions, plasma osmolality is the major controller of vasopressin release. A small rise in plasma osmolality causes release, probably mediated by special cells, "osmoreceptors", in axonal contact with the cell bodies of the neurosecretory cells of the hypothalamus. Secretion of vasopressin is also affected by changes in intravascular volume. This response is mediated via afferent nerve fibers from low-pressure (cardio-pulmonary) receptors and high-pressure (aortic and carotid) baroreceptors. Other factors such as pain, nausea, and hypoxia, as well as a number of hormones and drugs, stimulate secretion of vasopressin. Oxytocin may also be released in response to a number of stimuli, notably by reflexes originating from tactile stimulation of the nipple and from stretch receptors located in the lower genital tract. It has been established that co-secretion of oxytocin and vasopressin occurs in response to osmotic stimuli.

Once released into the systemic circulation, the two neurohypophysial hormones exhibit remarkably different patterns of effects. Vasopressin is a potent vasoconstrictor, as the name implies. The effect is mediated by stimulation of receptors (V_1 receptors) in vascular smooth muscle cells resulting in increased influx of Ca^{2+} which serve as second messenger. In particular, the flow resistance of the skin and the splanchnic vasculature is increased. Generally, the vascular effects of vasopressin have been demonstrated by use of supraphysiological plasma concentrations; however, it is possible that it participates in the regulation of vascular tone in relatively small concentrations. Secondly, vasopressin is the most important single factor controlling the renal excretion of water. The renal effects of vasopressin are mediated by adenyl cyclase-linked receptors (V_2 receptors) in the basolateral membranes of the cells of the collecting duct. Intracellular cyclic AMP activates protein kinase(s), as a result of which

	1 2 3 4 5 6 7 8 9
Arginine vasopressin	CYS–Tyr–Phe–Gln–Asn–Cys–Pro–Arg–Gly–NH_2 (SS bridge 1–6)
Lysine vasopressin	CYS–Tyr–Phe–Gln–Asn–Cys–Pro–Lys–Gly–NH_2 (SS bridge 1–6)
Oxytocin	CYS–Tyr– Ile– Gln–Asn–Cys–Pro–Leu–Gly–NH_2 (SS bridge 1–6)

FIGURE 1. Amino acid sequences of vasopressin and oxytocin.

water-permeable aggregates of membrane-associated proteins are inserted into the apical membrane. Incorporation of water-permeable patches allows water to be reabsorbed until osmotic equilibration between the tubular fluid and the hyperosmolar interstitium of the renal medulla occurs. Oxytocin seems to play important facilitatory roles in the complex processes of reproduction, particularly in labor and lactation (for a review, see Reference 2). It has been known since the 1920s that neurohypophysial extracts cause contraction of the uterus, particularly at term, and the myoepithelial cells of the lactating mammary gland. However, the hormone is not an essential requirement, as parturition may take place and lactation occur under conditions where changes in oxytocin secretion appear unlikely, and the precise role of circulating oxytocin in these processes has been the subject of numerous studies. A number of other effects have been investigated; for example, facilitation of sperm production and transport, renal water retention and sodium excretion, and influence on the secretion of other hypothalamo-hypophysial hormones or factors. Interestingly, both vasopressin and oxytocin stimulate ACTH secretion, most likely through the hypophysial portal veins.[3] However, the physiological significance of these other effects are at present uncertain.

The review will focus on the integrated aspects of the degradation and excretion of vasopressin and oxytocin. Hormone masses and activities not originally reported in SI units have been converted into moles (mol) and concentrations into moles per liter (mol/l, *M*), even though this represents an approximation in several situations. Molecular weights of arginine vasopressin and oxytocin are 1084 and 1007 g/mol, respectively. Conversion factors of 2.25 $\times 10^{-9}$ mol/IU for vasopressin and 1.70 $\times 10^{-9}$ mol/IU for oxytocin have been applied where necessary.[4]

II. VASOPRESSIN

A. CLEARANCE RATES, $T_{1/2}$ AND V_D

For a number of years, it has been possible to measure the concentration of vasopressin in plasma with an accuracy sufficient to allow reasonable estimates of vasopressin turnover. However, with the exception of very few papers (see below), the conceptual framework within which these results have been treated is rather conventional: the concentration in plasma is dependent on three factors: the rate of secretion, the apparent volume of distribution (V_d), and the rate of removal from plasma (see Chapter 2). In contrast to several other hormones, the rate of secretion of vasopressin has not been determined directly in any experimental set-up, but it is generally accepted that the rate of secretion is the regulated variable in the very precise control of body fluid osmolality. The V_d has been considered to be relatively constant and approximately equivalent to the volume of the extracellular fluid, but influenced by the degree of protein and receptor binding. The rate of removal from plasma has almost exclusively been described in terms of the plasma clearance of hormone. The plasma clearance rate is the

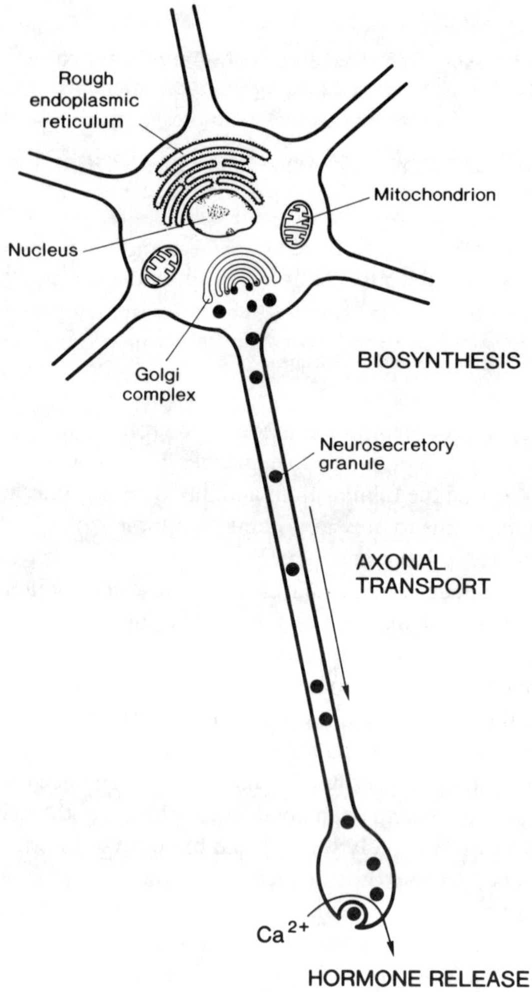

FIGURE 2. Schematic overview of neurosecretory neuron. (Simplified from Reference 1.)

removal of hormone per unit time (mol/min) divided by the plasma concentration (mol/l). Most frequently — and in this review — plasma clearance rate is given the dimension of $ml \cdot min^{-1} \cdot (kg\ b.wt)^{-1}$. Approximate values have been calculated under steady-state conditions during infusion experiments (i.e., constant plasma concentration and presumably constant V_d of hormone) in which the rate of removal must be equal to the rate of the infusion. Under these conditions, the plasma clearance rate equals the rate of infusion divided by the plasma concentration, assuming that the endogenous contribution to the latter variable is negligible.

The plasma clearance rate is the sum of all regional clearances and consists of metabolic clearance and urinary clearance. The metabolic clearance is mediated by enzymatic processes taking place mainly in the liver, splanchnic area, and kidney.[5,6] Some metabolic degradation of hormone takes place extracellularly by cystineaminopeptidase but the quantitative importance of this aspect is still unclear (see below). Since vasopressin is a relatively small molecule, it can easily pass the glomerular membrane in the kidney, limited only by the degree of protein binding. The magnitude of urinary clearance is determined by the balance between glomerular filtration and tubular reabsorption and secretion. By definition, urinary clearance is urinary rate of excretion (urine flow times urinary concentration) divided by plasma concentration.

Another point contributing to variation in plasma clearance rates deserves consideration. In a study by Os et al.[7] evidence for a substantial extraction of vasopressin by peripheral tissues was obtained in a study on 13 normotensive 50-year-old men. An arterio-venous difference of 3.2 pmol/l was observed over the forearm vascular bed in the supine position. After 30 min in the upright position, the arterio-venous difference decreased some 30% and was no longer statistically significant. Thus, application of peripheral venous plasma as reference in the clearance calculation, instead of (well-mixed) arterial plasma, may contribute to overestimation of the plasma clearance. In case of local tissue extraction, as suggested by the data of Os et al.,[7] peripheral venous plasma will reflect not only whole-body metabolism, but also the local degradation (see Chapters 2 and 3). In this way, a plasma sample obtained from a forearm vein may have a peptide concentration lower than the value presented to the metabolic sites. Consequently, the plasma clearance rate calculated from such data may constitute an overestimation.

Most frequently, the total elimination of hormone from plasma has been characterized only by the half-life ($t_{1/2}$); that is, the time it takes for the plasma concentration to be reduced by 50%. Vasopressin is usually considered to belong to the group of hormones for which the elimination within certain limits can be approximated by a monoexponential decay (first-order kinetics). It is a simple mathematical consequence of this approximation that the rate constant of elimination, k, is related to the half-life by $t_{1/2} = \ln 2/k$. Half-life is a function of both clearance and volume of distribution ($T_{1/2} \simeq \ln 2 \cdot V_d/\text{clearance}$) and is usually estimated directly from a multiple-point disappearance curve.

B. CONCENTRATION-DEPENDENT HORMONE CLEARANCE

A large number of reports deal with the factors controlling the rate of secretion of vasopressin. Plasma clearance rate and changes in catabolism have not received similar attention. However, it is quite possible that changes in the rate of removal of vasopressin might be associated with changes in plasma vasopressin, of importance to urinary concentration. There is evidence that the kidney is sensitive to extremely small changes in the plasma concentration of vasopressin[8-13] and, at least theoretically, such minute changes in the concentration of vasopressin might easily be caused by alterations in plasma clearance rate.

It is quite remarkable that there is still considerable disagreement with regard to the relation between plasma clearance rate and plasma concentration of vasopressin, although a number of experiments have been performed to investigate this relationship. Before the development of sensitive radioimmunoassays plasma clearance rate and V_d were usually estimated by use of monoiodinated vasopressin or by bioassay techniques. The relatively low sensitivity of bioassays necessitated the infusion of supraphysiological doses of vasopressin, and different laboratories have provided markedly discrepant results.[14] V_d varied from half to twice the plasma volume with a $t_{1/2}$ ranging from 1.2 to 42 min. More recently, Matsui et al.[15] infused vasopressin to anesthetized dogs at rates of 225, 900, and 1800 $\text{fmol} \cdot \text{min}^{-1} \cdot \text{kg}^{-1}$ and found that plasma clearance rate, urinary clearance, or renal organ clearance did not measurably change within this range of infusions. With regard to plasma clearance rate, this result is compatible with the data obtained by Bie et al.[16] from infusion of vasopressin at 92, 185, and 369 $\text{fmol} \cdot \text{min}^{-1} \cdot \text{kg}^{-1}$, and those of Montani et al.[17] who infused vasopressin in doses of 40 to 5000 $\text{fmol} \cdot \text{min}^{-1} \cdot \text{kg}^{-1}$. Similarly, Davison et al.,[18] in a recent study, did not observe changes in plasma clearance rate during a step-up infusion of vasopressin to hydrated women (doses of 22, 38, and 66 $\text{fmol} \cdot \text{min}^{-1} \cdot \text{kg}^{-1}$ preceded by priming doses). Likewise, Crofton et al.[19] found that the plasma clearance rate of vasopressin did not change in rats when vasopressin was infused stepwise in four rates ranging from 360 to 2800 $\text{fmol} \cdot \text{min}^{-1} \cdot \text{kg}^{-1}$ for 30 min periods. However, the results of several studies are in disagreement with these conclusions. In a recent study, Sondeen et al.[20] infused vasopressin in rates of 90, 225, and 450 $\text{fmol} \cdot \text{min}^{-1} \cdot \text{kg}^{-1}$ to anesthetized dogs and found that the nonurinary plasma clearance rate was markedly lower with the low dose than with the two higher doses which were associated with indistinguishable

TABLE 1
Total Clearance Rate (TCR) of Vasopressin

Species	Anesthesia	Hydration (ml/kg)	Infusion rate of vasopressin ($fmol\cdot min^{-1}\cdot kg^{-1}$)	TCR [ml/(min·kg)]	Ref.
Rat	Ether	None	360—2,800	40—60	19
Dog	Urethane/ chloralose	None	225	5—6	27
Dog	Pentobarbital	40—60	113—3,600	29—44	22
Dog	None	50	76—1,225[a]	20—37	21
Dog	None	None	40—5,000	12	17
Dog	Pentobarbital	None	1,125[a]	13	36
Dog	Pentobarbital	None	225—1,800	16	15
Dog	None	20	92—369	46	16
Dog	None	None	90—450	6—10	20
Human	None	20	8,700—87,000[d]	4	30
Human	None	20	1,800,000[d]	8	28
Human	None	20	22—788	7—18	23
Human	None	20	22—66[a]	750[c]	18

[a] Preceded by priming dose
[b] Single injection studies, fmol/kg.
[c] ml/min (body weight not available).
[d] ^{125}I-labeled vasopressin.

clearance rates of about 8 $ml\cdot^{-1}\cdot kg^{-1}$. In contrast, Weitzmann and Fisher[21] observed that vasopressin clearance *decreased* progressively with increasing infusion rates toward a plateau value of approximately 20 $ml\cdot min^{-1}\cdot kg^{-1}$ in conscious dogs. Similarly, Akatsuka et al.[22] concluded that in anesthetized dogs the clearance decreased with increasing plasma concentrations of vasopressin. Vasopressin was infused at rates of 113 to 3600 $fmol\cdot min^{-1}\cdot kg^{-1}$ and, when the dose was increased from 450 to 900 $fmol\cdot min^{-1}\cdot kg^{-1}$, the clearance decreased from approximately 40 to 30 $ml\cdot min^{-1}\cdot kg^{-1}$. In a study in normal human subjects, Moses and Steciak[23] concluded that both total metabolic and urinary clearance *increased* linearly when they infused vasopressin stepwise in doses of 22 to 788 $fmol\cdot min^{-1}\cdot kg^{-1}$. The very distinct discrepancies between these results are not immediately intelligible. In all the mentioned clearance studies (Table 1), vasopressin was infused for 60 min or more, except for the study by Weitzman and Fisher[21] who used 40-min infusions and Akatsuka et al.[22] who used 45-min infusions. These periods seem adequate for establishment of new steady states with regard to plasma concentration of vasopressin.[15,17,23] Absence of steady state is, therefore, an unlikely explanation of the observed differences. Moreover, the inconsistencies cannot be explained by species differences, since experiments on the same species turn out differently. As suggested by Robertson,[24] the wide variation in the results of half-life determinations could, at least partly, be due to a failure to distinguish between the phase of distribution and that of degradation. It is evident that $t_{1/2}$ estimated from data affected by the initial distribution will be smaller than the one measured from the elimination phase. From results in anesthetized rabbits, King et al.[25] recently concluded that the disappearance of vasopressin involved two components; a rapid component with a $t_{1/2}$ of 0.9 min and a slower component with a $t_{1/2}$ of 5.4 min, supporting the notion of Robertson. Another factor suggested by the latter and to some extent explaining the inconsistent results is the interindividual variation in turnover. In a group of normal adults studied under identical conditions $t_{1/2}$ varied between 8 and 32 min with a mean of 16.1 min. Data originating from the use of iodine-labeled vasopressin has been difficult to interpret. Klein and Roth[26] previously showed that the half-life of ^{131}I-vasopressin determined by single-injection studies in man and dog is longer than the half-life of unlabeled vasopressin. Later, Shade and Share[27] confirmed this finding and concluded that ^{131}I-vasopressin clearance calculated from data obtained by single injection cannot be used as a quantitative measure of the clearance of unlabeled vasopressin. Finally, Weitzman and Fisher[21] demonstrated that in conscious dogs the apparent volumes of distribution and metabolic clearance rates of monoiodinated and unlabeled vasopressin were markedly different, wherein unlabeled vasopressin was cleared more than five times faster than the iodinated hormone. The estimates of V_d for unlabeled vs. labeled vasopressin were approximately 13 and 7% of b.wt., respectively. Benmansour et al.[28] administered bolus injections of synthetic immunoreactive vasopressin (1.8 nmol) to 16 healthy volunteers and found that the elimination of vasopressin could be interpreted as being in accordance with a two-compartment system. They concluded that the volume of fast distribution was 219 ml·(kg b.wt.)$^{-1}$, that is, roughly equivalent to the extracellular fluid volume. Similar results in man were obtained by Maxwell et al.[29] who observed that the decline in plasma vasopressin after a bolus injection of 2.3 pmol/kg had two distinct phases. They calculated V_d to be of the order of 180 to 240 ml·(kg b.wt.)$^{-1}$.

V_d is dependent on protein and receptor binding. A variety of methods have been used to study this question, including ultrafiltration, dialysis, gel filtration, and electrophoresis. Older data ranging from complete to no plasma binding have been reported in humans (for a review, see Reference 14). In a later investigation, Baumann and Dingman[30] studied protein binding of 4.6 fmol of ^{125}I-vasopressin per milliliter plasma from hydrated normal subjects by gel chromatography. Their data suggest that monoiodinated vasopressin is not appreciably bound to plasma protein at physiological concentrations at 37°C. However, the degree of protein binding of ^{125}I-vasopressin may be quite different from that of genuine vasopressin. Using an experimental model where blood from anesthetized dogs was pumped through the dialysis coil of an artificial kidney, Share et al.[6,31] investigated whether plasma vasopressin is freely filterable. Binding of vasopressin to protein averaged 12% at plasma concentrations from 4.5 to 45 pmol/l, and 40% when plasma levels exceeded 45 pmol/l. They concluded that at normal plasma concentrations glomerular filtration of vasopressin is restricted only to a minimal extent by protein binding, but such binding at high plasma levels may result in a substantial reduction of glomerular filtration of the hormone.

C. SYSTEM DYNAMICS APPROACH

A more comprehensive mathematical treatment of pools and rate constants of vasopressin distribution, binding, and degradation was attempted more than a decade ago by Wilson et al.[32] They took advantage of the results, showing that ^{125}I-vasopressin does not have biological effects, most likely due to distortion at the molecular level because of the iodine moiety, and used ^{125}I-vasopressin to characterize the distribution and degradation of hormone occurring in the absence of binding to the high-affinity receptors, notably of the vascular endothelium and the basolateral membranes in various segments of the nephron. In accordance with the experimental data obtained by the same group, the system analysis indicated that plasma clearance was high when the ratio of free to occupied receptors was high; that is, at low plasma concentrations, and decreased to levels similar to those seen with ^{125}I-vasopressin when the receptors saturated.[32] Fitted with the data from conscious dogs, the model is compatible with a distribution of intact unlabeled vasopressin of 25% free in plasma, 19% on receptors, and 56% free in the extracellular volume. It also indicates that the plasma clearance rate is approximately 37 ml·min^{-1}·kg^{-1} at normal concentrations and that degradation of vasopressin in plasma and urinary excretion together accounts for less than 2% of total vasopressin disposal (Figures 3 and 4). However, it appears that the analysis with regard to urinary excretion was based exclusively on data from experiments with ^{125}I-vasopressin. It is conceivable that the renal handling of (genuine) vasopressin differs from that of iodine-labeled vasopressin; therefore, any model should preferably be based on data from both hormones.

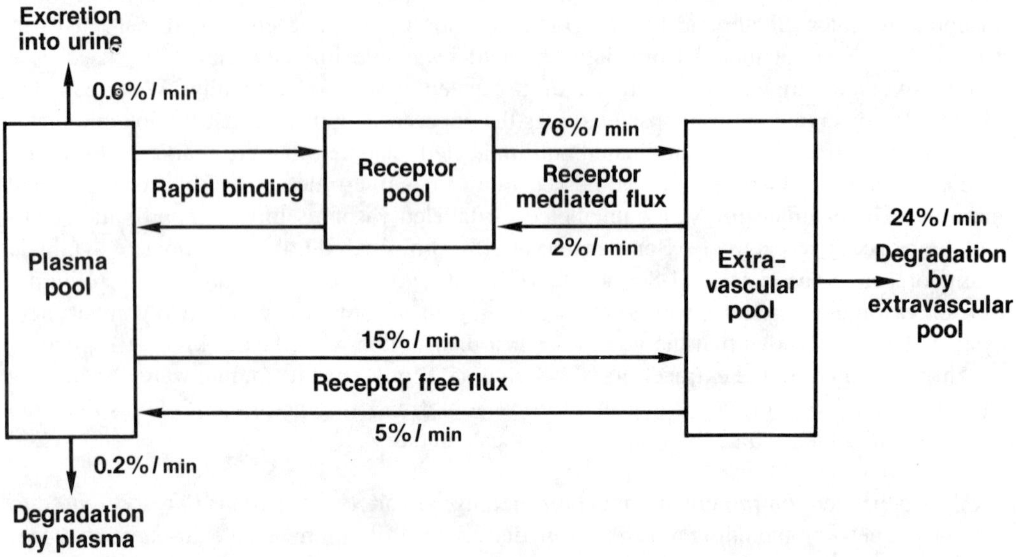

FIGURE 3. Model of vasopressin distribution and degradation. (Redrawn from Reference 32.)

Unfortunately, this potentially very powerful line of work does not appear to have been continued or taken up by others.

D. LIVER/SPLANCHNIC CIRCULATION

It has generally been assumed that vasopressin is cleared from the systemic blood mainly, if not totally, by the splanchnic viscera and the kidneys.[14,33] On the other hand, the relative contributions of the different organs or regions are not clarified, especially not in humans. Because of the technical difficulties in estimating the role of the liver and kidneys, most studies have been performed in anesthetized animals. General methods to measure organ clearance of vasopressin *in vivo* have been the direct and indirect application of the Fick principle. In the direct Fick method, extraction is calculated by simultaneous measurements of arterial and venous concentrations of hormone. The extraction fraction equals the difference between arterial and venous concentrations of hormone divided by the arterial concentration. Organ clearance is then calculated as the product of the extraction fraction and the organ plasma flow. This method requires high accuracy of the measurement of hormone concentration by either bio- or radioimmunoassay. In the indirect Fick method vasopressin is administered either to the systemic blood or to the blood supplying the organ of interest and pressor or antidiuretic responses are compared. The fraction removed by the organ can then be estimated by the decrease in the biological responses of the organ-administered dose. Finally, in the method of organ exclusion, clearance studies are made first in the intact animal and then after exclusion of the organ in question by ligation of the vessels. The latter two methods are relatively imprecise, and most organ clearance studies are now performed according to the direct Fick method.

To estimate the role of the liver in the metabolism of vasopressin, Lauson[14] reviewed data of hepatic extraction fractions and observed that in anesthetized rats values ranged from 0 to more than 50%. In dogs, hepatic removal of vasopressin was not observable, except in a single experiment on conscious animals where the hepatic extraction fraction was calculated to be 12%. Aziz et al.[34] studied the elimination of vasopressin by the normal and damaged liver in unanesthetized rats in water diuresis using the indirect Fick method; that is, caval vs. portal infusion of vasopressin, 36 fmol/min. In normal animals, extraction of vasopressin was not observed. In a subsequent study, Rabkin et al.[35] investigated the ability of the liver to remove

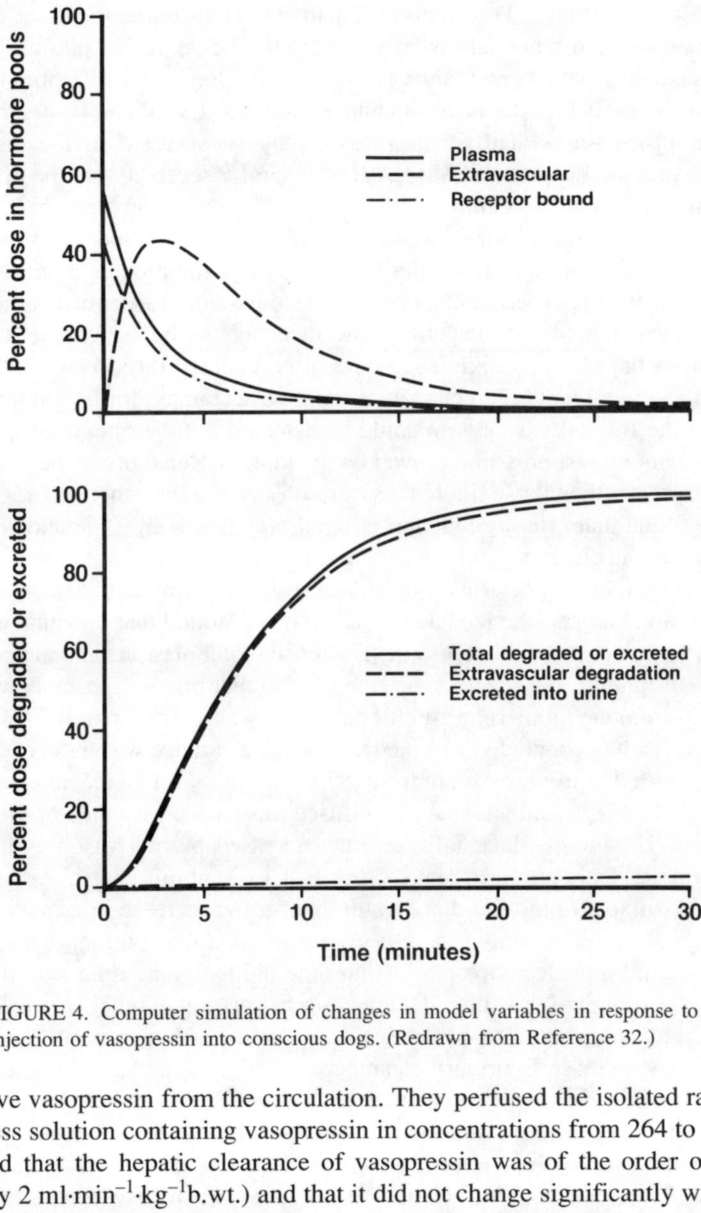

FIGURE 4. Computer simulation of changes in model variables in response to injection of vasopressin into conscious dogs. (Redrawn from Reference 32.)

immunoreactive vasopressin from the circulation. They perfused the isolated rat liver *in situ* with a bloodless solution containing vasopressin in concentrations from 264 to 20,315 pmol/l and observed that the hepatic clearance of vasopressin was of the order of 800 μl/min (approximately 2 $ml \cdot min^{-1} \cdot kg^{-1}$b.wt.) and that it did not change significantly with increasing inflow levels of vasopressin. They found no or only trace amounts of vasopressin in bile. Subsequently, Matsui et al.[5] investigated the total splanchnic clearance of vasopressin in the anesthetized dog and found evidence supporting prehepatic extraction. The prehepatic and total splanchnic clearances were approximately 1.6 and 3.0 $ml \cdot min^{-1} \cdot kg^{-1}$, respectively. The difference was attributed to removal by the liver. Vasopressin was not found in the bile. Further, they did not find it likely that the spleen contributes to the splanchnic extraction of vasopressin. Thus, according to these authors, a major part of the splanchnic extraction of vasopressin seems to occur in the mesenteric bed. Summarizing the role of the liver in the total removal of vasopressin, it must be concluded that our present knowledge is restricted to rather inconsistent results obtained from animal studies. Further investigations in this field are, therefore, desirable.

E. KIDNEYS

As mentioned above, the renal clearance of vasopressin accounts for a considerable fraction of the total plasma clearance. The relatively small size of vasopressin allows the molecule to pass the glomerular membrane and is only limited by the degree of plasma protein binding which, in most studies, have been found to be low. After glomerular filtration, vasopressin can either be excreted as such in the urine, it can be reabsorbed, or it can be degraded by tubular or peritubular processing. Finally, vasopressin may be secreted in the distal tubulus as indicated in a study by Kimura and Share.[36] It is generally accepted that the urinary clearance of intact hormone accounts for only a part of the total renal organ clearance. The role of the kidneys in the total removal of vasopressin was evaluated some years ago and it was concluded[14] to be of the order of one third to one half of the total clearance. However, the results from the different experiments were very inconsistent. Subsequently, additional information has been obtained with regard to the mechanisms by which the kidneys remove vasopressin from the blood. Rabkin et al.[37] perfused isolated rat kidneys and observed that only about 45% of the renal organ clearance could be accounted for by glomerular filtration. About 38% of the filtered vasopressin could be detected in the urine, corresponding to 18% of the total amount of vasopressin removed by the kidney. Renal organ clearance of amounts of vasopressin greater than those filtered was also observed in anesthetized dogs by Shade and Share.[38] They found that after a reduction in renal blood flow by 75%, glomerular filtration ceased by renal organ clearance did not decrease to zero but to 20% of control values; that is, from 2.8 to 0.6 $ml \cdot min^{-1} \cdot kg^{-1}$. Urinary clearance was approximately 70% of the renal organ clearance. In a study on anesthetized dogs, Matsui et al.[15] found that the renal organ clearance calculated for both kidneys made up some 27% of the total plasma clearance of vasopressin; that is, 4.4 of 16 $ml \cdot min^{-1} \cdot kg^{-1}$. Moreover, they found that urinary clearance was not significantly different from the glomerular filtration rate and was approximately 74% of renal organ clearance. Finally, they concluded that the renal organ clearance was independent of changes in plasma concentration of vasopressin from 9.7 to 123 pmol/l. In a recent clearance study on conscious dogs, Sondeen and Claybaugh[20] infused vasopressin in rates of 90, 225, and 450 $fmol \cdot min^{-1} \cdot kg^{-1}$. The intermediate infusion rate increased plasma vasopressin from 3 to 26 pmol/l and urinary clearance from approximately 1 to 2 $ml \cdot min^{-1} \cdot kg^{-1}$. Further elevation of plasma vasopressin to 49 pmol/l did not result in a further increase in urinary clearance. The plasma clearance showed a similar pattern. At all infusion rates, urinary clearance averaged 20 to 27% of the plasma clearance rate. At no time did the glomerular filtration rate change significantly. The design of this study did not allow estimation of renal organ clearance. Partly supporting results were obtained by Moses and Steciak[23] who, in normal human subjects, observed a steady increase in urinary clearance with increasing rates of infusion, elevating plasma vasopressin from approximately 6 to 55 pmol/l. Total metabolic clearance also increased with increasing doses and the fraction of total clearance accounted for by urinary excretion remained approximately constant (approximately 6%). Renal organ clearance was not measured. Benmansour et al.[28] compared the metabolism of vasopressin in a group of uremic patients with a group of normal subjects and found that the plasma clearance rate was 31% lower in the uremic patients. Their interpretation of this finding was that the kidneys are responsible for about two thirds of the plasma clearance rate. They calculated the renal organ clearance to be of the order of 333 ml/min; that is, a value considerably greater than the normal level of glomerular filtration rate (GFR). In a later study in humans from the same laboratory,[39] urinary clearance was estimated to be 7.5 ml/min or about 6% of GFR. Neither plasma clearance rate nor renal organ clearance was measured in this study.

Since the thorough review by Lauson in 1974,[14] most studies of the regional breakdown of vasopressin have focused on the renal clearance. However, despite the application of advanced technology (e.g., sensitive radioimmunoassays), the role of the kidneys in removal of vasopressin and particularly the mechanisms involved in this process are still not clarified. In the total removal of vasopressin, the kidneys have been found to account for about 60[28] and 30%[15] in man and dog, respectively. At least three studies[28,37,38] provide evidence that renal organ

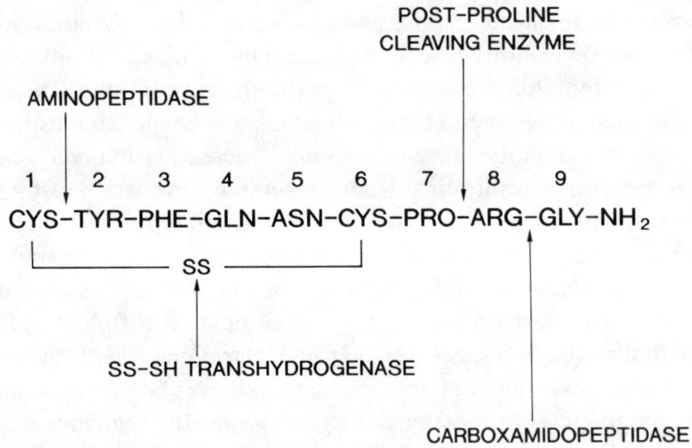

FIGURE 5. Sites of enzymatic cleavage of arginine vasopressin by renal enzymes. (Redrawn from Reference 40.)

clearance exceeds glomerular filtration, indicating a postglomerular degradation. Urinary clearance has been found to range between 18 and 74% of renal organ clearance[15,3738] and 5.6 and 27% of plasma clearance rate.[20,23,30] Moses and Steciak[23] and Sondeen et al.[20] observed in man and dog, respectively, that urinary clearance increased with increasing plasma levels of vasopressin without significant changes in GFR. In contrast, Matsui et al.[15] found that urinary clearance was independent of the plasma concentration of vasopressin. Recently, from a detailed study of renal degradation of vasopressin using ^{3}H-labeled vasopressin, intratubular microinjection, and HPLC,[40] it was concluded that luminally administered vasopressin was processed by two mechanisms, a major one involving hydrolysis by luminal aminopeptidase and a minor one involving endocytosis with transfer to lysosomes. Enzymatic cleavage of vasopressin was found to take place at several sites (Figure 5).

In conclusion, there is little doubt that the kidney plays an important role in the degradation of vasopressin, but the quantitative aspects are still rather poorly understood and need further investigation.

III. OXYTOCIN

A. QUANTITATIVE CONSIDERATIONS

As mentioned, the role of circulating oxytocin in parturition and lactation has been the subject of numerous investigations for decades. However, evaluation of the importance of oxytocin in normal pregnancy and parturition has been perplexed by the wide variation in the values reported for the concentration of oxytocin in plasma. Furthermore, physiological roles of oxytocin in the male and nonpregnant, nonlactating female have not been clearly defined. Plasma concentrations of oxytocin under basal conditions have now been measured by several laboratories using RIA techniques. The results are not very uniform. Amico et al. measured mean basal plasma concentrations of 3.0 ± 0.1 pmol/l in samples from normal men and 2.9 ± 0.1 pmol/l in samples from normal women.[41] Values were not significantly different. In a later study from the same laboratory, the plasma oxytocin in normal men was found to be slightly lower (0.3 to 1.7 pmol/l with a mean of 0.8 pmol/l),[42] possibly because of small changes in the assay technique. Leake and co-workers did not find significant differences between the plasma levels of oxytocin in normal male subjects, nonpregnant women, and pregnant women (greater than 37 weeks of gestation); corrected for recovery of the extraction procedure, values ranged between 1.0 and 1.4 pmol/1,[43] that is, similar to the latest results of

Amico et al. In contrast to these results, early radioimmunoassay studies of unextracted samples[44] had shown concentrations of several hundred picomoles per milliliter plasma during pregnancy. These data exemplify the variability in plasma concentration measurements most likely caused by differences between assay techniques. Especially in plasma from pregnant women, the high levels of oxytocin measured in unextracted plasma could be due to degradation of the tracer because of insufficient inactivation of cystine-aminopeptidase,[45] a plasma enzyme that inactivates oxytocin (see below). Recently, Amico et al.[46] have elucidated the importance of the specificity of the antibody against oxytocin used in the radioimmunoassay of extracted plasma and obtained results that indicate that skepticism is warranted. Plasma samples from 16 pregnant women were analyzed using two different antibodies against oxytocin. With the first antibody, a mean oxytocin concentration of 13.0 pmol/l was measured, whereas the corresponding value with the second antibody was one order of magnitude lower (1.5 pmol/l). Accordingly, it was suggested that in the plasma of pregnant women, there exists an oxytocin-like peptide that is recognized by the less-specific antibody only. Antibody specificity thus provides another technical reason for substantial differences between the results from different laboratories and this might, at least partly, explain the disparate results of plasma oxytocin measurements reported in the literature. The reliability of the information on the secretion, distribution, degradation, and excretion of oxytocin is crucially dependent on the specificity and precision by which oxytocin can be measured in plasma and urine. At present, the assay techniques cannot be assumed to provide reliable data on oxytocin metabolism without full documentation of assay performance. It appears that results obtained by radioimmunoassay of unextracted plasma should be evaluated with great caution and that antibody specificity should be thoroughly tested.

B. PROTEIN BINDING, V_d

Generally, the agreement between studies with regard to the degree to which plasma oxytocin is bound to protein is not impressive. Fabian et al.[47] found no evidence for protein binding of oxytocin in man at plasma concentrations between 85 and 680 pmol/l (whereas some 30% of vasopressin was bound). In rat plasma, the fraction of oxytocin bound to protein averaged 42% at plasma concentrations ranging from 85 to 28,900 pmol/l.[48] In the same study, the apparent V_d was calculated to be 73 ml/kg b.wt.; that is, a result between the plasma volume and the extracellular space. In a more recent study in man,[49] V_d was calculated to be approximately 300 ml/kg, a value considerably higher than the extracellular space and about fourfold higher than the results obtained in man by Fabian et al.[47] Again, differences in the experimental designs and assay techniques must be considered as the most probable reasons for these discrepancies. Fabian et al. used a relatively high dose, single injection, and bioassay, while in the study of Dawood et al.,[49] lower doses, constant infusion, and radioimmunoassay were applied. To fully elucidate the size of the apparent V_d and the degree of protein binding of oxytocin, further investigations seem to be required.

C. CLEARANCE RATE, $t_{1/2}$

Clearance studies of oxytocin are much less numerous than in the case of vasopressin. In the extensive review by Lauson,[14] a number of studies were evaluated but only two reports of plasma clearance rate of oxytocin in man obtained by the constant infusion method were mentioned. Fabian et al.[47] infused synthetic oxytocin (850 pmol/min) to 5 males and 2 females for a period of 25 min and calculated the plasma clearance rate on the basis of bioassay of plasma oxytocin. Plasma clearance rate was found to be of the order of 1500 ml/min (on the order of 20 ml/kg b.wt./min); however, as stated by Lauson this value may be somewhat overestimated since the samples for bioassay were collected only 22 min after the start of infusion; that is, it is likely that steady-state conditions were not achieved. In subsequent studies, it has been found[50,51] that steady-state conditions are obtained after approximately 40 min of constant infusion. In the other study by Chard et al.,[52] oxytocin was infused for 60 min at a rate of 680 pmol/min to 3 normal men, and plasma concentrations of oxytocin were measured by radioimmunoassay. According to Lauson,[14] the plasma clearance rate could be calculated to average 1000 ml/min when oxytocin concentrations were corrected for recovery. In contrast to the plasma clearance rate value obtained by Fabian et al., this might be slightly underestimated because of unspecific binding of the antibody used by Chard et al.

Apparently, only few clearance studies of oxytocin were performed on human subjects in later years. Dawood et al.[49] measured plasma clearance rates using the constant infusion method in eight normal adult males. Oxytocin was administered either as a step-up infusion for 1 h (54 to 435 pmol/min) or at a constant rate of 225 pmol/min. The mean plasma clearance rate of oxytocin was about 22 $ml \cdot min^{-1} \cdot kg^{-1}$. Similar results were obtained by Leake et al.[43] via infusion of synthetic oxytocin at a rate of 6.8 pmol/min to 6 men, and 6 nonpregnant and 10 pregnant women (greater than 37 weeks of gestation). Values were approximately 27, 21, and 23 $ml \cdot min^{-1} \cdot kg^{-1}$, respectively, and were not statistically different. Somewhat lower values have recently been obtained in a clearance study of men by Amico et al.[42] Oxytocin was infused in a stepwise fashion (approximately 5, 10, and 20 pmol/min), each dose being infused for 90 min. The plasma clearance rate was of the order of 16 $ml \cdot min^{-1} \cdot kg^{-1}$ and apparently independent of the rate of infusion. The notion that the plasma clearance rate over a broad range is independent of the plasma concentration of oxytocin is supported by the fact that, in the studies mentioned, oxytocin has been infused at rates from 5 to 850 pmol/min without the appearance of markedly different values of plasma clearance rate.

There is evidence that clearance of oxytocin in women may be dependent on the oestrogenic and gestagenic status. Years ago, Rydén et al.[53] investigated $t_{1/2}$ of oxytocin in three groups of women: in normally menstruating women during different phases of the menstrual cycle, in women using oral contraceptives, and in pregnant women in the first and second trimester. $t_{1/2}$ was found to be higher in the proliferative phase than in the secretory phase (4.5 ± 0.2 vs. 3.7 ± 0.2 min). In the group using oral contraceptives, $t_{1/2}$ was somewhat lower (3.3 ± 0.2 min) and in the group of pregnant women $t_{1/2}$ was low in the first trimester (3.0 ± 0.2 min) and increased markedly in the second trimester (4.9 ± 0.8 min, week 14 to 17). However, these results may not provide an accurate picture of the metabolism of endogenous oxytocin. Firstly, the values of $t_{1/2}$ were calculated from the disappearance curves of tritium as ^3H-labeled oxytocin was used; as described above, this may not provide pharmacokinetic data identical to those obtainable by use of the genuine hormone. Secondly, blood for oxytocin measurement was sampled within the first 5 min after injection; that is, most likely at a time where the plasma concentration was affected by the process of distribution. Results from studies on the relation between female hormonal status and the disposal of oxytocin seem to be very rare.

D. ORGAN CLEARANCE

The organs inactivating oxytocin in man have not been clearly defined but, like the degradation of vasopressin, the liver and the kidneys seem to play a major role. Most of the studies of organ clearance have been performed on rodents using the method of organ exclusion. Although the results are somewhat conflicting, it appears that about two thirds of the total clearance, at least in rodents, is accounted for by degradation in the liver and kidneys (for review, see Reference 14). In man, only a minor fraction of a given dose of oxytocin seems to be excreted in the urine.[14] In a recent study in man,[42] it was concluded that the absolute rates of urinary excretion were positively correlated with plasma levels of oxytocin but, in accordance with earlier reports, only a small fraction (<1%) of oxytocin was cleared into the urine as intact hormone. Years ago, the mammary glands in rats were found to be responsible for more than 20% of the total degradation (see Reference 14); therefore, it is possible that the mammary glands are not only the target organ but also contribute to the total clearance of oxytocin. In recent years, little progress seems to have been made with regard to elucidation of the regional clearance of oxytocin.

E. CYSTINEAMINOPEPTIDASE (OXYTOCINASE/VASOPRESSINASE)

Cystineaminopeptidase (CA-peptidase) is an enzyme which appears in maternal plasma during pregnancy; only extremely low levels of CA-peptidase are present in the circulation of men and nonpregnant women (for review, see Reference 14 and Chapter 4). The enzyme is found only in primates. It is able to inactivate the neurohypophysial hormones and is, therefore, also called oxytocinase or vasopressinase. The modes of degradation of oxytocin and vasopressin by CA-peptidase can be summarized as follows: initially, the Cys^1-Tyr^2 peptide bond is hydrolyzed, in this way breaking the hexapeptide ring of the hormone without preceding reduction of the disulfide bond. The enzyme then successively hydrolyzes the peptide bonds between residues in position 2 to 6, finally leaving a pentapeptide (original residues 1-S-S-6-7-8-9) which is resistant to CA-peptidase. The main source of the CA-peptidase is usually thought to be the placenta,[54] although small amounts have been isolated from a number of tissues (e.g., liver, kidney, pancreas, mammary gland, uterus, ovary, pituitary, and hypothalamus). Only trace amounts of CA-peptidase are found in fetal blood, indicating that the fetus is not the source of the enzyme (e.g., Reference 43).

The appearance of large amounts of CA-peptidase in the maternal plasma during pregnancy naturally raises the question of whether clearance of the neurohypophysial hormones is changed in this period. Controversial results have been obtained. In the study of Rydén,[54] CA-peptidase activity in about 700 blood samples from nonpregnant and pregnant women was investigated (by artificial substrate) and a 50-fold increase of the enzyme activity during pregnancy was observed. In accordance with this finding, Rosenbloom and co-workers, using vasopressin as substrate, observed a 30- to 60-fold increase in CA-peptidase activity from 7 weeks of gestation to near term.[55] Recently, Viinamäki et al.[56] also used vasopressin as substrate and found a 3- and 20-fold increase in enzyme activity in the first and third trimesters, respectively, compared to the nonpregnant level. Leake et al.[43] used oxytocin as substrate and found an 85% decrease in oxytocin concentration in plasma of pregnant women over 1 h of incubation (37°C). Significant degradation was not observed in plasma from normal men, nonpregnant women, or from cord blood. Despite the high activity of CA-peptidase in pregnancy plasma, the plasma clearance rates obtained during pregnancy were similar to those of nonpregnant women and men,[43] indicating somewhat unexpectedly that large amounts of CA-peptidase activity does not measurably affect the plasma clearance of oxytocin. This is supported by the results of Amico et al.[50] who were unable to detect any difference in plasma clearances in normal men compared to pregnant women during hypocontractile labor. In contrast to these findings indicating that clearance of *oxytocin* is relatively constant during pregnancy, clinical data indicate that disposal of vasopressin may be accelerated;[57] however, in this report, clearance rate of infused vasopressin was not available. Other studies provide evidence that the clearance of vasopressin is altered during pregnancy. Davison et al.[58] investigated women late in the gestational period and observed that the clearance of vasopressin was 4 to 5 times that observed 8 to 10 weeks postpartum. This observation was further investigated in a rather extensive study[18] in which vasopressin clearance was measured serially in five women before conception, during early, mid, and later pregnancy, and again post-partum. Plasma CA-peptidase activity, measured as degradation of ^{125}I-vasopressin tracer, was found to increase markedly by mid- and slightly more by late gestation from an undetectable level at gestational week 7 to 8. In accordance with the data of the previous investigation, clearance rates of vasopressin in mid- and late pregnancy were fourfold those measured either before conception, early in gestation, or post-partum. The increased plasma clearance rate of vasopressin observed in these studies is most likely due to the increased levels of CA-peptidase in plasma.[18,58] These elaborate studies thus show that vasopressin degradation is accelerated during pregnancy and seems to lead to the question of whether similar careful studies will be able to revise previous conclusions with regard to oxytocin metabolism.

IV. PATHOPHYSIOLOGICAL COMMENTS

A. GENERAL QUANTITATIVE ASPECTS

The control system regulating plasma osmolality and fluid volume through the renal actions of vasopressin includes several high-gain transfer functions and acts normally with remarkable precision. Plasma osmolality is mainly determined by the body contents of water and sodium. The regulation of water excretion is fast compared to the control of sodium excretion. In this setting even substantial disturbances in vasopressin disposal will hardly be noticeable; for example, if a sudden reduction occurred in the rate at which vasopressin is metabolized, this would elicit small and transient increases in plasma vasopressin and in renal water retention; the ensuing minute decreases in osmolality will reduce the rate of vasopressin secretion so that the increased plasma concentration is adjusted towards normal. *A priori,* it would thus be assumed that the appearance of symptoms from disruption in vasopressin disposal would require a massive change in the disposal machinery. Clinical conditions in which this is a prominent feature are rare (see below). On the other hand, if progressive decreases in the rate of degradation of vasopressin do occur to a significant extent, the dynamics of the system would be measurably changed; that is, half-life is increased prior to the appearance of significant steady-state deviations. In addition, radioimmunoassays are usually not designed to work with optimal precision at subnormal plasma concentrations. Disruption of vasopressin-degrading pathways could, therefore, participate in explaining clinical conditions including (1) inadequate decrease in plasma vasopressin within the time of a standard water-loading procedure of a few hours, (2) delayed excretion of the water load, and (3) normal sensitivity (threshold and slope) of vasopressin secretion in response to a positive osmotic stimulus.

With regard to oxytocin, the function of the regulatory mechanisms are less precisely described and knowledge of pathological conditions elicited by changes in oxytocin metabolism is even more scant than for vasopressin.

B. "INAPPROPRIATE" SECRETION OF VASOPRESSIN

It is important to realize in this context that it is not appropriate to label any rate of secretion of vasopressin as inappropriate unless secretion takes place against the concerted action of all major stimuli, notably plasma osmolality, cardiac filling pressures, and arterial pressure. Impairment of renal water excretion and reduction in dilution capability may be associated with a great number of clinical conditions, including sodium depletion, postsurgical stress, changes in glomerular filtration rate and renal hemodynamics, hypovolemia, use of diuretics and other drugs, glucocorticoid deficiency, chronic congestive heart failure, cirrhosis with ascites, and several carcinomas. In some cases, the mechanism is quite well elucidated. Hypovolemia and (relative) arterial hypotension normally cause secretion of vasopressin and may counterbalance osmotic stimuli. The hyponatremia and hypervolemia occurring with some cases of carcinoma was suspected to be due to an excess of vasopressin, as deduced from comparison with results of long term administration of vasopressin to normal subjects;[59] the condition is called Schwartz-Bartter syndrome or "syndrome of inappropriate antiduretic hormone secretion", abbreviated SIADH. In subsequent investigations, vasopressin was found in increased amounts in plasma and in carcinoma tissue[60] and changes in vasopressin degradation is not expected. Under other conditions, the relation between hormone administration/ secretion and elimination is less clear. Beginning in the 1960s, a number of reports have appeared on water intoxication (i.e., lack of diluting capability) as a complication to the use of oxytocin as a uterotonic agent. Even maternal death has been described.[61] The renal antidiuretic effect of oxytocin has usually been held responsible. However, it would not be surprising if the physiologically large doses of oxytocin administered to these patients interfered with the degradation of vasopressin liberated in unusual amounts due to stress and pain under these conditions. The kidney is so much more sensitive to vasopressin than to oxytocin[62] that it seems unlikely that oxytocin per se, even in the applied doses, exerts this dramatic antidiuretic effect. In chronic congestive heart failure, hyponatremia is quite commonly seen and it has been shown that plasma from the majority of these patients contain vasopressin in amounts inappropriate for *the concomitant plasma osmolality;*[63-66] generally, the authors agree that plasma vasopressin probably is controlled by nonosmolar factors, although the investigations failed to identify reasonable candidates. On one side, it is emphasized[63] that the data do not necessitate the assumption of changed functional characteristics of the osmoreceptor mechanism; on the other, Goldsmith and co-workers[65] found that plasma vasopressin did not change as expected in response to hemodynamic stimuli. Only one paper[65] mentions that the increased levels of vasopressin in congestive heart failure could (also) arise from decreased clearance of hormone secreted at a normal rate. However, vasopressin degradation was not quantified.

C. DIABETES INSIPIDUS

In diabetes insipidus, polyuria, excessive thirst, and polydipsia occur as a consequence of lack of renal concentrating ability. In central diabetes insipidus, this is caused by decreased amounts of vasopressin in plasma, secondary to an inadequate response of the neurosecretory cells to normal physiological stimuli. However, in a few cases, the vasopressin deficiency may be explained by a marked increase in vasopressin degradation so that the vasopressin secreted, even in increased amounts in response to augmented stimuli, does not reach the kidney in sufficient quantities. The transient diabetes insipidus sometimes seen during pregnancy[57,67] may represent an example of symptoms due to exaggerated hormone degradation. Dürr described a unique example of this condition.[68,69] The polyuria was apparently not due to lack of vasopressin, as urinary concentrating ability was adequate before and after the pregnancy. Neither were the symptoms due to lack of renal responsiveness, as urine osmolality promptly increased in response to dDAVP, which is a CA-peptidase-resistant analog of vasopressin with antiduretic agonist activity. In addition, it was shown[68,69] that the post-partum level of CA-peptidase was one order of magnitude higher than normally seen at term, and that injection of patient plasma to an assay rat elicited a decrease in urine osmolality and an increase in urine flow with no change in renal solute excretion (indicating inactivation or inhibition of endogenous vasopressin). This condition was, therefore, most likely due to exaggerated production of CA-peptidase and its existence indicates that the normal pregnancy may well include a latent vasopressin insufficiency due to accelerated vasopressin degradation. However, it is important to note that pregnancy-associated polyuria may also exhibit the characteristics of ordinary central diabetes insipidus[70] or may even appear to be caused by a primary change in the thirst threshold.[71] Evaluation of abnormal water handling during pregnancy is thus rather complex and must involve all aspects of fluid homeostasis.

V. CONCLUSION

The separate antidiuretic/vasopressor and oxytocic activities of neurophypophysial extract have been known since the late 1920s, synthetic hormones have been available since the 1950s, and highly sensitive radioimmunoassays capable of measuring plasma concentrations were developed in the early seventies. Vasopressin is universally recognized as the pivotal hormone of body water regulation, (i.e., volume control); oxytocin is an important and widely used tool for the obstetrician and likely to be important in other contexts as well. Based on this background, it is somewhat surprising that the rate of secretion, the distribution, and the degradation of these hormones are rather poorly elucidated. Estimates of the apparent volume of distribution in the normal mammalian organism are variable to an extent which makes correlations to physical features (fluid compartments, receptor pools, etc.) less meaningful. Metabolic clearance rates obtained in seemingly sound investigations are (mostly) within the wide range of 5 to 50 ml·min^{-1}·kg^{-1}. Disagreement prevails in the literature even with regard to qualitative issues. For example, the direction of change, if any, of metabolic clearance rate with increasing plasma concentration of hormone and the use of the more comprehensive techniques of system analysis to describe the complex dynamics of the neurohypophysial hormones is the exception rather than the rule. Apparently, the task of investigating the integrative aspects of vasopressin and oxytocin is demanding and requires special efforts of the clinician, the physiologist, the assayist, and the system analyst. However, for those capable of combining the skills of these persons, the field would seem to provide ample amounts of important, unanswered questions.

ABBREVIATIONS

IU	International unit
$t_{1/2}$	Half-life
V_d	Apparent volume of distribution
b. wt.	Body weight
k	Rate constant of monoexponential disappearance
CA-peptidase	Cystineaminopeptidase

REFERENCES

1. **Bennett, G. W. and Whitehead, S. A.,** *Mammalian Neuroendocrinology*, Oxford University Press, New York, 1983, 12.
2. **Chard, T.,** Oxytocin. Physiology and pathophysiology, in *The Posterior Pituitary. Hormone Secretion in Health and Disease,* Baylis, P. H. and Padfield, P. L., Eds., Marcel Dekker, New York, 1985, 361.
3. **Watanabe, T., Oki, U., and Orth, D. N.,** Kinetic actions and interactions of arginine vasopressin, angiotensin-II, and oxytocin on adrenocorticotropin secretion by rat anterior pituitary cells in the microperfusion system, *Endocrinology,* 125, 1921, 1989.
4. **Stürmer, E.,** Vasopressin and oxytocin and synthetic analogues: the use of bioassays, *J. Pharmaceut. Biomed. Anal.,* 7, 199, 1989.
5. **Matsui, K., Share, L., Brooks, D. P., Crofton, J. T., and Rockhold, R. W.,** Splanchnic clearance of plasma vasopressin in the dog: evidence for prehepatic extraction, *Am. J. Physiol.,* 245, E611, 1983.
6. **Share, L., Kimura, T., Matsui, K., Shade, R., and Crofton, J. T.,** Metabolism of vasopressin, *Fed. Proc.,* 44, 59, 1985.
7. **Os, I., Kjeldsen, S. E., Aakesson, I., Skjøtø, J., Westheim, A., Eide, I., and Leren, P.,** Arteriovenous difference of plasma vasopressin in normal man and effect of posture, *Acta Physiol. Scand.,* 122, 49, 1984.
8. **Shannon, J. A.,** The control of the renal excretion of water. II. The rate of liberation of the posterior pituitary antidiuretic hormone in the dog, *J. Exp. Med.,* 76, 387, 1942.
9. **Lauson, H. D.,** The problem of estimating the rate of secretion of antidiuretic hormone in man, *Am. J. Med.,* 11, 135, 1951.
10. **Hollander, W., Jr., Williams, T. F., Fordham, C. C., and Welt, L. G.,** A study of the quantitative relationship between antidiuretic hormone (vasopressin) and the renal tubular reabsorption of water, *J. Clin. Invest.,* 36, 1059, 1957.
11. **McMurray, J., Forsling, M. L., and Struthers, A. D.,** Arginine vasopressin dissociates the diuresis and natriuresis due to atrial natriuretic factor in man, *Eur. J. Clin. Invest.,* 18, 627, 1988.
12. **Baerwolff, M. and Bie, P.,** Effects of subpicomolar changes in vasopressin on urinary concentration, *Am. J. Physiol.,* 255, R940, 1988.
13. **Andersen, L. J., Andersen, J. L., Schütten, H. J., Warberg, J., and Bie, P.,** Antidiuretic effect of subnormal levels of arginine vasopressin in normal man, *Am. J. Physiol.,* 259, R53, 1990.
14. **Lauson, H. D.,** *Handbook of Physiology. Endocrinology, The Pituitary Gland and Its Neuroendocrine Control,* Sec. 7, Vol. IV, American Physiological Society, Washington, D.C., 1974, chap. 14.

15. **Matsui, K., Share, L., Wang, B. C., Crofton, J. T., and Brooks, D. P.,** Effects of changes in steady state plasma vasopressin levels on renal and urinary vasopressin clearances in the dog, *Endocrinology*, 112, 2107, 1983.
16. **Bie, P., Munksdorf, M., and Warberg, J.,** Renal effects of overhydration during vasopressin infusion in conscious dogs, *Am. J. Physiol.*, 247, F103, 1984.
17. **Montani, J., Liard, J., Schoun, J., and Möhring, J.,** Hemodynamic effects of exogenous and endogenous vasopressin at low plasma concentrations in conscious dogs, *Circ. Res.*, 47, 346, 1980.
18. **Davison, J. M., Sheills, E. A., Barron, W. M., Robinson, A. G., and Lindheimer, M. D.,** Changes in the metabolic clearance of vasopressin and in plasma vasopressinase throughout human pregnancy, *J. Clin. Invest.*, 83, 1313, 1989.
19. **Crofton, J. T., Ratliff, D. L., Brooks, D. P., and Share, L.,** The metabolic clearance rate of and pressor responses to vasopressin in male and female rats, *Endocrinology*, 118, 1777, 1986.
20. **Sondeen, J. L. and Claybaugh, J. R.,** Clearance and urinary excretion of vasopressin in conscious dogs, *Am. J. Physiol.*, 256, R291, 1989.
21. **Weitzman, R. E. and Fisher, D. A.,** Arginine vasopressin metabolism in dogs. I. Evidence for a receptor-mediated mechanism, *Am. J. Physiol,* 235, E591, 1978.
22. **Akatsuka, N., Moran, W. H., Morgan, M. L., and Wilson, M. F.,** Effects of steady-state plasma vasopressin levels on the distribution of intrarenal blood flow on electrolyte excretion, *J. Physiol.,* 266, 567, 1977.
23. **Moses, A. M. and Steciak, E.,** Urinary and metabolic clearance of arginine vasopressin in normal subjects, *Am. J. Physiol.*, 251, R365, 1986.
24. **Robertson, G. L.,** The regulation of vasopressin function in health and disease, *Rec. Prog. Horm. Res.*, 33, 333, 1977.
25. **King, K. A., Courneya, C. A., Tang, C., Wilson, N., and Ledsome, J. R.,** Pharmacokinetics of vasopressin and atrial natriuretic peptide in anesthetized rabbits, *Endocrinology*, 124, 77, 1989.
26. **Klein, L. A. and Roth, J.,** Antidiuretic hormone metabolism studied by radio-immunoassay, *Surg. Forum*, 18, 289, 1967.
27. **Shade, R. E. and Share, L.,** Metabolic clearance of immunoreactive vasopressin and immunoreactive ^{131}I-iodo vasopressin in the hypophysectomized dog, *Endocrinology,* 99, 1199, 1976.
28. **Benmansour, M., Rainfray, M., Paillard, F., and Ardaillou, R.,** Metabolic clearance rate of immunoreactive vasopressin in man, *Eur. J. Clin. Invest.*, 12, 475, 1982.
29. **Maxwell, D., McMurray, S., Szwed, J., Shelton, R., and Robertson, G.,** The effect of distribution and clearance on plasma vasopressin in man, *Clin. Res.,* 24, 407A, 1976.
30. **Baumann, G. and Dingman, J. F.,** Distribution, blood transport, and degradation of antidiuretic hormone in man, *J. Clin. Invest.,* 57, 1109, 1976.
31. **Share, L. and Crofton, J. T.,** Is vasopressin in the circulation of the dog freely filterable?, *J. Endocrinol.*, 86, 501, 1980.
32. **Wilson, K. C., Weitzman, R. E., and Fisher, D. A.,** Arginine vasopressin metabolism in dogs. II. Modeling and system analysis, *Am. J. Physiol.*, 235, E598, 1978.
33. **Ginsburg, M. and Heller, H.,** The clearance of injected vasopressin from the circulation and its fate in the body, *J. Endocrinol., 9, 283, 1953.*
34. **Aziz, O. and Schmidt, W.,** Elimination of vasopressin by the normal and the damaged liver, *Pflügers Archiv.,* 367, 165, 1976.
35. **Rabkin, R., Ghazeleh, S., Share, L., Crofton, J., and Unterhalter, S. A.,** Removal of immunoreactive arginine vasopressin by the perfused rat liver, *Endocrinology*, 106, 930, 1980.
36. **Kimura, T. and Share, L.,** Characterization of the renal handling of vasopressin in the dog by stop-flow analysis, *Endocrinology,* 109, 2089, 1981.
37. **Rabkin, R., Share, L., Payne, P. A., Young, J., and Crofton, J.,** The handling of immunoreactive vasopressin by the isolated perfused rat kidney, *J. Clin. Invest.*, 63, 6, 1979.
38. **Shade, R. E. and Share, L.,** Renal vasopressin clearance with reductions in renal blood flow in the dog, *Am. J. Physiol.,* 232, F341, 1977.
39. **Pruszczynski, W., Caillens, H., Drieu, L., Moulonguet-Doloris, L., and Ardaillou, R.,** Renal excretion of antidiuretic hormone in healthy subjects and patients with renal failure, *Clin. Sci.*, 67, 307, 1984.
40. **Carone, F. A., Christensen, E. I., and Flouret, G.,** Degradation and transport of AVP by proximal tubule, *Am. J. Physiol.*, 253, F1120, 1987.
41. **Amico, J. A., Seif, S. M., and Robinson, A. G.,** Oxytocin in human plasma: correlation with neurophysin and stimulation with estrogen, *J. Clin. Endocrinol. Metab.*, 52, 988, 1981.
42. **Amico, J. A., Ulbrecht, J. S., and Robinson, A. G.,** Clearance studies of oxytocin in humans using radioimmunoassay measurements of the hormone in plasma and urine, *J. Clin. Endocrinol. Metab.*, 64, 340, 1987.
43. **Leake, R. D., Weitzman, R. E., and Fisher, D. A.,** Pharmacokinetics of oxytocin in the human subject, *Obstet. Gynecol.*, 56, 701, 1980.

44. **Kumaresan, P., Anandarangam, P. B., Dianzon, W., and Vasicka, A.,** Plasma oxytocin levels during human pregnancy and labor as determined by radioimmunoassay, *Am. J. Obstet. Gynecol.,* 119, 215, 1974.
45. **Burd, J. M., Davison, J., Weightman, D. R., and Baylis, P. H.,** Evaluation of enzyme inhibitors of pregnancy associated oxytocinase: application to the measurement of plasma immunoreactive oxytocin during human labor, *Acta Endocrinol. (Copenhagen),* 114, 458, 1987.
46. **Amico, J. A., Ervin, M. G., Finn, F. M., Leake, R. D., Fisher, D. A., and Robinson, A. G.,** The plasma of pregnant women contains a novel oxytocin-vasotocin-like peptide, *Metabolism,* 35, 596, 1986.
47. **Fabian, M., Forsling, M. L., Jones, J. J., and Pryor, J. S.,** The clearance and antidiuretic potency of neurohypophysial hormones in man, and their plasma binding and stability, *J. Physiol.,* 204, 653, 1969.
48. **Fabian, M., Forsling, M. L., Jones, J. J., and Lee, J.,** The release, clearance and plasma protein binding of oxytocin in the anesthetized rat, *J. Endocrinol.,* 43, 175, 1969.
49. **Dawood, M. Y., Ylikorkali, O., Trivedi, D., and Gupta, R.,** Oxytocin levels and disappearance rate and plasma follicle-stimulating hormone and luteinizing hormone after oxytocin infusion in men, *J. Clin. Endocrinol. Metab.,* 50, 397, 1980.
50. **Amico, J. A., Seitchik, J., and Robinson, A. G.,** Studies of oxytocin in plasma of women during hypocontractile labor, *J. Clin. Endocrinol. Metab.,* 58, 274, 1984.
51. **Seitchik, J., Amico, J., Robinson, A. G., and Castillo, M.,** Oxytocin augmentation of dysfunctional labor. IV. Oxytocin pharmacokinetics, *Am. J. Obstet. Gynecol.,* 150, 225, 1984.
52. **Chard, T., Boyd, N. R. H., Forsling, M. L., McNeilly, A. S., and Landon, J.,** The development of a radioimmunoassay for oxytocin: the extraction of oxytocin from plasma, and its measurement during parturition in human and goat blood, *J. Endocrinol.,* 48, 223, 1970.
53. **Rydén, G. and Sjöholm, I.,** Half-life of oxytocin in blood of pregnant end non-pregnant women, *Acta Endocrinol.,* 61, 425, 1969.
54. **Ryden, G.,** Cystine aminopeptidase activity in pregnancy, *Acta Obstet. Gynecol. Scand.,* 50, 253, 1971.
55. **Rosenbloom, A. A., Sack, J., and Fisher, D. A.,** The circulating vasopressinase of pregnancy: species comparison with radioimmunoassay, *Am. J. Obstet. Gynecol.,* 121, 316, 1975.
56. **Viinamäki, O., Erkkola, R., and Kanto, J.,** Plasma vasopressin concentrations and serum vasopressinase activity in pregnant and nonpregnant women, *Biol. Res. Pregnancy,* 1, 17, 1986.
57. **Barron, W. M., Cohen, L. H., Uland, L. A., Lassiter, W. E., Fulghum, E. M., Emmanouel, D., Robertson, G., and Lindheimer, M. D.,** Transient vasopressin-resistant diabetes insipidus of pregnancy, *N. Engl. J. Med.,* 310, 442, 1984.
58. **Davison, J. M., Barron, W. M., and Lindheimer, M. D.,** Metabolic clearance rates of vasopressin increase markedly in late gestation: possible cause of polyuria in pregnant women, *Trans. Assoc. Am. Phys.,* 100, 91, 1987.
59. **Schwartz, W. B., Bennett, W., Curelop, S., and Bartter, F. C.,** A syndrome of renal sodium loss and hyponatremia probably resulting from inappropriate secretion of antidiuretic hormone, *Am. J. Med.,* 23, 529, 1957.
60. **Amatruda, T. T., Mulrow, P. J., Jr., Galagher, J. C., and Sawyer, W. H.,** Carcinoma of the lung with inappropriate diuresis. Demonstration of antidiuretic-hormone-like activity in tumor extract, *N. Engl. J. Med.,* 269, 544, 1963.
61. **Lillien, A. A.,** Oxytocin induced water intoxication, *Obstet. Gynecol.,* 32, 171, 1968.
62. **Bie, P. and Warberg, J.,** unpublished data, 1985.
63. **Szatalowicz, V. L., Arnold, P. E., Chaimovitz, C., Bichet, B., Berl, T., and Schrier, R. W.,** Radioimmunoassay of plasma arginine vasopressin in hyponatremic patients with congestive heart failure, *N. Engl. J. Med.,* 305, 263, 1981.
64. **Riegger, G. A. J., Liebau, G., and Kochsiek, K.,** Antidiuretic hormone in congestive heart failure, *Am. J. Med.,* 72, 49, 1982.
65. **Goldsmith, S. R., Francis, G. S., Cowley, A. W., Jr., Livine, T. B., and Cohn, J.,** Increased plasma arginine vasopressin levels in patients with congestive heart failure, *J. Am. Coll. Cardiol.,* 1, 1385, 1983.
66. **Pruszczynski, W., Vahanian, A., Ardaillou, R., and Acar, J.,** Role of antidiuretic hormone in impaired water excretion of patients with congestive heart failure, *J. Clin. Endocrinol. Metab.,* 58, 599, 1984.
67. **van der Weiden, R. M. F., Visser, W., Peeters, L. L. H., van Leeuwen, A. P., and Wallenburg, H. C. S.,** Transient diabetes insipidus of pregnancy, *Eur. J. Obstet. Gynecol. Reprod. Biol.,* 25, 331, 1987.
68. **Dürr, J. A.,** Diabetes insipidus in pregnancy, *Am. J. Kidney Dis.,* 9, 276, 1987.
69. **Dürr, J. A., Hoggard, J. G., Hunt, J. M., and Schrier, R. W.,** Diabetes insipidus in pregnancy associated with abnormally high circulating vasopressinase activity, *N. Engl. J. Med.,* 316, 1070, 1987.
70. **Hughes, J. M., Barron, W. A., and Vance, M. L.,** Recurrent diabetes insipidus associated with pregnancy: pathophysiology and therapy, *Obstet. Gynecol.,* 73, 462, 1989.
71. **Baylis, P. H., Thompson, C., Burd, J., Tunbridge, W. M. G., and Snodgrass, C. A.,** Recurrent pregnancy-induced polyuria and thirst due to hypothalamic diabetes insipidus: an investigation into possible mechanisms responsible for polyuria, *Clin. Endocrinol.,* 24, 459, 1986.

Chapter 14

ACTIVATION AND DEGRADATION OF COMPLEMENT: PHYSIOLOGY AND PATHOPHYSIOLOGY

Jens Møller Rasmussen, Børge Teisner, and Sven-Erik Svehag

TABLE OF CONTENTS

I. INTRODUCTION

The complement (C) system is an enzyme cascade system which acts as an integrated part of the immune system in initiating inflammation, in killing of bacteria and other invading pathogens, and in the clearance of microorganisms and immune complexes (IC).

Functionally, the C system is organized into two activation pathways, the classical pathway (CP) which is activated by soluble or surface-bound IC, and the alternative pathway (AP) which is activated by certain biological surfaces and macromolecules, independent of antibodies (Figure 1). Several of the components circulate in plasma as proenzymes and become enzymatically active through limited proteolysis by enzymes activated earlier in the cascade reaction. By this cleavage, biologically active peptides are released and new proteases generated, which in turn activate the next proenzyme in the cascade.[1,2]

The reaction is amplified by different mechanisms: (1) at each step in the sequence, one activated component can activate several proenzymes; (2) in the alternative pathway AP, positive feedback is obtained at the C3-level because one product of C3-cleavage (C3b) is an integral part of the enzyme complex (C3bBb), which cleaves C3; and (3) the spontaneous decay of the C3bBb-complex is retarded by binding of a plasma protein, factor P (Figure 1).[1,2]

Effective control mechanisms are necessary in order to regulate these amplification reactions. Unrestricted activation may lead to abrogation of biological function, due to exhaustion of C factors present in limiting concentrations and to severe side effects, due to overproduction of biologically active split-products. Therefore, several proteins in plasma and on cell surfaces regulate C activation at various stages.[1]

Activation by either pathway generates fragments (C4b, C3b, and C5b) which can bind to the activating substances. This alters the substances' fate in the organism and elicits biological responses by interaction with various cell types. In addition, small polypeptides are released, which act as inflammatory mediators by attracting myeloid cells and triggering their inflammatory potentials. Upon binding to a cell surface, the fragment C5b acts as the anchoring site for assembly of the lytic membrane attack complex (MAC), forming channels in the lipid bilayer of cell membranes.

Apart from the lytic capability, most of the biological effects of C activation are mediated through the interaction of cleavage products, especially from C3, with specific cell-surface receptors, inducing biological responses in a variety of cells. Virtually all cell types present in the circulation and the cells of the monocyte-macrophage system in various organs and tissues express one or several different types of C-receptors.

Inherited deficiencies have been described for most of the known plasma C components and C regulatory proteins in man. Deficiencies of precursor components lead to blockade of activation pathways and to impairment of the biological functions mediated by fragments of the components. Deficiencies of regulatory proteins may lead to secondary deficiencies, usually of several components, due to unregulated activation of the C cascade. Although many individuals with C-deficiencies are apparently healthy, others are predisposed to infections and/or to diseases of chronic inflammatory nature. Acquired deficiencies may occur transiently, due to consumption of C components during activation of the C system in chronic disease processes, or persistently, due to occurrence in plasma of factors causing deregulation of the activation pathways, such as autoantibodies to activated C components.

II. ACTIVATION OF THE COMPLEMENT SYSTEM

A. THE CLASSICAL PATHWAY

Activation of the classical pathway (CP) is initiated by the binding of C1q (Table 1) to Fc-regions of antibodies within IC.[3] Conformational changes in the C1q-$(C1rC1s)_2$ complex lead to autoactivation of Clr, exposing a proteolytic site which cleaves and activates C1s. Activated

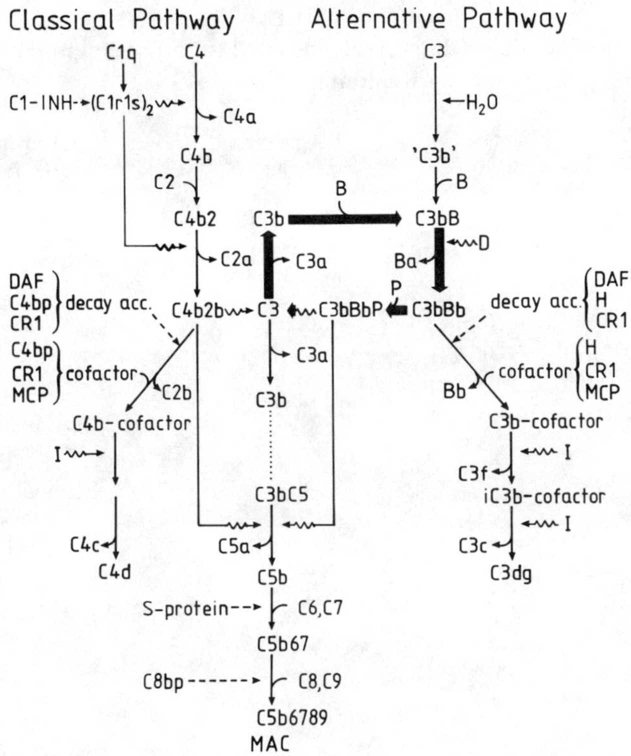

FIGURE 1. Activation of the complement system. 'C3b', C3 (H_2O) with C3b activity; ⇝, enzyme activity; - - ▸, inhibition; ➡, the amplification loop; and MAC, membrane attack complex.

C1rC1s cleaves C4 in the α-chain into two fragments, C4a and C4b. In the larger fragment, C4b, a reactive internal thiolester is exposed, which can form covalent bonds with nearby hydroxyl- or amino groups on cell surfaces or soluble macromolecules.[4,5] Covalently bound C4b can bind C2, which is cleaved by adjacent activated C1. A small fragment, C2a, is released, exposing a proteolytic site in the C2b part of C4b2b.[2] This complex is termed the CP C3/C5 convertase and has C3 as its natural substrate.

B. ACTIVATION AND BINDING OF C3

The C3-convertase cleaves C3 into two fragments, C3a and C3b. Analogous to C4b, a reactive thiolester is exposed in the α-chain[6,7] of the larger fragment, C3b. This thiolester reacts with free hydroxyl or amino radicals on cell surfaces or macromolecules in the immediate vicinity, forming hydroxyester or amide linkages.[8,9] The short half-life of the thiolester (≈ 60 μsec)[9] allows nascent C3b to diffuse only a short distance (≈ 30 nm) from the site of activation.[9] Consequently, only 5 to 10% of nascent C3b is actually bound in clusters around the activating convertase,[10] the remainder being hydrolyzed, to yield soluble, inactive C3bi.[11]

C. AMPLIFICATION OF THE ALTERNATIVE PATHWAY

Surface-bound C3b may bind factor B, forming a Mg^{2+}-dependent proenzyme complex, C3bB (reviewed in Reference 12). When bound to C3b, factor B is susceptible to cleavage by the serine esterase factor D and the AP C3/C5-convertase C3bBb is formed. C3bBb can cleave native C3, leading to the formation of more nascent C3b and hence C3bBb, thereby generating an amplification loop unique to the AP. C3bBb spontaneously dissociates with a half-life of

TABLE 1
Physicochemical Characteristics of Plasma C Components and Regulatory Proteins

Protein	Mol wt (kDa)	Polypeptide chains[a]	Approximate serum concentration (μg/ml)
C1q	410	A 23 (6) B 23 (6) C 23 (6)	180
C1r	85	—	34
C1s	85	—	30
C4	200	α 93 β 75 γ33	350
C2	110	—	15
C3	190	α 110 β 75	1300
B	90	—	200
D	24	—	1—5
C5	190	α 115 β 75	75
C6	105—128[b]	—	70
C7	92—121	—	60
C8	151	α 64 β 64 γ22	
C9	71	—	60
C1-INH	104	—	200
C4bp	550	70 (7)	250
I	88	Light 38 Heavy 50	35
H	160	—	360—550
P	220	53 (4)	20—30
S-protein	80	—	80

[a] Chain designation, molecular weight in kilodaltons, and number of chains (in parentheses) are indicated.
[b] Range reported.

≈ 90 s,[13,14] but binding of factor P prolongs the half-life of the C3bBbP-complex to about 10 min,[14,15] allowing further C3 cleavage and deposition.

The balance between amplification and inactivation allows the AP to continuously deposit small amounts of C3b on cells and molecules in solution, but the reaction is amplified only when C3b-molecules are bound to targets protecting C3b against inactivation. A wide variety of structures can offer such protection.[12] These include certain lipopolysaccharides on Gram-negative bacteria, surface molecules on fungi and yeast cells, some viruses and virus-infected cells, as well as malignantly transformed cells. Although these microorganisms and macromolecules differ widely in composition of surface carbohydrate moieties, one common feature is the relative lack of sialic acid. On surfaces with low sialic acid content, factor H has low affinity for C3b, allowing factor B to bind to C3b thereby amplifying the C activation.[16,17] On surfaces with high sialic acid content, the affinity of factor H for C3b is increased, allowing factor H to displace factor B from C3b, which renders C3b susceptible to proteolytic inactivation.[18] In recent years, the importance of cell-bound C3-binding proteins in regulation of surface activation has been recognized.[19]

D. ACTIVATION OF THE LYTIC TERMINAL PATHWAY

Assembly of the terminal lytic complex requires the presence of surface-bound C3bBb or

C4b2b plus excess C3b, deposited on the surface close by the C3-convertases (reviewed in Reference 2). Surface-bound C3b can bind C5 which is cleaved by the C3/C5-convertases into the C5a fragment, which is the major anaphylatoxin, and the larger C5b fragment. C5b expresses two binding sites, one for reversible binding to lipid or polysaccharide membranes and one for binding of C6, to form the complex C5b6. Binding of C7 stabilizes the C5b6 complex and facilitates its attachment to membranes. The C5b67 complex binds C8 and the C5b678 complex promotes the polymerization of C9, forming cylindrical, pore-forming structures consisting of 12 to 18 C9 subunits which transverse the lipid-bilayer membrane.[2]

III. REGULATION OF COMPLEMENT ACTIVATION

A. REGULATION OF C1 ACTIVATION

Activated C1 is regulated by C1-inhibitor (C1-INH), which binds to activated C1,[3] blocking the proteolytic sites of the enzymes and thus inhibiting further activation of C4.

B. REGULATION OF C3-CONVERTASE FORMATION

Central to the C system is activation and covalent attachment of C3b and C4b and formation of the C3 convertases. In order to control this process, a family of plasma and cell-bound proteins have evolved with the ability to interact with and modify the activity of C3b and C4b. These proteins, which constitute the *regulators of complement activation* (RCA) gene family, display primary structure and functional homologies[19] and they are encoded at a single chromosomal locus, the RCA gene cluster on chromosome 1.[20] The components of the RCA family include two plasma proteins with analogous functions in the AP and CP, factor H and C4-binding protein (4bp), two cell-associated proteins, decay-accelerating factor (DAF) and membrane cofactor protein (MCP), and two complement receptors (CR) – the C3b-receptor (CR1) and the C3dg-receptor (CR2).[20,21]

Three principles of C3-convertase inactivation have been recognized: (1) spontaneous dissociation of C3bBb and C4b2b, (2) acceleration of the rate of dissociation of C3bBb and C4b2b by the decay-accelerating factors, and (3) cleavage and inactivation of C3b and C4b by the plasma enzyme factor I, after binding of cofactors (Figure 1).

C4b2b spontaneously loses activity with a half-life of about 3 min at 37°C.[22] This rate may be significantly increased by C4 binding protein, C4bp.[23] Moreover, C4bp can bind directly to C4b, inhibiting the binding of C2 and formation of C4b2b. C4bp also has cofactor activity, rendering C4b susceptible to proteolytic cleavage by factor I (see below).[24,25] **Factor H**[26,27] competes with factor B for binding to C3b, thereby accelerating the dissociation of Bb from C3bBb, and renders C3b susceptible to proteolysis by factor I.[28]

Decay-accelerating factor, DAF,[29] is present on virtually all circulating blood and tissue cells[30] and in body fluids.[30] It binds to surface-bound C3b and C4b and accelerates the decay of both C3bBb and C4b2b.[29] **Membrane cofactor protein, MCP,** interacts with surface-bound C3b and C4b, rendering these molecules susceptible to proteolysis by factor I.[31]

Complement receptor 1, CR1, (= C3b/C4b receptor, CD35) inhibits the binding of factor B to C3b and C2 to C4b, accelerates the dissociation of Bb from C3bBb and C2b from C4b2b and is cofactor for factor I-mediated cleavage of surface-bound and soluble C3b to iC3b and of C4b to C4c and C4d.[21,32] CR1 may also be an essential cofactor for the cleavage of surface-bound and soluble iC3b to C3dg by factor I.[33-35] **Complement receptor 2, CR2,** (= C3dg/Epstein-Barr virus receptor, CD21) binds the iC3b and C3d fragments of C3[36] and the glycoprotein gp 350/220 of Epstein-Barr virus.[37] CR2 may participate in AP activation by B-lymphocytes as it binds C3b covalently.[38] Cofactor function of CR2 for factor I-mediated cleavage of iC3b to C3c and C3dg has also been reported.[39]

Factor I (C3b/C4b inactivator)[28,40,41] is a serine esterase occurring in active form in plasma. Following binding of the cofactors H[26,27] CR1,[32] or MCP,[31] C3b is cleaved by factor I. The iC3b fragment remains covalently attached to the activating surface but has lost the ability to amplify C-activation. Additional cleavage of iC3b releases the C3c-fragment to the fluid phase, leaving the 36-kDa C3dg fragment covalently attached to the surface.[42,43] Analogous to C3b, C4b is cleaved by factor I, leaving a 47-kDa C4d-fragment covalently bound to the activating immunoaggregate, requiring the cofactor activity of C4bp,[24] the CR,[44] or MCP.[31]

C. REGULATION OF MAC FORMATION

The binding of C5b67 to membranes is inhibited by the **S protein.**[45] S protein (vitronectin) binds to C5b-7, competing with membrane lipids for the membrane-binding site of C5b-7.[46] **Homologous restriction factor (HRF** = C8 binding protein (C8bp))[47] is a membrane protein (≈65 kDa) present on E, thrombocytes, neutrophils, monocytes, and lymphocytes. It inhibits channel-formation by homologous C5b-8, C5b-9, and by polymerized C9, thus protecting autologous cells against MAC attack.

IV. BIOLOGICAL FUNCTIONS OF THE COMPLEMENT SYSTEM

Activation of the complement (C) system by immune aggregates, cells, or microorganisms results in deposition on the activators of C fragments and C factor complexes which can then mediate cellular uptake of the former by binding to specific cell-surface C receptors. This type of interaction forms the basis for several of the C system's numerous biological functions.

A. MODULATION OF IMMUNE COMPLEXES

The interaction of the C-system with soluble IC is of major importance in controlling the clearance of foreign antigenic material from the circulation. Solubilization of IC refers to the process of deaggregation of preformed, precipitable IC after incubation with C.[48,49] The solubilization requires an intact AP. The CP contributes by accelerating the reaction,[49] but is not essential for the solubilization to occur. According to the "intercalation theory", solubilization arises from monovalent binding of C-fragments, especially C3b and C4b, to the antigen-antibody lattice, causing disruption of primary antibody-antigen bonds and rearrangement into smaller, more soluble ICs. However, other investigators have not observed disruption of specific antigen-antibody bonds.

B. OPSONIZATION AND CLEARANCE

After interaction with C and incorporation of C3b and C4b, IC may bind to C3/C4 receptors on various cell types. When ^{125}I-labeled IC are added to whole blood, the majority of the radioactivity rapidly becomes associated with the erythrocyte (E) fraction.[50,51] Although E has considerably fewer CR1 per cell than leukocytes, they nevertheless, because of their greater numbers, contribute with >90% of CR1 found in the circulation, and therefore bind the majority of IC.

In vivo studies have demonstrated that large, preformed IC infused into primates rapidly become bound to E and are subsequently deposited in the liver, without concomitant trapping of the E.[52] Complement depletion leads to impairment of IC uptake by E, accelerated clearance of IC from the circulation, and increased extrahepatic IC deposition.[53]

IC-C3b bound to CR1 is rapidly degraded to IC-iC3b and further to IC-C3dg[54] and may consequently react with iC3b-receptors (CR3) on the same or neighboring cells. Although both CR1 and CR3 participate in adhesion of opsonized particles and enhancement of phagocytosis, CR3 seems to be the receptor of greatest importance for phagocytic uptake by polymorphs, monocytes, and macrophages.[55]

C. INFLAMMATORY MEDIATORS

The major inflammatory mediators of the C system are the anaphylatoxins C3a, C4a, and

C5a, which are released to the fluid phase upon cleavage of native C3, C4, and C5 by the CP or AP C3/C5 convertases.[56] By binding to specific receptors, the anaphylatoxins may elicit inflammatory reactions, including aggregation and activation of thrombocytes, smooth muscle contraction, increased vascular permeability, and leucocyte chemotaxis. Several types of cells are responsive to the anaphylatoxin, including polymorphs, monocytes/macrophages, thrombocytes and mast cells. Responses to anaphylatoxin stimulation *in vitro* include release of cellular enzymes, activation and release of oxygen radicals, histamine and serotonin release, induction of arachidonic acid pathways, and interleukin 1 secretion. C5a is a major chemotactic and activating agent for neutrophilic and eosinophilic granulocytes and monocytes.

D. CYTOLYSIS

The mechanism of C-mediated lysis has been briefly discussed above (see Section II.D).

V. BIOSYNTHESIS OF COMPLEMENT COMPONENTS

A. CELLS AND SITES OF SYNTHESIS

The activity of the C system is regulated at least at three different points, at the level of biosynthesis, consumption, and catabolism.

Cell types shown to be capable of biosynthesis of C proteins include hepatocytes, monocytes, tissue macrophages, keratinocytes, endothelial cells, and renal mesangial cells. Only hepatocytes and macrophages/monocytes have been shown to synthesize the full range of functional C pathway proteins. C2, C3, C4, and C1 INH are all synthesized in human fetal liver.[56,57] The liver is the major site for synthesis and secretion of C proteins. By use of human hepatoma cell lines, the hepatocytes rather than the Kuppfer cells were indicated as the major source of hepatic C components.[59] In patients receiving liver transplants the allotypes of C3, C6, C8, and B converted to the donor types.[60,61]

Human monocytes/macrophages appear capable of synthesizing all functional components of the CP, AP, and MAC.[62-65] Resident macrophages from different compartments display differing rates of C biosynthesis which appear to be regulated at the pretranslational level.[66,67] Thus, C biosynthesis in different organs and by different cell populations is under separate regulation.[66] Macrophages participating in inflammatory processes secrete greater quantities of C proteins than do resident macrophages.[68] Thus, the C3 concentration in synovia of arthritis joints exceeds the plasma concentration of C3, indicating considerable local production in the joints.[69]

Extrahepatic synthesis of C proteins, particularly by tissue macrophages along mucosal surfaces, probably plays an important role for the secretion of these proteins locally and for the defense against invading microorganisms. The synthesis of functional C1 in the intestinal and genito-urinary epithelium has been reported.[70]

The tissue specificity in biosynthesis of C proteins is illustrated by the differential expression of the two C4 isotypes in hepatic cells and macrophages. Hepatocytes produced mostly C4A, whereas macrophages predominantly produced C4B.[71]

Peptide mapping of biosynthetically labeled C3 secreted from human monocytes, macrophages, the monocyte-like cell line U 937, a hepatoma derived cell line, and human fibroblasts indicated great similarity. However, the rate of production of C3 by hepatoma-derived cells exceeded that of monocytes or U 937 cells.[72]

Most C proteins are single gene products, consisting either of a single polypeptide chain or a multichain structure derived from post-translation cleavage of a single-chain precursor. However, two components, C1q and C8, are composed of three gene products. C1q consists of 6 copies of each of the 3 gene products where the 18 polypeptides combine to form the functional C1q molecule (Table 1).

TABLE 2
Metabolism of Selected Complement Factors in Healthy Individuals and Patients with Autoimmune Diseases

	Healthy individuals		Systemic lupus erythematosus		Rheumatoid arthritis	
	FCR[a]	SP[b]	FCR	SR	FCR	SR
C3	1.4—2.0; 1.7	0.76	3.1	0.82	2.5	1.01
C5	1.5—2.0; 1.7	0.09	2.4	0.11	2.8	0.14
B	1.7—2.2; 2.0	0.18	4.0	0.42	2.0	0.23
P	0.8—1.1; 0.7	0.09	1.7; 1.5	0.09	—	—
H	1.1—1.5; 1.3	0.37	1.8	0.63	1.6	0.95

Note: The table is compiled from data in References 76 to 78, 80, 81, and 83.

[a] FCR, fractional catabolic rate (% of plasma pool/hour), means, and/or ranges reported by different investigators.
[b] SR, synthetic rate (mg/kg/h), mean values reported by different investigators.

C3, C4, and C5 are synthesized as single polypeptide promolecules which undergo intracellular proteolytic cleavage by signal peptidases and other undefined enzyme(s) before assembly to their native functional form.

B. REGULATION OF BIOSYNTHESIS

Differences in net production of C proteins may be accounted for by differences in kinetics of secretion, differences in primary synthesis rate, or both. There is convincing evidence that the biosynthesis of C proteins is regulated by a multiplicity of hormones and interleukins. Soluble mediators reported to influence biosynthesis of C components include I1-1, I1-6, gamma-interferon, tumor necrosis factor, and PGE_2.[72] Gamma-interferon and soluble IC were reported to stimulate C2 and Factor B synthesis, whereas PGE_2, C3a, C5a, and histamine inhibited synthesis of C2. Lipopolysaccharide increased the synthesis of C3 and B, while the synthesis of C2 was unaffected.

Histamine can modify the biosynthesis of C proteins in human and murine hepatic cells and macrophages.[73] The histamine-induced changes appear dependent upon the type of histamine receptor involved. Thus, stimulation of H1 receptors has been observed to enhance biosynthesis of C proteins, whereas stimulation of the histamine receptor H2 pathway may down-regulate C protein synthesis.

During the I1-6-induced, acute-phase reaction, the plasma concentration of proteinase inhibitors and C proteins is increased. The plasma concentrations of the different C components show great variation, with C3 as the predominant component. The concentration of the different C proteins has to be carefully regulated and balanced, and control elements in the genes for these proteins are probably essential for this regulation.

VI. METABOLISM OF COMPLEMENT COMPONENTS

A. METHODOLOGICAL AND INTERPRETATIONAL PROBLEMS

Highly purified C components, trace labeled with radioactive iodine, have been used to study the catabolism of these proteins in healthy individuals and in patients with a variety of diseases (Table 2). The fractional catabolic rates (FCR) indicate that almost one half of the intravascular C proteins are being renewed each day. The turnover rates of C components reflect normal daily C activation, superimposed on the degradative processes that apply to all plasma proteins. Thus, for each C protein, there is a normal route of catabolism which probably is distinct from C consumption related to activation of the C system.

The measurement of catabolism of purified radioiodinated proteins is affected by the presence of aggregated or denatured protein. This is also a problem in assessing the fractional activity of trace-labeled C proteins. Artifacts may also arise due to deiodination processes *in vivo*.

Measurements of FCR are based on the assumption that the labeled protein is metabolically homogeneous and that the catabolism occurs in compartments in rapid equilibrium with the plasma pool. Ideally, the radioiodinated protein should behave *in vivo* like the native protein. The percentage of the labeled C proteins that remained biologically active has also been assessed in several studies by determining the "specific functional activity". To ascertain that a preparation is free of denatured protein is, however, impossible.

Plasma radioactivity curves for C-proteins have been analyzed by different methods. The integrated rate equations method of Nosslin[74] makes use of both plasma and urine data, which offers theoretical advantages over the multiexponential analysis of the plasma radioactivity disappearance curve,[75] particularly in studies of proteins with short half-lives. Discrepancies between estimates of FCR by the two methods may be due, in part, to the fact that extravascular catabolism is not accounted for in Matthews' method. Average daily FCRs have also been calculated by the metabolic clearance method in which the daily urinary radioactivity is divided by the mean total plasma radioactivity for that day.

Synthesis rates have been derived from the formula:

$$\frac{\text{FCR} \times \text{plasma volume} \times \text{plasma concentration}}{\text{kg}}$$

assuming that the synthetic rate is equal to the absolute catabolic rate, when the plasma concentration of the protein in question is stable throughout the study period.

B. METABOLISM IN HEALTHY INDIVIDUALS

The serum concentration of a C component represents the transient balance between synthesis, catabolism, and consumption. C3 and C4 are synthesized and catabolized rapidly, their normal catabolic rate being about 2% of the plasma pool per hour. Most C-proteins have a rapid turnover relative to other plasma proteins.

There are likely to be two separate mechanisms of C3 catabolism, one being degradation by C-activating enzymes and the other a physiological process of catabolism. The mean FCR of C3 in healthy subjects was 1.7%/h[76,77] (Table 1), the mean percent of C3 that was intravascular was 67% and the C3 synthetic rate averaged 0.76 mg/kg/h. A positive correlation between the synthetic rate of C3 and its serum concentration has been observed in healthy individuals. The degradation fragment C3a is cleared in the liver and the kidney. C3d may be removed from circulation by glomerular filtration with subsequent tubular degradation.

The fractional catabolic rate of C5 in normal subjects was 1.5 to 2.0% of the plasma pool per hour, the extra-/intravascular distribution ratio was 0.22 to 0.78.[78] The catabolic rate of C5 is similar to that of C3 in healthy subjects, but usually less than that of C3 in patients with C activation.[78] C5a and C5a des Arg are eliminated by neutrophils.[79] Neutrophils take up C5a and C5a des Arg which are then degraded by intracellular hydrolytic enzymes to peptides and amino acids.

The fractional catabolic rate of properdin in normal subjects was 0.8 to 1.1% of the plasma pool per hour.[80] This is about half of that found for C3. The FCR of C3, C4, C5, and factor B is similar. The FCR of factor B in healthy subjects was estimated to be 1.7 to 2.2% of the plasma pool/h. Half-life of factor H was 66 to 87 h with a FCR of 1.1 to 1.5%/h.[81] Both factors H and B showed significant extravascular catabolic rate.

Plasma levels of the nonlytic SC5b-9 complex normalize within 30 to 60 min after termination of dialysis, indicating an effective mechanism for elimination of small quantities of SC5b-9 in humans.[82] It is most probably trapped in the liver and spleen.

C. METABOLISM IN DISEASE

Catabolism of C components is increased in patients with a variety of diseases. Hypercatabolism of C components has been observed in both inborn and acquired diseases with C-activation. The greatest increase in FCR of C3 has been observed in patients with marked hypocomplementemia. The increased catabolism is usually compensated for by an increased rate of synthesis. However, hypercatabolism and depressed C synthesis (C3) has been observed in systemic lupus erythematosus patients with hypocomplementemia.[76-78,81]

Fractional catabolism and extravascular distribution of C3, C5, and factor H have been reported to be increased in systemic lupus erythematosus patients. Moreover, the extravascular catabolism of C3 and C5 seems to be increased in this disease.

Properdin was also hypercatabolized in systemic lupus erythematosus patients (mean FCR 1.7%).[80] The synthetic rate of properdin, however, was not reduced in these patients, indicating that the reduced serum levels of properdin are due to the increased catabolism.

Metabolic parameters such as FCR and synthetic rates for factor B have been determined in patients with systemic lupus erythematosus, rheumatoid arthritis, and other diseases which may be associated with hypocomplementemia.[76,83] An increased FCR of factor B was observed in most of the systemic lupus erythematosus patients, but they maintained normal plasma levels of factor B due to a compensatory increased rate of synthesis.

An increased rate of catabolism of factor B was shown also in patients with rheumatoid arthritis.[83] Serum levels of factor B were normal, the increased catabolism being matched by increased synthesis.

The mean FCR of C3 and C1q was higher in primary biliary cirrhosis than in controls.[84,85] This was probably due to chronic activation of the CP of complement. This also appears to be the case for hepatitis Bs-Ag positive patients with chronic active liver disease.

In primary biliary cirrhosis, C1q concentration and pool sizes were increased, indicating that synthesis had overcompensated for the increased catabolism.[85] Overcompensation was also noted in C3 metabolism in this disease.[84] C3 catabolism in alcoholic cirrhosis is normal and decreased synthesis appears to be responsible for low C3 levels in the patients' serum.

VII. COMPLEMENT DEFICIENCIES AFFECTING REGULATION AND DEGRADATION OF COMPLEMENT

A. INHERITED DEFICIENCIES

Homozygous inheritance of structurally abnormal C genes or abnormal regulatory genes leads to absence or reduced rates of synthesis of the individual C proteins.[86] This is manifested as undetectable or very low concentrations of the affected component in body fluids. In the heterozygous state, the affected component is usually present at a concentration corresponding to approximately 50% of the normal mean. Dependent on their location in the cascade sequence, deficiencies of components can result in blockade of C activation and/or impairment of a variety of C-mediated functions. These include derangement of mechanisms required for normal IC-clearance, such as opsonization, solubilization, and proper transport of intravascular IC, as well as diminished defense against microorganisms; in particular opsonization, extracellular lysis, and virus neutralization. Consequently, individuals with inherited C-deficiencies are predisposed to autoimmune IC-mediated diseases such as systemic lupus erythematosus-like syndromes[51] and to recurrent pyogenic infections.[87] The striking association of deficiencies of the MAC-components, especially C7 and C8, with recurrent, systemic Neisserial infections points to the importance of extracellular lysis in the primary defense against these organisms.[87]

Deficiencies of regulatory proteins may cause profound disturbances of the C system. Dysfunctional control mechanisms at various levels in the C-cascades may result in a prolonged half-life of activated enzyme complexes. This, in turn, leads to accelerated degradation

TABLE 3
Comparison of the Complement Profiles in Patients with AP Regulatory Protein Deficiencies and with C3 Nephritic Factor (C3-NeF)

	I-deficiency	H-deficiency	P-deficiency	C3-NeF
C3[a]	↓↓ (15—30%)	↓↓↓ (2—10%)	N	↓↓↓ (3—10%)
Native C3	↓↓↓ (5—10%)	↓↓↓	N	0
Native C3b	↑↑	—	—	—
Native C3c	0	N	—	↑
Native C3dg	0	↑↑	—	↑↑↑
B	↓↓↓ (0—10%)	↓↓ (6—30%)	N	N
Native B	0	↓↓ (6—30%)	N	N
Νατιϖε Βα	↑	—	—	↑
P	↓↓ (15—30%)	↓-N (18—60%)	0	N
H	↓ (25—60%)	0	N	N
I	0	N	N	N
C4	N	N	N	N
C4c	0	—	—	—
C4d	0	—	—	—
C5	↓ (30—70%)	↓↓-N (7—100%)	N	↓-N

Note: The complement components C1, C2, C1-INA, C4bp, D, C6, C7, C8, and C9 were found in normal concentrations in these four deficiencies. N, concentrations within normal range; %, percentage of the mean of the normal range.

[a] Measured using an anti-C3C antibody detecting C3, C3b, iC3b, as well as C3c.

and, hence, to more or less profound secondary depletion of C-components. It is characteristic that, in deficiencies of C3-convertase or MAC components, usually only one component is strongly affected, whereas deficiencies of regulatory proteins cause depletion of more than one C-protein (Table 3).

Deficiencies of regulatory proteins described in humans include C1-inhibitor deficiency in the CP and deficiency of factor I, factor H, and factor P in the alternative pathway.

1. C1-Inhibitor Deficiency

C1-INH deficiency was the first C-deficiency to be etiologically linked to a clinical syndrome, termed hereditary angioedema.[88] This syndrome usually manifests itself during the first decade of life. It is characterized by recurrent attacks of subcutaneous and submucosal edema, involving the skin and the respiratory and gastrointestinal tracts. C1-INH deficiency exists in two forms: type I (80% of cases), where the patients have low levels (10 to 35% of normal) of C1-INH, and type II (20%) where the patients have normal or elevated levels of dysfunctional C1-INH.[88]

The mechanisms of the paroxysmal angioedema in hereditary angioedema patients are not fully understood. Attacks are usually precipitated by minor trauma and physical or emotional stress. Inefficient inhibition of the spontaneous fluid-phase activation of the C1r-C1s complex due to low levels of functionally C1-INH leads to increased C1-esterase activity and degradation of C4 and C2. The angioedema is believed, in part, to be mediated by phlogistic peptides generated by cleavage of C4 and C2. C1-INH deficiency seems to be inherited as an autosomal dominant trait. Homozygosity is supposed to be incompatible with life. Several hundred kinships for inherited C1-INH deficiency are known. About 2% of the affected individuals develop systemic lupus erythematosus in the long term.[89]

Deficiencies of AP-regulator proteins can cause profound disturbances in AP homeostasis because of the potent nature of the pathway's amplification mechanism.

2. Factor I Deficiency

Since the description of the first case in 1970,[90] 13 patients from 12 different kinships have been described in the world.[91-96] These patients are usually profoundly immunologically compromised, predisposed to recurrent pyogenic infections, including neisserial meningitis and septicemia. One patient, without apparent immunodeficiency, developed a generalized vasculitis syndrome.[96]

Biochemically, profound disturbances of the AP of the complement system are seen (Table 3). Lack of factor I leads to impairment of the proteolytic inactivation of C3b and C3bBb in the AP, and of C4b in the CP. This results in a prolonged half-life for C3bBb which, in turn, cleaves and consumes native C3 and factor B. A dynamic equilibrium is established *in vivo*, characterized by low, but detectable, levels of native C3.[96] By contrast, factor B is totally consumed, indicating that this is the limiting factor for the accelerated AP-activity. The increased degradation of factor B is demonstrated by markedly elevated Ba and Bb levels.[97] Due to abolished enzymatic degradation of C3b and C4b, the fragments C3dg and C4d cannot be detected. The patients have reduced plasma concentration of factor H which has been demonstrated in a complex with C3b[93,98] in the fluid phase.

The metabolic abnormalities can be transiently normalized *in vivo* by infusion of purified factor I or normal human plasma. This is followed by a rapid increase in native factor B and a more gradual and prolonged normalization of C3.[96] The inhibitory effect of infused factor I on the accelerated degradation of factor B is evident from a fall in Ba levels.[97] In two patients the half-life of infused factor I was estimated to be 29 to 45 h,[96] corresponding to an average catabolic rate of about 1.4%/h. No data for the catabolism of factor I have been reported in healthy individuals.

Metabolic studies employing ^{125}I-labeled purified C3 in one patient[99] revealed a normal synthetic rate of 1.2 mg/kg/h but a great increase in FCR (10.1% of the plasma pool per hour). Feedback inhibition of C3-synthesis by plasma C3dg has been suggested in metabolic studies of patients with C3-hypercatabolism. This could not be confirmed in one factor I-deficient patient, in whom the SR of C3 increased from 1.2 to 1.6 mg/kg/h after plasma infusion,[99] which is known to generate large amounts of C3dg.[97]

3. Factor H Deficiency

Nine cases of factor H deficiency have been described to date.[100-103] Clinically, this deficiency is associated with renal disease, such as hemolytic uremic syndrome and membranoproliferative glomerulonephritis. It has also been described in patients with neisserial meningitis, either recurrent[104] or caused by low-virulent, rare serotypes.[105]

The C profile resembles that of factor I deficiency. An accelerated AP turnover is indicated by markedly reduced concentrations of native C3 (2 to 10% of normal) and elevated concentrations of C3dg (C. Koch, personal communication). The depletion of factor B is not as marked as in factor I deficiency, since the patients have 6 to 30% of the normal native B concentration. Factor P is only moderately affected. The concentrations of CP proteins are normal.[100,101]

Erythrocyte-CR1 expression was found normal in one patient (C. Koch, personal communication). Like factor H, CR1 exerts cofactor function for factor I-mediated degradation of surface-bound C3b and C4b as well as for soluble C3b. However, it seems that factor H has a critical role in maintaining homeostasis of the AP in the fluid phase, since CR1 in these patients is apparently unable to compensate for the lack of H cofactor function.

4. Factor P Deficiency

Deficiency of factor P is not associated with disturbed AP homeostasis, but rather with impaired amplification of AP activation, due to the inability to generate a stable C3 convertase.[106] Partial factor P deficiency has been described in patients with meningococcal meningitis. It seems to be associated with a severe course for the disease, since several cases of mortal disease have been reported in the families of patients with P deficiency.[107,108] Recently, several cases of factor P deficiency were identified in two studies of selected patients with meningococcal disease.[103,104] Except for factor P, normal concentrations of AP and CP proteins were found in these patients.

B. ACQUIRED DEFICIENCIES

In some acquired C deficiencies, the observed depletion of C components is due to generation of antibodies against C components or activated enzyme complexes.

1. C3 Nephritic Factor

A C3 cleaving activity in serum was first demonstrated by Spitzer et al.[109] in sera from patients with glomerulonephritis and the responsible factor was termed C3 nephritic factor (C3 NeF). Subsequently, C3 NeF was also demonstrated in patients with partial lipodystrophy, both with and without glomerulonephritis.[110] C3 NeF is particularly associated with type II membranoproliferative glomerulonephritis. More recently, C3 NeF has been demonstrated in sera from patients with recurrent[111] and sporadic meningococcal meningitis,[112] apparently without renal disease and/or lipodystrophy. C3 NeF is an IgG autoantibody,[113] which binds to, and increases, the half-life of solid-phase and fluid-phase C3bBb, leading to depletion of native C3.[114,115] This resembles the effect of factor P, except that C3 NeF-stabilized C3bBb is more resistant to the decay-accelerating activity of factor H and CR1 and to subsequent inactivation of C3b by factor I. In contrast to patients with factor I deficiency, C3 is completely degraded, giving rise to markedly elevated levels of C3dg and slightly elevated C3c.[111,116] Elevated Ba levels[97] indicate increased turnover of factor B, but the concentration of native B is within the normal range. Thus, in patients with C3 NeF, the rate-limiting factor for the excessive AP turnover seems to be C3.

2. C4 Nephritic Factor

A C4 nephritic factor has been isolated from sera of some patients with systemic lupus erythematosus.[117] This factor is the CP analogue to C3 NeF. C4 NeF is an IgG autoantibody which binds to, and prolongs the half-life of, activated C4b2b by rendering the enzyme relatively resistant to the decay-accelerating activity of C4bp.

3. Acquired C1-Inhibitor Deficiency

An acquired form of C1-INH deficiency has been described in association with B-cell malignancies. The identification of anti-idiotype antibodies to the malignant cells immunoglobulin lead to the hypothesis that formation of idiotype-anti-idiotype complexes and consumption of C1q and C1-INH was the pathological mechanism.[118] Recently, however, autoantibodies to C1-INH have been described in association with acquired C1-INH deficiency.[119]

VIII. CONCLUDING REMARKS

New, more specific and sensitive methods for measurement of C activation and degradation have been elaborated in the last decade. Measurements of native C components, using polyclonal antibodies, have been replaced by assays for C degradation products; more recently, employing monoclonal antibodies to neoepitopes expressed only on the degradation products. Furthermore, the characterization of a family of cell-bound regulator proteins has advanced our understanding of how control mechanisms of C activation have evolved.

Studies of the *in vivo* metabolism of complement components are complicated by the fact that degradation of native components by C-activating enzymes is superimposed on the physiological process of protein catabolism, which applies to all proteins in the circulation. Several of the C components exert their biological function only when activated by cleavage of the C-specific enzymes. These generated fragments of C proteins interact with other C components and C-activators, as well as with cellular receptors. Thus, cleavage products generated by the C-activating enzymes may be cleared by different mechanisms and at different catabolic rates.

In spite of these methodological and interpretational problems, a wealth of data indicate that most C proteins have a rapid turnover relative to other plasma proteins. Thus, FCR estimates for several C proteins indicate that almost 50% of the intravascular complement pool is being renewed each day.

Studies of complement deficiency states leading to dysregulation of the complement system have also given new insight into the mechanisms of homeostasis of the complement system. In particular, studies of patients with factor I deficiency have elucidated how the potent amplification mechanism in the alternative activation pathway is controlled, thereby emphasizing the role of the regulatory proteins in maintaining homeostasis of the complement system.

ABBREVIATIONS

AP	Alternative pathway
C4bp	C4 binding protein
C	Complement
CD	Cluster of differentiation
CP	Classical pathway
CR	Complement receptor
DAF	Decay accelerating factor
E	Erythrocyte
FCR	Fractional catabolic rate
HRF	Homologous restriction factor
IC	Immune complexes
Il	Interleukin
INH	Inhibitor
kDa	Kilodalton
MAC	Membrane attack complex
MCP	Membrane cofactor protein
NeF	Nephritic factor
PG	Prostaglandin
RCA	Regulators of complement activation
SR	Synthetic rate

REFERENCES

1. **Mollnes, T. E. and Lachmann, P. J.,** Regulation of complement, *Scand. J. Immunol.*, 27, 127, 1988.
2. **Müller-Eberhard, H. J.,** Molecular organization and function of the complement system, *Annu. Rev. Biochem.*, 57, 321, 1988.
3. **Ziccardi, R. J. and Cooper, N. R.,** Activation of C1 and the classical pathway of complement, in *Cytolytic Lymphocytes and Complement Effectors of the Immune System,* Vol. 1, Podack, E. R., Ed., CRC Press, Boca Raton, FL, 1988, 21.

4. **Law, S. K., Lichtenberg, N. A., Holcombe, F. H., and Levine, R. P.,** Interaction between the labile binding site of the fourth (C4) and fifth (C5) human complement proteins and erythrocyte cell membranes, *J. Immunol.*, 125, 634, 1980.
5. **Law, S. K., Lichtenberg, N. A., and Levine, R. P.,** Covalent binding and hemolytic activity of complement proteins, *Proc. Natl. Acad. Sci. U.S.A.*, 77, 7194, 1980.
6. **Pangburn, M. K. and Müller-Eberhard, H. J.,** Relation of a putative thiolester bond in C3 to activation of the alternative pathway and binding of C3b to biological targets of complement, *J. Exp. Med.*, 152, 1102, 1980.
7. **Tack, B. F., Harrison, R. A., Janatova, J., Thomas, M. L., and Prahl, J. W.,** Evidence for presence of an internal thiolester bond in the third component of human complement, *Proc. Natl. Acad. Sci. U.S.A.*, 77, 5764, 1980.
8. **Law, S. K., Lichtenberg, N. A., and Levine, R. P.,** Evidence for an ester linkage between the labile binding site of C3b and receptive surfaces, *J. Immunol.*, 123, 1388, 1979.
9. **Sim, R. B., Twose, T. M., Paterson, D. S., and Sim E.,** The covalent binding reaction of complement component C3, *Biochem. J.*, 193, 115, 1981.
10. **Mardiney, M. R., Müller-Eberhard, H. J., and Feldman, J. D.,** Ultrastructural localization of the third and fourth components of complement on complement-cell complexes, *Am. J. Pathol.*, 53, 253, 1968.
11. **Müller-Eberhard, H. J., Dalmasso, A. P., and Calcott, M. A.,** The reaction mechanism of β1C-globulin (C3) in immune hemolysis, *J. Exp. Med.*, 123, 33, 1966.
12. **Pangburn, M. K.,** Initiation and activation of the alternative pathway of complement, in *Cytolytic Lymphocytes and Complement Effectors of the Immune System,* Vol. 1, Podack, E. R., Ed., CRC Press, Boca Raton, FL, 1988, 41.
13. **Pangburn, M. K. and Müller-Eberhard, H. J.,** The C3 convertase of the alternative pathway of human complement. Enzymic properties of the bimolecular proteinase, *Biochem. J.*, 235, 723, 1986.
14. **Fearon, D. T. and Austen, K. F.,** Properdin: binding to C3b and stabilization of the C3b-dependent C3 convertase, *J. Exp. Med.*, 142, 856, 1975.
15. **Medicus, R. G., Götze, O., and Müller-Eberhard, H. J.,** Alternative pathway of complement: recruitment of precursor properdin by the labile C3/C5 convertase and potentiation of the pathway, *J. Exp. Med.*, 144, 1076, 1976.
16. **Pangburn, M. K. and Müller-Eberhard, H. J.,** Complement C3 convertase: cell surface restriction of β1H control and generation of restriction on neuraminidase treated cells, *Proc. Natl. Acad. Sci. U.S.A.*, 75, 2416, 1978.
17. **Fearon, D. T.,** Regulation by membrane sialic acid of β1H dependent decay dissociation of the amplification C3 convertase of the alternative complement pathway, *Proc. Natl. Acad. Sci. U.S.A.*, 75, 1971, 1979.
18. **Kazatchkine, M. D., Fearon, D. T., and Austen, K. F.,** Human alternative complement pathway: membrane associated sialic acid regulates the competition between B and β1H for cell bound C3b, *J. Immunol.*, 122, 75, 1979.
19. **Holers, V. M., Cole, J. L., Lublin, D. M., Seya, T., and Atkinson, J. P.,** Human C3b- and C4b-regulatory proteins: a new multi-gene family, *Immunol. Today,* 6, 188, 1985.
20. **Hourcade, D., Holers, V. M., and Atkinson, J. P.,** The regulators of complement activation (RCA) gene cluster, *Adv. Immunol.,* 45, 381, 1989.
21. **Ahearn, J. M. and Fearon, D. T.,** Structure and function of the complement receptors CR1 (CD35) and CR2 (CD21), *Adv. Immunol.*, 46, 183, 1989.
22. **Fishelson, Z. and Müller-Eberhard, H. J.,** C3 convertase of human complement: enhanced formation and stability of the enzyme generated with nickel instead of magnesium, *J. Immunol.,* 129, 2603, 1982.
23. **Scharfstein, J., Ferreira, A., Gigli, I., and Nussenzweig, V.,** Human C4-binding protein. I. Isolation and characterization, *J. Exp. Med.*, 148, 207, 1978.
24. **Fujita, T. and Tamura, N.,** Interaction of C4-binding protein with cell-bound C4b, *J. Exp. Med.*, 157, 1239, 1983.
25. **Fujita, T., Gigli, I., and Nussenzweig, V.,** Human C4-binding protein. II. Role in proteolysis of C4b by C3b inactivator, *J. Exp. Med.*, 123, 1195, 1979.
26. **Whaley, K. and Ryddy, S.,** Modulation of the alternative complement pathway by β1H globulin, *J. Exp. Med.,* 144, 1147, 1976.
27. **Weiler, J. M., Daha, M. R., Austen, K. F., and Fearon, D. T.,** Control of the amplification convertase of complement by the plasma protein β1H, *Proc. Natl. Acad. Sci. U.S.A.*, 73, 3268, 1976.
28. **Pangburn, M. K., Schreiber, R. D., and Müller-Eberhard, H. J.,** Human complement C3b inactivator: isolation, characterization and demonstration of an abolute requirement for the serum protein β1H for cleavage of C3b and C4b in solution, *J. Exp. Med.*, 146, 257, 1977.
29. **Nicholson-Weller, A., Burge, J., Fearon, D. T., Weler, P. F., and Austen, K. F.,** Isolation of a human erythrocyte membrane glycoprotein with decay accelerating activity for C3-convertases of the complement system, *J. Immunol.,* 129, 184, 1982.

30. **Lublin, D. M. and Atkinson, J. P.,** Decay-accelerating factor: biochemistry, molecular biology and function, *Annu. Rev. Immunol.*, 7, 35, 1989.
31. **Seya, T., Turner, J. R., and Atkinson, J. P.,** Purification and characterization of a membrane protein (gp45-70) that is a cofactor for cleavage of C3b and C4b, *J. Exp. Med.*, 163, 8376, 1986.
32. **Fearon, D. T.,** Regulation of the amplification C3 convertase of human complement by an inhibitory protein isolated form human erythrocyte membranes, *Proc. Natl. Acad. Sci. U.S.A.*, 76, 5867, 1979a.
33. **Medof, M. E., Prince, G. M., and Mold, C.,** Release of soluble immune complexes from immune adherence receptors on human erythrocytes is mediated by C3b inactivator independently of β1H and is accompanied by generation of C3c, *Proc. Natl. Acad. Sci. U.S.A.*, 79, 5047, 1982.
34. **Ross, G. D., Lambris, J. D., Cain, J. A., and Newman, S. L.,** Generation of three different fragments of bound C3 with purified factor I or serum. I. Requirements for factor H vs. CR1 cofactor activity, *J. Immunol.*, 129, 2051, 1982.
35. **Medicus, R. G., Melamed, J., and Arnout, M. A.,** Role of human factor I and C3b receptor in the cleavage of surface-bound C3bi molecules, *Eur. J. Immunol.*, 13, 465, 1983.
36. **Cooper, N. R., Moore, M. D., and Nemerow, G. R.,** Immunobiology of CR2, the B lymphocyte receptor for Epstein-Barr virus and the C3d complement fragment, *Annu. Rev. Immunol.*, 6, 85, 1988.
37. **Tanner, J., Weis, J., Fearon, D. T., Whang, Y., and Kieff, E.,** Epstein-Barr virus gp 350/220 binding to the B lymphocyte C3d receptor mediates adsorption, capping and endocytosis, *Cell*, 50, 203, 1987.
38. **Mold, C., Nemerow, G. R., Bradt, B. M., and Cooper, N. R.,** CR2 is a complement activator and the covalent binding site for C3 during alternative pathway activation by Raji cells, *J. Immunol.*, 140, 1923, 1988.
39. **Mitomoto, K., Fujita, T., and Iida, K.,** Functional and antigenic properties of complement receptor type 2, CR2, *J. Exp. Med.*, 165, 1424, 1987.
40. **Lachmann, P. J. and Müller-Eberhard, H. J.,** The demonstration in human serum of conglutinogen activity factor and its effect on the third component of complement, *J. Immunol.*, 100, 691, 1968.
41. **Fearon, D. T.,** Purification of C3b inactivator and demonstration of its two polypeptide chains, *J. Immunol.*, 119, 1248, 1977.
42. **Law, S. K., Fearon, D. T., and Levine, R. P.,** Action of the C3b-inactivator on cell bound C3b, *J. Immunol.*, 122, 759, 1979.
43. **Davis, A. E. and Harrison, R. A.,** Structural characterization of factor I mediated cleavage of the third component of complement, *Biochemistry*, 21, 5745, 1982.
44. **Iida, K. and Nussenzweig, V.,** Complement receptor is an inhibitor of the complement cascade, *J. Exp. Med.*, 153, 1138, 1981.
45. **Podack, E. R. and Müller-Eberhard, H. J.,** Isolation of human S-protein, an inhibitor of the membrane attack complex of complement, *J. Biol. Chem.*, 254, 9908, 1979.
46. **Podack, E. R., Preissner, K. T., and Müller-Eberhard, H. J.,** Inhibition of C9 polymerization within the SC5b-9 complex of complement by S-protein, *Acta Path. Microbiol. Scand.*, Suppl., 92, 89, 1984.
47. **Zalman, L. S., Wood, L. M., and Müller-Eberhard, H. J.,** Isolation of a human erythrocyte membrane protein capable of inhibiting expression of homologous complement transmembrane channels, *Proc. Natl. Acad. Sci. U.S.A.*, 83, 6975, 1986.
48. **Miller, G. W. and Nussenzweig, V.,** A new complement function: solubilization of antigen-antibody aggregates, *Proc. Natl. Acad. Sci. U.S.A.*, 72, 418, 1975.
49. **Petersen, I., Baatrup, G., Jepsen, H. H., and Svehag, S.-E.,** Complement-mediated solubilization of immune complexes and their interaction with complement C3 receptors, *Complement*, 2, 97, 1985.
50. **Medof, M. E. and Oger, J. J. F.,** Competition for immune complexes by red cells in human blood, *J. Clin. Lab. Immunol.*, 7, 7, 1982.
51. **Schifferly, J. A., Ng, Y. C., and Peters, D. K.,** The role of complement and its receptors in the elimination of immune complexes, *N. Engl. J. Med.*, 315, 488, 1986.
52. **Cornacoff, J. B., Hebert, L. A., Smead, W. L., VanAman, M. E., Birmingham, D. J., and Wazman, F. J.,** Primate erythrocyte-immune complex-clearing mechanism, *J. Clin. Invest.*, 71, 236, 1983.
53. **Waxman, F. J., Hebert, L. A., Cornacoff, J. B., VanAman, M. E., Smead, W. L., Kraut, E. H., Birmingham, D. J., and Taguiam, J. M.,** Complement depletion accelerates the clearance of immune complexes form the circulation of primates, *J. Clin. Invest.*, 74, 1329, 1984.
54. **Rasmussen, J. M., Jepsen, H. H., and Svehag, S.-E.,** Influence of processing by erythrocyte C3b/C4b receptors (CR1) on binding of immune complexes to Raji cells and polymorphonuclear cells, *Scand. J. Immunol.*, 26, 437, 1987.
55. **Griffin, F. M.,** Opsonization, phagocytosis and intracellular microbial killing, in *The Complement System*, Rother, K. and Till, G. O., Eds., Springer-Verlag, Berlin, 1988, 395.
56. **Bitter-Suermann, D.,** The anaphylatoxins, in *The Complement System*, Rother, K. and Till, G. O., Eds., Springer-Verlag, Berlin, 1988, 367.
57. **Colten, H. R.,** Ontogeny of the human complement system: *in vitro* biosynthesis of individual complement components by fetal tissues, *J. Clin. Invest.*, 51, 725, 1972.

58. **Colten, H. R.,** Biosynthesis of the fifth component of complement (C5) by human fetal tissues, *Clin. Immunol. Immunopathol.,* 1, 346, 1973.
59. **Morris, K. M., Aden, D. P., Knowles, B. B., and Colten, H. R.,** Complement biosynthesis by the human hepatoma-derived cell line Hep G2, *J. Clin. Invest.*, 70, 906, 1982.
60. **Alper, C. A. and Rosen, F. S.,** Genetics of the complement system, in *Advances in Human Genetics,* Harris, H. and Hirschhorn, K., Eds., Plenum Press, New York, 1976, 141.
61. **Hobart, M. J., Lachmann, P. J., and Calue, R. Y.,** C6: synthesis by the liver *in vivo*, *J. Exp. Med.*, 140, 629, 1977.
62. **Colten, H. R., Ooi, Y. M., and Edelson, P. J.,** Synthesis and secretion of complement proteins by macrophages, *Ann. N.Y. Acad. Sci.,* 332, 482, 1979.
63. **Whaley, K.,** Biosynthesis of the complement components and the regulatory proteins of the alternative complement pathway by human peripheral monocytes, *J. Exp. Med.*, 151, 501, 1980.
64. **Hetland, G., Johnson, E., Falk, R. J., and Eskeland, T.,** Synthesis of complement components C5, C6, C7, C8 and C9 *in vitro* by human monocytes, and assembly of the terminal complement complex, *Scand. J. Immunol.*, 24, 421, 1986.
65. **Johnson, E. and Hetland, G.,** Mononuclear phagocytes have the potential to synthesize the complete functional complement system, *Scand. J. Immunol.*, 27, 489, 1988.
66. **Cole, F. S., Auerbach, H. S., Goodberger, G., and Colten, H. R.,** Tissue-specific pretranslational regulation of complement production in human mononuclear phagocytes, *J. Immunol.,* 134, 2610, 1985.
67. **Falus, A., Benscher, H. U., Auerbach, H. S., and Colten, H. R.,** Constitutive and IL 1-regulated murine complement gene expression is strain and tissue specific, *J. Immunol.,* 138, 856, 1987.
68. **De Ceulaer, C., Papazoglou, S., and Whaley, K.,** Increased synthesis of complement components by monocytes, synovial fluid macrophages and synovial membrane cells from patients with rheumatoid arthritis, *Immunology*, 41, 37, 1980.
69. **Ruddy, S. and Colten, H. R.,** Rheumatoid arthritis. Biosynthesis of complement proteins by synovial tissues, *N. Engl. J. Med.*, 290, 1284, 1974.
70. **Morris, K. M., Colten, H. R., and Bing, D. H.** The first component of complement. A quantitative comparison of its biosynthesis in culture by human epithelial and mesenchymal cells, *J. Exp. Med.,* 148, 1007, 1978.
71. **Falus, A., Kramer, J., Walez, E., Varga, Z., Setalo, J., Jobst, K., Lakatos, T., and Meretey, K.,** Unequal expression of complement C4A and C4B genes in rheumatoid synovial cells, human monocytoid and hepatoma-derived cell lines, *Immunology*, 68, 133, 1989.
72. **Cole, F. S. and Colten, H. R.,** Complement biosynthesis, in *The Complement System*, Rother, K. and Till, G. O., Eds., Springer-Verlag, Berlin, 1988, 44.
73. **Lappin, D. and Whaley, K.,** Effect of histamine on monocyte complement production. I. Inhibition of C2 productive mediated by its action on H2 receptors, *Clin. Exp. Immunol.*, 41, 487, 1980.
74. **Nosslin, B.,** Analysis of disappearance time-curves after single injection of labeled proteins, in *Protein Turnover*, A Ciba Symposium, Association of Scientific Publishers, New York, 1973, 113.
75. **Matthews, C. M.,** The theory of tracer experiments with ^{131}I-labeled plasma proteins, *Phys. Med. Biol.,* 2, 36, 1957.
76. **Charlesworth, J. A., Williams, D. G., Sherington, E., Lachmann, P. J., and Peters, D. K.,** Metabolic studies of the third component of complement and the glycine-rich beta glycoprotein in patients with hypocomplementemia, *J. Clin. Invest.*, 53, 1578, 1974.
77. **Ruddy, S., Carpenter, C. B., Chin, K. W., Knostman, J. N., Soter, N. A., Götze, O., Müller-Eberhard, H. J., and Austen, K. F.,** Human complement metabolism: an analysis of 144 studies, *Medicine*, 54, 165, 1975.
78. **Peters, D. K., Sissons, J. G., and Liebowitch, J.,** Metabolism of C5 in health and disease, in *Clinical Aspects of the Complement System*, Opferkuch, W., Rother, K., and Schultz, D. R., Eds., Thieme, Stuttgart, 1978, 99.
79. **Chenoweth, D. E. and Hugli, T. E.,** Demonstration of specific C5a receptor on intact human polymorphonuclear leukocytes, *Proc. Natl. Acad. Sci., U.S.A.*, 75, 761, 1975.
80. **Ziegler, J. B., Rosen, F. S., Alper, C. A., Grupe, W., and Lepow, I. H.,** Metabolism of properdin in normal subjects and patients with renal disease, *J. Clin. Invest.*, 56, 761, 1975.
81. **Charlesworth, J. A., Scott, D. M., Pussel, B. A., and Peters, D. K.,** Metabolism of human β1H: studies in man and experimental animals, *Clin. Exp. Immunol.*, 38, 397, 1979.
82. **Salama, A., Hugo, F., Heinrich, D., Höge, R., Müller, R., Kiefel, V., Müller-Eckhardt, C., and Bhakdi, S.,** Deposition of terminal C56-9 complement complexes on erythrocytes and leukocytes during cardiopulmonary bypass, *N. Engl. J. Med.*, 318, 408, 1988.
83. **Krick, E. H., Deteer, D. H., Kaplan, R. A., Arrogave, C. M., and Vaughan, J. H.,** Metabolism of factor B of serum complement in rheumatoid arthritis, *Clin. Exp. Immunol.*, 34, 1, 1978.
84. **Potter, B. J., Elias, E., and Jones, E. A.,** Hypercatabolism of the third component of complement in patients with primary biliary cirrhosis, *J. Lab. Clin. Med.*, 88, 427, 1976.

85. **Potter, B. J., Elias, E., Thomas, H. C., and Sherlock, S.,** Complement metabolism in chronic liver disease: catabolism of C1q in chronic liver disease: catabolism of C1q in chronic active liver disease and primary bilary cirrhosis, *Gastroenterology*, 78, 1034, 1980.
86. **Lachmann, P. J. and Walport, M. J.,** Genetic deficiency diseases of the complement system, in *Immunobiology of the Complement System*, Ross, G. D., Academic Press, Orlando, 1986, 237.
87. **Petersen, B. H., Lee, T. J., Synderman, R., and Brooks, G. F.,** *Neisseria meningitis* and *Neisseria gonorrhoeae* bacteremia associated with C6, C7 or C8 deficiency, *Ann. Int. Med.*, 90, 917, 1979.
88. **Agostini, A.,** Inherited C1 inhibitor deficiency, *Complement*, 6, 112, 1989.
89. **Massa, M. C. and Conolly, M. D.,** An association between C1-esterase inhibitor deficiency and lupus erythematosus. Report of two cases and review of the literature, *J. Am. Acad. Dermatol.*, 7, 255, 1982.
90. **Alper, C. A., Abramson, N., Johnston, R. B., Jandl, J. H., and Rosen, F. S.,** Increased susceptibility to infection associated with abnormalities of complement-mediated functions and of the third component of complement (C3), *N. Engl. J. Med.*, 282, 349, 1970.
91. **Thompson, R. A. and Lachman, P. J.,** A second case of human C3b inhibitor (KAF) deficiency, *Clin. Exp. Immunol.*, 27, 23, 1977.
92. **Eng, R. H. K., Seligman, S. J., Arnout, M. A., and Alper, C. A.,** Variable expression of homozygous C3b inactivator deficiency, *Clin. Res.*, 26, 394, 1978 (Abstr).
93. **Wahn, V., Rother, U., Rauterberg, E. W., Day, N. K., and Laurell, A. B.,** C3b inactivator deficiency: association with an α-migrating factor H, *J. Clin. Immunol.*, 1, 228, 1984.
94. **Barrett, D. J. and Boyle, M. D. P.,** Restoration of complement function *in vivo* by plasma infusion in factor I (C3b inactivator) deficiency, *J. Pediatr.*, 104, 76, 1984.
95. **Solal-Celigny, P., Laviolette, M., Herbert, J., Atkins, P. C., Siriois, M., Brun, G., Lehner-Netsch, G., and Delage, J. M.,** C3b-inactivator deficiency with immune-complex manifestations, *Clin. Exp. Immunol.*, 47, 197, 1982.
96. **Rasmussen, J. M., Teisner, B., Jepsen, H. H., Svehag, S.-E., Knudsen, F., Kirstein, H., and Buhl, M.,** Three cases of congenital factor I deficiency: the effect of treatment with plasma, *Clin. Exp. Immunol.*, 74, 131, 1988.
97. **Rasmussen, J. M., Teisner, B., Brandt, J., Brandslund, I., and Gry, H.,** Metabolism of C3 and factor B in patients with congenital factor I deficiency, *J. Clin. Lab. Immunol.*, in press.
98. **Rasmussen, J. M., Teisner, B., Brandslund, I., and Svehag, S.-E.,** A family with complement factor I deficiency, *Scand. J. Immunol.*, 23, 711, 1986.
99. **Charlesworth, J. A., Gwyn-Williams, D., Sherington, E., Lachmann, P. J., and Peters, D. K.,** Metabolic studies of the third component of complement and the glycine-rich β-glycoprotein in patients with hypocomplementemia, *J. Clin. Invest.*, 53, 1578, 1974.
100. **Thompson, R. A. and Winterborn, M. H.,** Hypocomplementemia due to genetic deficiency of β1H globulin, *Clin. Exp. Immunol.*, 46, 110, 1981.
101. **Lopez-Larrea, C., Dieguez, M. A., Enguix, A., Dominiguez, O., Marin, B., and Gomez, E.,** A familial deficiency of complement factor H, *Biochem. Soc. Trans.*, 15, 648, 1987.
102. **Brai, M., Misiano, G., Maringhini, S., Cutaja, I., and Hauptmann, G.,** Combined homozygous factor H and heterozygous C2 deficiency in an Italian family, *J. Clin. Immunol.*, 8, 50, 1988.
103. **Wyatt, R. J., Julian, B. A., Weinstein, A., Rothfield, N. F., and McLean, R. H.,** Partial H (β1H) deficiency and glomerulonephritis in two families, *J. Clin. Immunol.*, 2, 110, 1982.
104. **Nielsen, H. E., Koch, C., Magnussen, P., and Lind, I.,** Complement deficiencies in selected groups of patients with meningoccoccal disease, *Scand. J. Infect. Dis.*, 21, 389, 1989.
105. **Fijen, C. A. P., Kuijper, E. J., Hannema, A. J., Sjöholm, A. G., and van Putten, J. P. M.,** Complement deficiencies in patients over ten years old with meningococcal disease due to uncommon serogroups, *Lancet*, 2, 585, 1989.
106. **Ruddy, S., Purkall, D. B., and Gelfand, E. W.,** Formation of unstable alternative pathway convertase in properdin deficient serum, *Complement*, 2, 68, 1985.
107. **Davis, C. A. and Forristal, J.,** Partial properdin deficiency, *J. Lab. Clin. Med.*, 96, 633, 1980.
108. **Sjöholm, A. G., Braconier, J. H., and Söderstrøm, C.,** Properdin deficiency in a family with fulminant meningococcal infections, *Clin. Exp. Immunol.*, 52, 291, 1982.
109. **Spitzer, R. E., Vallota, E. H., Forristal, J., Sudora, E., Stitzel, A., Davis, N. C., and West, C. D.,** Serum C′3 lytic system in patients with glomerulonephritis, *Science*, 164, 436, 1969.
110. **Sissons, J. G. P., West, R. J., Fallows, J., Williams, D. G., Boucher, B. J., Amos, N., and Peters, D. K.,** The complement abnormalities of lipodystrophy, *N. Engl. J. Med.*, 294, 461, 1976.
111. **Teisner, B., Elling, P., Svehag, S.-E., Poulsen, L., Lamm, L. U., and Sjöholm, A. G.,** C3 nephritic factor in a patient with recurrent *Neisseria meningitidis* infections, *Acta Path. Microbiol. Scand.*, Sect. C, 92, 341, 1984.
112. **Rasmussen, J. M., Weihe, P., Teisner, B., Mathiassen, B., Petersen, T., and Isager, H.,** Screening for complement deficiencies in patients surviving from epidemic meningococcal disease, *J. Clin. Lab. Immunol.*, 25, 161, 1988.

113. **Scott, D. M., Amos, N., Sissons, J. G. P., Lachmann, P. J., and Peters, D. K.,** The immunoglobulin nature of nephritic factor (NeF), *Clin. Exp. Immunol.*, 32, 12, 1978.
114. **Daha, M. R., Fearon, D. T., and Austen, K. F.,** C3 nephritic factor (C3NeF): stabilization of fluid phase and cell-bound alternative pathway convertase, *J. Immunol.*, 116, 1, 1976.
115. **Daha, M. R., Austen, K. F., and Fearon, D. T.,** The incorporation of C3 nephritic factor (C3NeF) into a stabilized C3 convertase, C3bBb(C3NeF), and its release after decay of convertase function, *J. Immunol.*, 119, 812, 1977.
116. **Teisner, B., Brandslund, I., Folkersen, J., Rasmussen, J. M., Poulsen, L. O., and Svehag, S.-E.,** Factor I deficiency and C3 nephritic factor: immunochemical findings and association with Neisseria meningitidis infection in two patients, *Scand. J. Immunol.*, 20, 291, 1984.
117. **Fujita, T., Sumita, T., Yoshida, S., Ito, S., and Tamura, N.,** C4 nephritic factor in patients with chronic glomerulonephritis, *J. Clin. Lab. Immunol.*, 22, 65, 1987.
118. **Geha, R. S., Quinti, I., Austen, K. F., Cicardi, M., Scheffer, A., and Rosen, F. S.,** Acquired C1-inhibitor deficiency associated with antiidiotypic antibody to monoclonal immunoglobulins, *N. Engl. J. Med.*, 312, 534, 1985.
119. **Jackson, J., Sim, R. B., Whelan, A., and Feighery, C.,** An IgG autoantibody which inactivates C1-inhibitor, *Nature (London)*, 323, 722, 1986.

Chapter 15

DEGRADATION OF CIRCULATING CONNECTIVE TISSUE COMPONENTS

Kirsten D. Bentsen

TABLE OF CONTENTS

I. INTRODUCTION

Connective tissue consists of several structurally and biochemically rather different components, such as fibroblasts, collagens, glycoproteins, and proteoglycans. These components are essential for the many functions of connective tissue, being maintenance of tissue strength and elasticity and regulation of extracellular transport, cell adhesion, and proliferation.

A constant metabolism of the different connective tissue elements takes place in the healthy organism during growth and remodeling processes. In the inflammatory process which follows tissue injury, increased connective tissue metablism is seen during repair. These processes comprise *in situ* degradation, as well as increased synthesis and deposition of connective tissue components.

Within recent years, there has been an increasing interest in the identification and determination of different connective tissue-related components in body fluids. This interest is based on the search for markers directly related to the repair processes in inflammation and, consequently, to disease activity. Different connective tissue-related components have now been identified in the circulation from healthy people and increased amounts are generally found in persons with inflammatory and fibrosing disorders. The presence of these components in serum reflects different mechanisms in relation to connective tissue metabolism. They can be degradation products from connective tissue molecules already present in tissue or components related to newly synthesized molecules.

This chapter comprises a survey of the connective tissue components, which currently can be detected in body fluids of man. An overview of the structure, antigenic components in serum, as well as the degradation and excretion of the aminoterminal propeptide of type III procollagen in the circulation is presented. Finally, present knowledge of other less well-investigated connective tissue components is reviewed.

II. CONNECTIVE TISSUE-RELATED COMPONENTS IN BODY FLUIDS

The quantitatively most important connective tissue elements are the fibrillar type I and III collagens. These collagens are secreted as precursor molecules, the procollagens, which, in addition to the collagen molecule, hold a propeptide in the aminoterminal and carboxyterminal ends. Extracellularly, these propeptides are cleaved from the collagen molecule by specific proteases, before or during formation and growth of the collagen fibers.[1,2] These propeptides are easily soluble and are released into tissue fluids and serum. A third collagen, type IV collagen, is a major component of basement membranes. The amino- and carboxyterminal extensions of this collagen are not released before an association with other type IV collagens takes place. Actually, both ends are involved in intermolecular aggregation.[3] Another basement membrane component is the glycoprotein, laminin, a large cross-shaped molecule, closely linked to type IV collagen in the basement membrane structure.[4] The methods described up to now for determination of connective tissue elements in body fluids comprise assays detecting the type III procollagen aminoterminal propeptide (PIIINP), the type I procollagen carboxyterminal propeptide (PICP), the aminoterminal part of type IV collagen (7S domain), the carboxyterminal part of type IV collagen (NC1 domain), and fragments of the laminin molecule (laminin P1 and laminin P2 fragments) (for review, see Reference 5). Assays detecting the glycosaminoglycan, hyaluronan, are also presently available and described elsewhere in this book. These methods for determination of connective tissue components in serum have been developed within the last decade and have been applied in several studies of fibrosing disorders. However, apart from hyaluronan, very few studies have dealt with investigations of the metabolism of the components.

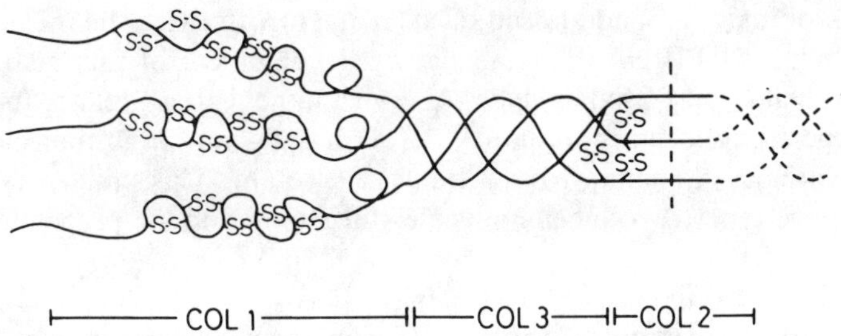

FIGURE 1. Schematic picture of the PIIINP molecule. The vertical stippled line indicates the physiological cleavage point between PIIINP and the type III collagen molecule.

III. TYPE III PROCOLLAGEN AMINOTERMINAL PROPEPTIDE

A. STRUCTURE

The conversion of type III procollagen to type III collagen is initiated by removal of the carboxyterminal propeptide. The remaining molecule (i.e., the collagen molecule still attached to the aminoterminal propeptide) is designated the pN-collagen. pN-collagen is found on the surface of growing collagen fibrils.[6] During further growth of the fibril, PIIINP is cleaved off and is released to tissue fluid and the blood stream. PIIINP is a peptide with a molecular weight of 42,000.[7] It is formed by the extensions of three polypeptide chains (α chains), which constitute the characteristic triple helix of the collagen molecule (Figure 1). PIIINP contains three domains. The aminoterminal end is a globular segment called the Col 1 domain, which contains five intra-chain bridges. Each of the amino acid chains in this domain is called the Col 1 fragment (MW, 9000). The Col 1 domain is attached to a collagenous triple helical sequence, Col 3. Between the Col 3 region and the triple helix of the collagen molecule is a short stretch of nonhelical structure, the Col 2 domain. The specific type III procollagen N-proteinase cleaves the propeptide from the collagen molecule in a certain region in the Col 2 domain.[8] However, the usual procedure to isolate the propeptide *in vitro* is by a bacterial collagenase. This procedure leaves the Col 2 domain intact. The physiologically released propeptide is therefore 17 amino acid residues shorter than the complete Col 1-3 molecule, which has a molecular weight of 45,000. However, the two molecules are immunologically indistinguishable. The isoelectric point of PIIINP is very low, about pH 3.0, in the human propeptide, probably due to the presence of sulfate, mainly located in the Col 1 domain.[9]

1. Antigenic Material in Body Fluids

Four PIIINP-related antigens have been demonstrated in body fluids from man, pig, and rat. The main antigen material in serum has a smaller molecular weight than the intact propeptide. This material consists of Col 1 fragments and somewhat larger molecules, probably complexes of Col 1 fragments.[10] It is still unknown whether these fragments are degradation products of tissue PIIINP or of circulating PIIINP. Col 1 fragments are always found in tissue fluids and their proportion decreases in relation to intact PIIINP with increasing fibrogenetic activity. This suggests that further degradation of recently synthesized PIIINP does not take place *in situ*. Only Col 1 fragments are found in urine. In serum from healthy adults, relatively small amounts of intact PIIINP are present, while an antigen of a molecular weight approximately two times that of the intact propeptide constitutes a larger fraction.[11] It may represent covalent cross-linked PIIINP molecules or complexes of PIIINP and other proteins.[12,13] The cross-linking is probably a process which takes place after the propeptide is released from tissue. Increased serum PIIINP values are generally associated with increased amounts of intact propeptide and of the antigen with twice the molecular weight of PIIINP.[11,14-16] Studies of PIIINP in serum during development of granulation tissue have shown that serum values of PIIINP and the somewhat larger PIIINP-related molecule reflects deposition of type III collagen in tissue.[14,17] A fourth antigen of high molecular weight has been found in variable amounts in body fluids and serum. This antigen is probably pN-collagen; that is, the type III collagen molecule still attached to the propeptide.

B. METHODS FOR PIIINP ANALYSES

Several assays for the determination of PIIINP, some of which are commercially available, have been developed.[5] In the studies of degradation and elimination of circulating PIIINP-related antigens, two assays have been of special benefit. In the PIIINP RIA Kit, the antibody has no affinity to the small PIIINP related antigens.[18,19] Thus, this assay measures the antigenecity of the molecules holding the intact PIIINP. No specific assay for PIIINP- related degradation products exists. In the PIIINP Fab assay, the antibody has an equal affinity to the molecules holding the intact propeptide and to the PIIINP fragments.[20,21] However, the PIIINP fragments constitute the main antigenecity in serum (at least 80 to 90%), which makes this assay sensitive to changes in the concentration of smaller PIIINP degradation products. In both assays, the reference antigen is the Col 1-3. In the PIIINP RIA Kit, the antigen is human, while it is bovine in the Fab-assay. Complete cross-reactivity exists between bovine and human PIIINP. The assays are suitable for analysis in man, pig, and sheep, but not for other laboratory animals, such as rats. The antibodies in both assays are raised in rabbits. Due to the use of Col 1-3 as a reference antigen, only approximate calculations of the concentrations of antigens larger or smaller than the intact PIIINP can be obtained.

C. DEGRADATION OF CIRCULATING PIIINP

Ultrastructural studies of normal rat liver have demonstrated PIIINP in the endothelial cells.[22] Intravenously injected radioactive-labeled bovine intact PIIINP in rat is mainly recovered in liver endothelial cells.[23] *In vitro* experiments using purified cultures of rat liver endothelial cells, Kuppfer cells, and hepatocytes have confirmed that a receptor-mediated uptake and lysosomal degradation of rat PIIINP takes place, mainly in the endothelial cells.[23] The receptors of the endothelial cells are specific and have a high affinity towards PIIINP. The number of receptors (1.6×10^5 PIIINP molecules were bound per cell) suggests that the normal liver has a potential for clearance of PIIINP, which is in excess of the uptake capacity needed in normal physiological situations. Investigations of intravenously administered radioactive-labeled bovine PIIINP in rat have suggeted a half-life in blood of about 2 min.[23] However, a recent tracer kinetic study in pigs has found the half-life of intact human and porcine PIIINP to be approximately 50 min.[24] The distribution volumen of PIIINP in pigs is 1.6-fold the plasma volume, which indicates a rapid initial extravascular distribution of PIIINP. The short half-life found in rats as compared to pigs could be due to an extravascular escape of injected PIIINP, as well as to the use of bovine PIIINP for elimination/degradation studies in rats. While a complete cross-reactivity exists between human, bovine, and procine PIIINP, rat PIIINP possesses somewhat different antigenic properties.[25] Major differences in PIIINP antigenetic properties between species could interfere with investigations of PIIINP metabolism.

Catheterization studies in pigs have confirmed that circulating intact PIIINP is extracted in the liver with an extraction ratio of 0.14, corresponding to a clearance of 90 ml/min.[26] Studies of the antigen profiles of the afferent and efferent vessels of the liver[27] as well as before and during liver transplantation,[28] indicate that the intact propeptide and pN-collagen are extracted in the liver. No hepatic extraction of the antigenic material with a molecular weight two times that of intact PIIINP is seen (Figure 2). In liver of healthy man, the hepatic extraction ratio is 0.17, which corresonds to a clearance of about 80 ml/min and a removal rate of 250 ng/min

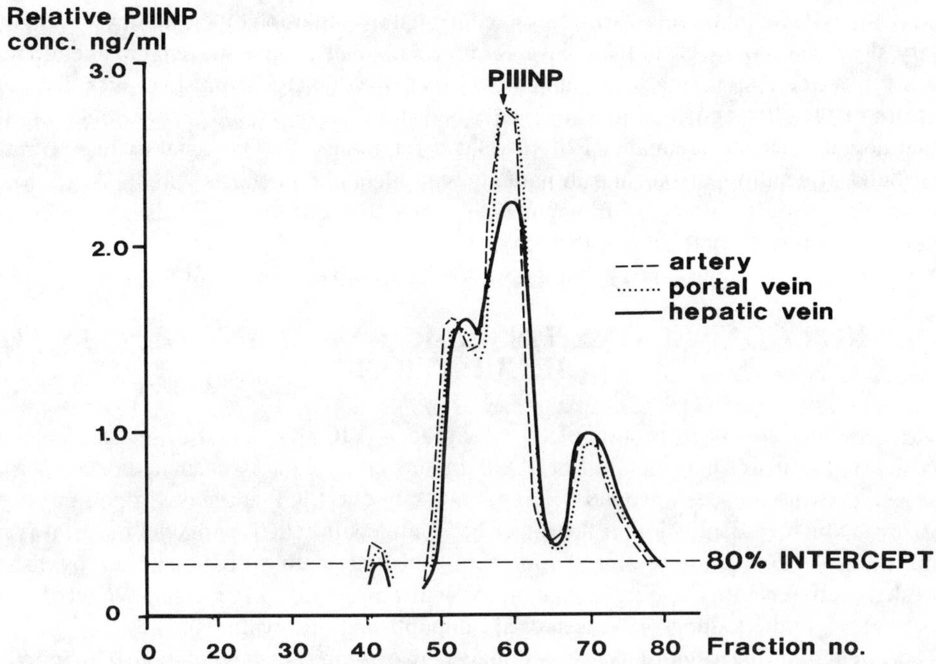

FIGURE 2. Antigen profile, showing PIIINP related antigens in serum from the artery, the hepatic vein and the portal vein in a pig (Reference 27). The antibody used for detecting PIIINP reactivity in these gelfiltrated samples, has a 10 to 20 times lower affinity to PIIINP fragments (fraction 66-78), as compared to the three other antigens holding the intact propeptide. The arrow indicates the elution position of the intact PIIINP.

in persons with serum PIIINP within the normal range.[29] The hepatic extraction of PIIINP and pN-collagen is in accordance with a capacity of the liver endothelial cells for capture and degradation of PIIINP. However, a transport of PIIINP from the circulation to the Disse space of the liver, through the fenestrated wall of the sinusoids, may also occur. No indications of hepatic extraction of PIIINP fragments have been found. This could suggest that the receptor specificity of the endothelial cells is related to the intact structure of the Col 1 domain or to the triple helix structure of the propeptide. Formation of complexes of PIIINP with other PIIINP molecules or proteins might interfere with the epitopes, reacting with specific PIIINP receptors on the endothelial cells, thus explaining the lack of a hepatic extraction of these PIIINP related antigens.

Development of alcoholic liver cirrhosis significantly decreases the hepatic extraction of circulating PIIINP with approximately 75%.[29] Development of alcoholic liver disease is characterized by capillarization of the sinusoidal wall (formation of a subendothelial basement membrane and defenestration of the endothelial cells) and development of perisinusoidal fibrosis. It is likely that the significant morphological changes of the endothelial cells in alcoholic liver diseases also reflect an impaired function of these cells and, thus, a decreased capacity for the capture and internalization of circulating PIIINP.

Intact PIIINP and Col 1 fragments are present in bile from liver-healthy persons and from patients with alcoholic cirrhosis.[30] Introducton of bile stasis in healthy pigs does not affect the hepatic extraction of circulating PIIINP, suggesting that PIIINP excretion in bile and the hepatic extraction of circulating PIIINP reflect different mechanism.[26] Biliary excretion of PIIINP increases during development of alcoholic cirrhosis, in accordance with an increased synthesis of type III collagen in the liver.[30] No data related to PIIINP metabolism in the intestine are currently available.

Only PIIINP related fragments are extracted from the circulation by the kidneys. In pig, the urinary clearance is only 2% of the renal clearance.[26] In healthy man, the renal extraction ratio is about 14%, corresponding to a net removal rate of 7000 μg/24 h.[29] However, the daily excretion of PIIINP fragments in urine is only between 100 and 350 μg,[10,30,31] which implies a renal degradation of the smaller PIIINP-related fragments. Similar renal extraction ratios were found in healthy persons and in patients with alcoholic cirrhosis,[29] in spite of a lower daily urinary excretion of PIIINP fragments in patients with cirrhosis.[20,30] This is probably due to a decreased renal blood flow in these patients.[32]

No indications of a pulmonary extraction of PIIINP have been found.[29]

IV. OTHER CONNECTIVE TISSUE-RELATED ANTIGENS IN THE CIRCULATION

The carboxyterminal propeptide of type I collagen (PICP) is found in serum of healthy individuals and increased amounts of PICP in the circulation is seen in bone diseases, alcoholic liver diseases, and wound healing.[5,33,34] Only one PICP antigen is found in serum, the size of which is identical with the intact PICP molecule.[5] PICP antigenic material is not found in urine. No data concerning the metabolism of circulating PICP has been published. However, preliminary investigations in man show that circulating PICP is not extracted either in the liver or in the kidneys. (Bentsen et al., unpublished observations).

The circulating antigens of type IV collagen may represent degradation of previously synthesized molecules or an overflow of newly synthesized molecules. Detection of type IV collagen in serum has been based on assays for the amino- and carboxyterminal ends of the molecule, the 7S and the NC1 domain, respectively.[5] In normal human serum, as well as in fibrosing disorders, the NC1 assay detects one type IV collagen fragment with a molecular size of 60,000.[35] In normal human serum, small amounts of antigenic material, corresponding to the aminoterminal part (the 7S domain) of type IV collagen, are present. Increased amounts of 7S immunoreactivity are found in liver diseases and are related to large homogenous molecules, possibly representing nearly intact type IV collagen molecules.[36] Some NC1 antigeneicity is found in urine. Further knowledge of the elimination routes and metabolism of type IV collagen in the circulation are currently unknown.

Laminin is a cross-shaped glycoprotein of high molecular weight. Like type IV collagen, laminin-related antigens in serum may represent degradation of basement membranes or laminin synthesis. Radioimmunoassays for two laminin related antigens (laminin P1 and laminin P2) have been developed.[5] The laminin P1 antigen is a globular domain present in the center of the molecule, while laminin P2 corresponds to globular domains of the two short arms of the molecule. In human serum, the laminin P1 antigen is found in several forms of different molecular sizes, of which the smallest corresponds to the size of the P1 fragment.[36,37] Laminin P2 has only been investigated in murine samples, which contain several P2-related antigens, some being intact laminin.[38] Only P2 antigenecity is found in urine.

Following intravenous injection of radioactive murine, intact laminin and laminin P1 in rat far most radioactivity is found in the liver endothelial cells.[39] A two-phase clearance of laminin from the circulation is observed. Initially, laminin is rapidly cleared from the circulation, which could indicate an initial extravascular distribution. A longer half-life of the second phase and the appearance of degraded laminin (acid-soluble radioactivity) during this phase could reflect capture, endocytosis, and degradation of laminin in liver endothelial cells. However, substantial data concerning presence, metabolism, and elimination of circulating laminin-related antigens are still lacking.

V. SUMMARY

Several connective tissue-related antigens are present in the circulation. Present knowledge of the degradation and elimination of these antigens is rather limited. The aminoterminal propeptide of type III procollagen (PIIINP) in the circulation is extracted and degraded in the liver. Hepatic extraction of circulating PIIINP is decreased in alcoholic liver diseases. Smaller PIIINP fragments are extracted and, to some extent, degraded in the kidneys. Preliminary results indicate that the liver is involved in the extraction and degradation of the type I procollagen carboxyterminal propeptide and of laminin.

REFERENCES

1. **Nimni, M. E.,** Collagen: structure and metabolism in normal and fibrotic tissues, *Sem. Arthr. Rheumat.*, 13, 1, 1983.
2. **Miller, E. J. and Gay, S.,** The collagens: an overview and update, *Methods Enzymol.*, 144, 3, 1987.
3. **Kühn, K., Glanville, R. W., Babel, W., Qian, R.-Q., Dieringer, H., Voss, T., Siebold, B., Oberbäumer, I., Schwartz, U., and Yamada, Y.,** The structure of type IV collagen, *Ann. N.Y. Acad. Sci.*, 460, 14, 1985.
4. **Martin, G. R.,** Laminin and other basement membrane components, *Annu. Rev. Cell Biol.*, 3, 57, 1987.
5. **Risteli, L. and Risteli, J.,** Noninvasive methods for detection of organ fibrosis, in *Connective Tissue in Health and Disease,* Vol. 1, Rojkind, M., Ed., CRC Press, Boca Raton, FL, 1989.
6. **Fleischmajer, R., Timpl, R., Tuderman, L., Raischer, L., Wiestner, M., Perlish, J., and Graves, P. N.,** Ultrastructural identification of extension aminopropeptides of type I and III collagens in human skin, *Proc. Natl. Acad. Sci. U.S.A.*, 78, 7360, 1981.
7. **Timpl, R. and Glanville, R. W.,** The aminopropeptide of collagen, Clin. Orthop. Relat. Res., 158, 224, 1981.
8. **Brandt, A., Glanville, R. W., Hörlein, D., Bruckner, P., Timpl, R., Fietzek, P., and Kühn, K.,** Complete amino acid sequence of the N-terminal extension of calf skin type III procollagen, *Biochem. J.*, 219, 625, 1984.
9. **Niemelä, O., Risteli, L., Parkkinen, J., and Risteli, J.,** Purification and characterization of the N-terminal propeptide of human type III procollagen, *Biochem. J.,* 232, 145, 1985.
10. **Niemelä, O.,** Radioimmunoassays for type III procollagen aminoterminal peptides in humans, *Clin. Chem.*, 31, 1301, 1985.
11. **Niemelä, O., Risteli, L., Sotaniemi, E. A., and Risteli, J.,** Heterogeneity of the antigens related to the aminoterminal propeptide of type III procollagen in human serum, *Clin. Chim. Acta,* 124, 39, 1982.
12. **Bowness, J. M., Folk, J. E., and Timpl, R.,** Identification of a substrate site for liver transglutaminase in the aminopropeptide of type III collagen, *J. Biol. Chem.*, 262, 1022, 1987.
13. **Bowness, J. M., Tarr, A. H., and Wiebe, R. I.,** Transglutaminase-catalysed cross-linking: a potential mechanism for the interaction of fibrinogen, low density lipoprotein and arterial type III procollagen, *Thromb. Res.*, 54, 357, 1989.
14. **Bentsen, K. D., Lanng, C., Hørslev-Petersen, K., and Risteli, J.,** The aminoterminal propeptide of type III procollagen and basement membrane components in serum during wound healing in man, *Acta Chir. Scand.*, 154, 97, 1988.
15. **Hørslev-Petersen, K., Bentsen, K. D., Engstrøm-Laurent, A., Junker, P., Halberg, P., and Lorenzen, I.,** Serum aminoterminal type III procollagen peptide and serum hyaluronan in rheumatoid arthritis: relation to clinical and serological parameters of inflammation during 8 and 24 months treatment with levamisole, penicillinamine or azathioprine, *Ann. Rheum. Dis.*, 47, 116, 1988.
16. **Jensen, L. T., Hørslev-Petersen, K., Toft, P., Bentsen, K. D., Grande, P., Simonsen, E. E., and Lorenzen, I.,** Serum aminoterminal type III procollagen peptide reflects repair after acute myocardial infarction. *Circulation,* 81, 52, 1990.
17. **Hørslev-Petersen, K., Kim, K. Y., Pedersen, L. R., Bentsen, K. D., Uldbjerg, N., Oxlund, H., Garbarsch, C., Hahn, E. G., Schuppan, D., and Lorenzen, I.,** Serum aminoterminal type III procollagen peptide. Relation to biosynthesis of collagen type III in experimentally induced granulation tissue in rats, *APMIS,* 96, 793, 1988.
18. **Risteli, J., Niemi, S., Trivedi, P., Mäentaustra, O., Mowat, A. P., and Risteli, L.,** Rapid equilibrium radioimmunoassay for the amino-terminal propeptide of human type III procollagen, *Clin. Chem.*, 34, 715, 1988.
19. Instruction booklet, *Procollagen PIIINP (^{125}I) Radioimmunoassay Kit*, Farmos Diagnostica, Farmos Group Ltd., Ouluensalo, Finland, 1987.
20. **Rohde, H., Langer, I., Krieg, T., and Timpl, R.,** Serum and urine analysis of the aminoterminal procollagen peptide type III by radioimmunoassay with antibody Fab fragments, *Coll. Relat. Res.*, 3, 371, 1983.
21. Instruction booklet, *RIA-Procollagen III Peptide (Fab),* Beehring AG, Marburg, Federal Republic of Germany, 1988.

22. **Geerst, A., Geuze, H. J., Slot, J. W., Voss, B., Schuppan, D., Schellinck, P., and Wisse, E.,** Immunogold localization of procollagen III, fibronectin and heparan sulfate proteoglycan on ultrathin frozen sections of normal rat liver, *Histochemistry*, 84, 355, 1986.
23. **Smedsröd, B.,** Aminoterminal propeptide of type III procollagen is cleared from the circulation by receptor-mediated endocytosis in liver endothelial cells, *Collagen Rel. Res.*, 8, 375, 1988.
24. **Jensen, L. T., Henriksen, J. H., Risteli, J., and Lorenzen, I.,** Metabolism of circulating aminoterminal propeptide of type III procollagen in pigs: a tracer kinetic study, manuscript in preparation.
25. **Schuppan, D., Dumont, J. M., Kim, K. Y., Hennings, G., and Hahn, E. G.,** Serum concentration of the aminoterminal procollagen type III peptide in the rat reflects early formation of connective tissue in experimental liver cirrhosis, *J. Hepatol.*, 3, 27, 1986.
26. **Bentsen, K. D., Henriksen, J. H., Boesby, S., Hørslev-Petersen, K., and Lorenzen, I.,** Hepatic and renal extraction of circulating type III procollagen amino-terminal propeptide and hyaluronan in pig, *J. Hepatol.*, 9, 177, 1989.
27. **Bentsen, K. D., Henriksen, J. H., Boesby, S., Jensen, L. T., Hørslev-Petersen, K., and Lorenzen, I.,** Hepatic extraction of the aminoterminal propeptide of type III procollagen during cholestasis in pigs, manuscript in preparation.
28. **Bentsen, K. D., Boesby, S., Kirkegaard, P., Hansen, C. P., Jensen, S. L., Hørslev-Petersen, K., and Lorenzen, I.,** Is the aminoterminal propeptide of type III procollagen degraded in the liver?, *J. Hepatol.*, 6, 144, 1988.
29. **Bentsen, K. D., Henriksen, J. H., Bendtsen, F., Hørslev-Petersen, K., and Lorenzen, I.,** Splanchnic and renal extraction of circulating type III procollagen aminoterminal propeptide in patients with normal liver function and in patients with alcoholic cirrhosis, *Hepatology*, 11, 957, 1990.
30. **Raedsch, R., Stiehl, A., Sieg, A., Walker, S., and Kommerell, B.,** Biliary excretion of procollagen type III peptide in healthy humans and in patients with alcoholic cirrhosis of the liver, *Gastroenterology*, 85, 1265, 1983.
31. **Hørslev-Petersen, K., Amnitzböll, T., Engström-Laurent, A., Bentsen, K. D., Jenker, P., Asboe-Hansen, G., and Lorenzen, I.,** Serum and urinary aminoterminal type III procollagen peptide in progressive systemic sclerosis: relationship to sclerodermal involvement, serum hyaluronan and urinary collagen metabolites, *J. Rheumatol.*, 15, 460, 1988.
32. **Ring-Larsen, H.,** Renal blood flow in cirrhosis: relation to systemic and portal haemodynamics and liver function, *Scand. J. Clin. Lab. Invest.*, 37, 635, 1977.
33. **Simon, L. S., Krane, S. M., Wortman, P. D., Krane, I. M., and Kovitz, K. L.,** Serum levels of type I and III procollagen fragments in Paget's disease of bone, *J. Clin. Endocrinol. Metab.*, 58, 110, 1984.
34. **Savolainen, E.-R., Goldberg, B., Leo, M. A., Velez, M., and Lieber, C. S.,** Diagnostic value of serum procollagen peptide measurements in alcoholic liver diseases, *Alcohol. Clin. Exp.*, 8, 384, 1984.
35. **Schuppan, D., Besser, M., Schwarting, R., and Hahn, E. G.,** Radioimmunoassay for the carboxyterminal cross-linking domain of type IV (basement membrane) procollagen in body fluids, *J. Clin. Invest.*, 78, 241, 1986.
36. **Niemelä, O., Risteli, L., Sotaniemi, E. A., and Risteli, J.,** Type IV collagen and laminin-related antigens in human serum in alcoholic liver disease, *Eur. J. Clin. Invest.*, 15, 132, 1985.
37. **Brocks, D. G., Strecker, H., Neubauer, H. P., and Timpl, R.,** Radioimmunoassay of laminin in serum and its application to cancer patients, *Clin. Chem.*, 32, 787, 1986.
38. **Risteli, J., Rohde, H., and Timpl, R.,** Sensitive radioimmunoassay for 7S collagen and laminin: Application to serum and tissue studies of basement membranes, *Anal. Biochem.*, 113, 372, 1981.
39. **Smedsröd, B., Paulsson, M., and Johansson, S.,** Uptake and degradation *in vivo* and *in vitro* of laminin and nidogen by rat liver cells, *Biochem. J.*, 261, 37, 1989.

Chapter 16

CATABOLISM OF HYALURONAN

Torvard C. Laurent and J. Robert E. Fraser

TABLE OF CONTENTS

I. INTRODUCTION

A. GENERAL

Hyaluronan (hyaluronic acid, hyaluronate, HYA; for nomenclature, see Balazs et al.[1]) is a polysaccharide found extracellularly in all tissues and tissue fluids but at the highest concentrations in soft connective tissue. It is therefore often regarded as a typical connective tissue component. Recent developments of new analytical techniques have made it possible to quantitate nanogram amounts of HYA and it has subsequently been found that increased concentration of the polysaccharide in blood and other body fluids is a sensitive marker of various pathological conditions.[2] This has given an impetus to the research on the catabolic pathways of HYA to be reviewed in this chapter.

The chemical structure of HYA is related to that of other connective tissue polysaccharides such as chondroitin sulphate and dermatan sulphate and they are collectively named glycosaminoglycans. Recent investigations have, however, shown that there are clear differences between HYA and the other glycosaminoglycans with regard to macromolecular properties, presence of core protein covalently bound to the polysaccharide, and site, mechanism, and regulation of biosynthesis.

B. STRUCTURE

HYA is a linear chain containing alternating residues of glucuronic acid and N-acetylglucosamine (Figure 1). It is polydisperse with regard to molecular weight, but the weight-average molecular weight is usually large when the polymer is extracted from tissues (of the order of several millions). The overall macromolecular structure of HYA is that of a coil extending over a large solvent domain. Its physiological function has often been connected with the unique physical-chemical properties of its solutions. Apparently, HYA plays an important role in the regulation of the water balance of tissues, transport through the extracellular matrix, lubrication of joints, etc. More recently, evidence has accumulated for direct interactions between cells and HYA. For reviews on the chemical and macromolecular structure, function, and biology of HYA, see References 3 through 7.

C. BIOSYNTHESIS

HYA is synthesized in the cell membrane.[8,9] The elongation takes place by addition of sugar residues to the reducing end of the chain.[10] The polymer is then presumably growing into the pericellular space, with the nonreducing end ahead. The cell surrounds itself with a HYA coat. This can be visualized by the exclusion of particles from the pericellular space.[11]

The molecular mechanism for the regulation of the synthesis of HYA is unknown. Certain evidence has been presented that phosphorylation is involved.[12] It has been known for a long time that a number of factors stimulate living cells to increase their production. These factors include various hormones, growth factors, inflammatory mediators, etc. A number of reports indicate that corticosteroids down regulate the synthesis. For references on various factors affecting the synthesis see Reference 5.

D. ANALYTICAL TECHNIQUES

The new methods which are now available to determine HYA on a nanogram scale are based on the specific affinity of certain proteins for the polysaccharide. Tengblad designed the first technique and he utilized the hyaluronan-binding region of the cartilage proteoglycan as a probe in a RIA-like assay.[13] His technique has subsequently been modified.[14-16] Delpech used the protein hyaluronectin in an ELISA-type assay.[17] Several other techniques based on similar principles have recently been published.[18-21]

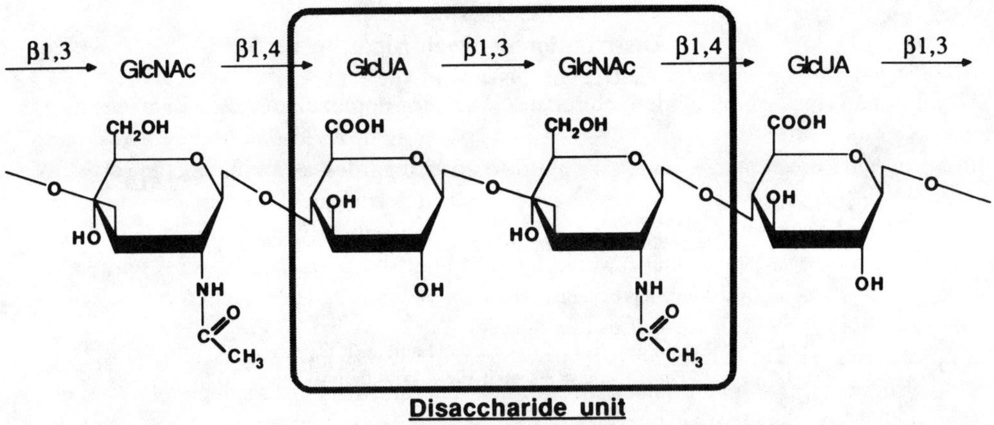

FIGURE 1. Structure of hyaluronan. GlcNAc, N-acetylglucosamine; GlcUA, glucuronic acid.

TABLE 1
Concentrations of Hyaluronan in Some Tissues and Tissue Fluids

	Conc (mg/l)
Human umbilical cord	4100
Human synovial fluid	1420—3600
Bovine nasal cartilage	1200
Human dermis	200
Rabbit brain	65
Rabbit muscle	27
Human thoracic lymph	8.5—18
Human urine	0.1—0.5
Human blood plasma	0.01—0.1

From Laurent, T. C. and Fraser, J. R. E., *Functions of the Proteoglycans,* Evered, D. and Whelan J., Eds., Ciba Foundation Symposium 124, John Wiley & Sons, Chichester, England, 1986, 9. With permission.

II. RATE OF TURNOVER OF HYALURONAN IN TISSUES

A. DISTRIBUTION OF HYALURONAN

HYA is found extracellularly in every tissue or body fluid that has been analyzed (Table 1). The highest concentrations are found in soft connective tissues such as synovial fluid, dermis, vitreous body, and umbilical cord, and the lowest levels are recorded in blood plasma or serum.

The total HYA content of a 250-g rat was determined by Reed et al. to be 40 to 60 mg.[22] A hypothetical extrapolation would give 11 to 17 g for the content of HYA in an adult human of 70 kg. The distribution of HYA between different tissues of the rat shows that a major part is deposited in skin and the musculoskeletal system (Table 2).

B. STUDIES ON THE LOCAL TURNOVER OF HYALURONAN

The following techniques have been used in studies on HYA turnover in a given tissue compartment:

TABLE 2
Distribution of Hyaluronan in Different Tissues of the Rat

	Percent
Skin	55.9
Muscles	7.8
Skeleton and supporting tissues	26.8
Intestines and stomach	0.8
Remaining internal organs	8.7

From Reed, R. K., Lilja, K., and Laurent, T. C., *Acta Physiol. Scand.*, 134, 405, 1988. With permission.

1. Schiller et al.[23-27] injected radioactive precursors (acetate, glucose, or glucosamine) into rabbits and rats. The animals were sacrificed at given time points after the injection, and HYA and chondroitin sulphate were isolated from skin and analyzed for radioactivity. Half-lives were calculated from the drop in specific radioactivity of the polysaccharide fractions with time. By this technique, the authors could measure the turnover of the total endogenous HYA pool in skin.
2. Denlinger[28] injected unlabeled HYA in joints and the vitreous body of the eye. She subsequently recovered the remaining material from the compartments after different time intervals and measured the amount of HYA. This technique requires that the concentration of exogenous HYA is considerably higher than the endogenous level, and the experiments are therefore carried out at nonphysiological conditions but are relevant for conditions when HYA is used clinically.
3. Labeled HYA can be injected in a compartment in trace amounts and its disappearance followed. This was first performed by injection of ^{3}H-labeled HYA in the blood stream[29] (see below) and subsequently used to study turnover in eye compartments.[30] In the latter study, the remaining radioactivity in the eye was determined after enucleation and combustion of the eye. A time study therefore required many animals. By this technique, it is possible to perform the studies at physiological concentrations but the injected HYA is in a soluble form and may not behave as HYA incorporated into structures in the tissue matrix.
4. The most recent technique[31] is a development of the previous one. Injected ^{3}H-labeled HYA, after having left the injection site, reaches the blood via the lymph (see below). It is rapidly taken up by the liver and degraded to $^3H\text{-}H_2O$. By following the appearance of tritiated water in serum and correcting for normal water turnover, one can determine the disappearance rate of radioactive HYA from the compartment where it has been injected.

The results of some investigations on the local turnover of HYA using these techniques have been summarized in Table 3. With the exception of the slow turnover rate in the vitreous body and the rapid turnover in the anterior chamber of the eye, all data indicate half-lives of the order of 0.5 to a few days. The somewhat slower turnover of endogenous HYA in skin, as recorded by Schiller et al.;[23-27] compared to the turnover of injected polysaccharide is not surprising in view of the fact that the endogenous polysaccharide is partly incorporated in structures and less mobile than the exogenous polymer.

The turnover is usually retarded when HYA is injected at higher concentrations and thus as more viscous solutions.

It is of interest that hormones may effect the turnover. Treatment of rats with hydrocortisone[25] decreases the turnover rate and, similarly, hypothyreotic rats exhibit a slow turnover.[27]

TABLE 3
Studies on the Half-Life ($t_{1/2}$) of Hyaluronan in Various Tissue Compartments

Species and tissue	Technique[a]	Conc of HYA injected (mg/ml)	$t_{1/2}$[b] (days)	Ref.
Rabbit skin	1	Endogenous	1.9—3.7	23, 24
Rat skin	1	Endogenous	2.6—4.5	24—27
Rabbit skin	4	Tracer	0.5	32
	4	10	2	31, 32
Rabbit muscle	4	10	1.25	31
Rhesus monkey vitreous body	2	10	20	33
Rabbit vitreous body	3	Tracer	70	30
Owl monkey anterior chamber	2	10	0.2—0.6	34
Cynomolgus monkey anterior chamber	4	10	0.8	35
Rabbit anterior chamber	3	Tracer	0.04—0.06	30
	4	10	0.5	31
Horse carpal and hock joints	2	10	1	28
Rabbit knee joint	2	20	0.5	28
	4	0.3	0.6	36
Sheep amniotic fluid	3			37
12 weeks of gestation		Tracer	3—8	
15—17 weeks of gestation		Tracer	0.5—0.8	
Rabbit and human blood circulation	3	Tracer	0.0014—0.0038	29, 50

[a] The techniques are described in detail in the text. 1. Incorporation of radioactive label in endogenous HYA and recordings of the fall in specific radioactivity. 2. Injection of unlabeled highly concentrated HYA and determination of absolute amounts in the tissues, which remain. 3. Injection of radioactive exogenous HYA and analysis of what remains in the tissue. 4. Injection of tritiated HYA locally and recording of tritiated water in blood.

[b] When the disappearance disobeys a first-order reaction, $t_{1/2}$ stands for the time when half of the material has left the compartment.

III. MECHANISM OF REMOVAL FROM TISSUES

A. LOCAL DEGRADATION

For a long time, it has been taken for granted that HYA is catabolized locally. It has been known that lysosomal hyaluronidase is present in many cells;[38] an abundant source is the liver from where it has been isolated.[39] The other lysosomal enzymes important in the degradation, β-glucuronidase and β-N-acetylglucosaminidase, are also widespread.[40]

Any direct evidence for a major local degradation is, however, lacking. It is of interest that adult skin fibroblasts seem to lack hyaluronidase[41] in spite of the fact that the skin is our major HYA depot. Furthermore, it was not possible to show degradation in synovial lining cells which form the border of a major HYA-containing compartment.[42] We can only conclude that the amount of HYA degraded in the peripheral tissues is unknown at present.

B. REMOVAL BY LYMPH DRAINAGE

A new aspect on the turnover of HYA was introduced by its identification in lymph in concentrations many-fold higher than in blood.[43] The presence of HYA in lymph has since then been verified in several different experimental systems.[44-49] The concentrations have varied between 0.2 and 50 μg/ml, depending on the site of collection and the species. The highest concentrations have been found in intestinal lymph. HYA in lymph is disperse with regard to molecular weight. Molecular weights of the order of 10^6 have been recorded in thoracic lymph, which is somewhat lower than that in most tissues but higher than that of HYA in serum (10^5).[44]

The discovery of HYA in lymph made it probable that a major route for its disappearance from the tissues could be via water flux, which washes out the polysaccharide from the extracellular matrix. Subsequent experiments on sheep lung[46,47] or dog paw[48] with increased transcapillary flow, due either to an increased hydrostatic pressure gradient or an increased capillary permeability, verified that HYA could be washed into the lymph in increasing amounts.

The total amount of HYA carried by lymph to blood in a healthy adult human has been estimated to be between 20 and 40 mg/day.[44]

It was recently shown that a considerable amount of HYA can be taken up and degraded in the lymph nodes.[45] Under certain conditions, greater than 95% of radioactive HYA was absorbed in the popliteal lymph node of sheep. The amount of HYA carried by afferent lymph to the nodes is therefore presumably larger than the amount registered in efferent lymph. This has been confirmed in studies on afferent and efferent popliteal lymph in sheep. Both the concentration and molecular weight of HYA drops when the lymph passes the node, due to a preferential uptake of the large HYA molecules.[51]

IV. TURNOVER OF HYALURONAN IN BLOOD

A. TURNOVER RATE

Fraser et al.[29] injected trace amounts of ^3H-labeled high-molecular weight HYA intravenously into rabbits and found that it disappeared with a $t_{1/2}$ of 2.5 to 5.5 min. The disappearance could be blocked by injection of milligram amounts of unlabeled HYA, indicating that the rapid uptake was a receptor-mediated process. Similar rapid turnover has since then also been noted in other species like man,[50] rat,[52] and sheep.[53]

The rate of disappearance from blood is a function of the molecular weight. In an experiment on rabbit,[42] two fractions of HYA with molecular weights of 5×10^6 and 10^5, respectively, were injected simultaneously; the former had a $t_{1/2}$ of 1.89 minutes and the latter a $t_{1/2}$ of 7 min.

From the turnover rate of tracers and the endogenous serum concentration of HYA, one can calculate that the total turnover in blood is of the order 10 to 100 mg/d,[15,50] in agreement with the supply from lymph. There seems to be an increase in turnover of HYA with age. The plasma concentration increases with age[15] from about 30 μg/l at 20 years to about 50 μg/l at age 50 and the endogenous turnover calculated from kinetic analyses increases from about 10 to 50 mg/d over the age interval 20 to 40 years.[54]

There is also a relatively larger turnover in the fetus compared with the adult individual. Experiments on adult and fetal sheep gave turnovers of 1.0 to 1.4 and 12.0 to 23.5 mg/d/kg body weight, respectively.[42,55]

B. SITES OF CLEARANCE

In the original paper by Fraser et al.,[29] it was shown that at least 90% of the intravenous radioactive HYA tracer ended up in the nonparenchymal cells of the rabbit liver and smaller amounts were found in the spleen. The major uptake in the liver has subsequently been confirmed in experiments on rats.[52] However, autoradiography on mice injected with high doses of ^{14}C-HYA indicated that, besides by liver, some material also was taken up by the kidney, spleen, lymph nodes, and bone marrow.[56] Analysis of blood obtained by catheterizations of humans has shown that the main extraction of the compound occurs in the splanch-

TABLE 4
The Michaelis-Menten Parameters Describing the Removal of Circulating Hyaluronan in Adult Sheep and Man[53, 54]

	Endogenous plasma conc (μg/ml)	K_m [a] (μg/ml)	V_{max} [b] μg/min/kg b.w.	V_{max} [b] mg/day	Endogenous turnover (mg/day)
Sheep	0.12	0.12	3.3	127	28
Man	0.031	0.34	3.5	350	34

[a] K_m = Michaelis-Menten constant.
[b] V_{max} = maximal metabolic rate.

nicus region (liver)[57] but that some HYA (as much as 10 to 20%) also is extracted in the kidneys[58] where it presumably is degraded. The amount of HYA appearing in urine is only about 1% of the total circulating polysaccharide and it has a molecular weight of less than 12,000, indicating that it has filtered through the glomeruli.[59]

The cells in the liver responsible for the clearance of HYA were identified both by *in vitro* experiments[60] and *in situ* autoradiography[61] as the endothelial cells lining the sinusoids. Neither hepatocytes nor Kupffer cells seemed to take part in the process.

The cells in spleen and lymph nodes, responsible for the uptake of radioactive HYA, are situated in the sinusoids of these organs.[42] The degradation of amniotic HYA takes place in the fetal intestine.[37]

C. A KINETIC MODEL FOR THE UPTAKE OF CIRCULATING HYALURONAN

Lebel et al.[53] used a computer program designed for studies on drug kinetics to describe the clearance of hyaluronan from blood. They injected intravenous bolus doses of HYA into sheep, followed by continuous infusions until plateaus above the normal endogenous level were reached. The disappearance rate of a radioactive tracer dose of HYA injected at a plateau level gave $t_{1/2}$ at that serum concentration. After termination of the infusion the HYA level fell towards the endogenous level. From the changes in concentration due to supply and removal of exogenous HYA, it was decided by the use of the computer program that the removal of HYA from blood is adequately described by a one-compartment, Michaelis-Menten kinetic model (see Chapter 3). The parameters describing the uptake, the Michaelis-Menten constant (K_m), and the maximal metabolic rate (V_{max}), are given in Table 4. The values of $t_{1/2}$ determined by the radioactive tracer at different serum levels agreed with the values calculated from the model parameters which gave a strong support for the model.

In similar experiments on man,[54] it could be shown that these data also could be described by Michaelis-Menten kinetics (Table 4). Although the maximal metabolic rate was approximately the same in man and sheep, 3.3 to 3.5 μg/min/kg b.w., there were distinct species differences. In man the Michaelis-Menten constant (0.34 μg/ml) was 10 times higher than the normal serum level (0.031 μg/ml). The maximal metabolic rate (350 mg/d) was also about ten times higher than the daily turnover (34 mg). In sheep, the Michaelis-Menten constant was equal to the plasma level (0.12 μg/ml) and the maximal metabolic rate 5 times the endogenous turnover, which means that the sheep has less reserve capacity.

The maximal metabolic rate per kilogram body weight has also been estimated for rabbit and rat to be of the same order as for man and sheep.[42]

The description of HYA uptake from plasma in terms of Michaelis-Menten kinetics verifies that the uptake is receptor mediated rather than occurring by unspecific pinocytosis. Assuming that the receptors are essentially localized in the liver, it follows that the turnover (as well as

TABLE 5
Binding Parameters for Hyaluronan Molecules to the Liver Endothelial Cell Receptor

Mol wt of hyaluronan	Binding constant K_d (μM)	Number of molecules bound per cell
2,400	1.0	200,000
7,000	0.2	120,000
400,000	6×10^{-5}	8,100
6,400,000	9×10^{-6}	900

From Laurent, T. C., Fraser, J. R. E., Pertoft, H., and Smedsröd, B., *Biochem. J.*, 234, 653, 1986. With permission.

"clearance" and "$t_{1/2}$") of HYA in the circulation is determined by the plasma concentration of HYA, the hepatic blood flow, and the number and activity of the hepatic receptors.

V. MOLECULAR ASPECTS ON UPTAKE AND DEGRADATION

A. RECEPTOR-MEDIATED UPTAKE

By *in vitro* experimentation, it became possible to demonstrate a receptor-mediated uptake of HYA in liver endothelial cells.[60,62-64] Only marginal uptake was recorded in hepatocytes and Kupffer cells.[60]

HYA shows saturated binding to the surface of liver endothelial cells in the cold. This binding is a function of the molecular weight of the polysaccharide[63] (Table 5). The larger the molecule, the higher the affinity. This can be explained by the increasing number of oligosaccharide sequences in the molecule that can bind to the receptor when the chain length increases. In contrast, the number of molecules that can bind at the same time to the cell surface decreases with increasing molecular size, due to molecular exclusion. The higher affinity of liver endothelial cells for large molecules explains the preference for removal of large molecules from the circulation.

The HYA receptor shows a high degree of specificity, but can also bind chondroitin sulphate.[63] The two polysaccharides have a very similar chain structure. As a matter of fact, when oligosaccharides of the same degree of polymerization were compared, chondroitin sulfate had a three times higher affinity for the receptor than HYA. Chondroitin sulfate in the circulation will thus compete with HYA for the receptor. There is conflicting evidence regarding the binding of heparin to the receptor.[60,64]

A protein from liver endothelial cell membranes, with a molecular weight of 100,000, which binds to a HYA-Sepharose column, was recently described by Forsberg and Gustafson.[65] It is presumably identical with the receptor because a polyvalent antiserum towards the protein inhibits binding of HYA to the cells.

When ^{3}H-labeled HYA is given to liver endothelial cells *in vitro* at 37°C, it becomes rapidly internalized.[60,62] The process is quite efficient and even at HYA concentrations up to 1 μg/ml each cell clears the polysaccharide from 2 pl medium/min. The process has been visualized in the electron microscope[66] with the aid of a gold-labeled ligand and the uptake is taking place in coated vesicles. McGary et al.[67] have shown that the receptors are rapidly recirculated and used for repeated uptake of HYA.

B. DEGRADATION

The internalized HYA is carried to lysosomes where it is degraded; this is a rapid process. After intravenous injection of labeled HYA, low molecular weight degradation products appear in the circulation already after 10 to 20 min. In the early experiments on rabbits by

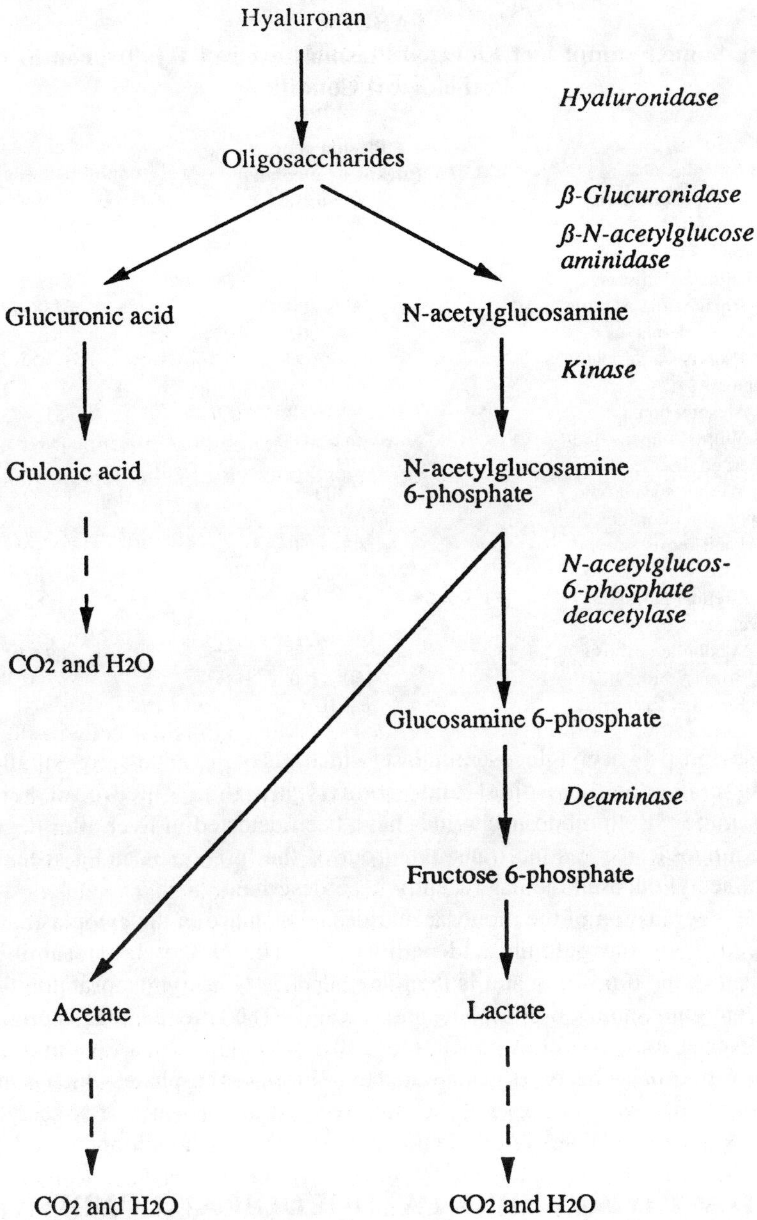

FIGURE 2. Scheme of hyaluronan degradation.

Fraser et al.,[29] the mean residence time of radioactivity from HYA in the liver was about 45 min. *In vitro* rat liver endothelial cells reach the steady state between uptake and degradation 75 min after addition of HYA to the culture.[60]

The end products of the degradation of HYA in a culture of liver endothelial cells are essentially acetate (from the N-acetyl group on glucosamine) and lactate (from the sugars).[60] These compounds are presumably taken up by the hepatocytes and degraded to carbon dioxide and water because *in vivo* ^3H-H_2O appears in the circulation and not ^3H-acetate.[29,50]

The general scheme for degradation of HYA is given in Figure 2. The polysaccharide is first acted upon by lysosomal hyaluronidase which is present in high concentrations in liver.[39] This endoenzyme degrades the polymer into shorter segments. These are acted upon by β-

TABLE 6
Some Examples of Elevated Plasma Levels of Hyaluronan in Pathological Conditions

Condition	Plasma conc (means or medians) (ng/ml)	Ref.
Normal (<60 years)	20—40	15, 17, 73
Inflammatory diseases		
Rheumatoid arthritis	165—400	73—77
Scleroderma	63—202	78, 79
Psoriasis/arthropathy	170	80
Tumors		
Mesothelioma	290—750	81—83
Wilms' tumor	Up to >20,000	84, 85
Inherited disease		
Werner's syndrome	300	59
Septic conditions, etc.		
Fatal sepsis	600—850	86
ARDS	620	87
Uremia/dialysis	150—210	88
Liver diseases		
Alcoholic cirrhosis	460—1225	89—91
Primary biliary cirrhosis	150—790	89, 90, 92
Rejected liver grafts	c:a 1000	93

glucuronidase and β-N-acetylglucosaminidase which cleave residues sequentially from the nonreducing end. The lysosomal end-products are thus glucuronic acid and N-acetylglucosamine.[40] Both monosaccharides have been detected in liver after degradation of HYA.[68] The monosaccharides are transported out of the lysosomes and a special transport system for N-acetylglucosamine has recently been described.[69]

The further degradation of the monosaccharides takes place in the cytoplasm. Glucuronic acid is degraded via the gulonic acid pathway.[40,68] The N-acetylglucosamine becomes phosphorylated in the 6-position and is then degraded by N-acetylglucosamine 6-phosphate deacetylase into glucosamine 6-phosphate and acetate.[70] The latter enzyme is present in large amounts in liver endothelial cells and Kuppfer cells, but is virtually absent in hepatocytes.[71] Glucosamine 6-phosphate becomes deaminated to fructose 6-phosphate, which is metabolized via the glycolytic pathway. Fraser et al.[45,55] have verified the presence of N-acetylglucosamine phosphates as intermediates in the degradation of HYA in lymph nodes and fetal liver.

VI. TURNOVER UNDER PATHOLOGICAL CONDITIONS

Our present knowledge on the metabolic routes of hyaluronan makes the use of hyaluronan as a clinical marker of a disease process more meaningful. It is now possible to relate pathological levels in blood and other fluids to specific deviations from the normal metabolism. Some examples will be given below (see also Table 6 and Reference 2).

A. INCREASED PLASMA LEVEL DUE TO INCREASED SYNTHESIS OF HYALURONAN

As mentioned in the introduction, various inflammatory mediators will activate the HYA synthesis, when given to fibroblasts *in vitro*. Also, inflammatory conditions *in vivo* may lead to increased synthesis and accumulation of HYA in tissues, the most striking example being the rheumatoid joint which may contain many-fold more than a normal joint.[72] This HYA can reach the circulation via lymph vessels.

Increased plasma levels of HYA have been recorded; for example, in rheumatoid arthritis, scleroderma, and psoriasis (Table 6). That this is due to an outflux of HYA from the tissues became apparent when the diurnal variation was investigated.[75,94] A rheumatoid patient often has a low plasma level of HYA after a night's rest and before the start of physical activities in the morning. One hour after rise from bed, the level has usually gone up to high values and it then falls towards the afternoon. The increase during the morning hour is correlated both to the extent of joint involvement in the disease and to the duration of morning stiffness. Apparently, HYA is accumulated in the tissues during the night, causing an edema and stiffness of the tissues. When the patient becomes physically active in the morning, HYA is pumped via lymph to the circulation, relieving the stiffness and increasing the plasma level.

The observations on humans with joint disease have recently been extended to animals with experimental arthritis, which also exhibit high plasma levels of HYA.[95,96]

There are other conditions in which an increased HYA synthesis results in increased plasma levels. Certain tumors produce HYA (e.g., mesotheliomas and Wilms' tumor) and, in some of these cases, extremely high levels of the polysaccharide are noted (up to 1000-fold the normal level). In these cases, the tumors must produce more circulating HYA than can be cleared by the liver endothelial cells.

There is also an inherited disease, Werner's syndrome, with a high plasma level.

B. ELEVATED PLASMA LEVEL DUE TO INCREASED INTERSTITAL WATER FLUX

The discovery that HYA is removed from the tissues by interstitial liquid flow offered another explanation for increased plasma levels; that is, a change in transvascular water flux through the tissues could effect the transport of HYA. This was tested experimentally by Lebel et al.,[46] who tapped lymph from sheep lung. Increase in the left atrial pressure resulted in a highly elevated lymph flow from the lung and an increased flux of HYA. Similarly, if the animals were made septic by the injection of *E. coli,* the lymph flow increased due to increased capillary permeability.[47] Also, in this case, HYA was washed out from the lungs in increasing amounts. An increase in the plasma level of HYA followed.

Although there has been no systematic study of the HYA plasma level in patients with disturbed water balance, there are reasons to believe that it is an important factor. This could be one of many factors which lead to increased HYA levels in septic patients (Table 6). Other factors could be increased synthesis of HYA or decreased clearance.

C. DECREASED CLEARANCE OF CIRCULATING HYALURONAN

A decreased capacity for uptake of HYA in the liver, and possibly other sites for its clearance, should also result in increased plasma levels. This hypothesis was tested at an early stage on patients with various liver diseases[89] and it was found that cases with liver cirrhosis showed markedly increased plasma HYA (Table 6). The observation has been verified in different laboratories and plasma HYA can be regarded as a very sensitive marker for liver cirrhosis. It was also shown that the clearance rate of HYA from blood is decreased in patients with liver cirrhosis.[97]

Different mechanisms can explain the high plasma level of HYA in liver cirrhosis. There could either be a decreased number of functional liver endothelial cells or there could be a decreased blood flow through the sinusoids. Our present techniques cannot distinguish between these alternatives. It is interesting that, in experiments on pigs with fecal peritonitis, there was a dramatic increase in plasma HYA which could be completely explained by decreased hepatic blood flow.[98] It has also been suggested that HYA synthesis in the cirrhotic liver causes increased plasma levels. Finally, one cannot exclude the possibility that increased splanchnic lymph flow due to the cirrhosis carries increasing amounts of HYA to the circulation.

TABLE 7
Some Hyaluronan Levels in Body Fluids other than Blood in Various Pathological Conditions

Fluid	Hyaluronan conc (mean or median)	Ref.
Urine (mg/24 h)		59
Healthy persons	330	
Rheumatoid arthritis	510	
Primary biliary cirrhosis	917	
Werner's syndrome	3313	
Lung lavage fluid (ng/ml)		
Healthy persons	<5	99
Sarcoidosis	16	99
Farmer's lung	547	100
Jejunal fluid (ng/ml)		101
Healthy persons	12	
Crohn's disease and celiac disease	34	
Skin blisters (μg/ml)		
Normal skin	1—10	102, 103
Active psoriasis	24—30	102
Scleroderma, etc.	2—55	103
Middle ear effusions (μg/ml)	3—42	104
Amniotic fluid (μg/ml)		105
Normal (16—20 weeks of gestation)	20	
Fetus with renal dysplasia	<0.2	

As mentioned above, there are indications that some HYA also is metabolized in the kidneys. In line with this observation, moderately elevated levels of HYA were observed in patients with kidney failure[88] (Table 6).

D. PATHOLOGICAL HYALURONAN LEVELS IN OTHER TISSUE FLUIDS

Increased HYA levels in other body fluids may also have diagnostic significance. Table 7 summarizes some studies of relevance. The increase of HYA in urine presumably mirrors an increased plasma level. Local inflammatory processes in lung, intestine, skin, and the middle ear are followed by sometimes highly increased HYA production. Oppositely, in a case of dysplastic kidneys in a fetus, no HYA was secreted via urine to the amniotic fluid, where the HYA concentration became abnormally low.

VII. SUMMARY

HYA is an extracellular polysaccharide produced continuously in the peripheral tissues. It is, to a significant extent, removed from tissues by lymph flow and degraded in lymph nodes or transported to the general circulation. HYA in plasma is cleared mainly by receptor-mediated uptake in liver endothelial cells.

A tentative scheme for the turnover of HYA in man, based on human data and extrapolations from animal experiments, is presented in Figure 3. A half-life of tissue HYA of 2 d has been assumed.

Pathological plasma levels are found both when there is an increased outflux of HYA from the tissues or an impaired clearance in the liver.

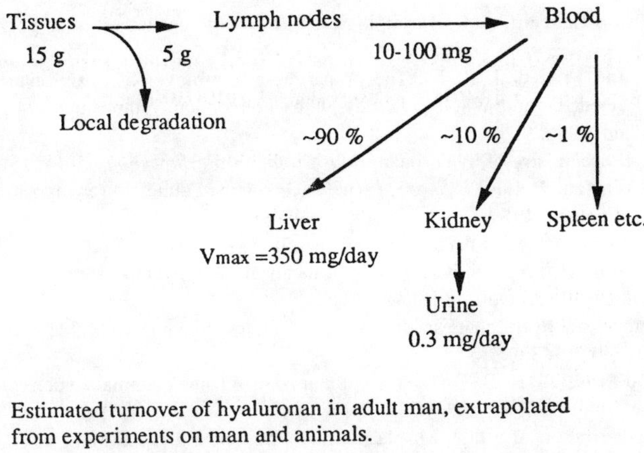

FIGURE 3. Estimated turnover of hyaluronan in adult man extrapolated from experiments on man and animals.

NOTE ADDED IN PROOF

After this article was submitted, it came to our attention that a series of articles published in Japanese[106-110] describe the fate of ^{14}C-labeled HYA injected in rabbits and rats. The HYA was injected intravenously, intraarticularly, subcutaneously, and intramuscularly. The results are compatible with those described above if one takes into consideration that the doses injected intravenously were 1 mg/kg body weight which is 1000-fold above the physiological level and therefore gives half-lives of the order of hours instead of minutes. High levels of radioactivity were found in liver, spleen, and lymph nodes, but the major part was recovered in ^{14}C-CO_2 in the respiratory gas.

ACKNOWLEDGMENTS

This project has been supported by the Swedish Medical Research Council (grant 03X-4), the Australian Rheumatism Foundation, Ingabritt and Arne Lundberg's Research Foundation, and Pharmacia AB.

ABBREVIATIONS

HYA Hyaluronan (hyaluronate, hyaluronic acid)
K_m Michaelis-Menten constant
V_{max} Maximal metabolic rate
$t_{1/2}$ Half-life of HYA in tissues or blood

REFERENCES

1. **Balazs, E. A., Laurent, T. C., and Jeanloz, R. W.,** Nomenclature of hyaluronic acid, *Biochem. J.*, 235, 903, 1986.
2. **Engström-Laurent, A. and Laurent, T. C.,** Hyaluronan as a clinical marker, in *Clinical Impact of Bone and Connective Tissue Markers*, Lindh, E. and Thorell, J. I., Eds., Academic Press, London, 1989, 235.
3. **Laurent, T. C.,** Structure of hyaluronic acid, in *Chemistry and Molecular Biology of the Intercellular Matrix*, Vol. 2, Balazs, E. A., Ed., Academic Press, London, 1970, 703.

4. **Comper, W. D. and Laurent, T. C.,** Physiological function of connective tissue polysaccharides, *Physiol. Rev.*, 58, 255, 1978.
5. **Laurent, T. C. and Fraser, J. R. E.,** The properties and turnover of hyaluronan, in *Functions of the Proteoglycans*, Evered, D. and Whelan, J., Eds., Ciba Foundation Symposium 124, John Wiley & Sons, Chichester, England, 1986, 9.
6. **Laurent, T. C.,** Biochemistry of hyaluronan, *Acta Otolaryngol. (Stockholm),* 442 (Suppl.), 7, 1987.
7. **Evered, D. and Whelan, J.,** Eds., *Biology of Hyaluronan,* Ciba Foundation Symposium 143, John Wiley & Sons, Chichester, England, 1989.
8. **Prehm, P.,** Hyaluronate is synthesized at plasma membranes, *Biochem. J.*, 220, 597, 1984.
9. **Philipson, L. H. and Schwartz, N. B.,** Subcellular localization of hyaluronate synthetase in oligodendroglioma cells, *J. Biol. Chem.,* 259, 5017, 1984.
10. **Prehm, P.,** Synthesis of hyaluronate in differentiated teratocarcinoma cells. Mechanism of chain growth, *Biochem. J.*, 211, 191, 1983.
11. **Clarris, B. J. and Fraser, J. R. E.,** On the pericellular zone of some mammalian cells *in vitro*, *Exp. Cell Res.*, 49, 181, 1968.
12. **Mian, N.,** Characterization of a high-M_r plasma-membrane-bound protein and assessment of its role as a constituent of hyaluronate synthase complex, *Biochem. J.,* 237, 343, 1986.
13. **Tengblad, A.,** Quantitative analysis of hyaluronate in nanogram amounts, *Biochem. J.*, 185, 101, 1980.
14. **Laurent, U. B. G. and Tengblad, A.,** Determination of hyaluronate in biological samples by a specific radioassay technique, *Anal. Biochem.*, 109, 386, 1980.
15. **Engström-Laurent, A., Laurent, U. B. G., Lilja, K. and Laurent, T. C.,** Concentration of sodium hyaluronate in serum, *Scand. J. Clin. Lab. Invest.*, 45, 497, 1985.
16. **Brandt, R., Hedlöf, E., Åsman, I., Bucht, A., and Tengblad, A.,** A convenient radiometric assay for hyaluronan, *Acta Otolaryngol. (Stockholm),* 442 (Suppl.), 31, 1987.
17. **Delpech, B., Bertrand, P., and Maingonnat, C.,** Immunoenzymoassay of the hyaluronic acid-hyaluronectin interaction: application to the detection of hyaluronic acid in serum of normal subjects and cancer patients, *Anal. Biochem.*, 149, 555, 1985.
18. **Goldberg, R. L.** Enzyme-linked immunosorbent assay for hyaluronate using cartilage proteoglycan and an antibody to keratan sulfate, *Anal. Biochem.*, 174, 448, 1988.
19. **Li, X.-Q., Thonar, E. J.-M. A., and Knudson, W.,** Accumulation of hyaluronate in human lung carcinoma as measured by a new hyaluronate ELISA, *Connective Tissue Res.*, 19, 243, 1989.
20. **Chichibu, K., Matsuura, T., Shichijo, S., and Yokoyama, M. M.,** Assay of serum hyaluronic acid in clinical application, *Clin. Chim. Acta,* 181, 317, 1989.
21. **Kongtawelert, P. and Ghosh, P.,** An enzyme-linked immunosorbent-inhibition assay for quantitation of hyaluronan (hyaluronic acid) in biological fluids, *Anal. Biochem.*, 178, 367, 1989.
22. **Reed, R. K., Lilja, K., and Laurent, T. C.,** Hyaluronan in the rat with special reference to the skin, *Acta Physiol. Scand.*, 134, 405, 1988.
23. **Schiller, S., Mathews, M. B., Goldfaber, L., Ludowieg, J., and Dorfman, A.,** The metabolism of mucopolysaccharides in animals. II. Studies in skin utilizing labeled acetate, *J. Biol. Chem.,* 212, 531, 1955.
24. **Schiller, S., Mathews, M. B., Cifonelli, J. A., and Dorfman, A.,** The metabolism of mucopolysaccharides in animals. III. Further studies on skin utilizing C^{14}-glucose, C^{14}-acetate and S^{35}-sodium sulfate, *J. Biol. Chem.*, 218, 139, 1956.
25. **Schiller, S. and Dorfman, A.,** The metabolism of mucopolysaccharides in animals: the effect of cortisone and hydrocortisone on rat skin, *Endocrinology,* 60, 376, 1957.
26. **Schiller, S. and Dorfman, A.,** The metabolism of mucopolysaccharides in animals. IV. The influence of insulin, *J. Biol. Chem.,* 227, 625, 1957.
27. **Schiller, S., Slover, G. A., and Dorfman, A.,** Effect of thyroid gland on metabolism of acid mucopolysaccharides in skin, *Biochim. Biophys. Acta,* 58, 27, 1962.
28. **Denlinger, J. L.,** Metabolism of Sodium Hyaluronate in Articular and Ocular Tissues, Ph.D. thesis, University of Lille, France, 1982.
29. **Fraser, J. R. E., Laurent, T. C., Pertoft, H., and Baxter, E.,** Plasma clearance, tissue distribution and metabolism of hyaluronic acid injected intravenously in the rabbit, *Biochem. J.*, 200, 415, 1981.
30. **Laurent, U. B. G. and Fraser, J. R. E.,** Turnover of hyaluronate in the aqueous humour and vitreous body of the rabbit, *Exp. Eye Res.,* 36, 493, 1983.
31. **Laurent, U. B. G., Fraser, J. R. E., and Laurent, T. C.,** An experimental technique to study the turnover of concentrated hyaluronan in the anterior chamber of the rabbit, *Exp. Eye Res.*, 46, 49, 1988.
32. **Reed, R. K., Laurent, U. B. G., Fraser, J. R. E., and Laurent, T. C.,** Removal rate of [^{3}H]-hyaluronan injected subcutaneously in the rabbit, *Am. J. Physiol.*, 259, H532, 1990.
33. **Denlinger, J. L. and Balazs, E. A.,** The fate of exogenous viscoelastic hyaluronan solutions in the primate eye, in *Viscoelastic Materials. Basic Science and Clinical Applications*, Rosen, E. S., Ed., Pergamon Press, Oxford, 1989, 185.

34. **Schubert, H. D., Denlinger, J. L., and Balazs, E. A.,** Exogenous Na-hyaluronate in the anterior chamber of the owl monkey and its effect on the intraocular pressure, *Exp. Eye Res.*, 39, 137, 1984.
35. **Laurent, U. B. G. and Fraser, J. R. E.,** Disappearance of concentrated hyaluronan from the anterior chamber of monkey eyes, *Exp. Eye Res.*, 51, 65, 1990.
36. **Brown, T. J., Laurent, U. B. G., and Fraser, J. R. E.,** Turnover of hyaluronan in synovial joint. Elimination of labelled hyaluronan from the knee joint of the rabbit, *Exp. Physiol.,* in press.
37. **Dahl, L. B., Kimpton, W. G., Cahill, R. N. P., Brown, T. J., and Fraser, J. R. E.,** The origin and fate of hyaluronan in amniotic fluid, *J. Dev. Physiol.*, 12, 209, 1989.
38. **Bollet, A. J., Bonner, W. M., Jr., and Nance, J. L.,** The presence of hyaluronidase in various mammalian tissues, *J. Biol. Chem.*, 238, 3522, 1963.
39. **Aronson Jr., N. N. and Davidson, E. A.,** Lysosomal hyaluronidase from rat liver, *J. Biol. Chem.,* 242, 437 and 441, 1967.
40. **Rodén, L., Campbell, P., Fraser, J. R. E., Laurent, T. C., Pertoft, H., and Thompson, J. N.,** Enzymic pathways of hyaluronan catabolism, in *The Biology of Hyaluronan,* Evered, D. and Whelan, J., Eds., Ciba Foundation Symposium 143, John Wiley & Sons, Chichester, England, 1989, 60.
41. **Arbogast, B., Hopwood, J. J., and Dorfman, A.,** Abscence of hyaluronidase in cultured human skin fibroblasts, *Biochem. Biophys. Res. Commun.*, 67, 376, 1975.
42. **Fraser, J. R. E. and Laurent, T. C.,** Turnover and metabolism of hyaluronan, in *The Biology of Hyaluronan*, Evered, D. and Whelan, J., Eds., Ciba Foundation Symposium No. 143, John Wiley & Sons, Chichester, England, 1989, 41.
43. **Laurent, U. B. G. and Laurent, T. C.,** On the origin of hyaluronate in blood, *Biochem. Int.,* 2, 195, 1981.
44. **Tengblad, A., Laurent, U. B. G., Lilja, K., Cahill, R. N. P., Engström-Laurent, A., Fraser, J. R. E., Hansson, H. E., and Laurent, T. C.,** Concentration and relative molecular mass of hyaluronate in lymph and blood, *Biochem. J.*, 236, 521, 1986.
45. **Fraser, J. R. E., Kimpton, W. G., Laurent, T. C., Cahill, R. N. P., and Vakakis, N.,** Uptake and degradation of hyaluronan in lymphatic tissue, *Biochem. J.*, 256, 153, 1988.
46. **Lebel, L., Smith, L., Risberg, B., Gerdin, B., and Laurent, T. C.,** Effect of increased hydrostatic pressure on lymphatic elimination of hyaluronan from sheep lung, *J. Appl. Physiol.*, 64, 1327, 1988.
47. **Lebel, L., Smith, L., Risberg, B., Gerdin, B., and Laurent, T. C.,** Increased lymphatic elimination of interstitial hyaluronan during *E. coli* sepsis in sheep, *Am. J. Physiol.*, 256, H1524, 1989.
48. **Reed, R. K., Laurent, T. C., and Taylor, A. E.,** Hyaluronan in prenodal lymph from skin: changes with lymph flow, *Am. J. Physiol.*, 259, H1097, 1990.
49. **Ferrara, J., Reed, R. K., Dyess, D. L., Townsle, M. I., Onarheim, H., Laurent, T. C., and Taylor, A. E.,** Increased hyaluronan flux from skin following burn injury, *J. Surg. Res.,* in press.
50. **Fraser, J. R. E., Laurent, T. C., Engström-Laurent, A., and Laurent, U. B. G.,** Elimination of hyaluronic acid from the blood stream in the human, *Clin. Exp. Pharmacol. Physiol.*, 11, 17, 1984.
51. **Fraser, J. R. E.,** Hyaluronan: sources, turnover and metabolism, in *Clinical Impact of Bone and Connective Tissue Markers,* Lindh, E. and Thorell, J. I., Eds., Academic Press, London, 1989, 31.
52. **Dahl, L. B., Laurent, T. C., and Smedsröd, B.,** Preparation of biologically intact radioiodinated hyaluronan of high specific radioactivity: coupling of ^{125}I-tyramine cellobiose to amino groups after partial N-deacetylation, *Anal. Biochem.,* 175, 397, 1988.
53. **Lebel, L., Fraser, J. R. E., Kimpton, W. S., Gabrielsson, J., Gerdin, B., and Laurent, T. C.,** A pharmacokinetic model of intravenously administered hyaluronan in sheep, *Pharm. Res.,* 6, 677, 1989.
54. **Lebel, L., Gabrielsson, J., Laurent, T. C., and Gerdin, B.,** Kinetics of circulating hyaluronan in man, in Turnover of Circulating Hyaluronan, Thesis, Lebel, L., Uppsala University, 1989.
55. **Fraser, J. R. E., Dahl, L. B., Kimpton, W. G., Cahill, R. N. P., Brown, T. J., and Vakakis, N.,** Elimination and subsequent metabolism of circulating hyaluronic acid in the foetus, *J. Dev. Physiol.,* 11, 235, 1989.
56. **Fraser, J. R. E., Appelgren, L.-E., and Laurent, T. C.,** Tissue uptake of circulatiing hyaluronic acid. A whole body autoradiographic study, *Cell Tissue Res.*, 233, 285, 1983.
57. **Henriksen, J. H., Bentsen, K. D., and Laurent, T. C.,** Splanchnic and renal extraction of circulating hyaluronan in patients with alcoholic liver disease, *J. Hepatol.*, 6, 158, 1988.
58. **Bentsen, K. D., Henriksen, J. H., and Laurent, T. C.,** Circulating hyaluronate: concentration in different vascular beds in man, *Clin. Sci.*, 71, 161, 1986.
59. **Laurent, T. C., Lilja, K., Brunnberg, L., Engström-Laurent, A., Laurent, U. B. G., Lindqvist, U., Murata, K., and Ytterberg, D.,** Urinary excretion of hyaluronan in man, *Scand. J. Clin. Lab. Invest.*, 47, 793, 1987.
60. **Smedsröd, B., Pertoft, H., Eriksson, S., Fraser, J. R. E., and Laurent, T. C.,** Studies *in vitro* on the uptake and degradation of sodium hyaluronate in rat liver endothelial cells, *Biochem. J.*, 223, 617, 1984.
61. **Fraser, J. R. E., Alcorn, D., Laurent, T. C., Robinson, A. D., and Ryan, G. B.,** Uptake of circulating hyaluronic acid by the rat liver. Cellular localization *in situ, Cell Tissue Res.*, 242, 505, 1985.

62. **Eriksson, S., Fraser, J. R. E., Laurent, T. C., Pertoft, H. and Smedsröd, B.,** Endothelial cells are a site of uptake and degradation of hyaluronic acid in the liver, *Exp. Cell Res.*, 144, 223, 1983.
63. **Laurent, T. C., Fraser, J. R. E., Pertoft, H., and Smedsröd, B.,** Binding of hyaluronate and chondroitin sulphate to liver endothelial cells, *Biochem. J.*, 234, 653, 1986.
64. **Raja, R. H., McGary, C. T., and Weigel, P. H.,** Affinity and distribution of surface and intracellular hyaluronic acid receptors in isolated rat liver endothelial cells, *J. Biol. Chem.*, 263, 16661, 1988.
65. **Forsberg, N. and Gustafson, S.,** Partial characterization of the receptor for hyaluronan on liver endothelial cells, to be submitted.
66. **Smedsröd, B., Malmgren, M., Ericsson, J., and Laurent, T. C.,** Morphological studies on endocytosis of chondroitin sulphate proteoglycan by rat liver endothelial cells, *Cell Tissue Res.*, 253, 39, 1988.
67. **McGary, C. T., Raja, R. H., and Weigel, P. H.,** Endocytosis of hyaluronic acid by rat liver endothelial cells. Evidence for receptor recycling, *Biochem. J.*, 257, 875, 1989.
68. **Fraser, J. R. E., Robinson, A. D., Holmes, M. F. W., Laurent, T. C., and Brown, T. J.,** Metabolic degradation of hyaluronan. Deacetylation follows the release of monosaccharides, to be submitted.
69. **Jonas, A. J., Speller, R. J., Conrad, P. B., and Dubinsky, W. P.,** Transport of *N*-acetyl-D-glucosamine and *N*-acetyl-D-galactosamine by rat liver lysosomes, *J. Biol. Chem.*, 264, 4953, 1989.
70. **Campbell, P., Laurent, T. C., and Rodén, L.,** Assay and properties of *N*-acetylglucosamine-6-phosphate deacetylase from rat liver, *Anal. Biochem.*, 166, 134, 1987.
71. **Campbell, P., Thompson, J. N., Fraser, J. R. E., Laurent, T. C., Pertoft, H., and Rodén, L.,** *N*-Acetylglucosamine-6-phosphate deacetylase in hepatocytes, Kupffer cells, and sinusoidal endothelial cells from rat liver, *Hepatology*, 11, 199, 1990.
72. **Dahl, L. B., Dahl, I. M. S., Engström-Laurent, A., and Granath, K.,** Concentration and molecular weight of sodium hyaluronate in synovial fluid from patients with rheumatoid arthritis and other arthropathies, *Ann. Rheum. Dis.*, 44, 817, 1985.
73. **Seibel, M. J., Raimann, S., Seglias, J., Lautenschläger, J., and Müller, W.,** Serum-Hyaluronsäure und aminoterminales Prokollagen-III-Peptid bei entzündlichen und degenerativen Gelenkerkrankungen, *Z. Rheumatol.*, 47, 98, 1988.
74. **Engström-Laurent, A. and Hällgren, R.,** Circulating hyaluronate in rheumatoid arthritis: relationship to inflammatory activity and the effect of corticosteroid therapy, *Ann. Rheum. Dis.*, 44, 83, 1985.
75. **Engström-Laurent, A. and Hällgren, R.,** Circulating hyaluronic acid levels vary with physical activity in healthy subjects and in rheumatoid arthritis patients, *Arthr. Rheum.*, 30, 1333, 1987.
76. **Hörslev-Petersen, K., Bentsen, K. D., Engström-Laurent, A., Junker, P., Halberg, P., and Lorenzen, I.,** Serum amino terminal type III procollagen peptide and serum hyaluronan in rheumatoid arthritis: relation to clinical and serological parameters of inflammation during 8 and 24 months' treatment with levamisole, penicillamine, or azathioprine, *Ann. Rheum. Dis.*, 47, 116, 1988.
77. **Levesque, H., Delpech, B., LeLoet, X., and Deshayes, P.,** Serum hyaluronate in rheumatoid arthritis: study by affino-immunoenzymatic assay, *Br. J. Rheumatol.*, 27, 445, 1988.
78. **Engström-Laurent, A., Feltelius, N., Hällgren, R., and Wasteson, Å.,** Raised serum hyaluronate levels in scleroderma: an effect of growth factor induced activation of connective tissue cells?, *Ann. Rheum. Dis.*, 44, 614, 1985.
79. **Hörslev-Petersen, K., Ammitzböll, T., Engström-Laurent, A., Bentsen, K., Junker, P., Asboe-Hansen, G., and Lorenzen, I.,** Serum and urinary aminoterminal type III procollagen peptide in progressive systemic sclerosis: relationship to sclerodermal involvement, serum hyaluronan and urinary collagen metabolites, *J. Rheum.*, 15, 460, 1988.
80. **Lundin, Å., Engström-Laurent, A., Hällgren, R., and Michaëlsson, G.,** Circulating hyaluronate in psoriasis, *Br. J. Dermatol.*, 112, 663, 1985.
81. **Frébourg, T., Lerebours, G., Delpech, B., Benhamou, D., Bertrand, P., Maingonnat, C., Boutin, C., and Nouvet, G.,** Serum hyaluronate in malignant pleural mesothelioma, *Cancer*, 59, 2104, 1987.
82. **Dahl, I. M. S. and Laurent, T. C.,** Concentration of hyaluronan in the serum of untreated cancer patients with special reference to patients with mesothelioma, *Cancer*, 62, 326, 1988.
83. **Dahl, I. M. S., Solheim, Ö. P., Erikstein, B., and Müller, E.,** A longitudinal study of the hyaluronan level in the serum of patients with malignant mesothelioma under treatment. Hyaluronan as an indicator of progressive disease, *Cancer*, 64, 68, 1989.
84. **Wu, A. H. B., Parker, O. S., and Ford, L.,** Hyperviscosity caused by hyaluronic acid in serum in a case of Wilms' tumor, *Clin. Chem.*, 30, 914, 1984.
85. **Cooper, E. H. and Forbes, M. A.,** Serum hyaluronic acid levels in cancer, *Br. J. Cancer*, 58, 668, 1988.
86. **Berg, S., Brodin, B., Hesselvik, F., Laurent, T. C., and Maller, R.,** Elevated levels of plasma hyaluronan in septicemia, *Scand. J. Clin. Lab. Invest.*, 48, 727, 1988.
87. **Hällgren, R., Samuelsson, T., Laurent, T. C., and Modig, J.,** Accumulation of hyaluronan (hyaluronic acid) in the lung in adult respiratory distress syndrome, *Am. Rev. Respir. Dis.*, 139, 682, 1989.
88. **Hällgren, R., Engström-Laurent, A., and Nisbeth, U.,** Circulating hyaluronate. A potential marker of altered metabolism of the connective tissue in uremia, *Nephron*, 46, 150, 1987.

89. **Engström-Laurent, A., Lööf, L., Nyberg, A., and Schröder, T.**, Increased serum levels of hyaluronate in liver disease, *Hepatology*, 5, 638, 1985.

90. **Frébourg, T., Delpech, B., Bercoff, E., Senant, J., Bertrand, P., Deugnier, Y., and Bourreille, J.**, Serum hyaluronate in liver diseases: study by enzymoimmunological assay, *Hepatology*, 6, 392, 1986.

91. **Bentsen, K. D., Horn, T., Risteli, J., Risteli, L., Engström-Laurent, A., Hörslev-Petersen, K., and Lorenzen, I.**, Serum aminoterminal type III procollagen peptide and the 7S domain of type IV collagen in patients with alcohol abuse. Relation to ultrastructural fibrosis in the acinar zone 3 and to serum hyaluronan, *Liver*, 7, 339, 1987.

92. **Nyberg, A., Engström-Laurent, A., and Lööf, L.**, Serum hyaluronate in primary biliary cirrhosis — a biochemical marker for progressive liver damage, *Hepatology*, 8, 142, 1988.

93. **Adams, D. H., Wang, L., Hubscher, S. G., and Neuberger, J. M.**, Hepatic endothelial cells. Targets in liver allograft rejection?, *Transplantation*, 47, 479, 1989.

94. **Lindqvist, U., Engström-Laurent, A., Laurent, U., Nyberg, A., Björklund, U., Eriksson, H., Pettersson, R., and Tengblad, A.**, The diurnal variation of serum hyaluronan in health and disease, *Scand. J. Clin. Lab. Invest.*, 48, 765, 1988.

95. **Björk, J., Kleinau, S., Tengblad, A., and Smedegård, G.**, Elevated levels of serum hyaluronate and correlation with disease activity in experimental models of arthritis, *Arthr. Rheum.*, 32, 306, 1989.

96. **Goldberg, R. L. and Rubin, A. S.**, Serum hyaluronate as a marker for disease severity in the *Lactobacillus casei* model of arthritis in the rat, *J. Rheumatol.*, 16, 92, 1989.

97. **Fraser, J. R. E., Engström-Laurent, A., Nyberg, A., and Laurent, T. C.**, Removal of hyaluronic acid from the circulation in rheumatoid disease and primary biliary cirrhosis, *J. Lab. Clin. Med.*, 107, 79, 1986.

98. **Arvidsson, D., Lebel, L., Rasmussen, I., Haglund, U., Laurent, T. C., and Gerdin, B.**, Hepatic extraction of hyaluronan in bacterial peritonitis in the pig, in Turnover of Circulating Hyaluronan, Thesis, Uppsala University, 1989.

99. **Hällgren, R., Eklund, A., Engström-Laurent, A., and Schmekel, B.**, Hyaluronate in brochoalveolar lavage fluid: a new marker in sarcoidosis reflecting pulmonary disease, *Br. Med. J.*, 290, 1778, 1985.

100. **Bjermer, L., Engström-Laurent, A., Lundgren, R., Rosenhall, L., and Hällgren, R.**, Hyaluronate and type III procollagen peptide concentrations in bronchoalveolar lavage fluid as markers of disease activity in farmer's lung, *Br. Med. J.*, 295, 803, 1987.

101. **Colombel, J. F., Hällgren, R., Engström-Laurent, A., and Rambaud, J. C.**, Hyaluronic acid and type III procollagen peptide in jejunal perfusion fluid as markers of connective tissue turnover, *Gastroenterology*, 96, 68, 1989.

102. **Lundin, Å., Engström-Laurent, A., Michaëlsson, G., and Tengblad, A.**, High levels of hyaluronate in suction blister fluid from active psoriatic lesions, *Br. J. Dermatol.*, 116, 335, 1987.

103. **Juhlin, L., Tengblad, A., Ortonne, J. P., and Lacour, J. P.**, Hyaluronate in suction blisters from patients with scleroderma and various skin disorders, *Acta Derm. Venereol. (Stockholm)*, 66, 409, 1986.

104. **Laurent, C., Tengblad, A., and Engström-Laurent, A.**, Hyaluronic acid in secretory otitis media. A biochemical study of mucoid middle-ear effusions and serum, *J. Laryngol. Otol.*, 100, 135, 1986.

105. **Dahl, L. B., Dahl, I. M. S., and Börresen, A.-L.**, The molecular weight of sodium hyaluronate in amniotic fluid, *Biochem. Med. Metabol. Biol.*, 35, 219, 1986.

106. **Sakamoto, T., Mizuno, S., Miyazaki, K., Yamaguchi, T., Toyoshima, H., and Namiki, O.**, Biological fate of sodium hyaluronate (SPH). I. Studies on distribution, metabolism and excretion of ^{14}C-SPH in rabbits after intra-articular administration, *Pharmacometrics*, 28, 375, 1984.

107. **Sakamoto, T., Mizuno, S., Miyazaki, K., Yamaguchi, T., Toyoshima, H., and Namiki, O.**, Biological fate of sodium hyaluronate (SPH). II. Studies on distribution, metabolism and excretion of ^{14}C-SPH in rabbits after intravenous administration, *Pharmacometrics*, 28, 421, 1984.

108. **Esumi, Y., Takaichi, M., Hasegawa, M., Sakamoto, T., Mizuno, S., and Yamaguchi, T.**, Biological fate of sodium hyaluronate (SPH). III. Distribution of ^{14}C-SPH revealed by autoradiography in rats after intravenous and in rabbits after intraarticular administration, *Pharmacometrics*, 28, 433, 1984.

109. **Mizuno, S., Tominaga, A., Sakamoto, T., and Tokuyasu, K.**, Biological fate of sodium hyaluronate (SPH). IV. Studies on absorption, distribution, metabolism and excretion of ^{14}C-SPH in rat after intravenous, subcutaneous and intramuscular administration, *Pharmacometrics*, 33, 849, 1987.

110. **Miyauchi, S., Mizuno, S., Onaya, J., Sakamoto, T., and Tokuyasu, J.**, Biological fate of sodium hyaluronate (SPH). V. Behaviour of ^{14}C-SPH in the serum, liver and spleen of rabbits after intravenous administration, *Pharmacometrics*, 33, 859, 1987.

Chapter 17

DEGRADATION OF CIRCULATING CORTICAL STEROIDS PHYSIOLOGY AND PATHOPHYSIOLOGY

Mauro Bernardi, Franco Trevisani, Antonio Gasbarrini, and Giovanni Gasbarrini

TABLE OF CONTENTS

I. PHYSIOLOGICAL ASPECTS

Steroidogenic cells are located in adrenal cortex, testis, and ovary. The enzymatic cascade leading to hormone biosynthesis, as well as the rate-limiting step, are similar in adrenocortical and gonadal tissues. The enzymes are distributed in the mitochondria, microsomes, and soluble cell fraction. The quantities of steroids synthesized in each gland vary according to the relative proportion of enzymatic activities. Steroid hormones can be classified as glucocorticoids, mineralocorticoids, progestogens (all containing 21 carbon atoms), androgens (19 carbon atoms), and estrogens (18 carbon atoms), depending on their main biological activity.

A. CHEMISTRY

The basic structure of steroid hormones consists of four carbon-containing rings which lie in a plane (Figure 1). Indeed, bond angles and ring fusions vary from one compound to another, due to different carbon-atom saturation and attached prostetic groups. The latter may extend above or below the plane. The degree of angulation between rings and the kind of prostetic groups are crucial determinants for the biological activity of the molecule.[1] Corticosteroid activity demands the presence of a double bond between carbons 4 and 5, hydroxyl groups at the 11β and 21 positions, and ketones at carbons 3 and 20. Mineralocorticoid activity is mainly determined by the presence of an aldehyde group in position 18, while glucocorticoid activity depends on the hydroxylation at carbon 11 and is potentiated by the presence of a hydroxyl group in position 17.[1,2]

B. BIOSYNTHESIS

Cholesterol is the common precursor of all steroid hormones. The cellular availability of this substrate derives from *de novo* synthesis or, to a greater extent, from the uptake of circulating cholesterol carried in LDL and HDL particles produced by the liver. LDL-cholesterol accounts for up to 70% of substrate for steroidogenesis.[3] Intracellular cholesterol can be immediately utilized or stored as esters in lipid droplets. These constitute a depot utilized during prolonged gland stimulation when facilitated uptake of LDL-cholesterol, due to an increased number of cellular receptors, raises the intracellular availability of substrate. These adaptative mechanisms succeed in increasing tenfold the normal production of cortisol in response to sustained stimulation of the adrenal cortex.[4]

Conversion to pregnenolone, by cleavage of the side chain from cholesterol, occurs in the mitochondrial membrane. This represents the rate-limiting step of steroidogenesis, since subsequent enzyme activities are present in relative excess.[4]

In the zonae reticularis and fasciculata of the adrenal cortex where 17α-hydroxylation can only occur, pregnenolone is sequentially converted to 17α-OH-pregnenolone, 17α-OH-progesterone, and then 17-hydroxyl-11-deoxycortisol by an enzyme complex located in the smooth endoplasmic reticulum.[5] The latter product flows back to mitochondria, where it is hydroxylated at position 11 to yield cortisol. Cortisol production can also follow an alternative pathway from pregnenolone to 17α-OH-progesterone through progesterone formation.

In the zona glomerulosa,[5,6] pregnenolone is preferentially converted to progesterone. Thereafter, this steroid is sequentially hydroxylated at carbons 21 and 11 to yield 11-deoxycorticosterone and corticosterone. The formation of the final product aldosterone is achieved through oxidation of the 18-hydroxyl group to an aldehyde by the enzyme 18-hydroxysteroid dehydrogenase, which is unique to the zona glomerulosa.

The zonae reticularis and fasciculata also secrete at least four compounds with androgenic activity:[5] dehydro-epiandrosterone (DHEA) and its sulfate form (DHEAS), which are produced in the greatest amount, androstenedione and testosterone. DHEA is mainly derived from 17α-OH-pregnenolone. Most of the androstenedione and testosterone derive from 17α-OH-progesterone and, to lesser extent, from DHEA.[7] The conversion of androstenedione to estrone

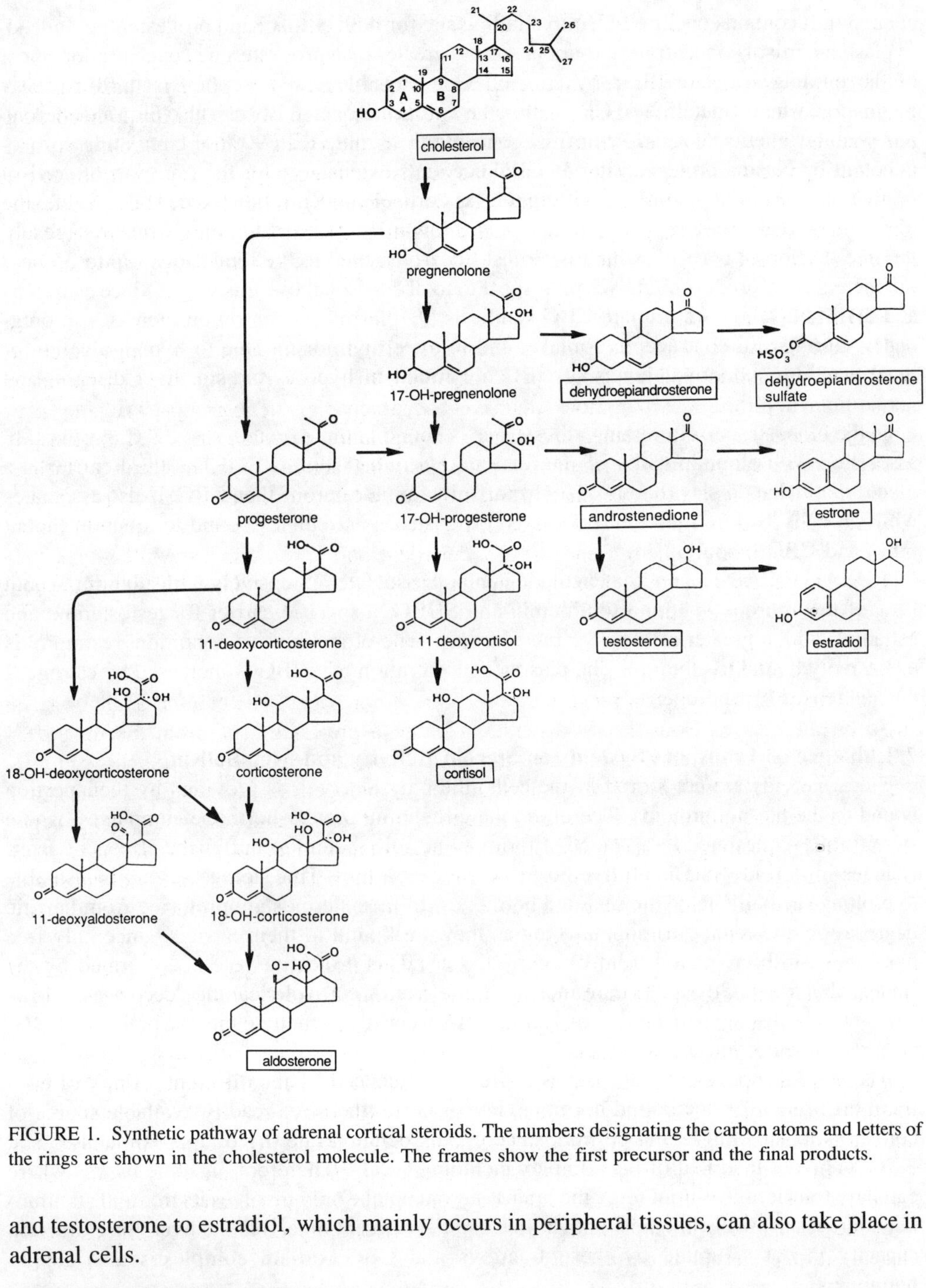

FIGURE 1. Synthetic pathway of adrenal cortical steroids. The numbers designating the carbon atoms and letters of the rings are shown in the cholesterol molecule. The frames show the first precursor and the final products.

and testosterone to estradiol, which mainly occurs in peripheral tissues, can also take place in adrenal cells.

C. TRANSPORT IN BLOOD

1. Transport Systems

The main transport systems of steroid hormones in blood are represented by two specific carriers, corticosteroid binding globulin (CBG, also called transcortin) and sex hormones binding globulin (SHBG, also named testosterone-estradiol binding globulin), and by albumin.[8,9]

CBG is an α-glycoprotein primarily synthesized by the liver, with a plasma half-life of about 5 d. It contains one high-affinity binding site for both cortisol and progesterone. Indeed, as plasma cortisol concentration is about tenfold greater than progesterone concentration, most of the binding sites are filled by cortisol. One physiological exception is third-trimester pregnancy, when about 25% of CBG is filled by progesterone.[8] CBG can also bind aldosterone and some synthetic steroids. At normal concentration, more than 90% of circulating cortisol is bound by plasma proteins. Human CBG accounts extensively for the transport of cortisol up to a plasma concentration of 250 ng/ml. As cortisol concentration exceeds this value, the hormone is weakly bound by albumin, that is, a low-affinity, nonsaturable carrier. As a result, the free portion of cortisol, which is almost linearly related to the total mount up to 250 ng/ml, increases disproportionately with respect to total cortisol above this value. Since estrogens and thyroid hormones stimulate CBG synthesis, its plasma concentration increases in pregnancy, during oral contraceptive intake, and in hyperthyroidism. Due to hypoproduction or renal wasting, subnormal levels of CBG are found in hypothyroidism, liver disease, and nephrotic syndrome.

In the bloodstream, circulating aldosterone is bound to three proteins. At 37°C, it is weakly associated with albumin (60% of the total amount) and CBG (20%). The third carrier is a glycoprotein that displays a tenfold higher affinity for aldosterone than CBG. It also associates with red cells.[10] More than 95% of deoxycorticosterone circulates bound to albumin (about 60%) and CBG (about 35%).

DHEA, DHEAS, and androstenedione in blood associates extensively with albumin. About 98% of testosterone is bound to albumin and SHBG, a specific carrier for testosterone and estradiol, with a greater affinity for the androgen. Therefore, in most conditions, estradiol is extensively bound to albumin. The plasma concentration of SHBG is increased by estrogens and decreased by androgens.

2. Influence of Transport Systems on Steroid Activity and Metabolism

Since steroids are not stored in the cells, the only reservoir is provided by their portion bound to the plasma proteins. Therefore, the circulating protein-hormone complexes, where steroids are sequestered in an inactive form, is the sole system to buffer the effects of sharp changes in steroidogenesis on free hormone concentrations. This storage system is also able to prolong the half-life of the secreted hormones. In fact, steroids are protected from hepatic degradation and renal clearance as long as they are bound to their carriers, since only free hormones can be removed from blood by tissues. This protective role is confirmed by the finding that when CBG levels are increased, the rate of cortisol clearance decreases.[8] However, it has been shown that albumin-bound cortisol can also penetrate into the brain and CBG-steroids can enter into hepatocytes.[11]

The steroid supply to target tissues is greatly influenced by the different affinity of each transport protein for the bound hormone. For instance, the more readily available source of cortisol is its albumin-bound portion, since the dissociation rate ($t_{1/2}$ of approximately 5 s at 37°C) is higher than that of the complex including CBG.[11] Therefore, in some tissues where capillary transit time is brief (e.g., the brain), steroid uptake only involves its free and albumin-bound fractions, while in others with a slow capillary transit time, like the liver, the extraction capacity is not saturated by these fractions and CBG-cortisol complexes also supply hormone.[12]

Since aldosterone rapidly dissociates from its plasma carriers, a large fraction of the total amount is biologically available as it flows through the target tissues.[12] The slower dissociation rate from CBG than from albumin determines a preferential liver uptake of the hormone bound to the latter. Changes in plasma cortisol concentration can modify the amount of CBG-bound aldosterone. Due to the higher affinity of this protein with respect to albumin, such changes can influence tissue removal of the hormone and then its metabolic clearance rate in spite of the small amount of aldosterone which is bound to CBG.[13]

D. DEGRADATION

It is the small fraction of "free" hormones, dynamically maintained in equilibrium with that bound to transport proteins,[12] which interacts with cellular receptors of the target tissues and induces biological effects. Internalization in target cells thus represents one of the metabolic pathways for steroids.

Since a large fraction of filtered hydrophobic compounds are actively reabsorbed by the renal tubule (e.g., 95% of cortisol and 85% of aldosterone),[14] renal excretion cannot represent an important clearance tool unless parent steroids are transformed into more hydrosoluble derivatives. The liver is the main site of this metabolic conversion, although many mammalian tissues, including the adrenal glands, can metabolize steroids.[14] Conversion to hydrophilic compounds involves the transformation of the ketone groups into hydroxyl groups and the groups in position 21 into acid forms.[1,2] The C–21 steroids are mainly esterified with glucuronic acid by the enzyme glucuronyl transferase, involving uridine-diphosphoglucuronic acid as donor. In normal adult subjects, this system is so active that almost all reduced corticosteroids reaching the liver are completely esterified in one passage.[2] The enzyme is not the same as that responsible for bilirubin conjugation, since steroid glucuro-conjugation is less severely affected in patients with altered bilirubin conjugation.[2] The C-19 steroids are extensively conjugated to sulfate by sulfotransferase.[1,2] The sulfate conjugates bind to plasma proteins to a greater extent and show a longer half-life than glucuronide conjugates. Liver, gonads, and placenta are able to hydrolyze steroid sulfates, producing free steroids.

The kidney eliminates over 90% of the steroid metabolites, the remainder being secreted in the gut lumen. Here they undergo further transformation by the bacterial flora (especially aldosterone derivatives) and are then reabsorbed.[1] Unlike the majority of steroids, a large fraction of corticosterone and deoxycorticosterone excreted into the gut is relevant (see Section I.D.1.c): more than 50 cortisol or aldosterone derivatives have been identified.

Although the changes in the molecular structure of steroids occurring during the metabolic pathway generally destroy their biological effectiveness, there are some exceptions. The activity of testosterone is enhanced by 5α-reductase, an enzyme located in the reproductive male tract and skin.[15] As a result, the active form 5α-dihydrotestosterone is generated in the target tissues. Moreover, circulating androstenedione can be converted into testosterone and both circulating androstenedione and testosterone are important sources of estradiol.[16]

1. Metabolic Pathways of the Single Adrenal Corticosteroids

a. Cortisol (Figure 2)

Radiolabeled cortisol disappears from the blood with a half-life of 80 to 120 min, according to various studies.[14,17] However, the half-life of endogenous cortisol at normal concentration has been estimated to be 66 min.[18] Over 90% of the injected radioactivity appears in the urine by 72 h, almost all as metabolites, since unchanged cortisol accounts for less than 1%. About 3% is recovered in the feces.[14] Therefore, in normal individuals, the amount of unmetabolized cortisol excreted in the urine is less than 100 μg/d and is reduced when renal function is impaired.[19]

The degradation of cortisol starts with an irreversible reduction of the 4,5 double bond of the A ring by at least two reductases, yielding the inactive molecule dihydrocortisol. The latter may be in α or β form, according to the orientation of the hydrogen atom in position 5. C-21 steroids are chiefly converted into the β-isomer. Both forms do not accumulate, since they are immediately reduced to tetrahydrocortisol by 3α-hydroxylases, located in both endoplasmic reticulum and cytosol.

Provided that the A ring is intact, cortisol can also be reversibly oxidized at the 11β-hydroxyl group to ketone, resulting in cortisone which is biologically inactive. As cortisol, cortisone is mainly metabolized via the dihydro to tetrahydro derivatives. Since there is a consistent conversion of cortisol to cortisone in physiological conditions, these molecules and

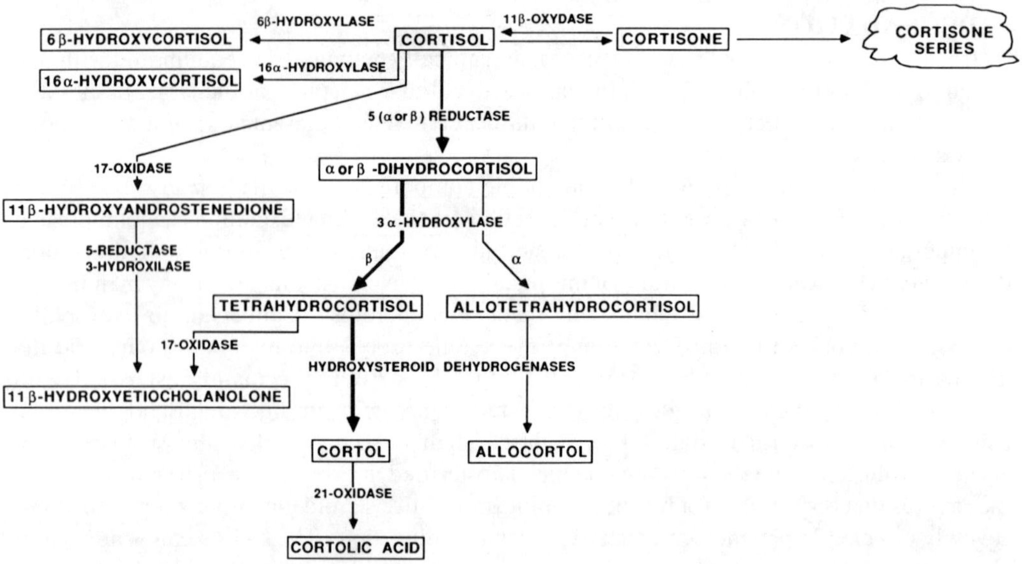

FIGURE 2. Pathways of cortisol degradation. The enzymes involved in the degradation steps are reported. Thick arrows show the main pathway.

their metabolites appear in the urine in similar amounts.[14,20] Cortisol, cortisone, and their derivatives can be further reduced to cortols and cortolones, respectively. After 5β-reduction of the A ring, an oxidation can occur at position 21, producing cortolic and cortolonic acids. A conversion to C-20 acids can also occur.[20] The transformation into acidic forms increases the water solubility of the steroid derivatives.

Other possible reactions involve the reduction of the C–20 keto-group of cortisol and cortisone. The liver can also hydroxylate cortisol at the 6β– and 16α-positions. Although these reactions are usually limited, β-hydroxylation becomes an important pathway for cortisol disposal in infants, whose esterification mechanisms are still immature, liver diseases, hyperestrogenic states, terminal illnesses, and during administration of drugs such as barbiturates, aminoglutethimide, phenytoin, and rifampin.[14,21–25] However, none of these drugs appear to affect the metabolic clearance rate of exogenous cortisol to a great extent.[14,26]

Finally, up to 10% of cortisol metabolites are C-19 molecules with a ketone group in position 17 (17-ketosteroids), that result from an oxidative cleavage of the side chain in this position.[2,14]

The majority of cortisol and cortisone derivatives are excreted by the kidney in esterified form. In contrast, both parent steroids appear largely in the urine as unconjugated molecules.[14]

b. Aldosterone (Figure 3)

Aldosterone half-life is about 15 min.[14,27] Hepatic extraction is extremely effective, so that the mineralocorticoid is almost completely removed from the blood by one passage through the liver.[2] Less than 0.5% of urinary aldosterone is in the free state.[14,27]

The metabolic pathways are similar to those of cortisol. The A ring and the ketone group in position 3 are reduced to form tetrahydroaldosterone. After glucuroconjugation, this metabolite is excreted by the kidney, and accounts for about 35% of the derivatives found in urine.[14] In normal subjects, up to 20% of aldosterone does not undergo reduction of 3-ketone and is conjugated in position 18 mainly in the renal tissue. This metabolite is called either the "acid-labile" conjugate, since it releases free steroid by hydrolysis at pH 1, or the "3-oxo" conjugate, due to the presence of an intact oxo group in position 3.[2,28] Since usual measurements of urinary aldosterone refer to this metabolite, it should be borne in mind that liver

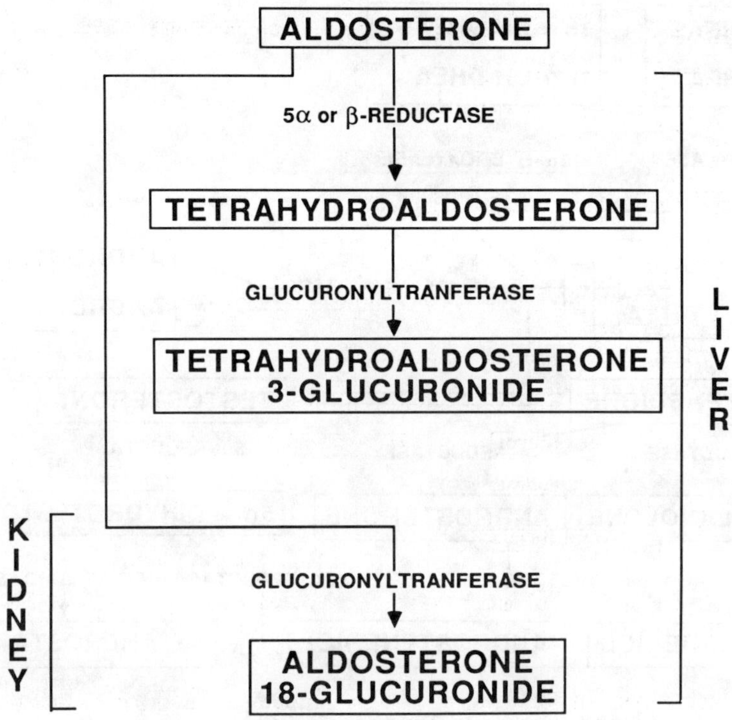

FIGURE 3. Pathways and sites of aldosterone degradation.

disease increases the fraction of steroid metabolized by the kidney and, consequently, the urinary fraction of 18-glucuronide.

In addition to the metabolites cited above, a number of other compounds can derive from aldosterone secreted by the adrenal cortex. They include polar derivatives and 21-deoxytetrahydroaldosterone, which is probably produced in the gut lumen by bacterial metabolism of the tetrahydroaldosterone excreted in the biliary tract and then reabsorbed.[27]

Potassium ion and spironolactone can affect aldosterone disposal.[27] Aldosterone clearance rate is accelerated by minoxidil, possibly through hemodynamic changes.[29]

c. Corticosterone and 11-Deoxycorticosterone

The half-life of corticosterone is shorter than that of the glucocorticoid.[14]

Although the metabolism of these mineralocorticoids is generally similar to that of cortisol, some peculiarities are noteworthy. In fact, lacking the 17-hydroxyl, they do not give rise to C-19 compounds (17-ketosteroids),[14] and a much greater fraction of metabolites is excreted into the gut (e.g., about 50% for 11-deoxycorticosterone).[20] Furthermore, a larger proportion of corticosterone than cortisol is excreted as 21-conjugate.[14]

d. Androgens (Figure 4)

DHEA can be irreversibly converted into androstenedione. DHEAS, which is also formed in liver and kidney, can be excreted unchanged. Both compounds can be further metabolized to androstenediol or its sulfate through hydroxylative processes.[1]

The hepatic clearing capacity for androstenedione is very high, so that more than 80% is extracted from the blood during one passage through the liver.[30] The steroid undergoes sequential reductions to androstanediol and etiocholanediol. Androstenedione can be converted to testosterone, chiefly in extrahepatic tissues.

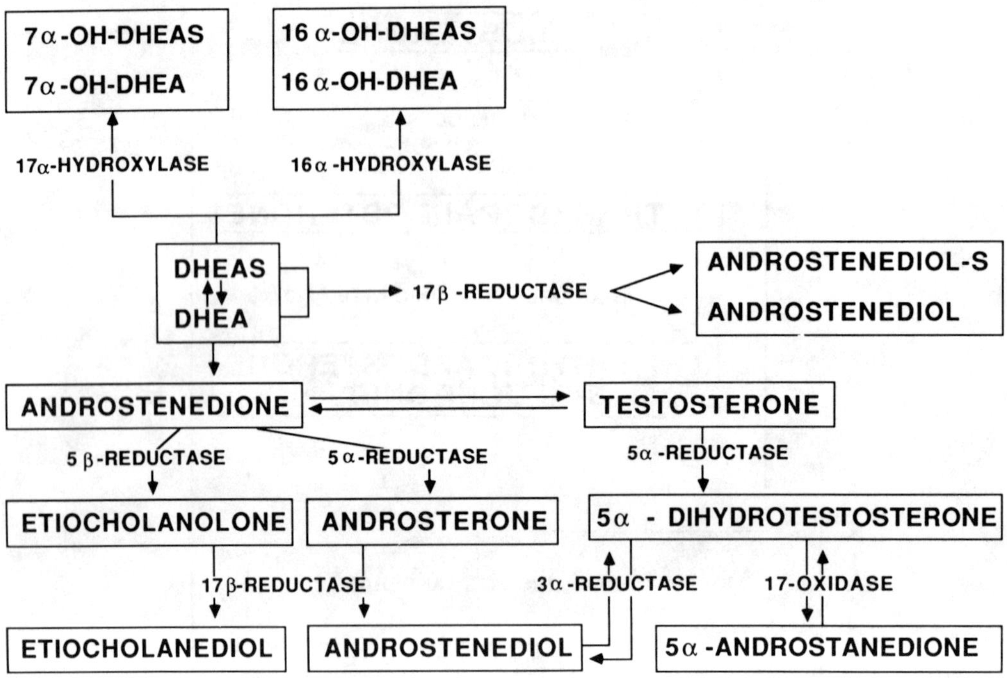

FIGURE 4. Pathways of androgen degradation. The α-reduction of testosterone producing the more active compound 5α-dihydrotestosterone occurs in the target tissues. DHEA(S), dehydroepiandrosterone (sulfate).

About 45% of testosterone is removed from the blood by a single liver passage.[30] In target tissues, it is converted to the active form 5α-dihydrotestosterone. This steroid gives rise to androstanediol (by irreversible 3α-reduction) or, to a lesser extent, 5α-androstanedione (by reversible 17-oxidation). These derivatives can be converted to androsterone.[1]

More metabolites of androgen steroids than cortisol derivatives are esterified with sulfate.[1,2] After conjugation, they are excreted with the urine and, in a small fraction, into the gut.[30]

The plasma clearance rate of testosterone and androstenedione is threefold greater in man than in woman, probably due to an accelerating effect of androgens on their own metabolism.[30]

II. PATHOPHYSIOLOGICAL ASPECTS IN LIVER DISEASE

As outlined in detail in the preceding section, the liver plays a central role in the degradation of adrenal corticosteroids. This is mainly due to its array of specific enzymatic and to the nonspecific glucuro- and sulfo-conjugating activity of liver cells. As a result, impaired degradation of these hormones can be expected to occur in those liver diseases characterized by extensive damage to the hepatic parenchyma. Moreover, cirrhosis, the liver disease in which corticosteroid metabolism has been most extensively investigated, is associated with striking derangements of splanchnic circulation: reduced liver blood flow and the opening of porta-systemic shunts diminishes the hormonal supply to parenchymal cells responsible for steroid degradation. Finally, liver disease affects protein synthesis and, therefore, albumin and CBG availability for hormonal transport. Although measurements of plasma CBG concentration in liver disease failed to yield concordant results, being found normal in some studies[14,31] and reduced in others,[32–34] serum albumin is invariably reduced in advanced cirrhosis. As stated above (see Section I.C.2), changes in protein binding can influence metabolic clearance rates of corticosteroids. Therefore, liver cirrhosis is the disease which exerts the most articulate effect on the metabolism of adrenocorticosteroids and represents a good natural model to investigate the clinical impact of their reduced degradation.

TABLE 1
Characteristics of Cortisol Metabolism in Liver Cirrhosis: Report from Various Studies

	Peterson[44]		Zumoff et al.[51]		Scavo et al.[47]	
	Cirrhosis	**Controls**	**Cirrhosis**	**Controls**	**Cirrhosis**	**Controls**
			Plasma concentration			
Range	6—21	6—25	—	—	5.6—11.0	2.5—17.5
Mean (±SEM)	13.5 ± 0.9	13.5 ± 0.3	—	—	7.9 ± 0.7	8.6 ± 1.3
Unit	μg/dl	μg/dl	—	—	μg/dl	μg/dl
			Half-Life			
Range	2.7—10.3	1.5—2.0	—	—	77.0—220	62.5—106
Mean (±SEM)	4.65 ± 0.88	1.83 ± 0.02	—	—	123.7 ± 16.6	78.2 ± 3.6
Unit	h	h	—	—	min	min
			Metabolic clearance rate			
Range	—	—	—	—	37.1—121.7	135.5—216.4
Mean (±SEM)	—	—	—	—	71.1 ± 7.9	174.1 ± 8.1
Unit	—	—	—	—	1/24 h	1/24 h
			Secretion rate			
Range	0.19—1.9	0.7—1.2	12—20	10—20	6.2—17.2	2.5—20.0
Mean (±SEM)	—	0.85 ± 0.02	16	15	12.1 ± 1.1	9.3 ± 1.7
Unit	mg/h	mg/h	mg/g^{cr}	mg/g^{cr}	mg/24 h	mg/24 h

Note: The values reported in this table are either derived directly from published data or are calculated from raw data, when available. Plasma half-life was calculated after the administration of 200 mg of cortisol (in Reference 44) and after a trace dose of tritiated cortisol in Reference 47. From the latter study, only the results derived from a two-compartment model are reported. Data reported under "secretion rate" in this table were calculated as "turnover rate" in Reference 44 and as "cortisol production" based on the specific activity of urinary metabolites tetrahydrocortisol and tetrahydrocortisone in Reference 51.

A. CORTISOL DEGRADATION (TABLE 1)

Seminal studies agreed in finding low levels of urinary 17-ketosteroids in various forms of liver disease, including acute viral hepatitis,[35–38] toxic hepatitis,[37] cirrhosis,[35-39] primary biliary cirrhosis,[40,41] and extrahepatic cholestasis.[35-38] However, since a small fraction of cortisol is metabolized to 17-ketosteroids, these results did not provide an adequate estimate of hormone metabolism. Urine cortisol and total 17-hydroxycorticosteroids are also excreted to a lower extent compared with normal subjects, even if broad overlap with the normal range has often been found.[31,37,38,42–45] Due to the influence of the glomerular filtration rate on urinary excretion of cortisol and its metabolites,[19] and the frequently occurring impairment of renal perfusion in cirrhosis,[46] even these results cannot be taken as definite evidence of reduced cortisol metabolism.

1. Cortisol Metabolic Clearance

Peterson[44] first attempted to gain a better insight into the pathophysiology of cortisol metabolism in liver disease by measuring the plasma disappearance rate of cortisol in eight patients with moderately severe cirrhosis of the liver and one with acute viral hepatitis. The biologic half-life of infused cortisol was prolonged in all the subjects tested, ranging from 2.67 to 10.3 h, with respect to the range found in 20 normal subjects, which varied from 1.5 to 2 h. Interestingly, the patient with acute viral hepatitis, whose cortisol half-life was 3.17 h, showed a reduction of the half-life to the upper limit of the normal range (2.17 h) after recovery. The disappearance rate of cortisol was heavily affected by the amount infused in both cirrhotics and controls, since the half-life of the hormone calculated after administration of pharmacological doses (up to 200 mg) was far greater than that observed with infusion of a trace (0.2 mg) quantity. This can be related to the fact that, while virtually all the circulating hormone is protein bound after a trace dose, only a portion (from 70 to 80%) is bound after the administration of a pharmacological dose. Cortisol can then diffuse in the extravascular spaces, which include ascites in cirrhotic patients, and reenter into the circulation in the following time.

These results have been confirmed more recently by the mathematical analysis of plasma 1,2-H^3-cortisol disappearance curves in a two-compartment model.[47] The half-life of the infused tracer hormone varied from 62.5 to 106 min (mean 78.2) in normal controls and from 77 to 220 min (mean 123.7) in cirrhotics without ascites. This was mainly due to the reduction of the cortisol metabolic clearance rate, which ranged from 37.1 to 121.7 l/24 h in cirrhotics and from 135.5 to 216.4 l/24 h in normal subjects.

Taken together, the available studies provide strong evidence for an important influence of liver disease on cortisol metabolism. This assumption is further confirmed by the finding that, following cortisol or ACTH infusion,[42,44,48,49] unconjugated 17-hydroxycorticosteroids remained elevated longer in plasma and enhanced amounts were excreted in urine of patients with liver disease compared with normal controls. In contrast, the rise of conjugated 17-hydroxycorticosteroids was delayed and less pronounced in plasma and urine recovery of total 17-hydroxycorticosteroids was lower.[50]

2. Factors Involved in the Altered Degradation of Cortisol

As stated above, several mechanisms may underlie the abnormality of cortisol degradation in liver cirrhosis.

a. Impairment of Specific Enzymatic Activities

Evidence of impaired enzymatic activity has been mainly provided by two studies. In Peterson's[44] experience, the plasma disappearance of infused cortisol was significantly prolonged in cirrhotics. In addition, the rate of urinary excretion of tetrahydrocortisol (the major glucuronide-conjugated metabolite of the hormone) following cortisol infusion was slower in a patient ($t_{1/2}$: 6 h) with respect to a control ($t_{1/2}$: 4.2 h). In contrast, the metabolism of infused cortisone and four major cortisol metabolites (dihydrocortisone, dihydrocortisol, tetrahydrocortisone, and tetrahydrocortisol) was not significantly affected. In fact, their plasma biologic half-life fell exactly within the range observed in normal controls. Such results suggested that the conversion of dihydrocortisol to tetrahydrocortisol occurred normally. This would imply a defect in the initial conversion of cortisol to dihydrocortisol, which depends on the activity of 5α- or 5β-reductase (Figure 2). A possible deficiency in the enzyme cofactor triphosphopyridine nucleotide (reduced form) was considered unlikely. This substance is needed for many different enzyme activities, and its deficiency would have involved the metabolism of most steroids. On the contrary, the metabolic defect found in patients with liver disease appeared to be a specific one.

In the second study, urine of cirrhotic patients given 4-^{14}C-cortisol was collected for prolonged periods (48 h).[51] Both the total conversion to 11-hydroxy metabolites and the conversion to each measured metabolite (i.e., tetrahydrocortisol and cortols) were comparable to that of controls. Such results apparently contrast with what has been reported above,[44] but this may be explained, at least in part, by differences in the protocols. In fact, in the second study,[51] a tracer dose of the hormone was given and urine was collected for 48 h. In Peterson's study,[44] a pharmacological dose of cortisol was given and urine collected for 24 h. Therefore, the defective cortisol-dihydrocortisol conversion may have emerged only after the administration of a pharmacological dose of the hormone; moreover, the recovery of a normal amount

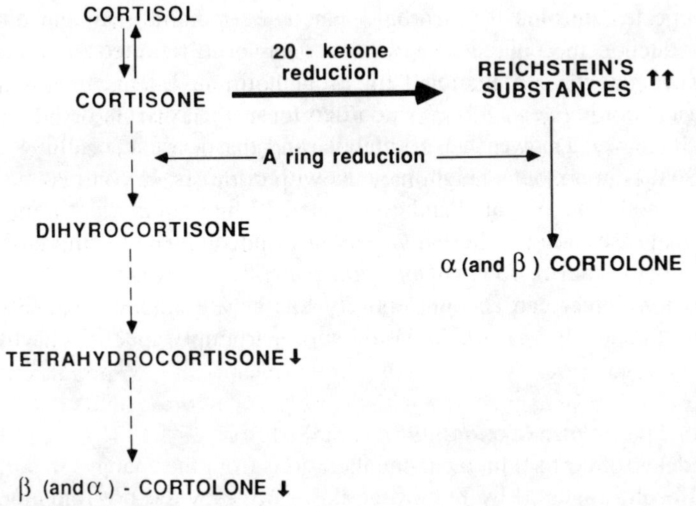

FIGURE 5. Proposed changes in cortisol metabolism in cirrhosis.[51] Due to a reduced conversion of cortisone to tetrahydrocortisone, reduction of the C-20 ketone of cortisone leads to an enhanced production of Reichsten's substance U and epi-U. As a result, the formation of α-cortolone is increased.

(as percentage of administered radioactivity) of total and individual 11-hydroxy metabolites might have been due to the prolonged urine collection.

A further alteration in cortisol metabolism, involving the 11-keto metabolites (tetrahydrocortisone and cortolones) has been disclosed in patients with cirrhosis.[51] Although the total amount recovered from patients' urine was normal after the injection of a tracer dose of labeled cortisol, the proportion of cortolones to tetrahydrocortisone was increased. This suggested a reduced conversion of cortisone to tetrahydrocortisone and an equivalent increase in conversion to cortolones. Since the main pathway leading to the conversion of cortisol to cortolones implies the formation of tetrahydrocortisone through cortisone and dihydrocortisone, the only explanation for such a finding was a diversion of the metabolic pathway to an alternative route. In fact, the urine of cirrhotic patients contained an increased amount of Reichstein's substance U (17,20β,21-trihydroxy-Δ^4-pregnene-3,11-dione) and epi-U (α-epimer of the former). These substances represent the intermediate metabolites of a pathway leading to the direct production of α-cortolones from cortisone (Figure 5). That such a metabolic pathway operates at higher than normal capacity in cirrhotic patients was confirmed by the increase of the α- and β-cortolones ratio. In fact, the former derive from the alternative pathway, and β–cortolones derive from tetrahydrocortisone. Such an abnormality of cortisol metabolism had also been disclosed in subjects with initially normal liver function given norethandrolone, a substance leading to pure intrahepatic cholestasis.[52] It was concluded that cholestasis was the main process responsible for the shift of cortisol metabolism towards the cortisone-Reichstein's substance U pathway found in cirrhotics.

The results of the cited studies are not in contrast with each other. Taken together, they suggest that, above all, liver disease can exert an adverse effect on the initial, crucial step of cortisol (and cortisone) degradation, consisting of A-ring reduction to dihydrocortisol (dihydrocortisone). In addition, involvement of the process of transport and for binding of cortisol to its intracellular metabolic sites cannot be ruled out.[51]

b. Impairment of Conjugating Activity

Conflicting results have been reported as far as hepatic glucuroconjugation of cortisol metabolites in cirrhosis is concerned. While normal controls and patients excreted the same fraction of the injected steroids as glucuronide-unconjugated and -conjugated metabolites in one study,[44] a reduction in conjugated metabolites has been reported in others.[51,53] Furthermore, a defect in glucuro-conjugation may account for a derangement in the circadian variation of plasma-conjugated 17-hydroxycorticosteroids in patients with cirrhosis. In fact, these compounds showed a slower than normal rise and broader peak, resulting in an inversion of the normal rhythm in six out of eight patients with cirrhosis. In contrast, morning plasma concentrations of both unconjugated and conjugated 17-hydroxycorticosteroids were within normal range in all cases and the diurnal variations of unconjugated compounds were within the broad limits of normal in four out of eight patients.[45] However, it is unlikely that the glucuro-conjugation defect can account entirely for the alterations of cortisol metabolism described above (Section II.2.a), which would appear to imply specific enzymatic deficiencies.

c. Alterations of Splanchnic Circulation

The independence of cortisol metabolism alterations from the changes in portal circulation of cirrhosis has been suggested by Zumoff et al.,[51] since they did not find any difference in cortisol degradation in cirrhotic patients with and without surgical portacaval shunts. It should be considered, however, that patients with severe portal hypertension can develop spontaneous porta-systemic shunting very close to what occurs after surgical diversion of portal blood flow.[54]

3. Consequences and Clinical Implications of the Altered Cortisol Metabolism

In the presence of reduced degradation, plasma levels of cortisol should tend to increase. Total plasma cortisol concentration, measured as cortisol itself or as hydroxycorticosteroids, have usually been found within the normal range in liver disease.[31,34,42,44,48,55–57] The explanation most often provided to explain such a finding was the feedback inhibition of ACTH secretion and, then, reduced cortisol production and secretion by the adrenal cortex. Indeed, the measurement of cortisol production rate by different authors yielded different results. Reduced cortisol secretion and turnover rates (0.19 to 0.78 mg/h; normal range 0.7 to 1.2 mg/h), unaffected by ascites removal, have been reported in one study.[44] This would fit with the above proposal, but others[31,51] failed to confirm such a finding, disclosing normal secretion rates. Another finding which militates against a preserved adrenal-hypophyseal feedback regulation is the increased plasma concentration of unbound 11-hydroxycorticosteroids, in the presence of low-normal total levels, found in a heterogeneous group of patients with liver diseases, including chronic persistent and active hepatitis and cirrhosis of various etiology.[31] Although a relative increase in unbound plasma cortisol may be simply explained by a reduced availability of albumin and transcortin,[32–34] the maintenance of elevated levels would imply a resetting of the negative feedback mechanism.[31] These data, however, have not been confirmed, since others reported normal plasma levels of both total and unbound cortisol in cirrhotic patients.[57] Thus, the characteristics of adrenal cortisol production and secretion in patients with liver disease still await a final clarification.

The circadian rhythm of plasma cortisol has been found by most authors to be preserved even in the advanced stages of the disease,[31,33,55,56] and suggestive of the preservation of the physiological interplay between the adrenal cortex and anterior hypophysis. The most consistent changes found in the circadian rhythm characteristics of total plasma cortisol were the reduction in amplitude and mean daily level, which would fit with a well-preserved feedback regulation of adrenal-hypophyseal axis, in the presence of an increased free cortisol fraction[33,55,56] (Figure 6). Further evidence in favor of this interpretation has recently been provided by finding reduced ACTH levels at 8:00 a.m. in a group of patients with post-hepatitic cirrhosis with and without ascites.[34]

In conclusion, even in the presence of reduced cortisol degradation, cirrhosis can be

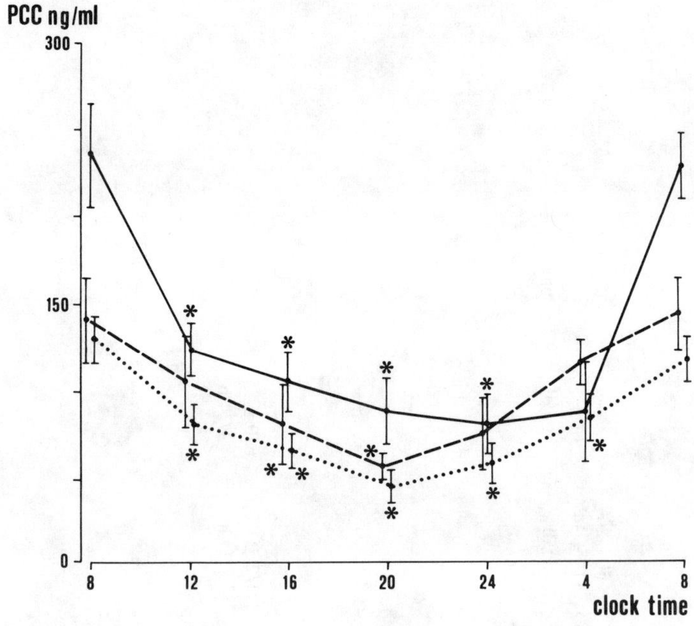

FIGURE 6. Circadian rhythms of plasma cortisol in normal subjects (—) and cirrhotics with (••••) or without (- - -) ascites. The circadian rhythmicity is preserved in patients, while amplitude and mean daily levels are reduced. These changes are statistically significant in ascitic patients ($p < 0.005$ with respect to controls). The asterisks indicate the statistical significance ($p < 0.05$ or lower) of changes with respect to 8:00 a.m. values within each group. PCC, plasma cortisol concentration. (From Reference 56, modified.)

considered a "eucorticoid" status, to use Peterson's[44] definition. Furthermore, the metabolism of exogenous cortisone and synthetic 9α-fluorocortisol and Δ^1-9α-fluorocortisol have been shown to occur normally in these patients.[44] Finally, adrenocortical responsiveness to ACTH[44] and suppressibility by dexamethasone test[34] are preserved. Nonetheless, corticosteroid therapy should be employed with caution in such patients, since the concomitant presence of severely reduced plasma protein binding and metabolic liver activity may result in excessive metabolic and immunological activity. This is particularly true in patients with severe hepatic encephalopathy and elevated serum levels of free triptophan, in whom plasma cortisol concentration has been found to be increased and steady throughout the day. In these cases, the pituitary-adrenal feedback interplay may be disrupted by enhanced CRF release, due to increased cerebral turnover of serotonin.[55]

B. ALDOSTERONE DEGRADATION (TABLE 2)

The finding of increased renal excretion of a sodium-retaining substance with the characteristics of aldosterone in patients with cirrhosis of the liver dates back to nearly 40 years.[40] This report was fully confirmed by several subsequent studies (for review, see Reference 58), and gave rise to the concept that liver cirrhosis represented a typical example of secondary hyperaldosteronism, whose clinical manifestations were ascites and edema formation. Later studies paid more attention to patient selection and to the control of experimental conditions, as far as posture, diet, and, above all, diuretic therapy were concerned. Moreover, methods for the measurement of plasma aldosterone concentration became readily available. As a result, the hyperaldosteronism of cirrhosis was put into perspective somewhat. In fact, up to 74% of untreated patients with ascites may show normal plasma aldosterone, whereas this percentage falls to 15% in cirrhotics with previous diuretic administration.[59] In other studies, such a

TABLE 2
Characteristics of Aldosterone Metabolism in Liver Cirrhosis: Report from Various Studies

	Coppage et al.[68]		Vecsei et al.[69]		Mitch et al.[63]		Rosoff et al.[72]		Wilkinson et al.[74]	
	Cirrhosis	Controls	Cirrhosis	Controls	Cirrhosis	Controls	Cirrhosis	Controls	Cirrhosis	Controls
Plasma Concentration										
Range	—	—	7.2—136	3.5—12.2	0.5—35.8	2.8—13.9	50—269	—	15—15,462	135—528
Mean (±SEM)	—	—	46.8	8.2	13.2 ± 1.9	8.1 ± 1.2	119 ± 34	23.7 ± 5.5	—	—
Unit	—	—	μg/dl	μg/dl	ng/dl	ng/dl	ng/dl	ng/dl	pmol/l	pmol/l
Half-Life										
Range	43—81	26—39	48—244	40—44	—	—	—	—	—	—
Mean (±SEM)	63	34	112.7	43	—	—	—	—	—	—
Unit	min	min	min	min	—	—	—	—	—	—
Metabolic Clearance rate										
Range	—	—	117—785	851—1,506	—	—	—	—	—	—
Mean (±SEM)	—	—	510	1155	—	—	1,072 ± 157	1,620 ± 85	—	—
Unit	—	—	ml/min	ml/min	—	—	1/24 h	1/24 h	—	—
HE										
Range	—	—	38—52	97—98	0—87	22—100	—	—	—	—
Mean (±SEM)	—	—	—	—	40.3 ± 6.7	64.8 ± 6.4	—	—	—	—
Unit	—	—	%	%	%	%	—	—	—	—
Secretion Rate										
Range	—	—	72—697	62—223	—	—	—	—	76—1369	≈16—291
Mean (±SEM)	—	—	299	134	—	—	1,281 ± 534	≈100	—	—
Unit	—	—	μg/24 h	μg/24 h	—	—	μg/24 h	μg/24 h	nmol/24 h	nmol/24 h

Note: HE, hepatic extraction rate. The values reported in this table are either derived directly from published data or are calculated from raw data, when available.

stratification has not been made, and the frequency of hyperaldosteronism in ascitic cirrhotics is found between the above values.[60-67] Moreover, plasma aldosterone concentration has been found to be in the low-normal range or reduced when patients are studied in the pre-ascitic stage.[59,62,65–67] In any case, there remains a large population of patients with cirrhosis showing an increased plasma aldosterone concentration, which is very elevated in the more advanced stages of the disease.

Due to the key role played by the liver in the degradation of aldosterone, one naturally speculates on how far a reduced metabolism of the hormone may account for the hyperaldosteronism of cirrhosis.

1. Aldosterone Metabolic Clearance

The initial observation that the plasma biological half-life of tritiated aldosterone was prolonged in one of two patients with liver cirrhosis[44] was confirmed by Coppage et al.[68] After injecting 7-^{3}H-*d*-aldosterone, they determined the plasma disappearance rate of aldosterone in 16 normal subjects and 9 patients showing "gross clinical evidence of hepatic decompensation" (7 with ascites). The prolonged plasma half-life of the administered aldosterone shown by the patients (43 to 81 min) was striking, since there was no overlap with the group of normal controls (26 to 39 min). Similar results were later obtained by Vecsei et al.,[69] who found $t_{1/2}$ values ranging from 48 to 244 min in cirrhotics and from 40 to 44 min in healthy controls. That such an abnormality was due to tracer hormone sequestration in the ascitic fluid during mixing with subsequent reentering into plasma appeared unlikely. In fact, a markedly prolonged disappearance rate was also found in patients without ascites.[68] The authors stated that the liver disease *per se*, rather than fluid retention, appeared to be responsible for such an abnormality. However, it should not be disregarded that subsequent studies have provided evidence for extracellular fluid expansion even in the pre-ascitic stage of cirrhosis.[59,65–67,70]

The increased biological half-life of aldosterone makes the reduced plasma clearance rate of 1,2-^{3}H-*d*-aldosterone observed in cirrhosis[69,71,72] an expected finding. This abnormality was also evident in those patients whose plasma aldosterone concentration was within the normal range.[69] The entity of the reduced plasma clearance rate was variable. It ranged from 117 to 785 ml/min with a mean value of 510 ml/min (normal 851 to 1506; mean 1115 ml/min) without appreciable differences between patients with or without ascites in one study.[69] In another study,[72] although the patients presumably had more severe disease, the aldosterone clearance rate was 1072 ± 157 (mean ± S.E.M.) l/d (normal 1620 ± 85 l/d).[72] These results suggest that the defect in hormonal disposal rate was not related to the severity of liver impairment.

The decrease in metabolic clearance rate appeared to be the major factor responsible for the prolonged biological half-life of the injected tracer hormone. A reduction in volume distribution of the compartment reflecting the "inner pool", which includes the liver along with the plasma and other extracellular volumes, may also have contributed.[69] That the liver damage was the main feature responsible for the decreased plasma clearance rate was confirmed by the finding that the hepatic extraction rate of aldosterone was strikingly reduced in patients with cirrhosis and other patients with liver impairment. Namely, the hepatic extraction ratio of arterial aldosterone dropped from the 95 to 98% of normal subjects to approximately 40% in cirrhotics and the hepatic clearance rate of the hormone was reduced to nearly one third of normal.[69] As a result, the relative proportion of the renal to the hepatic clearance rate of aldosterone rose from approximately 10 to 25%, even though kidney clearance was not increased in absolute terms. By splitting the renal clearance rate into its main components (i.e., "metabolic" and "excretory" rates), the former tended to be decreased and the latter to be increased.[69] A similar reduction in the hepatic extraction rate of aldosterone has also been reported by Mitch et al.,[63] who found a mean extraction rate of 40% (range 0 to 84) in cirrhotics. However, the difference between normal controls and cirrhotics was less striking, since normal aldosterone hepatic extraction ranged from 22 to 100% (mean 65). This abnormality of hepatic metabolism was not related to either the degree of portal hypertension, as evaluated by calculating the hepatic venous pressure gradient, or the sulfobromophtalein storage or transport maximum.

2. Factors Involved in the Altered Degradation of Aldosterone

As discussed for cortisol metabolism, several factors may account for the reduced capacity to degrade aldosterone in cirrhosis.

a. Impairment of Specific Enzymatic Activities

By confirming the first report by Hurter and Nabarro,[73] Coppage et al.[68] clearly demonstrated that the pattern of urinary excretion of 7-^{3}H-*d*-aldosterone was altered in patients with cirrhosis. A minute fraction of the tracer hormone was recovered unaltered in the urine of both controls and patients, even though the latter had a slightly higher range (0.08 to 0.8% vs. 0.14 to 0.17%). In contrast, clear differences in aldosterone metabolite excretion were disclosed. While normal subjects excreted from 5 to 11% (mean 8.5%) of the administered dose as the 3-oxo-conjugate (18-glucuronide aldosterone) and from 29 to 50% (mean 40%) as tetrahydroaldosterone, the ranges in patient urine samples were 5 to 25% (mean 16%) and 19 to 44% (mean 29%), respectively. It has been proposed that the reduction of the A ring, leading to tetrahydroaldosterone formation within the liver, is a rate-limiting step in the hormone metabolism and that this process was impaired in cirrhosis (Figure 7). Thus, 18-glucuronide production, which does not require A ring reduction, is enhanced.[68] This would imply an increased compensatory metabolism by the kidney, which only produces the 18-glucuronide form. According to these findings, in cirrhosis the correlation between the measurements of aldosterone secretion rate and the excretion rate of the 18-glucuronide metabolite was shifted to the left, when compared with predicted values derived from normal subjects.[74] Unless this is taken into account, values for 18-glucuronide excretion will give misleadingly high absolute values for the secretion rate.[74]

b. Alterations of Splanchnic Circulation

A factor of major importance in determining the rate of metabolic inactivation of aldosterone appears to be the liver blood flow. Such a key role was first suggested by Coppage et al.[68] who showed that the oral administration of tritiated aldosterone to normal subjects had no effect on urine electrolyte excretion and no detectable radioactivity appeared in the plasma up to 90 min. This result was achieved in spite of both complete, rapid intestinal absorption of labeled hormone and urine excretion of aldosterone metabolites in a proportion very close to that following intravenous administration. In contrast, when oral aldosterone was given to one cirrhotic patient with surgical porta-caval shunt, the labeled hormone was detectable in unaltered form in peripheral venous blood from 19 to 65 min after administration. Therefore, in this patient, much of whose portal blood bypassed the liver, oral administration simulated intravenous injection. Unfortunately, no patients with cirrhosis without surgical porta-systemic shunt were studied in such a way.

A more direct demonstration of the dependence of aldosterone metabolism on liver blood flow was subsequently provided by showing a very close relationship between hepatic clearance rate of the hormone and both hepatic plasma flow and hepatic extraction rate.[69]

Thus, contrary to what happens for cortisol, the circulatory derangement following post-sinusoidal portal hypertension, including reduction in total liver blood flow, changes in the relative hepatic artery/portal vein contribution and opening of spontaneous porta-systemic shunts, can be a major factor in conditioning an altered degradation of aldosterone in cirrhotic patients.

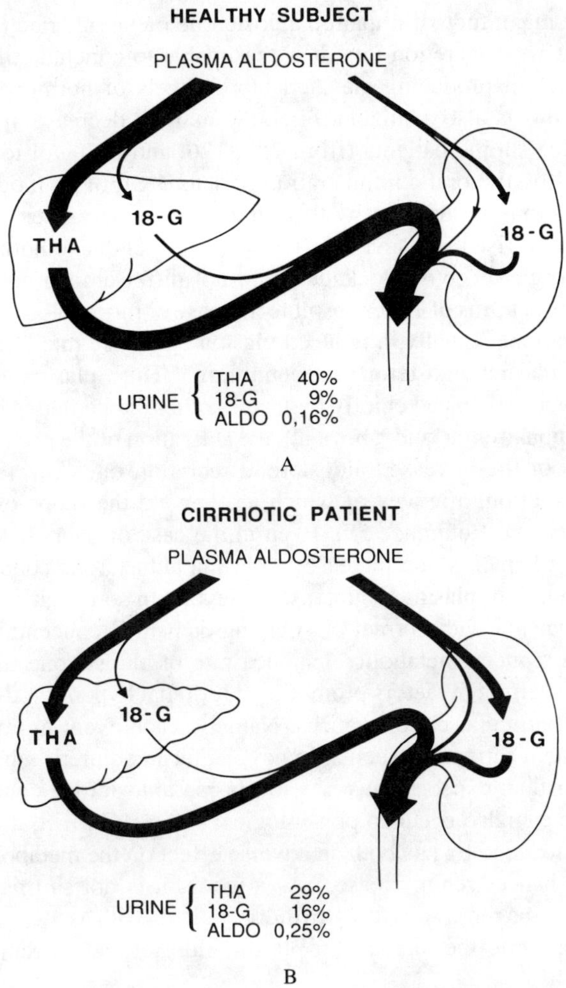

FIGURE 7. Alterations of aldosterone metabolism in cirrhosis, as shown for healthy (A) and cirrhotic (B) subjects. The impairment in liver-dependent reduction of A ring leads to lower tetrahydroaldosterone formation. As a result, a compensatory increase in metabolism by the kidney, with hyperproduction of 18-glucuronide, ensues. The thickness of the arrows is roughly proportional to the relative importance of the metabolic pathways. THA, tetrahydroaldosterone; 18-G, aldosterone 18-glucuronide; ALDO, free aldosterone. The percent urine recovery of the three substances are the mean values derived by Peterson.[44]

3. Consequences and Clinical Implications of the Altered Aldosterone Metabolism

Since hyperaldosteronism is a common feature in patients with advanced liver cirrhosis,[58–67] it is interesting to speculate how much this is due to the altered liver metabolism of the hormone. The aldosterone secretion rate has also been found to be increased in these patients.[68,74,75] In one study, the secretion rate ranged from 248 to 2080 μg/24 h (most patients being on a low-sodium diet), while that of normal subjects ranged from 95 to 249 μg/24 h on an unrestricted sodium diet and from 284 to 528 μg/24 h under sodium restriction.[68] An even more striking rise in aldosterone secretion rate in cirrhotics (mean 1281 ± 534 μg/24 h) with respect to normal controls (about 100 μg/d) was found by Rosoff et al.[72] Such a variation in results may be related to the differences in the patient populations under investigation: in the study by Rosoff et al.,[72] all cirrhotics had diuretic-resistant ascites, a condition usually appearing in the advanced stage of the disease.

By comparing the magnitude of impaired aldosterone metabolic rate (Section II.B.1) with the increase in aldosterone secretion rate, it is reasonable to conclude that hypersecretion is the predominant factor in producing elevated blood levels of hormone in ascitic patients. Indirect evidence of this is also available. First, the marked decrease in plasma aldosterone concentrations of three supine patients (from 282, 126, and 88 ng/dl to 20, 8, and 7 ng/dl, respectively) induced by the oral administration of 1 to 2 g/d of aminoglutethimide (a drug which reduces aldosterone production by the adrenal cortex) was very rapid.[72] Second, the metoclopramide-induced rise in plasma aldosterone (an effect due to hormone hypersecretion induced by dopaminergic receptor blockade[76]) did not differ among controls and cirrhotics with or without ascites in terms of either absolute increase or morphology of the plasma curves of hormone concentration.[77] Finally, a reduced plasma clearance rate has also been found in patients with normal plasma aldosterone concentration.[69] Thus, plasma levels of aldosterone in cirrhosis appear to be influenced chiefly by factors other than reduced hepatic degradation.

Hyperkalemia, hyponatremia, and, above all, the activation of the renin angiotensin system are the major causes of the increased aldosterone secretion rate. Several lines of evidence support such an assumption, a review of which are beyond the scope of this chapter (for a comprehensive review, see Reference 58). Even in the case of overwhelming liver damage, as occurs in fulminant hepatitis, the increase in plasma aldosterone concentration is, by and large, proportional to that in plasma angiotensin II levels. In some patients, however, hyper-aldosteronism is associated with normal plasma angiotensin II concentration.[78]

In conclusion, the reduced metabolic clearance rate of aldosterone found in severe liver disease exacerbates the effect of factors promoting its adrenal hypersecretion, and may assume clinical relevance in particular circumstances. Namely, each event promoting or enhancing renin-angiotensin system activation, such as hypovolemia and diuretic administration, may be followed by a greater than expected increase in plasma aldosterone concentration. Such an exaggerated response can also occur in physiological settings. In fact, the change in posture from supine position to standing has both an adverse effect on the metabolic clearance rate of the hormone[79] and enhances renin release.[80] As a result, it is not surprising that the tilting-induced increase in plasma renin activity in cirrhotic patients with ascites was 3.5 times greater than that of controls, while the increase in plasma aldosterone concentration was 7 times greater[81] (Figure 8).

III. CONCLUSIONS

The degradation of adrenal cortical steroids is a complex metabolic pathway which takes place mainly in the liver. The metabolic conversion of corticosteroids leads to the destruction of their biological activity (with few exceptions) and the transformation into more soluble derivatives. The latter are excreted by the kidney and, to a far lesser extent, by the gut. Any reduction in the metabolic clearance rate of corticosteroids cannot be without biological effects. Thus, changes in degradation can potentially assume clinical relevance.

Due to the key role played by liver cells in degradation and conjugation of corticosteroids and their derivatives, liver disease represents an ideal model to study the effects of a reduced degradation. A number of factors can affect corticosteroid degradation in liver disease, such as altered enzymatic activities, changes in liver and splanchnic blood flow, and impaired synthesis of protein involved in the hormone transport. Several studies have shown that both cortisol and aldosterone metabolic clearance rates are impaired in patients with liver cirrhosis. A defective reduction of the A ring is probably the principal metabolic alteration. Circulatory changes would be of scarce relevance in determining cortisol metabolism alterations, whereas they are crucial in contributing to the prolongation of aldosterone half-life. Even in the presence of a severe impairment of degradation, plasma levels of both cortisol and aldosterone

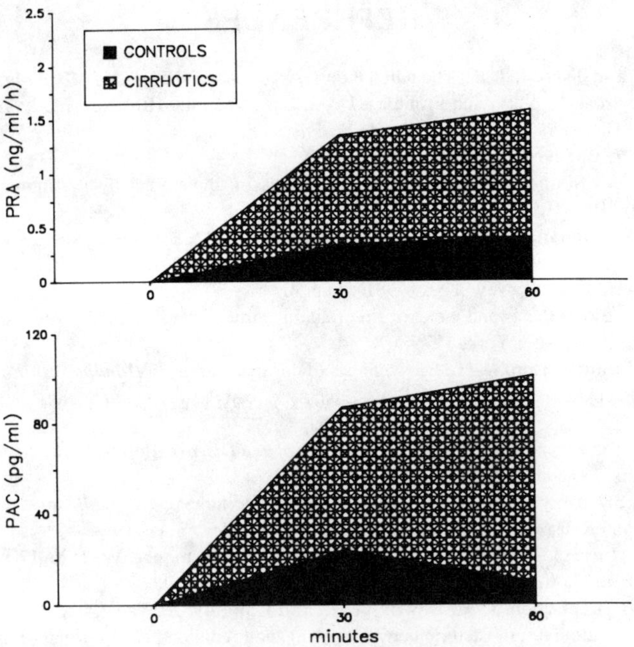

FIGURE 8. Effect of the change in posture (from supine to sitting) on plasma renin activity and aldosterone concentration in normal subjects (black area) and cirrhotics with ascites (shaded area). The incremental areas under the curves are shown. In patients, the maneuver produces a disproportionate increase in plasma aldosterone concentration (7 times greater than that of controls) with respect to the rise in PRA (3.5 times greater than that of normal subjects). This behavior is mainly due the lower metabolic clearance rate of aldosterone by the cirrhotic liver. PRA, plasma renin activity; PAC, plasma aldosterone concentration. The figure has been designed utilizing data published by Bernardi et al.[81]

are chiefly determined by their secretion rate and, therefore, by those factors/systems which regulate hormone secretion. As a result, most authors reported that plasma cortisol concentration of patients with cirrhosis is within the physiological range and show a preserved circadian variation. This is due to a normally operating feedback regulation of adrenal-hypophyseal axis, so that the tendency of plasma cortisol to increase is counterbalanced by a slight reduction in ACTH secretion. Such a formulation, however, deserves further investigation. Cirrhotic hyperaldosteronism is mainly attributable to hypersecretion which, in turn, is primed, among other factors, by the activation of the renin-angiotensin system. Based on present knowledge, the reduction in corticosteroid metabolism occurring in liver disease does not appear to assume a prevalent clinical relevance.

ABBREVIATIONS

LDL	Low-density lipoprotein
HDL	High-density lipoprotein
DHEA	Dehydroepiandrosterone
CBG	Corticosteroid binding globulin
SHBG	Sex hormone binding globulin
ACTH	Adrenocorticotropic hormone
DHEA(s)	Dehydro-epiandrosterone

REFERENCES

1. **Baxter, J. D. and Tyrrell, J. B.,** The adrenal cortex, in *Endocrinology and Metabolism,* 2nd ed., Feling, P., Baxter, J. D., Broadus, A. E., and Frohman, L. A., Eds., McGraw-Hill, New York, 1981, chap. 12.
2. **Bondy, P. K.,** Disorders of adrenal cortex, in *Williams Textbook of Endocrinology,* Wilson, J. D. and Foster, D. W., Eds., W. B. Saunders, Philadelphia, 1985, chap. 22.
3. **Carr, B. R. and Simpson, E. R.,** *De novo* synthesis of cholesterol by the human fetal adrenal gland, *Endocrinology,* 108, 2154, 1981.
4. **Baxter, J. D., Frohman, L. A., Broadus, A. E., and Feling, P.,** Biosynthesis, secretion, and metabolism of hormones, in *Endocrinology and Metabolism,* 2nd ed., Feling, P., Baxter, J. D., Broadus, A. E., and Frohman, L. A., Eds., McGraw-Hill, New York, 1981, chap. 4.
5. **Brooks, R. V.,** Biosynthesis and metabolism of adrenocortical steroids, in *The Adrenal Gland,* James, V. H. T., Ed., Raven Press, New York, 1979, 67.
6. **Samuels, L. T. and Nelson, D. H.,** Biosynthesis of corticosteroids, in *Handbook of Physiology: Endocrinology,* Vol. 6, Greep, R. O. and Astwood, E. B., Eds., American Physiological Society, Washington, D.C., 1975, 55.
7. **Bondy, P. K.,** The adrenal cortex, in *Metabolic Control and Disease,* Bondy, P. K. and Rosenberg, L. E., Eds., W. B. Saunders, Philadelphia, 1979, 1427.
8. **Ballard, P. L.,** Delivery and transport of glucocorticoids to target cells, in *Glucocorticoid Hormone Action,* Baxter, J. D. and Rousseau, G. G., Eds., Springer-Verlag, New York, 1979, 25.
9. **Siiteri, P. K., Murai, J. T., Hammond, G. L., Nisker, J. A., Raymoure, W. J., and Kuhn, R. W.,** The serum transport of steriod hormones, *Rec. Progr. Horm. Res.,* 38, 457, 1982.
10. **Channarri, M., Leutscher, J. A., Dowdy, A. J., and Ganguly, A.,** The effects of temperature and plasma cortisol on distribution of aldosterone between plasma and red blood cells: influence on metabolic clearance rate and on hepatic and renal extraction of aldosterone, *J. Clin. Endocrinol. Metab.,* 44, 752, 1977.
11. **Partridge, W. M., Sakiyama, R., and Judd, H. L.,** Protein-bound corticosterone in human serum is selectively transported into rat brain and liver *in vivo, J. Clin. Endocrinol. Metab.,* 57, 160, 1983.
12. **Partridge, W. M.,** Transport of protein-bound hormones into tissues *in vivo, Endocr. Rev.,* 2, 103, 1981.
13. **Zager, P. G., Burtis, W. J., Leutscher, J. A., Dowdy, A. J., and Sood, S.,** Increased plasma protein binding and low metabolic clearance rate of aldosterone in plasma of low cortisol concentration, *J. Clin. Endocrinol. Metab.,* 42, 207, 1976.
14. **Peterson, R. E.,** Metabolism of adrenal cortical steroids, in *The Human Adrenal Cortex,* Christy, N. P., Ed., Harper & Row, New York, 1971, 87.
15. **Wilson, J. D. and Walker, J. D.,** The conversion of testosterone to 5α-androstan-17β-ol-3-one (dihydrotestosterone) by skin slices of man, *J. Clin. Invest.,* 48, 371, 1969.
16. **Siiteri, P. K. and MacDonald, P. C.,** Role of extraglandular estrogen in human endocrinology, in *Handbook of Physiology: Endocrinology,* Vol. 2, Greep, R. O. and Astwood, E. B., Eds., Williams & Wilkins, Baltimore, 1973, 615.
17. **Kawai, S., Ichikawa, Y., and Homma, M.,** Differences in metabolic properties among cortisol, prednisolone, and dexamethasone in liver and renal disease: accelerated metabolism of dexamethasone in renal failure, *J. Clin. Endocrinol. Metab.,* 60, 848, 1985.
18. **Weitzman, E. D., Fukushima, D., Nogeire, C., Roffwang, H., Gallagher, T. F., and Hellman, L.,** Twenty-four hour pattern of the episodic secretion of cortisol in normal subjects, *J. Clin. Endocrinol. Metab.,* 33, 14, 1971.
19. **Sederberg-Olsen, P., Binder, C., and Kehlet, H.,** Urinary excretion of free cortisol in impaired renal function, *Acta Endocrinol.,* 78, 86, 1975.
20. **Monder, C. and Bradlow, L. H.,** Cortoid acids: explorations at the frontier of corticosteroid metabolism, *Rec. Progr. Horm. Res.,* 36, 345, 1980.
21. **Katz, F. H., Lipman, M. M., Frantz, A. G., and Jailer, J. W.,** The physiologic significance of 6β-hydroxycortisol in human corticoid metabolism, *J. Clin. Endocrinol. Metab.,* 22, 71, 1962.
22. **Werk, E. E., MacGee, J., and Sholiton, L. J.,** Altered cortisol metabolism in advanced cancer and other terminal illnesses: excretion of 6-hydroxycortisol, *Metabolism,* 13, 1425, 1964.
23. **Tamada, S. and Iwai, K.,** Induction of hepatic cortisol-6-hydroxylase by rifampicin, *Lancet,* 2, 366, 1976.
24. **Werk, E. E., MacGee, J., and Sholiton, L. J.,** Effect of dyphenylhydantoin on cortisol metabolism in man, *J. Clin. Invest.,* 43, 1824, 1964.
25. **Southren, A. L., Gordon, G. G., Tochimoto, S., Krikun, E., Krieger, D., Jacobson, M., and Kuntzman, R.,** Effect of *N*-phenylbarbital (petharbital) on the metabolism of testosterone and cortisol in man, *J. Clin. Endocrinol. Metab.,* 29, 251, 1969.
26. **Santer, R. J., Lipton, A., and Kendall, J.,** Successful medical adrenalectomy with aminoglutetimide: role of altered drug metabolism, *JAMA,* 230, 1661, 1974.
27. **Morris, D. J.,** The metabolism and mechanism of action of aldosterone, *Endocrinol. Rev.,* 2, 234, 1981.

28. **Leutscher, J. A., Hancock, E. W., Camargo, C. A., Dowdy, A. J., and Nokes, G. W.,** Conjugation of 1,2-^{3}H-aldosterone in human liver and kidney and renal extraction of aldosterone and labelled conjugates from blood plasma, *J. Clin. Endocrinol. Metab.*, 25, 628, 1965.
29. **Pratt, J. H., Grim, C. E., and Parkinson, C. A.,** Minoxidil increases aldosterone metabolic clearance in hypertensive patients, *J. Clin. Endocrinol. Metab.*, 49, 834, 1979.
30. **Migeon, C.,** Adrenal androgens in man, *Am. J. Med.*, 53, 606, 1972.
31. **McCann, V. J. and Fulton, T. T.,** Cortisol metabolism in chronic liver disease, *J. Clin. Endocrinol. Metab.*, 40, 1038, 1975.
32. **Doe, R. P., Fernandez, R., and Seal, U. S.,** Measurement of cortico-steroid-binding globulin in man, *J. Clin. Endocrinol. Metab.*, 24, 1029, 1964.
33. **Rosman, P. M., Farag, A., Benn, R., Tito, J., Mishik, A., and Wallace, E. Z.,** Modulation of pituitary-adrenocortical function: decreased secretory episodes and blunted circadian rhythmicity in patients with alcoholic liver disease, *J. Clin. Endocrinol. Metab.*, 55, 709, 1982.
34. **Demelia, L., Vallebona, E., Solinas, A., Carruxi, R., Sanna, G., and Pitzus, F.,** Hypothalamo-pituitary-adrenocortical axis in liver cirrhosis, *Ital. J. Gastroenterol.*, 21, 216, 1989.
35. **Fraser, R. W., Forbes, A. P., Albright, F., Sulkowitch, H., and Reifenstein, E. C., Jr.,** Colorimetric assay of 17-ketosteroids in urine, *J. Clin. Endocrinol.*, 1, 234, 1941.
36. **Gilder, H. and Hoagland, C. L.,.**Urinary excretion of estrogens and 17-ketosteroids in young adult males with infectous hepatitis, *Proc. Soc. Exp. Biol.*, 61, 62, 1946.
37. **Eisenberg, H. L., Kirshen, M. M., Atlas, D. H., and Gaberman, P.,** 17-Ketosteroids excretion in liver diseases, *Gastroenterology*, 18, 36, 1951.
38. **Peterson, R. E., Guerra, S., and Sborov, V. M.,** Steroids excretion patterns in acute viral hepatitis with and without adrenocorticotropin infusion, *J. Lab. Clin. Med.*, 43, 58, 1954.
39. **Dohan, F. C., Richardson, E. M., Bluemle, L. W., Jr., and Gyorgy, P.,** Hormone excretion in liver diseases, *J. Clin. Invest.*, 31, 481, 1952.
40. **Bongiovanni, A. M. and Eisenmenger, W. J.,** Adrenal cortical metabolism in chronic liver disease, *J. Clin. Endocrinol.*, 11, 152, 1951.
41. **Ahrens, E. H., Payne, M. A., Kunkel, H. G., Eisenmenger, W. J., and Blondheim, S. H.,** Primary biliary cirrhosis, *Medicine (Baltimore)*, 29, 299, 1950.
42. **Brown, H., Willardson, D. G., Samuels, L. T., and Tyler, F. H.,** 17-hydroxycorticosteroids metabolism in liver disease, *J. Clin. Invest.*, 33, 1524, 1954.
43. **Vermeulen, A. and Demeulenaere, L.,** Le role du foie dans le metabolisme des corticoides, *Rev. Fr. Etudes Clin. Biol.*, 1, 398, 1956.
44. **Peterson, R. E.,** Adrenocortical steroid metabolism and adrenal cortical function in liver disease, *J. Clin. Invest.*, 39, 320, 1960.
45. **Tucci, J. R., Rene, A., Albacete, M. D., and Martin, M. M.,** Effect of liver disease upon steroid circadian rhythms in man, *Gastroenterology*, 50, 637, 1966.
46. **Ring-Larsen, H.,** Hepatic nephropathy, related to haemodynamics, *Liver*, 3, 265, 1983.
47. **Scavo, D., Cugini, P., and Di Lascio, G.,** The metabolic clearance rate of cortisol in normal subjects and in various disease states, *J. Nucl. Med. All. Sci.*, 113, 22, 1978.
48. **Englert, E., Jr., Brown, H., Wallach, S., and Simons, E. L.,** Metabolism of free and conjugated 17-hydroxycorticosteroids in subjects with liver disease, *J. Clin. Endocrinol.*, 17, 1395, 1957.
49. **Christy, N. P., Wallace, E. Z., Gordon, W. E. L., and Jailer, J. W.,** On the rate of hydrocortisone clearance from plasma in pregnant women and in patients with Laennec's cirrhosis, *J. Clin. Invest.*, 38, 299, 1959.
50. **Sholiton, L. J., Werk, E. E., Jr., and Marnell, R. T.,** Diurnal variation of adrenocortical function in non-endocrine disease states, *Metabolism*, 10, 632, 1961.
51. **Zumoff, B., Bradlow, L., Gallagher, T. F., and Hellman, L.,** Cortisol metabolism in cirrhosis, *J. Clin. Invest.*, 46, 1735, 1967.
52. **Zumoff, B., Bradlow, L., and Hellman, L.,** Characterization of a defect in cortisol metabolism in cirrhosis and its reversible reproduction by norethandrolone (Nilevar), *J. Clin. Invest.*, 45, 1091, 1966 (Abstract).
53. **Peterson, R. E., Wyngaarden, J. B., Guerra, S. L., Brodie, B. B., and Bunim, J. J.,** The physiological disposition and metabolic fate of hydrocortisone in man, *J. Clin. Invest.*, 34, 1779, 1955.
54. **Groszmann, R., Kotelanski, B., Cohn, J. N., and Khatri, I. M.,** Quantitation of portasystemic shunting from the splenic and mesenteric beds in alcoholic liver disease, *Am. J. Med.*, 53, 715, 1972.
55. **Ventura, E., Zeneroli, M. L., Tampieri, A., Pinelli, G., Licari, F., Gollini, C., Cremonini, C., Ricci, P., Casalgrandi, Baragnini, G., and Marrama, P.,** Chronobiological study of circadian variations of plasma free tryptophan and of cortisol in liver cirrhosis with severe degree of encephalopathy, *It. J. Gastroenterol.*, 16, 1, 1984.
56. **Bernardi, M., De Palma, R., Trevisani, F., Santini, C., Capani, F., Baraldini, M., and Gasbarrini, G.,** Chronobiological study of factors affecting plasma aldosterone concentration in cirrhosis, *Gastroenterology*, 91, 683, 1986.

57. **Kley, H. K., Nieschlag, E., Wiegelmann, W., Solbach, H. G., and Kruskemper, H. L.,** Steroid hormones and their binding in plasma of male patients with fatty liver, chronic hepatitis and liver cirrhosis, *Acta Endocrinol.,* 79, 275, 1975.
58. **Bernardi, M., Trevisani, F., and Gasbarrini, G.,** The renin-angiotensin-aldosterone sytem in liver disease, in *Cardiovascular Complications of Liver Disease,* Bomzon A. and Blendis, L. M., Eds., CRC Press, Boca Raton, FL, 1990.
59. **Wernze, H., Spech, H. G., and Muller, G.,** Studies on the activity of the renin-angiotensin-aldosterone system (RAAs) in patients with cirrhosis of the liver, *Klin. Wschr.,* 56, 389, 1978.
60. **Epstein, M., Levinson, R., Sancho, J., Haber, E., and Re, R.,** Characterization of the renin-aldosterone system in decompensated cirrhosis, *Circ. Res.,* 41, 819, 1977.
61. **Chonko, A. M., Bay, W. H., Stein, J. H., and Ferris, T. F.,** The role of renin and aldosterone in the salt retention of edema, *Am. J. Med.,* 63, 881, 1977.
62. **Wilkinson, S. P., Jowett, T. P., Slater, J. D. H., Arroyo, V. Moodie, H., and Williams, R.,** Renal sodium retention in cirrhosis: relation to aldosterone and nephron site, *Clin. Sci.,* 56, 169, 1979.
63. **Mitch, W. E., Whelton, P. K., Cooke, C. L., Walker, W. G., and Maddrey, W. C.,** Plasma levels and hepatic extraction of renin and aldosterone in alcoholic liver disease, *Am. J. Med.,* 66, 804, 1979.
64. **Arroyo, V., Bosch, J., Mauri, M., Viver, J., Mas, A., Rivera, F., and Rodés, J.,** Renin, aldosterone and renal haemodynamics in cirrhosis with ascites, *Eur. J. Clin. Invest.,* 9, 69, 1979.
65. **Bosch, J., Arroyo, V., Betriu, A., Mas, A., Carrilho, F., Rivera, F., Navarro-Lopez, F., and Rodés, J.,** Hepatic hemodynamics and the renin-angiotensin-aldosterone system in cirrhosis, *Gastroenterology,* 78, 92, 1980.
66. **Bernardi, M., Trevisani, F., Santini, C., De Palma, R., and Gasbarrini, G.,** Aldosterone related blood volume expansion in cirrhosis before and after the early phase of ascites formation, *Gut,* 24, 761, 1983.
67. **Bernardi, M., De Palma, R., Trevisani, F., Santini, C., Capani, F., Baraldini, M., and Gasbarrini, G.,** Chronobiological study of factors affecting plasma aldosterone concentration in cirrhosis, *Gastroenterology,* 91, 683, 1986.
68. **Coppage, W. S., Island, D. P., Cooner, A. E., and Liddle, G. W.,** The metabolism of aldosterone in normal subjects and in patients with hepatic cirrhosis, *J. Clin. Invest.,* 41, 1672, 1962.
69. **Vecsei, P., Dusterdieck, G., Jahnecke, J., Lommer, D., and Wolff, H. P.,** Secretion and turnover of aldosterone in various pathological states, *Clin. Sci.,* 36, 341, 1969.
70. **Lieberman, F. L. and Reynolds, T. B.,** Plasma volume in cirrhosis of the liver: its relation to portal hypertension, ascites and renal failure, *J. Clin. Invest.,* 46, 1297, 1967.
71. **Wolff, H. P., Bette, L., Blaise, H., Düsterdieck, G., Jahnecke, J., Kobayashi, T., Krück, F., Lommer, D., and Schieffer, H.,** Role of aldosterone in edema formation, *Ann. N.Y. Acad. Sci.,* 139, 285, 1966.
72. **Rosoff, L., Jr., Zia, P., Reynolds, T. B., and Horton, R.,** Studies of renin and aldosterone in cirrhotic patients with ascites, *Gastroenterology,* 69, 698, 1975.
73. **Hurter, R. and Nabarro, J. D. N.,** Aldosterone metabolism in liver disease, *Acta Endocrinol.,* (Kbh), 33, 168, 1960.
74. **Wilkinson, S. P., Wheeler, P. G., Jowett, T. P., Smith, I. K., Keenan, J., Slater, J. D. H., and Williams, R.,** Factors relating to aldosterone secretion rate, the excretion of aldosterone 18-glucuronide, and the plasma aldosterone concentration in cirrhosis, *Clin. Endocrinol.,* 14, 355, 1981.
75. **Ulick, S., Laragh, J. H., and Lieberman, S.,** The isolation of a urinary metabolite of aldosterone and its use to measure the rate of secretion of aldosterone by the adrenal cortex of man, *Trans. Assoc. Am. Phys.,* 71, 225, 1958.
76. **Noth, R. H., McCallum, R. W., Contino, C., and Havelick, J.,** Tonic dopaminergic suppression of plasma aldosterone, *J. Clin. Endocrinol. Metab.,* 51, 64, 1980.
77. **Bernardi, M., De Palma, R., Trevisani, F., Malatesta, R., Baraldini, M., Cursaro, C., and Gasbarrini, G.,** Unaltered dopaminergic secretion of aldosterone secretion in cirrhosis, *Clin. Sci.,* 74, 137, 1988.
78. **Bernardi, M., Wilkinson, S. P., Wernze, H., Spech, H. J., Müller, G., Poston, L., and Williams, R.,** The renin-angiotensin-aldosterone system in fulminant hepatic failure, *Scand. J. Gastroenterol.,* 18, 369, 1983.
79. **Balikian, H. M., Brodie, A. H., Dale, S. L., Melby, J. C., and Tait, J. F.,** Effect of posture on the metabolic clearance rate, plasma concentration, and blood production rate of aldosterone in man, *J. Clin. Endocrinol. Metab.,* 28, 1630, 1968.
80. **Ciér, J. F.,** La physiologie du système rénine-angiotensine, *J. Physiol. (Paris),* 75, 179, 1979.
81. **Bernardi, M., Santini, C., Trevisani, F., Baraldini, M., Ligabue, A., and Gasbarrini, G.,** Renal function impairment induced by change in posture in patients with cirrhosis and ascites, *Gut,* 26, 629, 1985.

Chapter 18

DEGRADATION OF ENDOGENOUS CATECHOLAMINES

Niels Juel Christensen and Jens H. Henriksen

TABLE OF CONTENTS

I. INTRODUCTION

This chapter deals with kinetics and degradation of circulating catecholamines. The main stress is laid on the quantitative aspects of norepinephrine (NE) uptake and release, whereas the biochemistry will only be reviewed briefly. Clearance studies may elucidate the mechanism by which catecholamines are metabolically degraded and, furthermore, such studies also make it possible to calculate catecholamine turnover rates.

The methods of studying the clearance rate of catecholamines are broadly divided into whole-body clearance studies and organ degradation or clearance studies. We consider factors which may influence catecholamine kinetics. Finally, we discuss how pathophysiological states may influence catecholamine degradation rates.

II. THE ASSESSMENT OF SYMPATHOADRENAL ACTIVITY

The sympathoadrenal system is one of the major pathways mediating physiological responses and psychological influences on different organs and tissues in the body. In pathophysiological states, the sympathoadrenal system defends steady states of the body, homeostasis, by corrective physiological mechanisms. Moreover, chemical agents that selectively block or stimulate adrenergic receptors are used extensively in clinical medicine. The sympathetic system works in a highly differentiated way and with subdivisions specialized to regulate different organ functions according to their demand. Sympathetic activity, sensitivity, and responsiveness to catecholamines, as well as degradation rate, should therefore preferably be measured in single organs or tissues.[1]

Classically, the function of the sympathetic nervous system has been evaluated indirectly by measuring changes in physiological variables (blood flow, blood pressure, etc.) considered to be controlled by sympathetic activity. More direct approaches of studying sympathetic activity are available: by measuring the release of NE from organs and tissues into the veins that selectively drain them, or by recording impulses in sympathetic nerves to skin or muscles. Radioenzymatic techniques or HPLC (high-performance liquid chromatography with electrochemical determination) permit precise measurements of catecholamines in small volumes. Furthermore, tungsten microelectrodes may be used to record sympathetic axon potentials in human peripheral nerves at rest and during various maneuvers. Recordings can be made from fibers in muscle and skin nerves, whereas nerves to other regions are inaccessible. Close correlations have been observed between forearm venous plasma NE and muscle sympathetic nerve activity during most conditions; that is, in the resting state, during exercise, during mental stress, as well as after raised and lowered arterial blood pressure with intravenous infusion of phenylephrine and nitroprusside.[2–4] This relationship may be explained by three factors: (1) skeletal muscle constitutes a major tissue, (2) approximately 50% of NE in forearm venous blood is derived locally from forearm tissue, mainly muscle, and (3) NE from the gastrointestinal tract is largely eliminated in the liver.

Plasma epinephrine (E) concentration should preferably be measured in arterial blood because the extraction of E across the forearm may increase considerably when plasma E concentration increases. This has been observed during infusion of E, during mental stress, as well as in the post-prandial period.[5-7]

Plasma NE and E do not necessarily change in parallel: during mental stress, sympathetic nerve activity to visceral organs is increased and plasma E is increased, whereas muscle sympathetic nerve activity and forearm venous plasma NE remain relatively unchanged during a brief mental stress period, provided that muscular activity is not increased. After an oral glucose load, plasma E is initially suppressed, whereas plasma NE is increased.[6,7]

A. METABOLISM: BIOCHEMICAL ASPECTS

NE and E are rapidly inactivated after their release (Figure 1). Three enzymes, catechol-

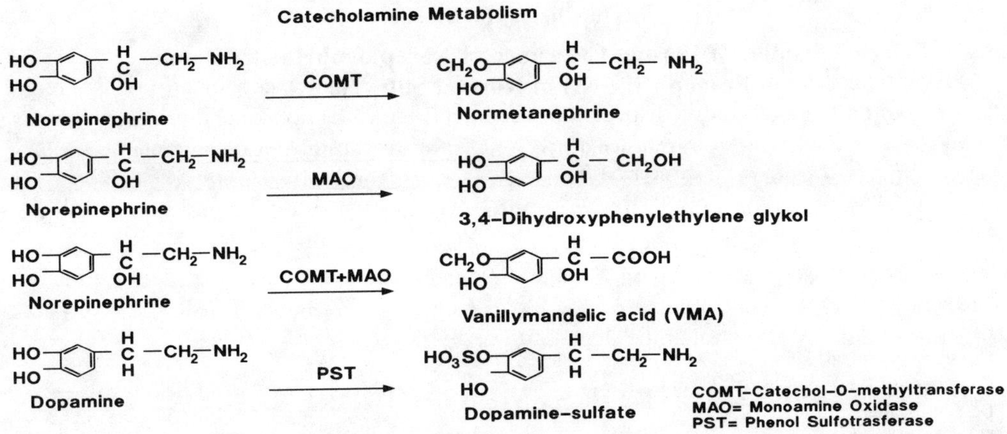

FIGURE 1. Catecholamine metabolism.

O-methyltransferase (COMT), monoamine oxidase (MAO), and phenol sulphotransferase (PST), as well as reuptake in sympathetic nerve endings, are important for the metabolism and inactivation of catecholamines.

NE which is taken up by axon terminals (uptake$_1$) is either incorporated into the vesicles and reused or deaminated by MAO in the cytoplasma. Deamination by MAO results in the formation of aldehydes which are further reduced to the glycol derivative or oxidized extra-neuronally to the corresponding acid.

Catecholamines are also taken up by nonneuronal tissues (uptake$_2$) and methylated by COMT and, in most tissues, also deaminated by MAO. Thus, MAO is present in both neuronal and nonneuronal tissues, whereas COMT is mainly present in nonneuronal tissues. The deaminated products thus reflect total NE metabolism (extra- and intraneuronal metabolism), whereas the methylated product normethanephrine is an index of the fraction of NE which is released from the sympathetic nerves. Catecholamines present in the food, especially DOPA and dopamine, are sulfated in the liver and in the gut by PST, which is also present in many other tissues. The major end-product of NE and E is vanillylmandelic acid (VMA), which is both deaminated and methylated (3-methoxy-4-hydroxy-mandelic acid).

There are substantial amounts of dopamine in the blood of which 99% is sulphoconjugated and biologically inactive. Sulphoconjugated dopamine in the blood is derived from the food and probably also from decarboxylation of circulating DOPA, as well as from other sources. The origin of free DOPA in plasma is not known. The urine contains free dopamine, which is derived from decarboxylation of plasma DOPA in the kidney. Sulphoconjugated dopamine is also excreted in the urine. The major dopamine metabolite is homovanillic acid, which is derived from the brain and from intraneuronal metabolism of dopamine and subsequent methylation.

B. WHOLE-BODY NOREPINEPHRINE TURNOVER AND SPILLOVER TO PLASMA

With the advent of precise and sensitive techniques for measuring plasma NE and other catecholamines, several authors have attempted to measure NE spillover to plasma. The plasma NE concentration is dependent not only on the rate of NE release but also on the rate at which NE is removed from the circulation. It has therefore been assumed that the product of plasma NE and plasma clearance is a better index of sympathetic nerve activity than plasma NE concentration alone. As pointed out above, the sympathetic nervous system works in a differentiated way and sympathetic nervous activity should therefore preferably be measured in single organs or tissues. Furthermore, plasma NE spillover values have been difficult to

TABLE 1
Selected Studies of Plasma Clearance of Norepinephrine (NE) and Epinephrine (E) in Normal Subjects

Authors	Infusion of tracer NE/E unlabeled NE/E	Plasma clearance $1 \cdot min^{-1}$ NE Cl_v	NE Cl_a	E Cl_v	E Cl_a
Vendsalu, 1960[8]	NE/E	3.800	—	9.800	—
Silverberg et al., 1978[18]	NE	3.100	—	—	—
FitzGerald et al., 1979[19]	NE	4.350	—	—	—
Esler et al., 1979[20]	tNEb	2.900	—	—	—
Christensen et al., 1979[21]	E	—	—	5.260	—
Clutter et al., 1980[22]	E	—	—	6.680	—
Best and Halter, 1982[5]	E	—	—	5.250	2.620
Hilsted et al., 1983[23]	tNE	2.470	1.400	—	—
Kjær et al., 1985[24]	tE	—	—	—	2.100

[a] Fluorimetric technique.
[b] tNE = tracer NE; tE = tracer E, Cl_v and Cl_a = whole-body plasma clearances assessed by venous and arterial samples, respectively.

interpret. Some of the problems which have arisen will be discussed in the following: (1) historical aspects, (2) basic concept, (3) sampling site, (4) recycling and the length of the infusion period, and (5) modeling.

1. Historical Aspects

In 1960, Vendsalu[8] studied the plasma clearance of NE and E by infusing unlabeled NE and E and measuring steady-state concentrations of the catecholamines in venous blood (see Table 1). Gitlow et al.[9] studied the plasma clearance of tritiated NE in normal human subjects and in patients with essential hypertension. DeQuattro et al.[10,11] studied NE turnover in normotensive and hypertensive subjects using tritium-labeled DOPA. Maas et al.[12,13] studied the metabolism of circulating NE by human subjects. Tritium-labeled NE was infused for a 48-h period to obtain an isotopic steady state. It was found, however, that specific activities of both urinary 3-methoxy-4-hydroxymandelic acid and 3-methoxy-4-hydroxyphenolglycol were lowered by at least a factor of ten than the specific activities of urinary NE and normetanephrine. They concluded that most of the endogenous urinary 3-methoxy-4-hydroxymandelic acid and 3-methoxy-4-hydroxy-phenoglycol originated in pools which, relative to their total rates of NE synthesis or size, are poorly penetrated or not penetrated at all by circulating NE. Coulombe et al.[14] studied catecholamine metabolism in thyroid diseases. Plasma secretion rates of E and urinary secretion rates were not different in patients with hyperthyroidism compared to normal subjects.

2. Basic Concept

As pointed out by Lassen and Perl and by others,[15–17] the basic assumption for the total NE spillover concept — equal specific activity (tritium-labeled NE dpm per nanomole endogenous NE) in plasma and at breakdown sites — is violated, even in the perfect steady state.

Endogenous NE is synthesized and metabolized in the same sites; that is, in sympathetic axon terminals. When tracer NE is infused intravenously, the specific activity of tracer NE in these terminals will never reach that of plasma despite neuronal tracer uptake. Moreover, there is no way to estimate the discrepancy between the plasma appearance rate and the total spillover rate of NE:

$$\text{Plasma appearance rate} = Cl_{[^3H]NE} \times X_{NE}$$

Total spillover rate =

$$Cl_{[^3H]NE} \times C_{NE} \times (a_{plasma} / a_{breakdown\ site})$$

where $Cl_{[3H]NE}$ is the plasma clearance of tracer NE, C_{NE} is the concentration of endogenous NE in plasma, and a_{plasma} and $a_{breakdown\ site}$ are specific activities of tracer NE in plasma and breakdown sites, respectively.

In contrast to NE, E is released into the circulation from the adrenal medulla and broken down at sites of no E synthesis. Therefore, intravenous infusion of tracer E fulfills basic assumptions of the kinetic theory, and it can be predicted with confidence that the specific activity of tracer E in plasma is the same as that at the breakdown sites. This statement implies that total spillover of E into plasma can be assessed during steady-state conditions as the concentration of endogenous E in plasma multiplied by the plasma clearance of tracer E.

3. Sampling Site

Table 1 presents a selected number of studies of plasma clearance of NE and E in normal human subjects. It appears from the table that (1) the calculated plasma clearance of E is higher than that of NE, (2) the clearance based on arterial sampling is considerably lower than that based on venous sampling, and (3) plasma clearance values are lower after tracer E and NE infusion as compared to infusion of nonlabeled catecholamines. Esler et al.[25] considered no sampling site ideal, but this is not correct. The correct sampling site depends on the site of infusion of the tracer. If tracer E is infused into a peripheral vein, the correct sampling site is the pulmonary artery or, alternatively, the systemic arteries because only a small extraction of E or NE takes place across the lungs.[17,26,27] If, on the other hand, the tracer concentration is measured in peripheral venous plasma, the calculated clearance value will be erroneously high because of the local uptake of catecholamines in the forearm tissues. As can be seen from Table 1, in many studies peripheral venous sampling has been used and the reported plasma clearance values are close to or exceed the cardiac output of plasma. These values can be explained by the effect of a local extraction of tracer NE or E in addition to the systemic clearance. Consequently, pulmonary or systemic arterial samples should be applied as references.

4. Recycling and the Length of the Infusion Period

Whole-body plasma clearance of NE depends on the length of the tracer infusion period and how the clearance value is calculated. Several authors have found that the plasma clearance of NE, determined as infusion rate divided by plasma concentration of tracer after 20-min infusion, was significantly higher than clearance values obtained after 40 or 80 min of tracer infusion, indicating that steady state with respect to plasma specific activity was not achieved after 20 min (see Table 2). Furthermore, Henriksen and Christensen[28] showed that, in contrast to tritium-labeled isoproterenol which has no neuronal uptake, plasma NE clearance as obtained from infusion rate divided by plasma concentration of tracer was significantly higher than the value obtained by total tracer infusion divided by total plasma area of tracer (Table 2). The mean residence time in plasma after stopping the infusion of tracer NE increased along an almost straight line with progressive infusion time, whereas constant values were recorded after 40-min infusion of tracer isoproterenol.[28] These results suggest the presence of a large cellular pool from which a reversible transport of tracer NE back into plasma takes place. The plasma clearance of tracer NE, as determined from infusion rate and plasma concentration of the tracer, includes transport to and accumulation in this large store. Thus, the final metabolic clearance reflecting irreversible removal of NE is smaller than previously estimated due to recycling through the plasma space.

TABLE 2
Whole-Body Clearance of Tracers Determined from Infusion Rate and Plasma Concentration and Infused Dose and Plasma Concentration-Time Area in Normal Volunteers

Tracer	Whole-body plasma clearance (l/min)					
	20 min		40 min		80 min	
	I/C	D/A	I/C	D/A	I/C	D/A
L-[^{3}H]norepinephrine	2.1 ± 0.5[a]	1.79 ± 0.78[f]	1.50 ± 0.48	1.35 ± 0.44[f]	1.74 ± 0.64	1.27 ± 0.51[e]
DL-[^{3}H]isoproterenol	2.5 ± 0.9[b]	1.25 ± 0.33[f]	1.56 ± 0.26	1.42 ± 0.14	1.34 ± 0.07	1.31 ± 0.14
^{131}I-hippurate	0.84 ± 0.07[c]	0.63 ± 0.10[d]	0.61 ± 0.08	0.54 ± 0.10[d]	0.56 ± 0.04	0.52 ± 0.14[d]

Note: Values are means ±SD. I/C, infusion rate/plasma concentration; D/A, infused dose/area underneath plasma concentration curve.

[a] Significantly different from 40-min value: $P < 0.005$
[b] Significantly different from 40-min value: $P < 0.01$.
[c] Significantly different from 40-min value: $P < 0.001$
[d] Significantly different from L-[^{3}H]norepinephrine and DL-[^{3}H]isoproterenol, $P < 0.001$.
[e] Significantly different from I/C values, $P < 0.01$.
[f] Significantly different from I/C values, $P < 0.05$.

From Henriksen, J. H. and Christensen, N. J., *Am. J. Physiol.*, 257, E743, 1989. With permission.

5. Modeling

In the studies summarized in Table 1, the plasma clearance was calculated from the infusion rate divided by the steady-state plasma NE and E concentrations. This model can be considered to be either a simple form of compartmental modeling or black-box analysis. As regards the post-infusion decay, both compartmental analysis and black-box analysis have been applied.[28–31] Linares et al.[31] devised a two-pool model of NE disposition, whereas Henriksen and Christensen[28] used black-box analysis. The major advantage of a noncompartmental model is that any number of recirculations and exchanges can occur with any number of noncentral pools, none of which has to be identified with specific physiological structures. A noncompartmental analysis is not, however, model independent. Thus, the noncompartmental model requires that the substance under study is directly eliminated from the central pool and that all sources of the substance under study must be directed into the same pool into which the tracer is introduced. It should be noticed that the post-infusion decay rate calculated either as the half-life or as the residence time in plasma is dependent not only on the plasma clearance, but also on the volume of distribution. More information can thus be obtained by studying both the plasma clearance rate and the post-infusion decay rate, and this will be considered in more detail in relation to autonomic neuropathy.

C. PLASMA EPINEPHRINE CLEARANCE

As pointed out previously, total spillover of E into plasma can be assessed with confidence during steady-state conditions as the product of the endogenous E concentration in plasma and the plasma clearance of tracer E. Kjœ et al.[24] studied the effect of exercise and E turnover in trained and untrained male subjects. E clearance was closely related to relative work load. It increased approximately 15% above basal at 30% of maximal O_2 uptake to 22% below at 76% of maximal O_2 uptake. E concentrations increased much more with work intensity that could be accounted for by changes in clearance. The authors concluded that during exercise changes in E concentrations reflect changes in secretion rather than in clearance.

D. FACTORS INFLUENCING THE WHOLE-BODY CLEARANCE OF NE AND E

It follows from the above-mentioned considerations that the concept of clearance cannot readily be applied to NE and, therefore, literature studies of total NE spillover will not be extensively discussed. Furthermore, since sympathetic outflow is highly differentiated, NE turnover rate should preferably be measured in single organs and tissues. Thus, the present discussion will be limited to three problems: (1) the effect of desipramine and autonomic neuropathy on the plasma clearance of NE and E, (2) the effect of propranolol on the plasma clearance of NE and E, and (3) age and plasma clearance of NE.

Esler et al.[32] reported that desipramine, which blocks neuronal uptake, decreased plasma NE clearance by approximately 20%. The mean plasma NE clearance was found to be 25% lower in patients with idiopathic autonomic insufficiency as compared to controls. Goldstein et al.[30] reported that about 15% of infused NE, which was removed in the arm, was removed by $uptake_1$. Hoeldtke and Cilmi[33] observed that plasma clearance of NE was not different in patients with diabetic autonomic neuropathy compared to normal subjects. Dejgaard et al.[34] observed that the plasma clearance of NE and isoproterenol was approximately similar, and not influenced by, the presence of autonomic failure in diabetic patients, whereas the plasma appearance rate of NE was significantly reduced in patients with neuropathy as compared to patients without neuropathy. The post-infusion decay was also similar in patients with and without diabetic autonomic neuropathy. The authors concluded that the extraneuronal uptake of isoproterenol was greater than that of NE. Furthermore, it was possible that a relatively small decrease in neuronal uptake of NE in diabetic patients with autonomic neuropathy was outweighed by a slightly higher nonneuronal uptake in these patients. In a subsequent study, Dejgaard et al.[35] studied plasma E kinetics (clearance, extraction across the forearm, initial plasma disappearance rate, mean residence time in plasma, and volume of distribution) in diabetic patients with and without autonomic neuropathy. The most marked abnormality was a decrease in the post-infusion decay rate (initial slope and residence time) in patients with autonomic neuropathy as compared to those without. The plasma clearance was not significantly different in the two groups, suggesting that E in patients with autonomic neuropathy is distributed in a larger volume as compared to control subjects. These results may indicate that patients with diabetic neuropathy are not able to adjust the plasma E concentration to changes in E secretion rate as rapidly as those without neuropathy; that is, the effect of an elevated E secretion rate may be prolonged in patients with diabetic autonomic neuropathy.

Taken together, these studies indicate that neuronal uptake removes a significant but relatively small proportion of circulating catecholamines. Furthermore, in patients with autonomic failure, the post-infusion residence time in plasma may be prolonged. The plasma clearance of NE or E may or may not be decreased in patients with autonomic neuropathy, probably dependent on the severity of the autonomic neuropathy. The studies also indicate that measurements of the plasma residence time may be a more sensitive index of a reduced neuronal uptake than measurements of the plasma clearance of NE or E.

Several studies have shown that propranolol decreased the plasma clearance of E and NE.[5,36–38] Propranolol may decrease the plasma clearance of E by approximately 75%, whereas the effect on NE is smaller, around 25%. Propranolol also decreased the post-infusion decay of isoproterenol in plasma.[39] The effect of propranolol on plasma E clearance is probably mediated by β_1-adrenergic mechanisms. It is not entirely clear whether this action of propranolol is directly at the sites of catecholamine removal or secondary to the vasoconstriction and reduced cardiac output that accompany β_1-adrenergic blockade. The results do suggest, however, that propranolol acts on nonneuronal processes rather than on neuronal uptake mechanisms. Interestingly, increments in plasma E increases the extraction of E across the forearm.[5–7] Furthermore, a high cardiac output may explain the negative relationship observed between the forearm extraction ratio of NE and E and the whole-body plasma clearance of NE and E, respectively.[23,35] Best and Halter[5] observed, however, no increase in total body clearance of E during E infusion, whereas Kjœr et al.[24] observed a slight increase in plasma E clearance during light exercise. Finally, Linares et al.[40] reported that the plasma clearance of NE decreased markedly in the upright position.

These results suggest that the plasma clearance of both NE and E may be dependent on the level of the cardiac output. Furthermore, the fact that plasma E increases its own extraction across the forearm makes it of paramount importance that changes in plasma E and plasma E kinetics are evaluated on the basis of arterial rather than venous measurements. Although propranolol in the studies mentioned above decreased the plasma clearance of NE, the plasma NE concentration was also increased and, thus, the calculated appearance rate was approximately unchanged. It would be very unwise, however, to draw any conclusions from these findings regarding sympathetic activity after administration of propranolol in view of the many problems with the interpretation of total NE spillover data. Microneurographic recordings suggest, as one would expect, that muscle sympathetic activity is increased after acute β-adrenergic receptor blockade.[41]

It is generally accepted that plasma NE concentration increases with age (Table 3). Furthermore, circulating levels of NE increased more in old subjects than in young subjects, in response to various maneuvers that activate the sympathetic nervous system (for a review, see Reference 42). Esler et al.[43] reported that the increase in plasma NE with age was due to a decreased plasma clearance of NE in elderly subjects. This finding has not, however, been confirmed by other investigators (Table 3) who found no differences in the plasma clearance rate of NE between young and old subjects. The clearance values reported in most of the studies are erroneously high because venous blood was used for determination of steady-state concentrations of the infused NE.[23,26] When arterial sampling was used, the whole-body

TABLE 3
Plasma Clearance of Norepinephrine in Young and Elderly Subjects

	Young subjects (l/min·1.8 m²)	Elderly subjects (l/min·1.8 m²)	Arterial or venous sampling	Change[a]
Esler et al., 1981[43]	2.7	2.2	Venous	S
Young et al., 1980[44]	3.0	2.7	Venous	NS
Rubin et al., 1982[45]	4.1	3.8	Venous	NS
Hoeldtke and Cilmi, 1985[46]	2.6	2.3	Venous	NS
Hilsted et al., 1985[47]	1.0	1.0	Arterial	NS
MacGilchrist et al., 1989[48]	6.0	5.0	Venous	NS

[a] S and NS = significant and not significant, respectively.

clearance of NE was, however, also similar in young and old subjects.[47] That sympathetic nervous activity increases with age is also supported by Walin et al.[4] who observed that the number of impulses in sympathetic nerves to muscle increased with age. Furthermore, renal venous-arterial differences of NE have been shown to increase with age.[49]

E. OTHER TECHNIQUES OF STUDYING NOREPINEPHRINE TURNOVER

Hoeldtke et al.[50] recommended that total NE production is estimated in addition to NE secretion rate from integrated NE metabolite secretion. Apparent secretion rate was only 22% of total NE production. According to Maas et al.,[12] the bulk of the total pool of metabolites excreted in the urine has its origin in intracellular metabolites without prior release at nerve endings. Plasma DHPG (3,4-dihydroxyphenylglycol) is probably derived exclusively from sympathetic nerves. Plasma DHPG reflects intraneuronal deamination of NE. An increase in plasma DHPG indicates neuronal uptake of NE, which may either be taken up in the vesicles and reused or deaminated to DHPG. Most of the basal plasma DHPG is probably derived from intraneuronal deamination of NE, which has not been released at nerve endings.[51-53]

Finally, intraneuronal NE turnover rates may be calculated from the increase in plasma DOPA after administration of a DOPA decarboxylase inhibitor (Eldrup et al., manuscript in preparation).

The techniques mentioned above may offer additional information about the sympathetic nervous system, although they do not directly estimate the amounts of NE released at nerve endings.

F. CATECHOLAMINE TURNOVER IN SINGLE ORGANS

In animal experiments, it is possible to study the turnover of NE in single organs: after isotopically labeled NE is taken up by the organ's nerve terminals or after NE synthesis is blocked, the disappearance of isotopic label and the decrease in tissue concentration of NE, respectively, is followed to drive a rate of turnover. These parameters are not, however, identical to NE release because only a fraction of NE synthesized in the axon terminals is released into the synaptic cleft.

In humans, organ degradation can be assessed by determination of the excretion rate (bile or urine), by the Fick principle, or by kinetic analysis of outflow profiles after tracer injection into the artery.[54,55] Catecholamines with a tissue uptake, but no release (like E or isoproterenol), can be assessed directly from the arterio-venous concentration difference and determination of the plasma flow rate. In case of NE, the vascular areas will, as previously mentioned, release as well as remove NE. Quantification of NE kinetics, therefore, demands a method by which one is able to distinguish both flux rates. Studies with ^{3}H-NE have provided valuable results in organs and tissues of normal humans as well as in patients with different diseases.[17,30,31,38,54,55] Some fundamental assumptions and limitations of organ NE kinetics will be considered briefly: (1) steady state, (2) neural uptake and re-release of tracer, (3) inhomogeneous specific activity, (4) tissue pools, and (5) reference samples.

In order to apply the Fick principle, a steady state with respect to concentrations and flux rates for endogenous and tracer catecholamines must exist. Most studies are carried out under standardized conditions, either in supine, resting subjects, or during an exercise protocol.[1,24,25,54] It seems reasonable to assume that steady-state conditions are fulfilled for the endogenous catecholamines, at least in rest. Tracer NE or tracer E have a high overall turnover rate with a short plasma half-life. Therefore, a primed continuous intravenous infusion may give an almost constant plasma concentration within approximately 20 min.[30,31] Thus, steady-state conditions seem to be fulfilled. However, recent experiments in man have shown that the different organs do achieve steady state within very different times.[54] Thus, for example, the kidneys obtain steady state with respect to renal venous NE, inflow, and outflow of tracer and extraction ratios within approximately 25 min.[54] In contrast, other organs like the hepato-splanchnic system and the lower limbs seem to achieve a steady state very slowly. This must be considered before one applies the Fick principle. Another possibility is to use the extraction of endogenous E as an index of tissue uptake of NE.[56,57] In most studies, the extraction fraction of E is somewhat higher compared to that of ^{3}H-NE.[54,58,59] Whether this is due to a different tissue uptake of E and NE or to re-release of ^{3}H-NE is not settled at present. However, from a pragmatic point of view, assessment of the extraction of E seems to be a reasonable alternative to the use of ^{3}H-NE.

Another way to assess catecholamine kinetics in organs has been described by Goresky et al.,[60] Cousineau et al.,[61] and Rose et al.[55] They have developed a highly sophisticated multiindicator technique with injection into an organ artery and repeated sampling from the organ venous outlet. By determination of endogenous NE in artery and vein, injection of, for example, ^{125}I-albumin, ^{14}C-sucrose, and ^{3}H-NE, and application of a multicompartment iterative computer technique, it is possible to estimate NE uptake, release, extraction, transcapillary passage, and average interstitial space concentration. Another advantage of this method is that repeated injections can be performed under different conditions; for example, before and after administration of pharmacological agents. The drawbacks are, along with injection into selective arteries, that a number of assumptions regarding capillary dispersion, longitudinal and radial diffusion, transit times, interstitial space/capillary volume ratios, and recirculation correction must be fulfilled. Moreover, the model does not take into account that the concentration of NE in the interstitial space is most likely very heterogeneous with high concentration near the axon terminals of post-ganglionic fibers. This method has been applied in the liver and heart of dogs and in the human heart.[55,60,61]

1. Lungs

With respect to catecholamine extraction, the lungs are remarkable because they have a very high perfusion relative to their tissue mass. Earlier investigations have suggested a relatively high catecholamine uptake in the lungs.[62-64] This conception was based on animal experiments and studies in humans undergoing major surgery. However, recent studies applying labeled and endogenous NE have disclosed a relatively modest extraction of tracer and endogenous NE.[27] The pulmonary clearance of NE has been estimated to be of the order of 150 ml/min. Extraction of E is small and insignificant in normal lungs;[27] see Table 4.

2. Heart

In the heart, a substantial extraction of E has been found, as evaluated by catheterization of the coronary sinus.[65,66] ^{3}H-NE has an extraction ratio of 0.54 to 0.70[25,55] and the normal clearance of E and tracer NE is about 85 and 90 ml/min, respectively.[25,55,57,67] Normally, the concentration of endogenous NE in the coronary sinus is slightly above the arterial level, indicating a spillover of NE in the heart.[55,57]

TABLE 4
Organ Disposal of Circulating Endogenous Norepinephrine (NE), Epinephrine (E), and Tracer NE (^{3}H-NE)

	Extraction ratio			Clearance		Spillover of NE (ng/min)
	NE	3H-NE	E	E (ml/min)	NE (ml/min)	
Lung	0.08[a]	0.07[a]	0.03[a]	—	150[a]	—
Heart	–0.05[b]	0.73[b]	0.52[c]	85[d]	90[d]	25[d]
Kidney	–0.09[a]	0.35[f]	0.47[f]	253[f]	187[f]	49[f]
Liver-	0.63[f]	0.59[f]	0.67[f]	587[f]	421[f]	–5[f]
intestines	0.68[g]		0.78[h]	740[h]	540[g]	
			0.90[g]			
Limb	–0.09[f]	0.40[f]	0.63[b]	185 pr. 30 kg[f]	120 pr. 30 kg[f]	30 pr. 30 kg[f]

[a] From Reference 27.
[b] From Reference 55.
[c] From Reference 57.
[d] Calculated under the assumption of 275/ml/min coronary flow and 0.40 hematocrit fraction.
[e] From Reference 25.
[f] From Reference 54.
[g] From Reference 74.
[h] From Reference 56.

3. Kidney

In addition to urinary excretion, the kidney can degrade a number of bioactive peptides and amines.[56,68,69] Animal and human studies have shown that the extraction ratio of E is around 0.5 in normal kidneys.[57,58,65,70] As the kidney receives a relatively large proportion of the cardiac output, this organ will contribute substantially to E disposal. It should be recalled that the left adrenal gland has its venous outlet directly into the left renal vein in most cases. Thus, it may be difficult to obtain representative blood samples from the left kidney without adreno-medullary contamination. Most studies have therefore applied the right renal vein for determination of kidney extraction of E.

By applying ^{3}H-NE, it has been shown that the normal kidney extracts about 40% of the tracer.[54,58] Endogenous NE in the renal vein of supine, relaxing subjects without sympathoadrenal stimulation is similar to, or slightly above, the concentration in the artery.[49,54] Kostraeva[71] and Kopp[72] and co-workers have shown an almost linear relationship between burst frequency in stimulated sympathoadrenal nerves to the kidney and renal venous NE spillover. The renal venous plasma NE concentration is thus a reliable index of renal sympathetic nervous activity. As explained later, further evidence of that conception has been obtained from pathophysiological studies in patients with heart and liver disease. By applying tracer NE, it has been shown that even in supine resting humans, there is a significant sympathetic tone in this organ.[58] Moreover, recent tracer studies have shown that a steady state is easily obtained after a relatively short time of intravenous tracer infusion.[54] This may be due to the relatively high blood flow compared to the relatively low kidney weight.

The urinary excretion of NE and E is small compared to the overall disposal of these catecholamines by the kidney. This indicates that, in addition to filtration and reabsorption, there is a substantial deamination, methylation, and neuronal re-uptake.[1,73]

4. Liver-Intestines

Blood flow in the splanchnic organs is complex because the gastrointestinal tract, spleen, and pancreas are inserted in parallel with a common outlet through the liver veins. In humans, numerous studies have confirmed a substantial extraction of both NE and E in these organs.[1,54,56,74,75] A few studies performed under anesthesia, or invasive procedures like trans-hepatic portal vein catheterization, have shown a relatively high content of NE in the portal vein.[76,77] Portal venous E is below that of the artery.[57,76] Thus, intestines may have a net spillover of NE followed by extraction in the liver. The reason for the very low content of E in hepatic veins may be: (1) two tissue systems are connected in series, each having a major E extraction with an overall arterio-hepatic venous extraction ratio of approximately 70% or (2) COMT may be very effective in the liver. Bile secretion of catecholamines constitutes only a very small fraction of overall disposal of catecholamines in the liver.[78]

In the normal condition, tracer studies have not given any indication of NE release in the hepato-splanchnic system.[54] There may be some problems, however, when applying tracer NE in the splanchnic system, due to insufficient steady state.[54]

5. Skeletal Muscle, Cutaneous, and Subcutaneous Tissues

Comparison of femoral venous and cubital venous NE has shown a somewhat higher level in the latter vessel.[79] Most often, NE concentration in arterial and peripheral venous plasma are similar, indicating that net uptake equals net spillover.[23] Extraction of tracer NE is about 40% in normal humans, but problems of non-steady-state exist because of the small blood flow relative to the large tissue mass.[54]

Extraction of E, as evaluated from femoral and cubital venous samples, lies in the interval from 30 to 60%.[79] Under some conditions, for example, after food intake, a highly increased extraction of E can be observed.[1,6] This is important to notice because a simultaneous increase in arterial E may be overlooked under such circumstances if only venous samples are drawn. By heating hand and forearm (45 to 60°C), one may obtain venous samples with an oxygen content above 95% saturation.[1] In this case, venous E will increase towards the arterial level due to the highly increased peripheral blood flow. Sampling sites and the relationship between cubital venous NE and skeletal muscle sympathetic nerve activity have been considered previously.

III. PATHOPHYSIOLOGY OF CATECHOLAMINE KINETICS AND CLINICAL APPLICATION

A number of major diseases show substantial derangement of sympathetic nervous activity and catecholamine kinetics; for example, heart failure, thyroid disorders, diabetes mellitus, liver cirrhosis, gastro-duodenal ulcers, and surgical and mental stress.[1] It is beyond the scope of this review to give a comprehensive survey of catecholamine degradation in all these diseases. A few examples will be considered.

Patients with *chronic obstructive lung disease* have elevated circulating levels of catecholamines and NE is inversely correlated to the arterial oxygen tension.[80] This is most likely due to an enhanced sympathoadrenal activity. Very little is known of tissue uptake and pulmonary degradation of catecholamines in lung disease, but the relatively modest clearance of catecholamines in normal lungs[27] suggests that a compromised disposal in patients with lung disease will affect the circulating level of endogenous catecholamines only minimally.

In *heart failure* and *cirrhosis*, circulating levels of NE are increased.[1] Most investigators have pointed to an increased appearance rate, due to enhanced sympathetic nervous tone, as the main cause of increased plasma NE and in some cases increased E in cirrhosis.[59,74,81,82] The role of catecholamine degradation in various organs and tissues has been investigated extensively in patients with cirrhosis. Only a minor fraction of the increased plasma NE and E can be explained by decreased disposal in the liver.[74,75]

Tracer studies have indicated that, in patients with cirrhosis, intestines have an enhanced neuronal release of NE, giving a slightly decreased overall net hepato-intestinal extraction of circulating NE[59] No significantly reduced whole-body degradation has been described.[54] In this disease, highly increased spillover of NE has been traced to the azygos vein and the renal vein (Bendtsen et al., unpublished results). A number of studies indicate that enhanced sympathetic nervous activity is of importance for the deranged sodium and water homeostasis.[58,75,81,82] This will not be considered in detail here, but it is important to notice that the overall clearance of catecholamines probably is normal in a number of diseases with decreased organ function. The elevated plasma concentrations of NE and E indicate, therefore, enhanced sympathoadrenal activity.

Controversy exists as regards sympathetic nerves in *heart failure*. Tissue levels of NE are reduced in cardiac muscle and plasma levels are increased.[83,84] Rose et al.[55] found a marked reduction in cardiac NE release and uptake in patients with left ventricular failure secondary to mechanic overload. In patients with heart failure, the circulating levels of NE and E are increased and inversely correlated to the oxygen tension in the pulmonary artery.[85] This probably reflects a relationship between sympathetic tone on one side and cardiac output relative to oxygen uptake on the other.

Different drugs (e.g., clonidin, beta-adrenergic blocking agents, alpha-adrenergic blocking agents, and ganglionic blockers) may affect the circulating level of catecholamines. The increased NE and E concentration after, for example, beta-adrenergic blockage which is seen in normal subjects, patients with heart failure and cirrhosis of the liver is, in part, caused by a decreased overall clearance of catecholamines (see earlier). Recent investigations in patients with cirrhosis may suggest that the clearance in major organs is relatively unchanged after propranol intake. Thus, in these patients, the increased plasma level of NE and E in part reflects increased appearance and thereby enhanced sympathetic nervous activity.

IV. CONCLUSION

Due to the advent of precise and sensitive techniques of measuring catecholamines, much new information has been obtained in recent years regarding whole-body and regional clearances of NE and E. Clearance studies are important because they may elucidate the mechanism by which catecholamines are degraded and such studies also make it possible to calculate catecholamine turnover rates. Changes in plasma NE and E concentrations are predominantly due to changes in NE release rate and E secretion rate because clearance values are relatively constant during physiological conditions, but may be changed by a number of pharmacological substances. Catecholamines are removed from the circulation by nonneuronal and, to a lesser extent, neuronal uptake, mainly in liver-intestines, kidney, and muscles. Lung clearance is insignificant for E and small for NE. The extraction ratios are approximately 0.50 in most organs and slightly larger for E than for NE. The extraction ratio decreases with increasing blood flow. Moreover, increments in plasma E increases the extraction of E across the organ, which makes it difficult to evaluate changes in plasma E concentrations from venous measurements.

The total spillover rate of E into plasma can be assessed with confidence during steady-state conditions as the product of endogenous E in arterial plasma and the plasma clearance of E. In contrast, the total spillover of NE to plasma cannot, for a number of reasons, be assessed with confidence. Furthermore, the sympathetic nervous system works in a highly differentiated state and NE release should therefore preferably be measured in single organs and tissues. Whereas muscle sympathetic nerve activity can be assessed directly by microneurography, NE release in internal organs can only be assessed by measuring the release of NE from these organs into the veins that selectively drain them. Fundamental assumptions and limitations of studying organ NE kinetics have also briefly been summarized.

REFERENCES

1. **Christensen, N. J., Henriksen, O., and Lassen, N. A., Eds.,** *The Sympathoadrenal System. Physiology and Pathophysiology,* Raven Press, New York, 1986, 1.
2. **Christensen, N. J.,** Discussion, in *The Sympathoadrenal System. Physiology and Pathophysiology,* Christensen, N. J., Henriksen, O., and Lassen, N. A., Eds., Raven Press, New York, 1986, 232 and 246.
3. **Eckberg, D. L., Rea, R. F., Andersson, O. K., Hedner, T., pernow, J., Lundberg, J. M., and Wallin, B. G.,** Baroreflex modulation of sympathetic activity and sympathetic neurotransmitters in humans, *Acta Physiol. Scand.,* 133, 221, 1988.
4. **Wallin, B. G., Sundlöf, G., Eriksson, B. M., Dominiak, P., Grobecker, H., and Lindblad, L. E.,** Plasma noradrenaline correlates to sympathetic muscle nerve activity in normotensive man, *Acta Physiol. Scand.,* 111, 69, 1981.
5. **Best, J. D. and Halter, J. B.,** Release and clearance rates of epinephrine in man: importance of arterial measurements, *J. Clin. Endocrinol. Metab.,* 55, 263, 1982.
6. **Astrup, A. Bülow, J., Christensen, N. J., Madsen, J., and Quaade, F.,** Facultative thermogenesis induced by carbohydrate: a skeletal muscle component mediated by epinephrine, *Am. J. Physiol.,* 13, E226, 1986.
7. **Jørgensen, L. S., Bönlökke, L., and Christensen, N. J.,** Plasma adrenaline and noradrenaline during mental stress and isometric exercise in man. The role of arterial sampling, *Scand. J. Clin. Lab. Invest.,* 45, 447, 1985.
8. **Vendsalu, A.,** Studies on adrenaline and noradrenaline in human plasma, *Acta Physiol. Scand.,* 49 (Suppl.) 173, 7, 1960.
9. **Gitlow, S. E., Mendlowitz, M., Wilk, E. K., Wilk, S., Wolf, R. L., and Naftchi, N. E.,** Plasma clearance of dl-β-^{3}H norepinephrine in normal human subjects and patients with essential hypertension, *J. Clin. Invest.,* 43, 2009, 1964.
10. **DeQuattro, V. and Sjoerdsma, A.,** Catecholamine turnover in normotensive and hypertensive man: effects of antiadrenergic drugs, *J. Clin. Invest.,* 47, 2359, 1968.
11. **DeQuattro, V.,** Evaluation of increased norepinephrine excretion in hypertension using L-Dopa-^{3}H, *Circ. Res.,* 28, 84, 1971.
12. **Maas, J. W., Benesohn, H., and Landis, D. H.,** A kinetic study of the disposition of circulating norepinephrine in normal male subjects, *J. Pharmacol. Exp. Ther.,* 174, 381, 1970.
13. **Maas, J. W. and Landis, D. H.,** The metabolism of circulating norepinephrine by human subjects, *J. Pharmacol. Exp. Ther.,* 177, 600, 1971.
14. **Coulombe, P., Dussault, J. H., Letarte, J., and Simard, S. J.,** Catecholamines metabolism in thyroid diseases. I. Epinephrine secretion rate in hyperthyroidism and hypothyroidism, *J. Clin. Endocrinol. Metab.,* 42, 125, 1976.
15. **Lassen, N. A. and Perl, W.,** *Tracer Kinetic Methods in Medical Physiology,* Raven Presss, New York, 1979, 1.
16. **Lassen, N. A.,** The concept of clearance as applied to noradrenaline, in *The Sympathoadrenal System, Physiology and Pathophysiology,* Christensen, N. J., Henriksen, O., and Lassen, N. A., Eds., Raven Press, New York, 1986, 48.
17. **Christensen, N. J. and Henriksen, J. H.,** Total norepinephrine spillover, *Hypertension,* 12, 468, 1988.
18. **Silverberg, A. B., Shah, S. D., Haymond, M. W., and Cryer, P. E.,** Norepinephrine: hormone and neurotransmitter in man., *Am. J. Physiol.,* 234, E252, 1978.
19. **FitzGerald, G. A., Hossmann, V., Hamilton, C. A., Reid, J. L., Davies, D. S., and Dollery, C. T.,** Interindividual variation in kinetics of infused epinephrine, *Clin. Pharmacol. Ther.,* 26, 669, 1979.
20. **Esler, M., Jackman, G., Bobik, A., Kelleher, D., Jennings, G., Leonard, P., Skews, H., and Korner, P.,** Determination of norepinephrine apparent release rate and clearance in humans, *Life Sci.,* 25, 1461, 1979.
21. **Christensen, N. J., Brandsborg, O., Løvgreen, N. A., and Brandsborg, M.,** Elevated plasma noradrenaline concentrations in duodenal ulcer patients are not normalized by vogotomy, *J. Clin. Endocrinol. Metab.,* 49, 331, 1979.
22. **Clutter, W. E., Bier, D. M., Shah, S. D., and Cryer, P. E.,** Epinephrine plasma metabolic clearance rates and physiologic thresholds for metabolic and hemodynamic actions in man, *J. Clin. Invest.,* 66, 94, 1980.
23. **Hilsted, J., Christensen, N. J., and Madsbad, S.,** Whole body clearance of norepinephrine. The signficance of arterial sampling and of surgical stress, *J. Clin. Invest.,* 71, 500, 1983.
24. **Kjœr, M., Christensen, N. J., Sonne, B., Richter, E. A., and Galbo, H.,** Effect of exercise on epinephrine turnover in trained and untrained male subjects, *J. Appl. Physiol.,* 59, 1061, 1985.
25. **Esler, M., Jennings, G., Korner, P., Willett, I., Dudley, F., Hasking, G., Anderson, W., and Lambert, G.,** Assessment of human sympathetic nervous system activity from measurements of norepinephrine turnover, *Hypertension,* 11, 3, 1988.
26. **Christensen, N. J.,** Sympathetic nervous activity and age, *Eur. J. Clin. Invest.,* 12, 91, 1982.
27. **Henriksen, J. H., Christensen, N. J., and Ring-Larsen, H.,** Pulmonary extraction of circulating norepinephrine in man, *Eur. J. Clin. Invest.,* 16, 423, 1986.

28. **Henriksen, J. H. and Christensen, N. J.,** Plasma norepinephrine in humans: limitations in assessment of whole-body norepinephrine kinetics and plasma clearance, *Am. J. Physiol.,* 257, E743, 1989.
29. **Esler, M., Jackman, G., Bobik, A., Leonard, P., Kelleher, D., Skews, H., Jennings, G., and Korner, P.,** Norepinephrine kinetics in essential hypertension. Defective neuronal uptake of norepinephrine in some patients, *Hypertension,* 3, 149, 1981.
30. **Goldstein, D. S., Zimlichman, R., Stull, R., Folio, J., Levinson, P. D., Keiser, H. R., and Kopin, I. J.,** Measurement of regional neuronal removal of norepinephrine in man, *J. Clin. Invest.,* 76, 15, 1985.
31. **Linares, O. A., Jacquez, J. A., Zech, L. A., Smith, M. J., Sanfield, J. A., Morrow, L. A., Rosen, S. G., and Halter, J. B.,** Norepinephrine metabolism in humans. Kinetic analysis and model, *J. Clin. Invest.,* 80, 1332, 1987.
32. **Esler, M., Jackman, G., Leonard, P., Skews, H., Bobik, A., and Korner, P.,** Effect of norepinephrine uptake blockers on norepinephrine kinetics, *Clin. Pharmacol. Ther.,* 29, 12, 1981.
33. **Hoeldtke, R. D. and Cilmi, K. M.,** Norepinephrine secretion and production in diabetic autonomic neuropathy, *J. Clin. Endocrinol. Metab.,* 59, 246, 1984.
34. **Dejgaard, A., Hilsted, J., and Christensen, N. J.,** Noradrenaline and isoproterenol kinetics in diabetic patients with and without autonomic neuropathy, *Diabetologia,* 29, 773, 1986.
35. **Dejgaard, A., Hilsted, J., Henriksen, J. H., and Christensen, N. J.,** Plasma adrenaline kinetics in Type 1 (insulin-dependent) diabetic patients with and without autonomic neuropathy, *Diabetologia,* 32, 810, 1989.
36. **Cryer, P. E., Rizza, R. A., Haymond, M. W., and Gerich, J. E.,** Epinephrine and norepinephrine are cleared through β-adrenergic, but not α-adrenergic mechanisms in man, *Metabolism,* 29, 1114, 1980.
37. **Esler, M., Jackman, G., Leonard, P., Skews, H., Bobik, A., and Jennings, G.,** Effect of propranolol on noradrenaline kinetics in patients with essential hypertension, *Br. J. Pharmacol.,* 12, 375, 1981.
38. **Best, J. D. and Halter, J. B.,** Blood pressure and norepinephrine spillover during propranolol infusion in humans, *Am. J. Physiol.,* 248, R400, 1985.
39. **Ziegler, M. G., Chernow, B., Woodson, L. C., Coyle, J., Cruess, D., and Lake, C. R.,** The effect of propranolol on catecholamine clearance, *Clin. Pharmacol. Ther.,* 40, 116, 1986.
40. **Linares, Q. A., Zech, L. A., Jacquez, J. A., Rosen, S. G., Sanfield, J. A., Morrow, L. A., Supiano, M. A., and Halter, J. B.,** Effect of sodium-restricted diet and posture on norepinephrine kinetics in humans, *Am. J. Physiol.,* 254, E222, 1988.
41. **Sundlöf, G., Wallin, B. G., Strömgren, E., and Nerhed, C.,** Acute effects of metoprolol on muscle sympathetic activity in hypertensive humans, *Hypertension,* 5, 749, 1983.
42. **Christensen, N. J.,** Is plasma norepinephrine an index of biologic age?, in *The Sympathoadrenal System. Physiology and Pathophysiology,* Christensen, N. J., Henriksen, O., and Lassen, N. A., Eds., Raven Press, New York, 1986, 266.
43. **Esler, M., Skews, H., Leonard, P., Jackman, G., Bobik, A., and Korner, P.,** Age-dependence of noradrenaline kinetics in normal subjects, *Clin. Sci.,* 60, 217, 1981.
44. **Young, J. B., Rowe, J. W., Palotta, J. A., Sparrow, D., and Landsberg, L.,** Enhanced plasma norepinephrine response to upright posture and oral glucose administration in elderly human subjects, *Metabolism,* 29, 532, 1980.
45. **Rubin, P. C., Scott, P. J. W., McLean, K., and Reid, J. L.,** Noradrenaline release and clearance in relation to age and blood pressure in man, *Eur. J. Clin. Invest.,* 12, 121, 1982.
46. **Hoeldtke, R. D. and Cilmi, K. M.,** Effects of ageing on catecholamine metabolism, *J. Clin. Endocrinol. Metab.,* 60, 479, 1985.
47. **Hilsted, J., Christensen, N. J., and Larsen, S.,** Plasma clearance of noradrenaline does not change with age in normal subjects, *Clin. Physiol.,* 5, 443, 1985.
48. **MacGilchrist, A. J., Hawksby, C., Howes, L. G., and Reid, J. L.,** Rise in plasma noradrenaline with age results from an increase in spillover rate, *Gerontology,* 35, 7, 1989.
49. **Kjeldsen, S. E., Eide, I., Christensen, C., Westhein, A., and Müller, O.,** Renal contribution to plasma catecholamines — effect of age, *Scand. J. Clin. Lab. Invest.,* 42, 461, 1982.
50. **Hoeldtke, R. D., Cilmi, K. M., Reichard, G. A., Jr., Boden G., and Owen, O. E.,** Assessment of norepinephrine secretion and production, *J. Lab. Clin. Med.,* 101, 772, 1983.
51. **Christensen, N. J., Dejgaard, A., and Hilsted, J.,** Plasma dihydroxyphenylglykol (DHPG) as an index of diabetic autonomic neuropathy, *Clin. Physiol.,* 8, 577, 1988.
52. **Ludwig, J., Gerhardt, T., Halbrügge, T., Walter, J., and Graefe, K.-H.,** Plasma concentrations of noradrenaline and 3,4-dihydroxyphenylethyleneglycol under conditions of enhanced sympathetic activity, *Eur. J. Clin. Pharmacol.,* 35, 261, 1988.
53. **Goldstein, D. S., Eisenhofer, G., Stull, R., Folio, C. J., Keiser, H. R., and Kopin, I. J.,** Plasma dihydroxyphenylglykol and the intraneuronal disposition of norepinephrine in humans, *J. Clin. Invest.,* 81, 213, 1988.
54. **Henriksen, J. H., Christensen, N. J., and Ring-Larsen, H.,** Continuous infusion of tracer norepinephrine may miscalculate unidirectional nerve uptake of norepinephrine in humans, *Circ. Res.,* 65, 388, 1989.

55. **Rose, C. P., Burgess, J. H., and Cousineau, D.,** Tracer norepinephrine kinetics in coronary circulation of patients with heart failure secondary to chronic pressure and volume overload, *J. Clin. Invest.*, 76, 1740, 1985.
56. **Brown, M. J., Jenner, D. A., Allison, D. J., and Dollery, C. T.,** Variations in individual organ release of noradrenaline measured by an improved radioenzymatic technique; limitations of peripheral venous measurments in the assessment of sympathetic nervous activity, *Clin. Sci.*, 61, 585, 1981.
57. **Péronnet, F., Béliveau, L., Boudreau, G., Trudeau, F., Brisson, G., and Nadeau, R.,** Regional plasma catecholamine removal and release at rest and exercise in dogs, *Am. J. Physiol.*, 254, R663, 1988.
58. **Henriksen, J. H., Ring-Larsen, H., and Christensen, N. J.,** Kidney, lower limb and whole-body uptake and release of catecholamines in alcoholic liver diseases, *Clin. Physiol.*, 8, 203, 1988.
59. **Henriksen, J. H., Ring-Larsen, H., and Christensen, N. J.,** Hepatic intestinal uptake and release of catecholamines in alcoholic cirrhosis. Evidence of enhanced hepatic intestinal sympathetic nervous acivity, *Gut*, 28, 1637, 1987.
60. **Goresky, C. A., Bach, G. G., Cousineau, D., Schwab, A. J., Rose, C., Lee, S., and Goresky, S.,** Handling of tracer norepinephrine by the dog liver, *Am. J. Physiol.*, 256, G107, 1989.
61. **Cousineau, D., Goresky, C. A., and Rose, C. P.,** Decreased basal cardiac interstitial norepinephrine release after neuronal uptake inhibition in dogs, *Circ. Res.*, 58, 859, 1986.
62. **Esler, M., Willet, I., Leonard, P., Hasking, G., Johns, J., Little, P., and Jennings, G.,** Plasma noradrenaline kinetics in humans, *J. Auton. Nerv. Syst.*, 11, 125, 1984.
63. **Gillis, C. N., Greene, N. M., Cronou, L. H., and Hammond, B. L.,** Pulmonary extraction of 5-hydroxytryptamine and norepinephrine before and after cardiopulmonary bypass in man., *Circ. Res.*, 30, 666, 1972.
64. **Sole, M. J., Drobac, M., Schwartz, L., Hussain, M. N., and Waughan-Neil, E. F.,** The extraction of circulating catecholamines by the lungs in normal man and in patients with pulmonary hypertension, *Circulation*, 60, 160, 1979.
65. **Manhem, P., Lecerof, H., and Hökfelt, B.,** Plasma catecholamine levels in the coronary sinus, the left renal vein and peripheral vessels in healthy males at rest and during exercise, *Acta Physiol. Scand.*, 104, 346, 1978.
66. **Miura, Y., Haneda, T., Sato, T., Miyazawa, K., Sakuma, H., Kobayashi, K., Minai, K., Shirato, K., Honna, T., Takishima, T., and Yoshinaga, K.,** Plasma catecholamine levels in the coronary sinus, aorta and femoral vein of subjects undergoing cardiac catheterization at rest and during exercise, *Jpn. Circ. J.*, 40, 929, 1976.
67. **Cousineau, D., Rose, C. P., and Goresky, C. A.,** Labeled catecholamine uptake in the dog heart: interactions between capillary wall and sympathetic nerve uptake, *Circ. Res.*, 47, 329, 1980.
68. **Hollister, A. S., Rodeheffer, R. J., White, F. J., Potts, J. R., Imada, T., and Inagami, T.,** Clearance of atrial natriuretic factor by lung, liver, and kidney in human subjects and the dog, *J. Clin. Invest.*, 83, 623, 1989.
69. **Henriksen, J. H., Tronier, B., and Bülow, J. B.,** Kinetics of circulating endogenous insulin, C-peptide, and proinsulin in fasting nondiabetic man, *Metabolism*, 36, 463,1987.
70. **Unger, T., Buu, N. T., and Kuchel, O.,** Renal handling of free and conjugated catecholamines following surgical stress in the dog, *Am. J. Physiol.*, 235, F542, 1978.
71. **Kostreva, D. R., Castaner, A., and Kampine, J. P.,** Reflex effect on hepatic baroreceptors on renal and cardiac sympathetic nerve activity, *Am. J. Physiol.*, 238, R390, 1980.
72. **Kopp, U., Bradley, T., and Hjemdahl, P.,** Renal venous outflow and urinary excretion of norepinephrine, epinephrine and dopamine during graded renal nerve stimulation, *Am. J. Physiol.*, 244, E52, 1983.
73. **Pedersen, E. B. and Christensen, N. J.,** Catecholamines in plasma and urine in patients with essential hypertension determined by double-isotope derivative techniques, *Acta Med. Scand.*, 198, 373, 1985.
74. **Keller, U., Gerberg, P. P. G., Buhler, F. R., and Stauffacher, W.,** Role of the splanchnic bed in extracting circulating adrenaline and noradrenaline in normal subjects and in patients with cirrhosis of the liver, *Clin. Sci.*, 67, 45, 1984.
75. **Henriksen, J. H., Ring-Larsen, H.,, Kanstrup, I.-L., and Christensen, N. J.,** Splanchnic and renal elimination and release of catecholamines in cirrhosis. Evidence of enhanced sympathetic nervous activity in patients with cirrhosis, *Gut*, 25, 1034, 1984.
76. **Shaldon, C., Peacock, J. H., Walker, R. M., Palmer, D. B., and Badrick, F. E.,** The portal venous content of adrenaline and noradrenaline in portal hypertension, *Lancet*, i, 957, 1961.
77. **Joly, J., Leduc, J., Berner, J., Lavoie, P., and Viallet, A.,** Catecholamine levels in portal, hepatic and systemic venous blood in portal hypertension, *Lancet*, ii, 121, 1967.
78. **Schanker, L. S.,** Secretion of organic compounds in bile, in *Handbook of Physiology, Section 6, Vol. V, Alimentary Canal*, Code, C. F., Ed., American Physiological Society, Washington, D.C., 2433, 1968.
79. **Henriksen, J. H., Ring-Larsen, H., and Christensen, N. J.,** Catecholamines in plasma from artery, cubital vein, and femoral vein in patients with cirrhosis. Significance of sampling site, *Scand. J. Clin. Lab. Invest.*, 46, 39, 1986.
80. **Henriksen, J. H., Christensen, N. J., Kok-Jensen, A., and Christiansen, I.,** Increased plasma noradrenaline concentration in patients with chronic obstructive lung disease: relation to haemodynamics and blood gases, *Scand. J. Clin. Lab. Invest.*, 40, 419, 1980.

81. **Ring-Larsen, H., Hesse, B., henriksen, J. H., and Christensen, N. J.,** Sympathetic nervous activity and renal and systemic hemodynamics in cirrhosis: plasma norepinephrine concentration, hepatic extraction, and renal disease, *Hepatology,* 2, 304, 1982.
82. **Bichet, D. G., Van Putten, V. J., and Schrier, R. W.,** Potential role of increased sympathetic activity in impaired sodium and water excretion in cirrhosis, *N. Engl. J. Med.,* 307, 1552, 1982.
83. **Chidsey, C. A., Sonneblick, E. H., Morrow, A. G. J., and Braunwald, E.,** Norepinephrine stores and contractile force of papillary muscle from the failing human heart, *Circulation,* 33, 43, 1966.
84. **Spann, J. F., Chidsey, C. A., Pool, P. E., and Braunwald, E.,** Mechanism of norepinephrine depletion in experimental heart failure produced by aortic constriction in the guinea pig, *Circ. Res.,* 17, 312, 1965.
85. **Hansen, J. F., Hesse, B., and Christensen, N. J.,** Enhanced sympathetic, nervous activity after intravenous propranolol in ischaemic heart disease: plasma noradrenaline, splanchnic blood flow and mixed venous oxygen saturation at rest and during exercise, *Eur. J. Clin. Invest.,* 8, 31, 1978.

INDEX

A

B

C

D

E

I

K

L

O

P

Q

R

S

T

U

V

W

X

Z